# Geospace Mass and Energy Flow

## Results From the International Solar-Terrestrial Physics Program

# Geophysical Monograph Series

Including
**IUGG Volumes**
**Maurice Ewing Volumes**
**Mineral Physics Volumes**

# Geophysical Monograph Series

70 **Synthesis of Results from Scientific Drilling in the Indian Ocean** *Robert A. Duncan, David K. Rea, Robert B. Kidd, Ulrich von Rad, and Jeffrey K. Weissel (Eds.)*

71 **Mantle Flow and Melt Generation at Mid-Ocean Ridges** *Jason Phipps Morgan, Donna K. Blackman, and John M. Sinton (Eds.)*

72 **Dynamics of Earth's Deep Interior and Earth Rotation (IUGG Volume 12)** *Jean-Louis Le Mouël, D.E. Smylie, and Thomas Herring (Eds.)*

73 **Environmental Effects on Spacecraft Positioning and Trajectories (IUGG Volume 13)** *A. Vallance Jones (Ed.)*

74 **Evolution of the Earth and Planets (IUGG Volume 14)** *E. Takahashi, Raymond Jeanloz, and David Rubie (Eds.)*

75 **Interactions Between Global Climate Subsystems: The Legacy of Hann (IUGG Volume 15)** *G. A. McBean and M. Hantel (Eds.)*

76 **Relating Geophysical Structures and Processes: The Jeffreys Volume (IUGG Volume 16)** *K. Aki and R. Dmowska (Eds.)*

77 **The Mesozoic Pacific: Geology, Tectonics, and Volcanism** *Malcolm S. Pringle, William W. Sager, William V. Sliter, and Seth Stein (Eds.)*

78 **Climate Change in Continental Isotopic Records** *P. K. Swart, K. C. Lohmann, J. McKenzie, and S. Savin (Eds.)*

79 **The Tornado: Its Structure, Dynamics, Prediction, and Hazards** *C. Church, D. Burgess, C. Doswell, R. Davies-Jones (Eds.)*

80 **Auroral Plasma Dynamics** *R. L. Lysak (Ed.)*

81 **Solar Wind Sources of Magnetospheric Ultra-Low Frequency Waves** *M. J. Engebretson, K. Takahashi, and M. Scholer (Eds.)*

82 **Gravimetry and Space Techniques Applied to Geodynamics and Ocean Dynamics (IUGG Volume 17)** *Bob E. Schutz, Allen Anderson, Claude Froidevaux, and Michael Parke (Eds.)*

83 **Nonlinear Dynamics and Predictability of Geophysical Phenomena (IUGG Volume 18)** *William I. Newman, Andrei Gabrielov, and Donald L. Turcotte (Eds.)*

84 **Solar System Plasmas in Space and Time** *J. Burch, J. H. Waite, Jr. (Eds.)*

85 **The Polar Oceans and Their Role in Shaping the Global Environment** *O. M. Johannessen, R. D. Muench, and J. E. Overland (Eds.)*

86 **Space Plasmas: Coupling Between Small and Medium Scale Processes** *Maha Ashour-Abdalla, Tom Chang, and Paul Dusenbery (Eds.)*

87 **The Upper Mesosphere and Lower Thermosphere: A Review of Experiment and Theory** *R. M. Johnson and T. L. Killeen (Eds.)*

88 **Active Margins and Marginal Basins of the Western Pacific** *Brian Taylor and James Natland (Eds.)*

89 **Natural and Anthropogenic Influences in Fluvial Geomorphology** *John E. Costa, Andrew J. Miller, Kenneth W. Potter, and Peter R. Wilcock (Eds.)*

90 **Physics of the Magnetopause** *Paul Song, B.U.Ö. Sonnerup, and M.F. Thomsen (Eds.)*

91 **Seafloor Hydrothermal Systems: Physical, Chemical, Biological, and Geological Interactions** *Susan E. Humphris, Robert A. Zierenberg, Lauren S. Mullineaux, and Richard E. Thomson (Eds.)*

92 **Mauna Loa Revealed: Structure, Composition, History, and Hazards** *J. M. Rhodes and John P. Lockwood (Eds.)*

93 **Cross-Scale Coupling in Space Plasmas** *James L. Horwitz, Nagendra Singh, and James L. Burch (Eds.)*

94 **Double-Diffusive Convection** *Alan Brandt and H.J.S. Fernando (Eds.)*

95 **Earth Processes: Reading the Isotopic Code** *Asish Basu and Stan Hart (Eds.)*

96 **Subduction Top to Bottom** *Gray E. Bebout, David Scholl, Stephen Kirby, and John Platt (Eds.)*

97 **Radiation Belts: Models and Standards** *J. F. Lemaire, D. Heynderickx, and D. N. Baker (Eds.)*

98 **Magnetic Storms** *Bruce T. Tsurutani, Walter D. Gonzalez, Yohsuke Kamide, and John K. Arballo (Eds.)*

99 **Coronal Mass Ejections** *Nancy Crooker, Jo Ann Joselyn, and Joan Feynman (Eds.)*

100 **Large Igneous Provinces** *John J. Mahoney and Millard F. Coffin (Eds.)*

101 **Properties of Earth and Planetary Matierials at High Pressure and Temperature** *Murli Manghnani and Takehiki Yagi (Eds.)*

102 **Measurement Techniques in Space Plasmas: Particles** *Robert F. Pfaff, Joseph E. Borovsky, and David T. Young (Eds.)*

103 **Measurement Techniques in Space Plasmas: Fields** *Robert F. Pfaff, Joseph E. Borovsky, and David T. Young (Eds.)*

Geophysical Monograph 104

# Geospace Mass and Energy Flow

## Results From the International Solar-Terrestrial Physics Program

James L. Horwitz
Dennis L. Gallagher
William K. Peterson
*Editors*

American Geophysical Union
Washington, D.C.

Published under the aegis of the AGU Books Board

**Library of Congress Cataloging-in-Publication Data**
Geospace mass and energy flow : results from the International Solar
 -Terrestrial Physics Program / James L. Horwitz, Dennis L.
 Gallagher, William K. Peterson.
   p. cm. -- (Geophysical monograph series ; 104)
 Includes bibliographical references.
 ISBN 0-87590-087-9
  1. Magnetosphere. 2. Ionosphere. 3. Cosmic physics.
4. International Solar-Terrestrial Program. I. Horwitz, James L.
II. Gallagher, Dennis L. III. Peterson, William M. IV. Series.
QC809.M35G49   1998                                          98-36221
551.51'4--dc21                                                  CIP

ISBN 0-87590-087-9
ISSN 0065-8448

Copyright 1998 by the American Geophysical Union
2000 Florida Ave., N.W.
Washington, DC 20009

Figures, tables, and short excerpts may be reprinted in scientific books and journals if the source is properly cited.

Authorization to photocopy items for internal or personal use, or the internal or personal use of specific clients, is granted by the American Geophysical Union for libraries and other users registered with the Copyright Clearance Center (CCC) Transactional Reporting Service, provided that the base fee of $1.50 per copy plus $0.35 per page is paid directly to CCC, 222 Rosewood Dr., Danvers, MA 01923. 0065-8448/98/$1.50+0.35.

This consent does not extend to other kinds of copying, such as copying for creating new collective works for resale. The reproduction of multiple copies and the use of full articles or the use of extracts, including figures and tables, for commercial purposes requires permission from AGU.

Printed in the United States of America

# Contents

**Preface**
*James L. Horwitz, Dennis L. Gallagher, and William K. Peterson* ............................................................... xi

## Solar Wind

**WIND Observations of Suprathermal Particles in the Solar Wind**
*R. P. Lin, D. Larson, T. Phan, R. Ergun, J. McFadden, K. Anderson, C. Carlson, M. McCarthy, G. K. Parks, R. Skoug, R. Winglee, H. Rème, N. Lormant, J. M. Bosqued, C. d'Uston, T. R. Sanderson, and K.-P. Wenzel* ................................................................................................................................................. 1

## Dayside Magnetosphere

**Dayside Electrodynamics Observed by POLAR with Northward IMF**
*N. C. Maynard, W. J. Burke, D. R. Weimer, F. S. Mozer, J. D. Scudder, C. T. Russell, and W. K. Peterson* ........ 13

**Interball Tail Probe Measurements in Outer Cusp and Boundary Layers**
*S. P. Savin, N. L. Borodkova, E. Yu. Budnik, A. O. Fedorov, S. I. Klimov, M. N. Nozdrachev, I. E. Morozova, N. S. Nikolaeva, A. A. Petrukovich, N. F. Pissarenko, V. I. Prokhorenko, S. A. Romanov, A. A. Skalsky, Yu. I. Yermolaev, G. N. Zastenker, L. M. Zelenyi, P. Triska, E. Amata, J. Blecki, J. Juchniewicz, J. Buechner, M. Ciobanu, R. Grard, G. Haerendel, V. E. Korepanov, R. Lundin, I. Sandahl, U. Eklund, Z. Nemecek, J. Safrankova, J. A. Sauvaud, J. Rustenbach, and J. L. Rauch* ..................................................................... 25

**Convection of Plasmaspheric Plasma into the Outer Magnetosphere and Boundary Layer Region: Initial Results**
*Daniel M. Ober, J. L. Horwitz, and D. L. Gallagher* ................................................................................ 45

**AMPTE-IRM Observations of Particles and Fields at the Dayside Low-Latitude Magnetopause**
*T. M. Bauer, G. Paschmann, N. Sckopke, W. Baumjohann, R. A. Treumann, and T.-D. Phan* ..................... 51

**The Effect of the Interplanetary Magnetic Field on the Dayside Magnetosphere**
*Daniel W. Swift* ................................................................................................................................ 67

**Entry of Solar Wind Plasma into the Magnetosphere: Observations Encounter Simulation**
*Patrick T. Newell and Simon Wing* ..................................................................................................... 73

## High-Latitude Ionosphere

**Under What Conditions Will Ionospheric Molecular Ion Outflow Occur?**
*G. R. Wilson and P. D. Craven* ........................................................................................................... 85

**Day-Night Asymmetry Of Polar Outflow Due to the Kinetic Effects of Anisotropic Photoelectrons**
*Sunny W. Y. Tam, Fareed Yasseen, and Tom Chang* ............................................................................. 97

**POLAR Observations of Properties of $H^+$ and $O^+$ Conics in the Cusp Near ~5300 km Altitude**
*M. Hirahara, J. L. Horwitz, T. E. Moore, M. O. Chandler, B. L. Giles, P. D. Craven, and C. L. Pollock* ............ 107

**Recent Developments in Ion Acceleration in the Auroral Zone**
*Tom Chang and Mats André* .............................................................................................................. 115

## Aurora

**High-Latitude Auroral Boundaries Compared With GEOTAIL Measurements During Two Substorms**
*W. J. Burke, N. C. Maynard, G. M. Erickson, M. Nakamura, S. Kokubun, B. Jacobsen, and R. W. Smith* ....... 129

**Energy Characterization of a Dynamic Auroral Event Using GGS UVI Images**
*G. A. Germany, G. K. Parks, M. J. Brittnacher, J. F. Spann, J. Cumnock, D. Lummerzheim, F. Rich, and P. G. Richards*................................................................................................................................143

**Auroral Observations from the POLAR Ultraviolet Imager (UVI)**
*G. A. Germany, J. F. Spann, G. K. Parks, M. J. Brittnacher, R. Elsen, L. Chen, D. Lummerzheim, and M. H. Rees* 149

**Field Line Resonances, Auroral Arcs, and Substorm Intensifications**
*J. C. Samson, R. Rankin, and I. Voronkov*................................................................................................161

## Magnetotail/Plasma Sheet/Substorms

**A New Approach to the Future of Space Physics**
*S.-I. Akasofu and W. Sun*............................................................................................................................169

**Reconnections at Near-Earth Magnetotail and Substorms Studied by a 3-D EM Particle Code**
*K.-I. Nishikawa*............................................................................................................................................175

**Magnetospheric-Ionospheric Activity During an Isolated Substorm: A Comparison Between WIND/Geotail/IMP 8/CANOPUS Observations and Modeling**
*R. M. Winglee, A. T. Y. Lui, R. P. Lin, R. P. Lepping, S. Kokubun, G. Rostoker, and J. C. Samson*..............181

**Sporadic Localized Reconnections and Multiscale Intermittent Turbulence in the Magnetotail**
*Tom Chang*..................................................................................................................................................193

**Geotail Observations of Current Systems in the Plasma Sheet**
*W. R. Paterson, L. A. Frank, S. Kokubun, and T. Yamamoto*....................................................................201

**Near-Earth Plasma Sheet Behavior During Substorms**
*T. Nagai, R. Nakamura, S. Kokubun, Y. Saito, T. Yamamoto, T. Mukai, and A. Nishida*........................213

**Separation of Directly-Driven and Unloading Components in the Ionospheric Equivalent Currents During Substorms by the Method of Natural Orthogonal Components**
*W. Sun, W.-Y. Xu, and S. I. Akasofu*..........................................................................................................227

**Simulation of the March 9, 1995 Substorm and Initial Comparison to Data**
*R. E. Lopez, C. C. Goodrich, M. Wiltberger, K. Papadopoulos, and J. G. Lyon*.....................................237

**Large-Scale Dynamics of the Magnetospheric Boundary: Comparisons Between Global MHD Simulation Results and ISTP Observations**
*J. Berchem, J. Raeder, M. Ashour-Abdalla, L. A. Frank, W. R. Paterson, K. L. Ackerson, S. Kokubun, T. Yamamoto, and R. P. Lepping*................................................................................................................247

**Study of an Isolated Substorm with ISTP Data**
*A. T. Y. Lui, D. J. Williams, R. W. McEntire, S. Ohtani, L. J. Zanetti, W. A. Bristow, R. A. Greenwald, P. T. Newell, S. P. Christon, T. Mukai, K. Tsurada, T. Yamamoto, S. Kokubun, H. Matsumoto, H. Kojima, T. Murata, D. H. Fairfield, R. P. Lepping, J. C. Samson, G. Rostoker, G. D. Reeves, A. L. Rodger, and H. J. Singer*................261

**Structure of the Magnetotail Reconnection Layer in 2-D Ideal MHD Model**
*Y. Lin and X. X. Zhang*................................................................................................................................275

**Particle Acceleration Due to Magnetic Field Reconnection in a Model Current Sheet**
*Victor M. Vazquez, Maha Ashour-Abdalla, and Robert L. Richard*.........................................................287

**Modeling Magnetotail Ion Distributions with Global Magnetohydrodynamic and Ion Trajectory Calculations**
*M. El-Alaoui, M. Ashour-Abdalla, J. Raeder, V. Peroomian, L. A. Frank, W. R. Paterson, and J. M. Bosqued*..........291

**Determination of Particle Sources for a Geotail Distribution Function Observed on May 23, 1995**
*M. Ashour-Abdalla, M. El-Alaoui, V. Peroomian, J. Raeder, R. L. Richard, R. J. Walker, L. M. Zelenyi,
L. A. Frank, W. R. Paterson, J. M. Bosqued, R. P. Lepping, K. Oglivie, S. Kokubun, and T. Yamamoto* ...................297

**Three-Dimensional Reconnection in the Earth's Magnetotail: Simulations and Observations**
*J. Büchner, J. P. Kuska, B. Nikutowski, H. Wiechen, J. Rustenbach, U. Auster, K. H. Fornacon, S. Klimov,
A. Petrukovich, and S. Savin* ...................................................................................................................313

## Inner Magnetosphere

**Drift-Shell Splitting in an Asymmetric Magnetic Field**
*Mei-Ching Fok and Thomas E. Moore* ...............................................................................................327

**Comparison of Photoelectron Theory Against Observations**
*G. V. Khazanov and M. W. Liemohn* ..................................................................................................333

**Determining the Significance of Electrodynamic Coupling Between Superthermal Electrons
and Thermal Plasma**
*M. W. Liemohn and G. V. Khazanov* ..................................................................................................343

**A New Magnetic Storm Model**
*Robert B. Sheldon and Harlan E. Spence* ...........................................................................................349

**The Magnetospheric Trough**
*M. F. Thomsen, D. J. McComas, J. E. Borovsky, and R. C. Elphic* .......................................................355

## Magnetic Field Models

**Data-Based Models of the Global Geospace Magnetic Field: Challenges and Prospects of the ISTP Era**
*Nikolai A. Tsyganenko* .........................................................................................................................371

**Improvements to the Source Surface Model of the Magnetosphere**
*Vahé Peroomian, Larry R. Lyons, and Michael Schulz* ......................................................................383

# PREFACE

The International Solar-Terrestrial Program (ISTP) was conceived by the space agencies of numerous countries as a coordinated effort to determine the global flow of mass, energy and momentum through the solar-terrestrial system. The physical region of interest extends from the Sun through the solar wind, to the terrestrial magnetosphere, down to the ionosphere, with the principal focus on the magnetosphere and its coupling with the solar wind and ionosphere. With the launch of NASA's POLAR spacecraft in February 1996, the major elements of the ISTP program were in place. This volume is one of the very first compendiums of both new observations and new modeling results either directly or indirectly deriving from this major ISTP undertaking.

This monograph is organized into sections associated with the following regions and phenomena, which comprise the major topics impacted by the ISTP program: the solar wind, the dayside magnetosphere, the high-latitude ionosphere, the aurora, the magnetotail/plasma sheet/substorms, the inner magnetosphere, and models of the magnetic field. Among the major highlights are research papers describing new global auroral imaging results from POLAR and comparisons of these images with particle and field data from POLAR and other ISTP spacecraft; multiple ISTP spacecraft and ground observations of magnetospheric substorms; and global scale magnetohydrodynamic and particle trajectory simulations of ISTP observations of substorms and particle distributions.

The impetus for this monograph grew out of a workshop in Guntersville, Alabama, in September 1996. The workshop, entitled "Encounter between Global Observations and Models in the ISTP Era," was originally conceived by Tom Moore, and was convened by D. L. Gallagher and J. L. Horwitz. The success of this volume is due in part to the members of the workshop organizing committee: T. E. Moore, R. H. Comfort, R. Carovillano, M. Mellott, R. Robinson, W. Burke, M. Acuna, R. Hoffman, J. Slavin, A. Nishida, L. Zelenyi, R. Schmidt, R. Greenwald, S. Curtis, R. Zwickl, and R. Lopez. We also must thank M. Hargrave, A. Haller, P. Moss, and J. Christensen of the Center for Space Plasma and Aeronomic Research for their assistance with the conference and this volume. Finally, financial support from NASA Headquarters and the Alabama Space Grant consortium is gratefully acknowledged.

James L. Horwitz
Dennis L. Gallagher
William K. Peterson
*Editors*

# WIND Observations of Suprathermal Particles in the Solar Wind

R. P. Lin[1*], D. Larson[1], T. Phan[1], R. Ergun[1], J. McFadden[1], K. Anderson[1*], C. Carslon[1], M. McCarthy[2], G. K. Parks[2], R. Skoug[2], R. Winglee[2], H. Rème,[3] N. Lormant[3], J. M. Bosqued[3], C. d'Uston[3], T. R. Sanderson[4], and K.-P. Wenzel[4]

The 3-D Plasma and Energetic Particle Instrument on the WIND spacecraft provides high sensitivity electron and ion measurements from solar wind thermal plasma up to ~300 keV energy. Some of the many new results are reviewed here, including: 1) the remote sensing of the jump in magnetic field and the electric potential of the Earth's bow shock, using angular distributions of backstreaming electrons; 2) the observation of solar impulsive electron events extending down to ~0.5 keV energy; 3) the detection of a quiet-time population (the "superhalo") of electrons extending up to ~$10^2$ keV energy; and 4) the probing of the magnetic topology and source region for magnetic clouds.

## INTRODUCTION

For many years the solar wind thermal plasma has been regarded as a regime essentially independent from the energetic particle populations found in interplanetary space. Because of dynamic range considerations, previous instruments designed to measure the solar wind plasma ions and electrons lack the sensitivity to detect the suprathermal particles from just above solar wind plasma to a few hundred keV, except during highly disturbed times. These suprathermals, play a key role in the varied plasma and energetic particle phenomena observed to occur in the interplanetary medium (IPM) and upstream from the Earth's magnetosphere.

The 3D Plasma and Energetic Particles Experiment [*Lin et al.*, 1995] on the WIND spacecraft is designed to bridge the gap between solar wind plasma and energetic particle measurements by providing high sensitivity, wide dynamic range, good energy and angular resolution, full 3-D coverage, and high time resolution over the energy range from a few eV to ≳ 300 keV for electrons and ≳ 6 MeV for ions. Here some of the results from the first year of operation are reviewed: 1) the remote probing of the Earth's bow shock by upstream electrons; 2) the discovery that the spectrum of impulsive solar electron events extend down to as low as ~0.5 keV energy; 3) the detection of a quiet-time "superhalo" component of interplanetary electrons from ~1 keV to ~$10^2$ keV; and 4) the probing of the field line topology of magnetic clouds ejected from the Sun.

## PROBING THE EARTH'S BOW SHOCK

To illustrate the type of 3-D data obtained by this experiment, we use backstreaming electrons in the Earth's foreshock region as an example. Plate 1 [from *Larson et al.*, 1996b] shows observations obtained with the Electron ElectroStatic Analyzer-Low (EESA-L) of the WIND 3-D Plasma and Energetic Particles instrument (see Lin et al.

[1]Space Sciences Laboratory, University of California, Berkeley, California
[2]Geophysics Program AK50, University of Washington, Seattle, Washington
[3]Centre d'Etude Spatiale des Rayonnements, Universite Paul Sabatier, Toulouse, France
[4]Space Science Department of ESA, European Space Research and Technology Centre, Noordwijk, Netherlands
*also at Physics Dept., U.C. Berkeley, California

Geospace Mass and Energy Flow: Results From the International Solar-Terrestrial Physics Program
Geophysical Monograph 104
Copyright 1998 by the American Geophysical Union

1995 for description) in the Earth's foreshock. This detector measures electrons from ~10 eV to 1.1 keV with 15 logarithmically spaced energy steps and full $4\pi$ angular coverage in one spacecraft rotation (3 seconds). The data are sorted on board the spacecraft into 88 angular bins, each with roughly $22° \times 22°$ resolution. These measurements are converted into units of distribution function and transformed into the solar wind rest frame using the solar wind velocity obtained from the Proton ElectroStatic Analyzer-Low (PESA-L) ion detector. A correction is made for an estimated spacecraft potential of ~8 V.

The top panel of Plate 1 shows $\Theta_{BN}$, the angle between the magnetic field and shock normal, calculated extrapolating the locally measured magnetic field vector (from the WIND Magnetic Field Instrument (MFI) [Lepping et al., 1995]) in a straight line to the model bow shock [Slavin and Holzer, 1981] scaled to match the bowshock crossing observed at ~0046 UT. Essentially all connections to the shock have $\Theta_{BN}$ greater than 90° (**B** points into the bow shock surface). WIND is located at $X_{GSE} \sim 22\ R_E$ and $Y_{GSE} \sim -21\ R_E$, about 15 $R_E$ upstream of the bow shock, but the features observed here are representative of other times.

The following panel, labeled "Solar Flux", shows that the flux of electrons flowing outward from the Sun (the halo or solar heat flux electrons), measured in the solar wind plasma rest frame, is relatively stable over this time period. The next panel, labeled "Backstreaming Flux", shows that the electron flux streaming back toward the Sun turns on and off abruptly. The bottom three color panels present the pitch angle spectrograms for 780, 56, and 13 eV electrons. Backstreaming flux at 780 and 56 eV are detected when $\Theta_{BN}$ indicates a magnetic connection to the shock. During the disconnected periods, the 13 eV flux is greater at 180° than at 0° pitch angle, because the solar wind core electrons which dominate the 13 eV flux must have a small net motion back toward the Sun to balance the current of halo electrons moving away from the Sun [Feldman et al., 1975]. Spikes of backstreaming electrons, often extending to high energies ( ~150° - 180° flux at 780 eV), are observed near the foreshock boundary, when the field line is nearly tangent to the shock surface.

Plate 2 displays the full 3-D electron distribution function when WIND was deep in the foreshock (1801 UT), well away from the foreshock boundary. Backstreaming electrons appear in the center of each plot whereas the solar wind heat flux will appear split between the left and right edges of each oval. The distributions are essentially gyrotropic; the asymmetries in gyro-angle are due to instrumental effects or statistical fluctuations. The most striking feature of each plot is the enhanced ring of flux centered about the anti-magnetic field direction. The angular diameter of this ring increases with decreasing energy.

The distributions are most easily analyzed by transforming to the deHoffman-Teller reference frame (dHTF), where the upstream plasma flow is parallel to the magnetic field direction and the motional electric field is zero. Electrons with a sufficiently large pitch angle will be magnetically mirrored by the increased magnetic field in the shock, and travel back upstream with a parallel velocity equal in magnitude to its incident value in the dHTF.

The shock velocity, $V_{SB}$, in the dHTF is parallel to **B** by construction, and given by $V_{SB} = V_{sw} - V_{HT}$, where $V_{HT} = \mathbf{n} \times (V_{sw} \times B)/B \cdot \mathbf{n}$, and **n** is the shock normal. The parallel velocity of the reflected electrons in the solar wind plasma rest frame is then given by $v_{\parallel r} = -v_{\parallel i} + 2V_{SB}$, where $v_{\parallel i}$ is the parallel velocity of the incoming electrons. Assuming Liouville's Theorem holds, the phase space density of the reflected population can be written in terms of the phase space density of the incident (solar wind) population: $f_r(v_\parallel, v_\perp) = f_{sw}(-v_\parallel + 2V_{SB}, v_\perp)$, where the values of $v_\parallel$ and $v_\perp$ are limited to those outside the loss cone. Electrons with pitch angles less than the mirror angle will pass through the shock, thus the reflected population should exhibit a loss cone. Electrons escaping from the magnetosheath would populate this loss cone. The two black circles shown in each panel divide angular space into three regions: escaping magnetosheath electrons in the center, reflected electrons in the annulus, and solar wind electrons outside.

Assuming gyrotopy, the 3-D distributions shown in Plate 2 are used to calculate the phase space density $f(v_\parallel, v_\perp)$ shown in Figure 1. The angular width of the loss cone varies as a function of electron energy due to the effect of the cross-shock potential. By assuming conservation of energy and magnetic moment, it can be shown [Feldman et al. 1983, Fitzenreiter et al. 1990] that electrons with pitch angles less than the critical angle $\alpha_m$, defined by

$$\sin^2(\alpha_m) = B_0/B_{max}(1 + e\Delta\Phi/E)$$

will not be mirrored at the shock surface, whereas electrons with pitch angles greater than $\alpha_m$ will be reflected and return. Here $B_0$ is the locally measured magnetic field strength and $B_{max}$ is the maximum field strength in the shock. $\Delta\Phi$ is the electrostatic potential across the shock (as measured in the dHT frame) and $E$ is the particle's initial energy. For the example of Figure 1, the data fit to approximate values: $B_{max}/B_0 = 5.0$ and $e\Delta\Phi = 86$ eV. These values are consistent with direct measurements at the bow shock [see Scudder, 1995, for review].

Thus remote monitoring of the shock using 3D measurements of electron distribution can provide these parameters. Correlation to solar wind parameters may help to achieve a detailed understanding of the shock formation process.

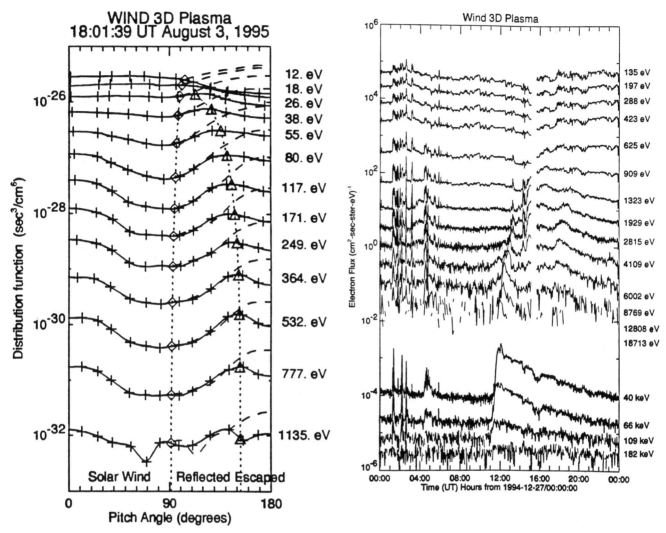

**Figure 1.** Pitch angle distribution for the sample shown in Plate 2. Diamonds connected by dots show the approximate boundary between solar wind and backstreaming electrons. Triangles connected by dots separate flux due to reflected solar wind electrons from escaping magnetosheath electrons. Dashed lines show the predictions for reflected electrons assuming no loss cone. [from *Larson et al.* 1996b]

**Figure 2.** Electron fluxes from ~100 eV to ≥100 keV for 27 December 1994. The solar electron event begins at ~1100 UT at ~100 keV, with velocity dispersion evident down to 624 eV. Two smaller events, at ~1100 UT and 1620 UT, are visible below ~6 keV. The dip at ~1500 UT at low energies is due to the close approach to the Moon, resulting in a plasma shadow. [from *Lin et al.* 1996]

## SOLAR IMPULSIVE ELECTRON EVENTS

The steady-state solar wind electron population is dominated by a core with temperature $kT \sim 10$ eV, containing ≥95% of the plasma density and moving at about the solar wind bulk velocity, plus ~5% in a hot, $kT \sim 80$ eV, halo population carrying heat flux outward from the Sun, often in the form of highly collimated strahl [*Feldman et al.*, 1975]. At energies of ~ keV and above, impulsively accelerated solar electron events occur at the Sun, on average, several times a day or more during solar maximum. As these electrons escape they produce solar and interplanetary type III radio bursts through beam-plasma interactions [see *Lin*, 1990 for review].

Figure 2 shows the first observation [*Lin et al.*, 1996] of solar impulsive electron events spanning the entire energy range from solar wind to suprathermal particle (few eV to hundreds of keV). A solar impulsive electron event begins at 1100 UT, easily identified by its velocity dispersion, e.g., the faster electrons arriving earlier, as expected if the elec-

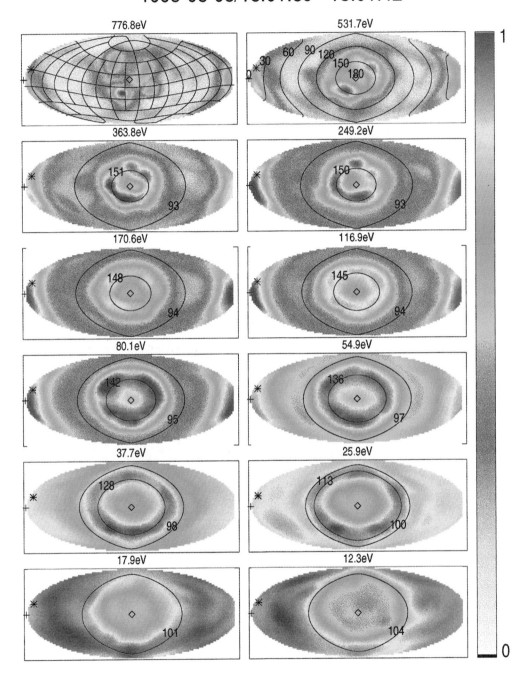

**Plate 2.** Full 3-D angular distribution for electrons in the deep foreshock. Each plot represents the normalized flux on a surface of constant energy in the solar wind plasma rest frame. The Hammer-Aitoff equal area projection is used to display $4\pi$ steradians angular coverage. The true bin resolution is shown for the highest energy step (top left). Interpolation is used to smooth between the bin centers. The projection is rotated to place the **-B** direction (diamond) in the center of each plot. Pitch angles are displayed in the next plot (top right). The circles in each successive energy step separate the populations of escaping magnetosheath electrons (inner region), reflected electrons (within annulus) and solar wind electrons (outer region). [from *Larson et al.* 1996b]

## QUIET TIME ELECTRONS

Unlike ISEE-3, the WIND spacecraft was launched near the minimum in the solar activity cycle. Thus, there are substantial periods free from energetic solar particle events and streams. Figure 4 shows the WIND omnidirectional electron spectrum from ~ 5 eV to ~ $10^2$ keV measured during such a quiet period on February 22, 1995. Because of the extremely wide range of electron fluxes, three separate detectors--EESA-L, EESA-H, and SST--are required to measure this spectrum. The solar wind plasma Maxwellian core dominates from ~5 to ~50 eV; the solar wind halo takes over from ~ $10^2$ eV to ~ 1 keV. The halo is believed to be due to the escape of coronal thermal electrons which have a temperature of ~ $10^6$ K. Note, however, that the halo spectrum departs significantly from isothermal at energies above ~ 0.7 keV.

A third, much harder component has been discovered in the WIND observations, beginning above ~2 keV and extends to $\gtrsim 10^2$ keV, which we have denoted the "super-halo" [*Larson et al.*, 1996a]. The spectrum of the "super-halo" appears to be approximately power-law with exponent $\delta \sim 2.5$. If this "super-halo" is solar in origin, (a more detailed analysis will be required to confirm this), it would imply that electrons of such energies must be continuously present at, and escaping from, the Sun. It should be noted here that for exospheric models of the solar wind [e.g. *Maksimovic et al.*, 1996] the presence of a significant non-thermal tail to the coronal electron population is sufficient by itself to accelerate the solar wind. The mechanism for producing such a non-thermal population, if solar, is unknown, but it is tempting to speculate that it is related to the mechanism which heats the corona [see, for example, *Scudder*, 1992].

Preliminary analysis of the angular distribution of these "super-halo" electrons at very quiet times, however, shows it to be nearly isotropic, with a very slight anisotropy flowing toward the Sun. Such angular distributions would be expected for sources beyond 1 AU; the near isotropy would be due to mirroring of this incoming population in the stronger magnetic fields of the inner heliosphere. Possibly, acceleration by Co-rotating Interaction Regions (CIRs) beyond 1 AU may be responsible for the superhalo [see *Gosling et al.*, 1993], but a preliminary analysis shows no obvious one-to-one correlations with CIRs or with active regions.

## MAGNETIC CLOUDS

Magnetic clouds are a subclass of Coronal Mass Ejection (CME) characterized by relatively strong magnetic fields

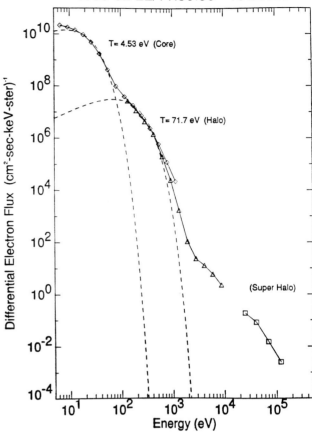

**Figure 4**. Electron differential flux spectrum from ~5 eV to $\gtrsim 100$ keV, measured at a very quiet time, in the absence of any solar particle events, by the Wind 3D Plasma and Energetic Particle experiment [*Larson et al.*, 1996a]. The diamonds, triangles, and squares indicate the three different detectors used to accommodate the wide range of fluxes over this energy range. The dashed lines give fits to Maxwellians for the solar wind core and halo.

(low plasma beta) and a smooth rotation of the magnetic field direction over a ~1 day period at 1 AU, consistent with the passage of an approximately force-free helical magnetic flux rope of diameter ~0.25 AU [*Burlaga et al.* 1988]. They often exhibit CME characteristics such as enriched alpha particle content, low proton and electron temperatures and bidirectional electron streaming [*Gosling, 1996*]. The extended periods of southward (and northward) magnetic field provided by the smooth rotation make clouds important for geomagnetic activity. As discussed by *Gosling* [1975] and *MacQueen* [1980] CMEs, including magnetic clouds, expel new magnetic flux, still connected to the Sun, into the interplanetary medium, so to maintain the IMF magnitude at a roughly constant level [*King, 1979*] compensating disconnection of magnetic flux must occur.

Solar electrons from ~0.1 to ~$10^2$ keV are excellent tracers of the structure and topology of interplanetary magnetic field (IMF) lines since they are fast and have very small gyroradii. At low energies, ~0.1 - 1 keV, the interplanetary electron fluxes are normally dominated by the continuous outflow of hot coronal electrons--the solar wind halo and strahl which provide an outward heat flux from the Sun. At higher energies ( $\gtrsim$1 to ~$10^2$ keV), the Sun often impulsively accelerates electrons in flare or flare-like event. As these electrons escape into the interplanetary medium they produce solar type III radio bursts, which can be tracked by radio observations back to the Sun. Here the WIND electron observations [*Larson et al.* 1997] are used to trace the October 18-20, 1995 cloud magnetic field back to a the Sun, measure the lengths of the field lines, and determine when magnetic disconnection from the Sun occurs.

The top three panels (a, b, c) of Plate 3 show the magnetic field strength (|B|) and direction ($\Theta$, $\Phi$) for October 18-20, 1995 magnetic cloud, from the magnetometer experiment on WIND [*Lepping et al.*, 1995]. The field rotates smoothly from south to north during the 30 hour passage of the cloud. *Lepping et al.* [1996] have fit this signature to a force-free flux-rope geometry and estimated the axis of the cloud to be nearly in the ecliptic plane ($\Theta$ = -12°), close to the Parker spiral angle ($\Phi$ = 291°), and with a right handed helicity. The cloud diameter is estimated to be about 0.27 AU and WIND passed very close to the central axis, ($y_0/R_0$=0.087).

Plate 3d is a color spectrogram of the radio observations provided by the WAVES instrument on WIND [*Bougeret et al.* 1995]. The numerous fast drift bursts, from high (14 MHz) to low (tens of kHz) frequency, are solar type III radio bursts. produced by fast electrons propagating from near the Sun (few solar radii) to ~1 AU [see *Lin* 1990 for review].

Panel g of Plate 3 shows the flux of electrons streaming away from the Sun (135° - 180° pitch angles since $\Phi$ = 291°) in 17 energy channels logarithmically spaced between ~0.1 and $10^2$ keV, respectively. At high energies, $\gtrsim$20 keV several impulsive electron events with rapid rise and slow decay can be seen, for example, at ~2100 UT Oct 18, ~0600, ~1130 and ~1800UT Oct 19. These events are also evident at lower electron energies, and generally show velocity dispersion, e.g. the faster electrons arriving earlier, typical of impulsive injections of the electrons at the Sun simultaneous at all energies, followed by travel along the same magnetic field lines out to 1 AU. A large solar energetic particle event started at ~0600 UT Oct 20 after the cloud's passage.

Panel h of Plate 3 shows a color spectrogram of the ratio $F/F_0$ of the outward streaming electron flux, $F$ (pitch angles 135° - 180°) divided by an average background flux ($F_0$) in the absence of impulsive events, to illustrate the velocity dispersion in the arrival of energetic electrons impulsively accelerated at the Sun. Each of the impulsive electron events can be traced back, with the type III radio burst produced by these electrons, to a flare observed by the Yohkoh Soft X-ray Telescope within solar active region 7912. Clearly, electrons accelerated in these flares escape outward along open field lines and are detected by the 3D Plasma experiment if the WIND spacecraft is on a field line connected to the flare site.

The crosses in Plate 3h mark the initial onset (at various energies) of the injected electrons. Using the start of the solar type III burst (red arrows in Plate 3d)--identified by the sharp increase in >5 MHz wave power measured by the WAVES experiment--as the electron injection time at the sun, the field line length ($L = v_{el} \cdot \Delta t$) traveled by the electrons was determined and plotted in Plate 3j. The field line length varies from ~4 AU near the leading edge of the cloud to ~1.5 AU near the center and rises again on the trailing part.

Panels e and f of Plate 3 show pitch angle spectrograms of 118 eV and 289 eV halo electrons. Early in the cloud (~0000-0700 UT October 19), bidirectional streaming of halo electrons is observed, indicating that both ends of the cloud field line are connected back to the solar corona. Later, after ~0700 UT October 19, the electron pitch angle distributions are uni-directional, indicating magnetic connection to the Sun on only one leg of the cloud. In addition, the electron fluxes exhibit many abrupt, discontinuous drops in level, occurring simultaneously at all energies (see panel g of Plate 3). The pitch angle spectrograms of Plate 3, panel e & f, identify these as heat flux dropouts (HFDs). *McComas et al.* [1989] suggested that HFDs indicate the disconnection of the IMF from the corona, but *Lin and Kahler* [1992] examined higher energy, >2 keV solar electrons at the times of *McComas et al.*'s HFDs and found that in most of them the IMF was still connected to the Sun. Almost all the HFDs for this cloud, however, extend up to high energies, and are observed in the impulsive events as well, indicating that these are times when the magnetic field is truly completely disconnected on both ends from the Sun. Plate 3i plots the connection to the Sun. Note that the disconnected regions range from a few minutes to hours long and they are intertwined with connected regions.

Plate 4 is a schematic representation of a flux rope model consistent with the observations. The correlation of impulsive solar electron events with x-ray flares observed by Yohkoh in active region 7912 provide the first direct identi-

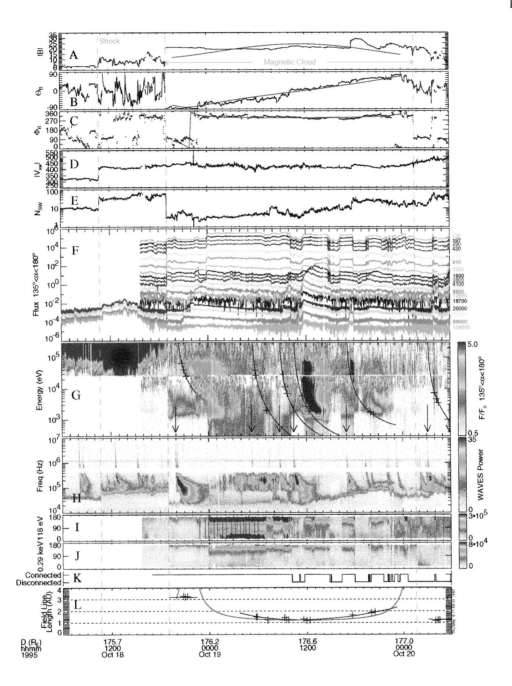

**Plate 3.** Summary plot of cloud event showing the magnetic field magnitude (|B|) and direction (Θ, Φ) in the top three panels (a, b, c). The following panel (d) is a spectrogram of the radio observations from 14 MHz to 10 kHz from the WIND Waves experiment [*Bougeret et al.* 1995]. Panel e and f show pitch angle spectrograms of 118 and 289 eV halo electrons. Panel g show the electron flux for electrons traveling outward away from the sun, anti-parallel (pitch angles, α, from 135° to 180°) to the magnetic field direction for 17 logarithmically-spaced energy channels ranging in energy from 100 eV to 60 keV. The following spectrogram (panel h), presents the ratio, $F/F_0$, electron flux ($F$) divided by an average flux ($F_0$) measured during a stable period, to show the impulsive solar electron events. Magnetic connection to or disconnection from the Sun from the heat flux dropouts is indicated in panel i. From the flare time (marked by red arrows) and the arrival time of electrons as a function of energy, the field line length at each point was determined and is shown in the panel j labeled 'Field Line Length'.

10 WIND OBSERVATIONS OF SUPRATHERMAL PARTICLES

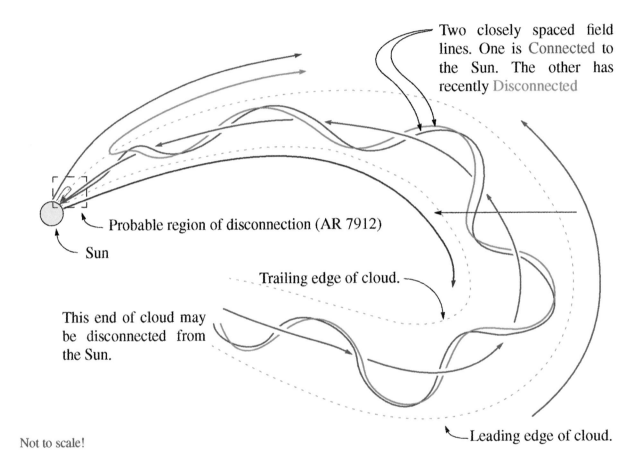

**Plate 4**. Schematic picture of possible magnetic cloud topology. In a magnetic flux rope with constant alpha, magnetic field lines wind around a central core axis in a helix pattern. Two closely spaced field lines of a magnetic flux rope are shown. The two field lines are nearly parallel but one is connected to the shown helically wound around a central core fieldline.

fication of the source region of a magnetic cloud. The field line length determined from electron arrival times are the shortest near the middle of the cloud and the longest near the edges, qualitatively supporting a flux rope model (smooth curve drawn in Plate 3j) with field lines twisted about a central core.

As the magnetic cloud passes by, the spacecraft crosses field lines with different connectivity (Plate 3i). Assuming the disconnections are the result of reconnection near the Sun, an important question is when these reconnection events occurred. It is unlikely that the disconnections occurred before the ejection of the cloud from the Sun ~4 days earlier, since then the magnetic tension forces would most likely have significantly distorted the cloud. Since the travel time from the Sun to the spacecraft at 1 AU varies from ~20 minutes for the fastest electrons to ~6 hours for the slowest, and the HFDs are seen simultaneously across all energies, the disconnections must have occurred ~>6 hours earlier.

*Burlaga* [1991] and *Rust* [1994] suggest that magnetic clouds are the interplanetary manifestation of solar filaments (prominences on the limb), which are often observed to have helical magnetic topology and to be embedded in CMEs. Alternatively *Gosling* [1990] suggested that 3-D magnetic reconnection across the legs of sheared arcades of coronal loops will result, after relaxation, in magnetic cloud flux rope structures when the plasma beta is low. *Gosling et al.* [1995] have argued this process could produce field lines that are connected at one or both ends to the Sun and even completely disconnected from the Sun. However, the cloud could also have emerged from the solar surface as a fully evolved flux rope, and parts of the flux rope could have become disconnected from the solar surface as the cloud moves through the interplanetary medium. As mentioned earlier, such a process must occur eventually for all CMEs that expel magnetic flux into the interplanetary medium.

*Acknowledgments.* This research was supported in part by NASA grant NAG5-2815.

# REFERENCES

Bougeret, J.-L., M.L. Kaiser, P.J. Kellogg, R. Manning, K. Goetz, S.J. Monson, N. Monge, L. Friel, C.A. Meetre, C. Perche, L. Sitruk, and S. Hoang, WAVES: the radio and plasma wave investigation on the WIND spacecraft, *Space Sci. Rev., 71*, 231, 1995.

Burlaga, L. F., Magnetic clouds and force-free fields with constant alpha, *J. Geophys. Res., 93*, 7217, 1988.

Burlaga, L. F., Magnetic clouds, in *Physics of the Inner Heliosphere II*, ed. R. Schwenn and E. Marsch, Springer-Verlag, Berlin, 1, 1991.

Dulk, G.A., and D.J. McLean, Coronal magnetic fields, *Solar Physics, 57*, 279, 1978.

Feldman, W. C., J. R. Asbridge, S. J. Bame, M. D. Montgomery, and S. P. Gary, Solar wind electrons, *J. Geophys. Res., 80*, 4181, 1975.

Feldman, W. C., R. C. Anderson, S. J. Bame, S. P. Gary, J. T. Gosling, D. J. McComas, M. F. Thomsen, G. Paschmann, and M. M. Hoppe, Electron velocity distributions near the Earth's bow shock, *J. Geophys. Res., 88*, 96, 1983.

Gosling, J. T., Large-scale inhomogeneities in the solar wind of solar origin, *Rev. Geophys., 13*, 1053, 1975.

Gosling, J. T., Coronal mass ejections and magnetic flux ropes in interplanetary space, in *Physics of Magnetic Flux Ropes*, ed. C. T. Russell, E.R. Priest, and L.C. Lee, *Geophys. Monogr.* **58** Amer. Geophys. Union, 343, 1990.

Gosling, J. T., S. J. Bame, W. C. Feldman, D. J. McComas, J. L. Phillips, B. E. Goldstein, Counterstreaming suprathermal electron events upstream of corotating shocks in the solar wind beyond ~2 AU: Ulysses, *Geophys. Res. Lett. 20*, 2335, 1993.

Gosling J. T., J. Birn, and M. Hesse, Three-dimensional magnetic reconnection and the magnetic topology of coronal mass ejection events, *Geophys. Res. Lett., 22*, (8)869-872, 1995.

Fitzenreiter, R. J., J. D. Scudder, and A. J. Klimas, Three- dimensional analytical model for the spatial variation of the foreshock electron distribution function: Systematics and comparisons with ISEE observations, *J. Geophys. Res., 95*, 4155, 1990.

King, J. H., Solar cycle variations in IMF intensity, *J. Geophys. Res., 84*, 5938, 1979.

Larson, D.E., R.P. Lin, K.A. Anderson, S. Ashford, C.W. Carlson, D. Curtis, R.E. Ergun, J.P. McFadden, M. McCarthy, G.K. Parks, H. Reme, J. M. Bosqued, J. Coutelier, F. Cotin, C. d'Uston, K.P. Wenzel, T.R. Sanderson, J. Henrion, and J.C. Ronnet, *Solar Wind 8 Conf. Proc.*, in press, 1996a.

Larson, D.E., R.P. Lin, J.P. McFadden, C.W. Carlson, K.A. Anderson, L. Muschietti, M. McCarthy, G. Parks, H. Reme, J. M. Bosqued, C. d'Uston, T.R. Sanderson, K.P. Wenzel, R.P. Lepping, Probing the Earth's bow shock with upstream electrons, *Geophys. Res. Lett., 23*, 2203, 1996b.

Larson, D.E., R.P. Lin, R.E. Ergun, J. McTiernan, J.P. McFadden, C.W. Carlson, K.A. Anderson, M. McCarthy, G.K. Parks, H. Reme, T.R. Sanderson, M. Kaiser, and R.P. Lepping, Using electronic electrons to Probe the Topology of the October 18-20, 1995 Magnetic Cloud, *Geophys. Res. Lett.*, in preparation 1997.

Lepping, R. P. et al., The WIND magnetic field investigation, *Space Sci. Rev., 71*, 207, 1995.

Lepping, R. P., L. F. Burlaga, A. Szabo, K. W. Ogilvie, W. H. mish, D. Vassiliadis, A. J. Lazarus, J. T. Steinberg, C. J. Farrugia, L. Janoo, and F. Mariani, The WIND magnetic cloud and events of October 18-20, 1995: Interplanetary properties and as triggers for geomagnetic activity, *J. Geophys. Res.*, submitted, 1996.

Lin, R.P., Non-relativstic solar electrons, *Space Sci. Rev., 16*, 189, 1974.

Lin, R.P., Electron beams and Langmuir turbulence in solar Type III radio bursts observed in the interplanetary medium, in *Basic Plasma Processes on the Sun*, ed. E. Priest and V. Krishan, p.

467, International Astronomical Union, The Netherlands, 1990.

Lin, R. P.; S. W. Kahler, Interplanetary magnetic field connection to the Sun during electron heat flux dropouts in the solar wind, *J. Geophys. Res., 97,* 8203, 1995.

Lin, R. P., K. A. Anderson, S. Ashford, C. Carlson, D. Curtis, R. Ergun, D. Larson, J. McFadden, M. McCarthy, G. K. Parks, H. Rème, J.M. Bosqued, J. Coutelier, F. Cotin, C. d'Uston, K.-P. Wenzel, T.R. Sanderson, J. Henrion, and J.C. Ronnet, A three-dimensional plasma and energetic particle investigation for the Wind spacecraft, *Space Sci. Rev., 71,* 125, 1995.

Lin, R. P., Solar impulsive electron events, *CESRA Proc.*, submitted, 1996.

Maksimovic, M., V. Pierrard, and J. Lemaire, *Astron. Astrophys.*, submitted, 1996.

MacQueen, R. M., Coronal transients: A summary, *Philos. Trans. R. Soc. London, Ser. A., 297,* 605, 1980.

McComas, D. J., J. T. Gosling, J. L. Philips, S. J. Bame, J. G. Luhmann, and E. J. Smith, Electron heat flux dropouts in the solar wind: Evidence for interplanetary magnetic field reconnection?, *J. Geophys. Res., 94,* 6907, 1989.

Pannekoek, A., Ionization in stellar atmospheres, *Bull. Astron. Inst. Neth., 1,* 107, 1922.

Potter, D. W., R. P. Lin and K. A. Anderson, Impulsive 2-10 keV solar electron events not associated with flares, *Astrophys. J., 236,* L97, 1980.

Rosseland, S., Electrical state of a star, *Monthly Notices Royal Astron. Soc., 84,* 720, 1924.

Rust, D. W., Spawning and shedding helical magnetic fields in the solar atmosphere, *Geophys. Res. Lett., 21,* 241, 1994.

Saito, K., A. I. Poland, and R. H. Munro, A study of the background corona near solar minimum, *Solar Phys., 55,* 121, 1977.

Scudder, J. D., On the causes of temperature change in inhomogeneous, low-density astrophysical plasmas, *Astrophys. J., 398,* 299, 1992.

Scudder, J. D., A review of the physics of electron heating at collisionless shocks, *Adv. Space Res. 15(8/9),* 181, 1995.

Scudder, J. D., Electron and ion temperature gradients and suprathermal tail strengths at Parker's solar wind sonic critical point, *J. Geophys. Res. 101,* 11039, 1996.

Slavin, J. A., and R. E. Holzer, Solar wind flow about the terrestrial planets: 1. Modeling bow shock postition and shape, *J. Geophys. Res., 86,* 11401, 1981.

---

R.P. Lin, D. Larson, T. Phan, R. Ergun, J. McFadden, K. Anderson, and C. Carlson, Space Sciences Laboratory, University of California, Berkeley, CA 94720-7450

M. McCarthy, G. K. Parks, R. Skoug, and R. Winglee, Geophysics Program AK50, Box 351650, University of Washington, Seattle, WA 98195-1650

H. Rème, N. Lormant, J. M. Bosqued, and C. d'Uston, Centre d'Etude Spatiale des Rayonnements, Universite Paul Sabatier, B.P. 4346, 9 avenue du Colonel Roche, F-31029 Toulouse, France

T. R. Sanderson and K.-P. Wenzel, Space Science Department of ESA, European Space Research and Technology Centre, Postbus 229, Domeinweg, 2200 AG Noordwijk, Netherlands

# Dayside Electrodynamics Observed by Polar with Northward IMF

N. C. Maynard[1], W. J. Burke[2], D. R. Weimer[1], F. S. Mozer[3], J. D. Scudder[4], C. T. Russell[5], and W. K. Peterson[6]

We present measurements of electric/magnetic fields and energetic particle fluxes acquired during two Polar passes above the northern dayside ionosphere. The WIND satellite determined that in both instances the interplanetary magnetic field (IMF) had northward components, similar clock angles in the $Y_{GSM}$-$Z_{GSM}$ plane, and comparable intensities. At different times during the passes Polar encountered particle fluxes with central plasma sheet (CPS), boundary layer, cusp and polar cap spectral characteristics. Electric fields measured during cusp/boundary layer passages are marked by very large variability with variations extending from the PC 1 through the PC 4 ranges. Although the orbital paths of Polar during the two passes followed similar trajectories, the large-scale dynamics encountered were different. In one instance the downward moving ions displayed a "reverse", velocity-dispersion feature, with the highest energies detected near the poleward boundary of the cusp, indicating that Polar crossed magnetic field lines that map upward to a merging region poleward of the cusp. The cusp was part of a counter-clockwise rotating (positive potential) cell, adjacent to a negative potential afternoon cell, which included precipitation from CPS and boundary layer sources. In the other case, Polar exited a wide boundary layer directly into the polar cap, where it detected a clockwise-rotating (negative potential) lobe cell, poleward of its merging line. We compare electric potential distributions detected at altitudes near 5 $R_E$ with observations from satellites in low-Earth orbit and predictions of a model derived from them, under prevailing solar wind conditions.

[1] Mission Research Corporation, Nashua, New Hampshire
[2] Phillips Laboratory, Hanscom Air Force Base, Massachusetts
[3] Space Sciences Laboratory, University of California, Berkeley
[4] Department of Physics and Astronomy, University of Iowa, Iowa City
[5] Institute for Geophysics and Planetary Physics, University of California, Los Angeles
[6] Lockheed Martin Space Sciences Laboratory, Palo Alto, California

Geospace Mass and Energy Flow: Results From the International Solar-Terrestrial Physics Program
Geophysical Monograph 104
Copyright 1998 by the American Geophysical Union

## INTRODUCTION

The entry of particles and electric fields from the interplanetary medium into the Earth's magnetosphere during periods of northward IMF has been studied using a broad spectrum of ground- and space-based sensors. The dayside magnetosphere quickly responds to changes in the polarities of IMF $B_Y$ and/or $B_Z$. Within a few minutes (approximately an Alfvén travel time) of the changes reaching the magnetopause, characteristic optical [*Sandholt*, 1991; *Sandholt et al.*, 1996; *Murphree et al.*, 1990] and plasma convection signatures [*Clauer and Friis-Christensen*, 1988] appear in the ionospheric projection of the cusp. The changes are believed to result from a shift in the site of magnetic merging between the IMF and the Earth's field from equatorward of the cusp with IMF $B_Z < 0$ to the poleward boundary of the cusp with $B_Z > 0$ [*Russell*, 1972].

If the IMF retains a northward orientation for longer than a half hour significant changes appear in global ionospheric convection patterns. *Maezawa* [1976] reported detecting magnetic perturbations during the Antarctic summer, indicating that regions of sunward convection develop in the polar cap when the IMF is northward. Electric and magnetic fields measured by the S3-2 [*Burke et al.*, 1979], Atmospheric Explorer [*Reiff and Heelis*, 1994] and MAGSAT [*Iijima and Shibaji*, 1984] satellites suggested that with IMF $B_Z > 0$ and $B_Y \sim 0$, a four-cell convection patterns evolves. The pattern consists of two cells in the central polar cap, which are of opposite polarity to the standard negative afternoon cell and positive morning cell and are driven by merging at the poleward boundary of the cusp [*Dungey*, 1961; *Russell*, 1972], and a weak but standard-polarity pair of cells in the auroral oval, possibly driven by the low latitude boundary layer [*Eastman et al.*, 1976]. Using electric field measurements of the DE 2 satellite, *Heppner and Maynard* [1987] showed that during periods of northward IMF in which $B_Y$ has large values, two distorted convection cells operate at high latitudes. The sense of cell rotation is the same as that observed with IMF $B_Z < 0$, but the axis of symmetry for the two cells is rotated far from the noon-midnight meridian. The orbits used by *Heppner and Maynard* [1987] to infer the convection pattern with $B_Z > 0$ and $B_Y > 0$ were reexamined by *Burke et al.* [1994] using simultaneous measurements from the retarding potential analyzer [*Hanson et al.*, 1981] and electron spectrometer [*Winningham et al.*, 1981] on DE 2. These provided supplementary information about the directions of convection flow lines (equipotentials) crossing the selected DE 2 trajectories and about the source regions of precipitating electrons. They concluded that the afternoon (negative potential) cell is rotated into the prenoon sector and consists of two parts: (1) equipotentials whose associated magnetic flux is always open, and (2) equipotentials whose associated flux is both open and closed. Equipotentials with circulating open flux, hereafter referred to as lobe cells [*Reiff and Burch*, 1985], are embedded within the afternoon cell and have the same sense of rotation.

Passages of satellites through the dayside high-latitude ionosphere are marked by encounters with distinctive particle and field characteristics. *Newell et al.* [1991a, b] have identified the spectral properties of electrons and ions in the dayside ionosphere whose magnetospheric sources are the central plasma sheet (CPS), the low-latitude boundary layer (LLBL), the cusp and the mantle. In energy-versus-time spectrograms the cusp is marked by intense fluxes of low energy (<100 eV) electrons and energy-dispersed ions. The latter signature is a time-of-flight, velocity-filter effect. During periods of northward (southward) IMF, the highest energy ions are detected near the poleward (equatorward) boundary of cusp precipitation [*Reiff et al.*, 1980; *Burch et al.*, 1980]. *Maynard* [1985] showed that cusp entry is also marked by a significant increase in the level of low-frequency noise measured by electric field sensors. The rapid traversal of the cusp by satellites in low-altitude orbits only allows for the detection of waves in the Pc-1 band [*Maynard*, 1985; *Basinska et al.*, 1992]. Satellite crossings of the ionospheric projections of magnetopause merging sites are frequently marked by electromagnetic spikes at the equatorward [*Maynard et al.*, 1991] and poleward [*Basinska et al.*, 1992] boundaries of cusp precipitation during periods of southward and northward IMF, respectively. On larger scales, ionospheric plasma convection in the cusp has a sunward (poleward) component when the IMF has a northward (southward) component. The azimuthal component of convection is controlled by the polarity of IMF $B_Y$. Convection is westward in the northern hemispheric cusp when IMF $B_Y > 0$. The opposite polarity relationship maintains in the southern hemisphere cusp. Finally we note that besides the large scale, Region 1/Region 2 systems, the dayside ionosphere is marked by cusp and mantle field-aligned current (FAC) systems, often referred to as Region 0 [*Iijima and Potemra*, 1976; *Erlandson et al.*, 1988].

Empirical models for patterns of high-latitude potentials (convection) have been constructed for various solar wind/IMF conditions. These have been based on pattern recognition normalization technique using the DE-2 data set [*Heppner and Maynard*, 1987] and average values from the large DMSP data bases [*Rich and Hairston*, 1994]. Recently *Weimer* [1995, 1996] has developed a technique that uses least-squares fits of spherical harmonics that also includes effects of the dipole tilt angle. Published representations of the modeled convection patterns appear to be quite complex, especially for periods with IMF $B_Z > 0$ and $|B_Y| \leq B_Z$. Multiple small cells evolve within larger convection cells.

This paper presents electric/magnetic field and particle measurements acquired by instrumentation on the Polar satellite during two passes at dayside, high latitudes in extended periods of northward IMF in which $|B_Y| \leq B_Z$. The following section contains brief descriptions of Polar sensors used in this study. The observations section describes particle and field measurements taken at middle altitudes by Polar on April 3, and 8, 1996. Interpretations of the observations in terms of Polar encounters with middle altitude projections of previously identified magnetospheric plasma regimes are presented in the final section. We also compare Polar measurements of high-latitude convection with the predictions of the *Weimer* [1995, 1996] model under prevailing interplanetary conditions.

## INSTRUMENTATION

Polar was launched into a 90° inclination orbit from the Western Test Range at Vandenburg, California on February 24, 1996. The initial orbital plane was in the early post-noon/midnight local time sectors. Apogee (above the northern polar cap) and perigee are at geocentric distances of 9 and 1.8 $R_E$, respectively. The spacecraft is spin stabilized at 10 rpm, with its spin axis perpendicular to the orbital plane. The three instruments whose measurements are used in this study are the electric field instrument (EFI), the magnetic field investigation (MFI), and the HYDRA electron/ion spectrometer. Comprehensive descriptions of these sensors are provided by *Harvey et al.* [1994], *Russell et al.* [1994], and *Scudder et al.* [1994], respectively. Only brief summaries of their relevant capabilities are provided below.

EFI consists of three dipoles to measure vector electric fields from potential differences between three pairs of spherical sensors. Two of the pairs of sensors are held at separation distances of 100 m and 130 m by wire booms that move in the spacecraft's spin plane. The third pair is held at a separation of 14 m by a pair of rigid booms, aligned with the spin axis. The two spin-plane components of the electric field are represented by the symbols $E_{X-Y}$ and $E_Z$. $E_{X-Y}$ is the projection of the spin-plane component of the electric field onto the GSE (geocentric solar ecliptic) plane. It is positive whenever the unit vector has a component in the $-X_{GSE}$ direction. $E_Z$ is positive toward the GSE north pole. The third component, $E_{56}$, points along the spin axis, positive in the sense that completes an orthogonal, right-hand coordinate system. Thus with Polar orbiting in the noon meridian, components of the vector ($E_{X-Y}$, $E_Z$, $E_{56}$) are positive in the ($-X_{GSE}$, $+Z_{GSE}$, $+Y_{GSE}$) directions. Data are sampled at a rate of 40 s$^{-1}$ by all three sensors. Corrections must be made for dc offsets and fields induced by satellite's motion ($-\mathbf{V_s} \times \mathbf{B}$) and by transformation of the measurements into the corotating frame. In addition, to eliminate magnetic wake effects at lower densities, data taken when a sensor axis was within an avoidance angle to the magnetic field were eliminated. The remaining data have been fit to a sine wave. The optimum avoidance angle value varies with density and is given for each data set below. Spin-fit measurements are presented below every 6 s for $E_{X-Y}$ and $E_Z$. Measurements of $E_{56}$ by the short booms are contaminated by unknown levels of dc offsets. To estimate their values we normally make independent calculations of $E_{56}$ using the $\mathbf{E} \cdot \mathbf{B} = 0$ condition. However, in the two cases chosen for analysis, the magnetic field vector lay very close the spin plane. This extrapolation then causes small uncertainties in the spin plane to be multiplied by large factors which creates large uncertainties in the extrapolated $E_{56}$ value. For this reason we present in this paper only the electric field's spin-plane components.

MFI consists of two orthogonal, triaxial fluxgate magnetometers that are mounted on a nonconducting boom at separation distances from the nearest satellite surface of 5.97 m and 4.75 m. the outer sensor operates in two ranges ±5525 nT and ±694 nT; the inner sensor in the ranges ±46,700 nT and ±5860 nT. Data are sampled at a rate of 500 s$^{-1}$ in all three components and averaged with a recursive filter to provide data at 100 samples per second and 8 samples per second. Only 8 samples per second data are available for the times under study here. In this study we are mostly concerned with magnetic perturbations produced by currents that couple the high-latitude to the magnetosphere or magnetosheath along closed or open magnetic field lines, respectively. Since these large-scale FAC systems generally extend much further in longitude than latitude [*Iijima and Potemra*, 1976], associated magnetic perturbations are mostly in the azimuthal component $B_{56}$. Positive-slope deflections in $B_{56}$ with time (latitude) are detected as the northward-moving Polar crosses FAC sheets directed into the ionosphere.

The HYDRA instrument consists of two pairs of electron and ion/electron spectrometers, each mounted 180° apart on the spacecraft body. In this paper we only use ion and electron measurements from the Duo Deca Electron Ion Spectrometer (DDEIS). As the name suggests, DDEIS is made up of six pairs of 127° electrostatic analyzers looking in different directions outward on a unit sphere. Each measures the fluxes of ions and electrons in 55 energy steps between 10 eV/q to 10 keV/q with a resolution in energy of $\Delta E/E = 6\%$, and angle of 10°. Fully three-dimension distribution functions for electrons and ions are acquired in 0.5 s.

## OBSERVATIONS

This section presents particles and fields measured by Polar during dayside passes on April 3, and April 8, 1996.

## Table 1. Interplanetary conditions

| Date | SW Speed | $B_X$ | $B_Y$ | $B_Z$ |
|---|---|---|---|---|
| 4/3/96 | 353 km/s | -0.3 ±0.5 nT | 3.1 ±0.8 nT | 5.1 ±1.0 nT |
| 4/8/96 | 312 km/s | 0.9 ±0.8 nT | 6.4 ±0.5 nT | 7.0 ±0.4 nT |

In both cases the satellite was moving toward higher altitudes and northern-hemisphere latitudes. Average solar wind speeds (km/s) and the interplanetary magnetic field (nT) $X_{GSM}$, $Y_{GSM}$ and $Z_{GSM}$ components along with the variability over 40 min are summarized for the periods of interest in Table 1. The data come from the WIND satellite located near ($X_{GSM}$, $Y_{GSM}$, $Z_{GSM}$) of (76.9, 2.1, 0.3) $R_E$ and (82.1, 27.5, -2.8) $R_E$ on April 3, and April 8, respectively. Since interplanetary conditions were stable across the studied interval, and we are primarily interested in steady-state responses of the magnetosphere, knowing exact propagation times between WIND and the Earth are not critical for this study. Table 1 shows that in both cases the solar wind speed was in the low-to-moderate range. The IMF had substantial, positive $Y_{GSM}$ and $Z_{GSM}$ components, with $B_Y < B_Z$, and with $B_X \sim 0$. on April 3. It had comparable values on April 8. We note that the level of geomagnetic activity was low on both days, during the times of interest, with $Kp$ indices of 2- and 2, respectively.

The remainder of this section is divided into two parts which present Polar observations first from April 8, then April 3. Here our task is to present physical quantities measured by the three sensors and point out empirical relationships between them. Interpretations, especially of potentially controversial aspects, are relegated to the discussion section. Because of their greater familiarity for identifying source regions, we show HYDRA measurements before those from EFI/MFI.

*April 8, 1996*

Plate 1A displays electron (top) and ion (bottom) differential energy flux accumulated in 24 ms intervals by DDEIS from 1400 to 1700 UT on April 8, 1996, in an energy-versus-time spectrogram format. Averages over 13.8 s of data acquisition of these fluxes per 24 ms values are displayed, with each spectra representing the average over 72 such readings of the fluxes at the indicated energy. The color bar indicates flux intensities. The dashed lines crossing the two spectrograms represent the time dependence of the average energies of the sampled populations. The measurements are presented as functions of universal time (UT), invariant latitude (ILT), magnetic local time (MLT) and geocentric distance in $R_E$. Values of the ILT, determined using IGRF 95, are related to the magnetic L shell by the familiar relationship

$$\text{ILT} = \cos^{-1}(1/L^{-\frac{1}{2}})$$

On a purely empirical basis, the particle measurements from April 8 divide into four time intervals. (1) Prior to 1500 UT the highest ion count rates were at energies in the 8 to 3 keV/q range. The energy of peak flux decreased with increasing ILT. Electron fluxes were high (>100) at energies below 60 eV and low (<10) at energies above 100 eV. Their sharp spectral cutoff suggests that low-energy population mostly consists of atmospheric photoelectrons [*Doering et al.*, 1976]. (2) The period between 1500 and 1505 UT is marked by a burst of electrons with energies between 200 eV and 2 keV. A similar increase in the flux of ions with ~5 keV/q was detected simultaneously. (3) The third period extends from 1505 to 1540 UT and is marked by nearly constant fluxes of electrons with energies <200 eV and dynamic variations in the ion measurements. From about 1505 to 1515 UT the energies of highest ion fluxes decreased from 4 to 0.2 keV/q. With notable small-scale variations, average ion energies remained within 0.2 - 0.3 keV/q until 1526 UT when they began to increase systematically to 2 keV/q at 1537 UT. Phenomenologically, we describe the changing ion spectral characteristics at 1501 - 1508 UT and 1526 - 1537 UT as "standard" and "inverse" ion dispersion events. (4) After 1540 UT, DDEIS detected electron fluxes whose intensities and energies were significantly reduced, and ion count rates near background levels. Based on experience with the spectral characteristics of ion and electron fluxes observed at low altitudes [*Newell et al.*, 1991a, b], we identify the source regions encountered by Polar during the first interval as the central plasma sheet (before 1500 UT), and during the fourth interval as the polar cap (after 1540 UT), respectively. Interpretation of the source regions for particles observed during the second and third segments is deferred.

Figure 1 shows the spin-plane components of the electric field, the electric potential distribution along Polar's trajectory, and the spin-axis components of the plasma drift $V_{56}$ and magnetic field $B_{56}$ with the T96 model field [*Tsyganenko*, 1996] subtracted. A magnetic

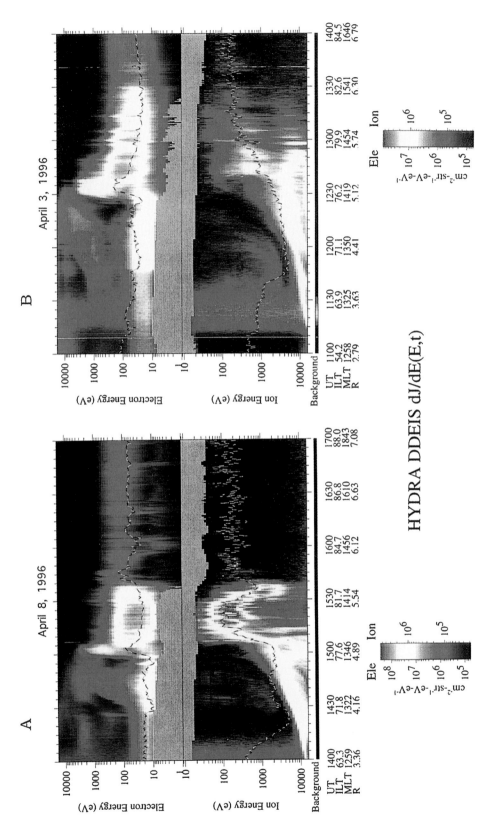

Plate 1. Energy-versus time spectrogram showing differential energy flux detected by HYDRA of electrons (top) and ions (bottom) with energies between 10 eV/q and 20 keV/q from (A) 1400–1700 UT on April 8, 1996 and from (B) 1100–1400 UT on April 3, 1996 (see text). The lowest energies are displayed at the center of the spectrogram. The measured energies have been adjusted in accordance with the spacecraft potential determined by the electric field instrument (gray areas at the top of the ion spectrogram and the bottom of the electron spectrogram). Color bars are different and are given under each spectrogram.

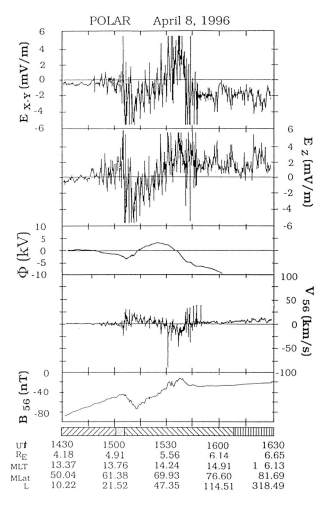

**Figure 1.** Electric and magnetic field measurements from 1430 – 1630 UT on April 8, 1996. From top to bottom the panels give the electric field spin-plane components $E_{X-Y}$ and $E_Z$, the electric potential $\phi$ derived from an integration along the Polar trajectory, the plasma-drift $V_{56}$ and magnetic field $B_{56}$ component transverse to the orbital plane. We have subtracted the *Tsyganenko* [1996] T96 model field from the $B_{56}$ measurement to obtain the values plotted. Data are displayed as functions of universal time, the geocentric distance of the satellite, magnetic local time, magnetic latitude and L shell. The four particle regions described in the text are marked at the bottom of the figure for reference}

field avoidance angle of 0° up through 1550 UT and 45° thereafter was used for the electric field spin fits. Potential values are for the corotating reference frame for easy comparison to the ionospheric patterns. Since the satellite velocity $\mathbf{V}_s$ lies very close to the spin plane, estimates of the potential distribution along the trajectory

$$\phi = \int -\mathbf{E} \cdot \mathbf{V}_s \, dx$$

should be accurate, despite our lack of knowledge about $E_{56}$. Its utility for understanding geomagnetic processes requires that the convection pattern be relatively steady over the interval in which the measurements were acquired. The steady conditions observed by WIND in the interplanetary medium provides justification for this an assumption. Likewise our calculation of the $V_{56}$ component of

$$\mathbf{V} = (\mathbf{E} \times \mathbf{B})/|B|^2$$

requires no knowledge of $E_{56}$. For reference, we have marked the four different particle-flux segments at the bottom of the figure.

Attention is directed to the following seven points. (1) Prior to 1452 UT, the absolute values and variability of electric field was <1 mV/m. After this the amplitudes of variations grew. (2) From 1503 to 1542 UT the amplitude of electric field variations assumed values <5 mV/m. The periods of the variations ranged from a few tens of seconds to a few minutes (Pc 1 to Pc 4). (3) Average (quasi dc) values of the $E_{X-Y}$ and $E_Z$ show similar variations, with the polarity of both components reversing near 1525 UT. (4) The third panel shows Polar crossed two regions of negative potential that bound a region of positive potential (1512 - 1534 UT). Viewed from above the north pole, the sense of rotation for plasma convection is clockwise/counterclockwise in regions of negative/positive potential (for instance, see patterns of *Heppner and Maynard* [1987]). In this case the potential of the counterclockwise rotating cell is ~3 kV. (5) The east-west component $V_{56}$ had low values prior to 1504 UT. From 1504 to 1526 UT Polar detected eastward flow (positive $V_{56}$), with an average value of ~10 km/s, as it crossed the region of "standard" ion dispersion and of low-energy, ion/electron fluxes. An average westward flow (negative $V_{56}$) of ~10 km/s to the west marked the "inverse" dispersion event. This is consistent with counterclockwise plasma rotation in the cusp derived from $\phi$ measurements. In the polar cap $V_{56}$ assumed low, steadily eastward values. (6) Variations in the trace of $B_{56}$ mimic those of quasi-dc values of $E_{X-Y}$ and $E_Z$. This indicates that Polar crossed several large-scale FAC sheets which close via Pedersen currents in the ionosphere [*Smiddy et al.*, 1980]. (7) Positive (negative) slopes in the $B_{56}$ trace indicate FACs into (out of) the ionosphere. Consistent with a postnoon MLT trajectory, the positive (before 1505 UT) and negative (1505 - 1512 UT) slopes in $B_{56}$ appear to be encounters with the afternoon Region 2 and Region 1 systems [*Iijima and Potemra*, 1976], respectively. The remaining FACs belong to the Region 0 system

[*Erlandson, et al.*, 1988]. As expected, the main Region 0 current has a polarity opposite to that of the adjacent Region 1. A small region of upward current is seen poleward of the main Region 0 current.

*April 3, 1996*

Plate 1B presents HYDRA measurements from 1200 to 1345 UT on April 3, 1996. Again, we divide the measurements as coming from four phenomenologically distinct regions. (1) Prior to ~1229 UT two populations of low- and high-energy electrons are sampled. From spectral similarities to fluxes detected at the same invariant latitudes during the previous pass, we identify them as photoelectrons and electrons from the CPS. Over the same period energetic (>1 keV) ion fluxes, whose mean energies decreased with increasing invariant latitude, were observed. (2) Fluxes of electrons with energies between about 500 eV and 4 keV underwent rapid increases and decreases at 1229 and 1238 UT, respectively. In this interval fluxes of ions with energies <200 eV rose above background. (3) From about 1238 to 1330 UT Polar crossed a region of high fluxes of electrons whose mean energies, except for a few small-scale structures, decreased smoothly with invariant latitude. This region is marked by two ion populations with energies centered near 500 eV and 7 keV. (4) After 1330 UT HYDRA detected polar rain electron and low ion flux levels, more typical of the polar cap. Again we describe particle from the first and fourth intervals as marking the CPS and polar cap, respectively. Sources for particles of the second and third will be discussed later.

Electric and magnetic field measurements from the same period are given in Figure 2. A magnetic field avoidance angle of 30° was used throughout for the electric field spin fits. Variations with time-scales of a few minutes or more in the spin-plane components of the electric field, $E_{X-Y}$ and $E_Z$ track well with each other and with those of $B_{56}$, with the model field subtracted, over the entire interval. We see that large-amplitude fluctuations with periods near 10 s appear continuously between 1233 and 1300 UT, and in bursts near 1330 UT. The potential distribution detected by EFI, indicates that Polar was in a region of negative potential, characteristic of the afternoon convection cell. The plasma flow component $V_{56} = (E_{X-Y} B_Z - E_Z B_{X-Y})/B^2$ was weaker than observed on April 8, and after 1235 UT was predominantly toward the east. On large-scale, the trace of $B_{56}$ appears similar to that observed on April 8. Positive slopes, indicating FACs into the ionosphere were observed before 1236 UT (region 2) and from 1248 to 1308 UT; a negative trend in the slope of $B_{56}$ (Region 1)

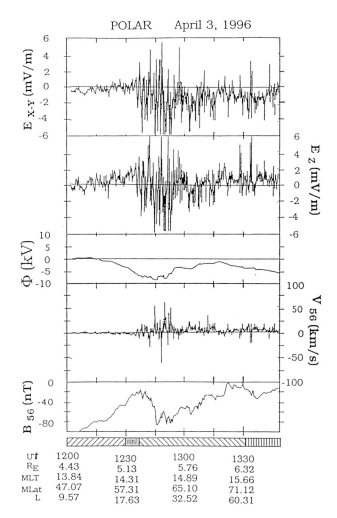

**Figure 2**. Electric and magnetic field measurements from 1200 – 1345 UT on April 3, 1996, in the same format as Figure 1.

appeared between these intervals. A significant difference between the $B_{56}$ traces of Figures 1 and 2 is the abundance of low-frequency fluctuations with amplitudes of a few nT. Their correlation with electric field variations indicates that Polar encountered Alfvén waves in the Pc 4 and 5 bands, as well as the large-scale FAC systems and suggests a more dynamic environment than that of the April 8 pass. Finally we note that the sharp negative turns in the $B_{56}$ trace at 1300 and 1304 UT coincide with small-scale bursts of electron found in Plate 1B.

## DISCUSSION

Data from the two Polar passes with northward IMF exhibit both similarities and differences. The magnetic field variations are characteristic of the afternoon Regions 0, 1 and 2 systems of FACs. At the highest and lowest

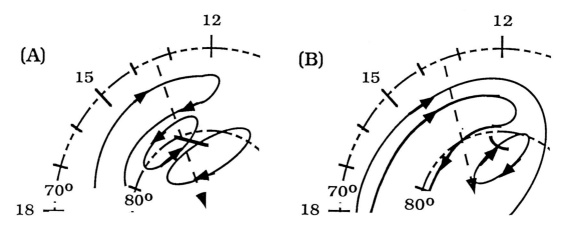

**Figure 3.** Schematic representation of ionospheric projection of potential distributions detected by Polar near (A) 1515 UT on April 8, 1996 and (B) 1300 UT on April 3, 1996. Directions of plasma flow are indicated by arrows.

invariant latitudes for which HYDRA spectral measurements are shown in Plates 1 and 2, one finds familiar characteristics of the CPS and polar cap. Localized bursts of kilovolt electrons are found in both instances near the poleward boundary of Region 2 currents. Since Region 2 currents are driven by pressure gradients in the plasma sheet [*Harel et al.*, 1981], and particle spectral characteristics do not point elsewhere, we suggest that the bursts originate in the outer CPS. Physical processes responsible for the observed flux increases must operate deeper in the magnetospheric than the Polar orbit. The most intriguing aspects of the two data sets were observed in the third particle segments (1505 – 1540 UT, April 8, and 1330 UT, April 3), which roughly spans the Regions 1 and 0 FACs. In both cases the segments of interest began as Polar entered the Region 1 current sheet and ended as it entered the polar cap. Between these boundaries significantly different observations were made. Our analysis and interpretation of the observations from this portion of the Polar orbits follows.

Figure 3a schematically represents the data constraints on the convection pattern encountered by Polar on April 8, projected onto the ionosphere. From 1505 to 1512 UT the plasma had an eastward drift component and the ion spectrogram displayed a "standard" dispersion signature. During periods of southward IMF, such ion-dispersion structures are normally regarded as time-of-flight effects as ions move from magnetic-merging sites on the magnetopause toward the ionosphere. With a northward IMF, the inner (low-latitude) portion of the LLBL is dominated by high-energy ions from the plasma sheet and the outer (high-latitude) portion by low-energy ions from the magnetosheath [*Eastman et al.*, 1976]. The apparent "standard" dispersion results from the superposition of these two populations. The separation of the two sources is even more evident in the second event (Plate 1B). During the period of the apparent "standard" dispersion, Polar moved about 1.1° in invariant latitude. This is the approximately the same latitudinal width of the LLBL observed above the *ionosphere* [*Smiddy et al.*, 1982]. An examination of the potential distribution measured between 1452 and 1512 UT, indicates that equipotentials in the CPS continue into the LLBL. Finally, our interpretation of the particles detected by HYDRA as coming from the LLBL is consistent with theoretical arguments developed by *Siscoe et al.* [1991] that the LLBL is the current generator for the dayside Region 1.

From 1512 to 1537 UT Polar crossed a positive potential cell in which plasma convection had an eastward (westward) drift component in its equatorward (poleward) parts. The "inverse" ion dispersion structure appeared in the poleward part. The simplest interpretation of these observations is that Polar went through the mid-altitude projection of the cusp. Magnetic merging with the northward IMF occurred at the cusp's poleward boundary. In this scenario, Polar crossed the magnetic mapping of the merging line (the heavy line in Figure 3a) where it encountered the highest energy ions of the "inverse" dispersion structure. After entering the polar cap, Polar again found itself in a region of negative potential. The question arises as to whether equipotentials with values >– 2 kV in the polar cap connect with those in the LLBL. The answer appears to be no. From the Polar trajectory in Figure 3a, it is clear that plasma drifting along such equipotentials would have to move with a west component. In fact, data in Figure 1 show that the plasma had an eastward drift component. For this reason we have represented the second negative potential cell as being made up of open flux circulating clockwise in the polar cap. HYDRA measurements of polar rain fluxes indicate

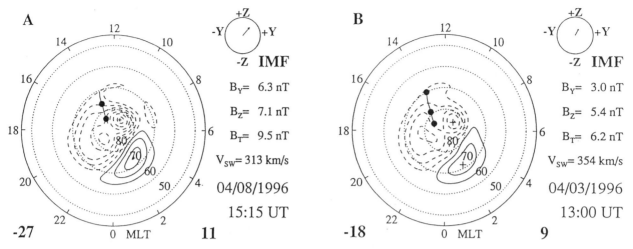

**Figure 4.** High-latitude potential distributions predicted by model of *Weimer* [1996] for interplanetary conditions prevailing near (A) 1515 UT on April 8, 1996, and (B) 1300 UT on April 3, 1996. The IMF values are averages over 40 minutes prior to the stated time of the data projected to the magnetopause in accordance with the measured solar wind velocity. The projected Polar orbit tracks using the *Tsyganenko* [1996] T-96 model are shown. The dots represent satellite projected positions at 1500 and 1600 UT on April 8, 1996 (A) and at 1200, 1300 and 1400 UT on April 3, 1996 (B).

that the satellite must have passed to the poleward side of this cell's merging line.

Figure 3b provides a sketch of the data constraints on the convection directions in the pattern crossed by Polar on April 3, based on the potential distribution found in the third panel of Figure 2. There are two classes of equipotentials with westward drifting plasma in the region of CPS fluxes. Equipotentials crossed at the lowest latitudes are not intersected again by the orbit. At higher latitudes within the CPS, we find equipotentials that turn and are intersected for a second time between 1240 and 1330 UT as the potential increased from about -8 to -1.5 kV. It encompasses the Region 1 and Region 0 current systems and spans the invariant-latitude range 76.5° to 82.6°. Ion fluxes observed across the entire interval are generally weak with broad distributions in energy. In the equatorwardmost portion (1240 – 1255 UT) they appear bimodal in energy indicating that they are of LLBL origin [*Eastman et al.*, 1976].

Electric potential and particle measurements are most simply explained if Polar remained within a projection of the boundary layer until it entered the polar cap. First, the electron fluxes observed in the early part of the boundary layer are higher in energy than those observed in the cusp on April 8, consistent with the findings of *Newell et al.* [1991b]. Second, the observed ion fluxes are an order of magnitude less than those observed in the cusp on April 8 and are more characteristic of boundary layer fluxes [*Haerendel et al.*, 1978]. Third, throughout the 1255 – 1330 UT period EFI measurements indicate that the plasma drift had an eastward component. During periods of IMF $B_Y > 0$, independent of the sign of IMF $B_Z$, magnetic tension forces on newly opened flux causes plasma to drift toward noon (west of the Polar orbit) then into the morning side of the polar cap. This is opposite to the observed eastward direction. Indeed the *Tsyganenko* [1996] model predicts that field lines near 80) and 1500 MLT map to the night rather than to the dayside of the magnetosphere. The observed fluxes must be coming from boundary layers somewhere back in the tail. Fourth, electron and ion fluxes observed between 1255 and 1330 UT show gradual, rather than abrupt, transitions to polar cap levels. This is seen most clearly in the electrons, whose average energies monotonically decreased with increasing invariant latitude. It may be viewed as due to an adiabatic cooling of electrons as flux tubes grow in volume and convect down stream along the flank of the magnetotail. The latitudinal width of the projection of this boundary layer is larger than that reported by *Smiddy et al.* [1982]. We note, however, that potential drop of ~8 kV measured across the boundary layer is similar to potentials reported by *Smiddy et al.* [1982] and in ISEE 1 observations of potentials across the low latitude boundary layer on the dusk side [*Mozer*, 1984]. Independent evidence of the boundary layer hypothesis is provided by the TIMAS ion mass spectrometer on Polar [*Shelley et al.*, 1995] which detected weak, variable low-energy fluxes of $He^{++}$ during the interval which must have entered the boundary layer some distance down the tail.

Figures 4a and 4b present ionospheric potential distributions in ILT/MLT predicted by the model of *Weimer*, [1995, 1996] for interplanetary conditions prevailing during the periods of interest on April 8 and April 3, respectively. Below the circular grids are listed minimum and maximum potentials predicted for the afternoon and morning cells; −28 and +12 kV on April 8, −18 and +9 kV on April 3. Polar trajectories, mapped to the ionosphere using the T-96 magnetic field model [*Tsyganenko*, 1996], are provided for reference. In both cases the model correctly predicts that Polar should detect measurable convection poleward of 72° ILT, and that over the studied periods, it should remain within the afternoon cell. The morning (positive potential) cell is restricted to the postmidnight – predawn quadrant. In both cases the model predicts that small lobe cells should develop in the noon sector, poleward of 80° ILT. The sense of rotation for convection in these cells is clockwise, similar to that reported by *Burke et al.*, [1994] for northward IMF with $B_Y > 0$. Attention is directed to the fact that even though IMF clock angles for the two events are similar, their differences, along with dipole tilt angle distinctions, are sufficient to produce clear differences in predicted convection patterns for the early afternoon sector. The model correctly predicted that Polar would pass through a more stagnant region near 80° ILT on April 3 than on April 8.

A comparison of the potential distributions shown in Figures 3 and 4 also reveals some significant differences. Obviously the model did not predict Polar's encounter with a positive potential cell during the April 8, pass. The data are more suggestive of a four-cell pattern [*Burke et al.*, 1979; *Reiff and Heelis* 1994] than a distorted two-cell pattern with an embedded lobe cell [*Burke et al.*, 1994]. *Rich and Hairston* [1994], suggested that the four-cell pattern may occur so rarely that it is not reproduced by empirical models developed from large data bases. However, the model of *Weimer* [1995, 1996] does produce the positive potential lobe cell on the dusk side of noon for pure northward IMF in a structured four-cell configuration, and this may be the start of the development of that cell.

*Acknowledgments.* This work was performed under funding from the NASA GGS Program (part of the International Solar Terrestrial Physics Program). Work at UC Berkeley and Mission Research Corporation was supported under contract number NAS5-30367 and grant number NAG5-3182. Work at UCLA was supported under grant number NAG5-3171. Work at Iowa was supported under grant NAG S-2231. WKP was supported by contract NAS5-33032. WJB was supported in part by NASA and in part by the U. S. Air Force Office of Scientific Research task 2311PL014. We thank R. Lepping and K. W. Ogilvie for the use of the WIND magnetic field and plasma data.

## REFERENCES

Basinska, E. M., W. J. Burke, N. C. Maynard, W. J. Hughes, J. D. Winningham, and W. B. Hanson, Small-scale electrodynamics of the cusp with northward interplanetary magnetic field, *J. Geophys. Res.*, *97*, 6369, 1992.

Burch, J. L., P.H. Reiff, R. A. Heelis, R. W. Spiro, and S. A. Fields, Cusp region particle precipitation and ion convection for northward interplanetary field, *Geophys. Res. Lett.*, *7*, 393, 1980.

Burke, W. J., E. M. Basinska, N. C. Maynard, W. B. Hanson, J. P. Slaven, and J. D. Winningham, Polar cap potential distributions during periods of positive IMF $B_Y$ and $B_Z$, *J. Atmos. Terres. Phys.*, *56*, 209, 1994.

Burke, W. J., M. C. Kelley, R. C. Sagalyn, M. Smiddy, and S. T. Lai, Polar cap electric field structures with northward interplanetary magnetic field, *Geophys. Res. Lett.*, *6*, 21, 1979.

Clauer, C. R., and E. Friis-Christensen, High-latitude electric fields and currents during strongly northward magnetic field: observations and model simulations, *J. Geophys. Res.*, *91*, 6959, 1986.

Doering, J. P., W. K. Peterson, C. O. Bostrom and T. A. Potemra, High resolution daytime photoelectron energy spectra from AE-E, *Geophys. Res. Lett.*, *3*, 129, 1976.

Dungey, J. W., Interplanetary magnetic field and the auroral zones, *Phys. Rev. Lett.*, *6*, 47, 1961.

Erlandson, R. E., L. J. Zanetti, T. A. Potemra, P. F. Bythrow, and R. Lundin, IMF $B_Y$ dependence of region 1 Birkeland currents near noon, *J. Geophys. Res.*, *93*, 9804, 1988.

Eastman T. E., E. W. Hones Jr., S. J. Bame, and J. R. Asbridge, The magnetospheric boundary layer: site of plasma, momentum and energy transfer from the magnetosheath into the magnetosphere, *Geophys. Res. Lett.*, *3*, 685, 1976.

Haerendel, G., G. Paschmann, N. Sckopke, H. Rosenbauer, and P. C. Hedgecock, The frontside boundary layer of the magnetosphere and the problem of reconnection, *J. Geophys. Res.*, *83*, 3195, 1978.

Hanson, W. B., R A Heelis, R. A. Power, C. R. Lippincott, D. R. Zuccaro, B. J. Holt, L. H. Harmon, and S. Sanatani, The retarding potential analyzer for Dynamics Explorer-B, *Space Sci. Instru.*, *5*, 503, 1981.

Harel, M., R. A. Wolf, P. H. Reiff, R. W. Spiro, W. J. Burke, F. J. Rich, and M. Smiddy, Quantitative simulation of a magnetospheric substorm 1. Model logic and overview, *J. Geophys. Res.*, *86*, 2217, 1981.

Harvey, P., et al., The electric field instrument on the Polar satellite, in *The Global Geospace Mission*, ed. by C. T. Russell, p. 583, Kluwer Academic Publishers, Dordrecht, The Netherlands, 1995.

Heppner J.P., and N.C. Maynard, Empirical high-latitude electric field models, *J. Geophys. Res.*, *92*, 4467, 1987.

Iijima, T., and T. A. Potemra, Field aligned currents in the dayside cusp observed by TRIAD, *J. Geophys. Res.*, *81*, 5971, 1976.

Iijima, T., and T. Shibaji, Global characteristics of northward IMF-associated (NBZ) field-aligned currents, *J. Geophys. Res.*, *89*, 2408, 1984.

Maezawa, K., Magnetospheric convection induced by positive and negative Z components of the interplanetary magnetic field: Quantitative analysis using polar cap magnetic records, *J. Geophys. Res.*, *81*, 2289, 1976.

Maynard, N. C., Structure in the dc and ac electric fields associated with the dayside cusp region, in *The Polar Cusp*, edited by J.A. Holtet and A. Egeland, p. 305, D. Reidel, Hingham, MA, 1985.

Maynard, N. C., T. L. Aggson, E. M. Basinska, W. J. Burke, P. Craven, W. K. Peterson, M. Sugiura and D. R. Weimer, Magnetospheric boundary dynamics: De-1 and DE-2 observations near the magnetopause and cusp, *J. Geophys. Res.*, *96*, 3505, 1991.

Mozer, F. S., Electric field evidence on the viscous interaction at the magnetopause, *Geophys. Res. Lett*, *11*, 135, 1984.

Murphree, J. S., R. D. Elphinstone, D. Hearn and L. L. Cogger, Large-scale high-latitude dayside auroral emissions, *J. Geophys. Res.*, *95*, 2345, 1990.

Newell, P. T., W. J. Burke, C. -I. Meng, E. R. Sanchez, and M. E. Greenspan, Identification and observation of plasma mantle flow at low altitude, *J. Geophys. Res.*, *96*, 21,013, 1991a.

Newell, P. T., W. J. Burke, E. R. Sanchez, C. -I. Meng, M. E. Greenspan, and C. R. Clauer, The low-latitude boundary layer and the boundary plasma sheet at low altitude: prenoon precipitation regions and convection reversal boundaries, *J. Geophys. Res.*, *96*, 35, 1991b.

Reiff, P. H. and R. A. Heelis, Four cells or two? Are four cells really necessary?, *J. Geophys. Res.*, *99*, 3955, 1994.

Reiff, P. H., and J.L. Burch, IMF $B_y$-dependent plasma flow and Birkeland currents in the dayside magnetosphere, 2. A global model for northward and southward IMF, *J. Geophys. Res.*, *90* 1595, 1985.

Reiff, P. H., J. L. Burch, and R. W. Spiro, Cusp proton signatures and the interplanetary magnetic field, *J. Geophys. Res.*, *85*, 5997, 1980.

Rich, F. J., and M. Hairston, Large-scale convection patterns observed by DMSP, *J. Geophys. Res.*, *99*, 3827, 1994.

Russell, C. T., The configuration of the magnetosphere, *in Critical Problems of Magnetospheric Physics*, edited by E. R. Dryer, p. 1, IUCSTP, National Academy of Sciences, Washington, DC, 1972.

Russell, C. T., R. C. Snare, J. D. Means, D. Pierce, D. Dearborn, M. Larson, G. Barr, and G. Le, The GGS/Polar magnetic field investigation, in *The Global Geospace Mission*, edited. by C. T. Russell, p. 563, Kluwer Academic Publishers, Dordrecht, The Netherlands, 1995.

Sandholt, P. E., Auroral electrodynamics at the cusp/cleft poleward boundary during northward interplanetary magnetic field, *Geophys. Res. Lett*, *18*, 805, 1991.

Sandholt, P. E., C. J. Faruggia, M. Øieroset, P. Staunning, and S. W. H. Cowley, Auroral signature of lobe reconnection, *Geophys. Res. Lett*, *23*, 1725, 1996.

Scudder, et al., HYDRA – a 3 dimensional electron and ion hot plasma instrument for the Polar spacecraft of the GGS mission, in *The Global Geospace Mission*, edited by C. T. Russell, p. 459, Kluwer Academic Publishers, Dordrecht, The Netherlands, 1995.

Shelley, et al., The toroidal imaging ion mass spectrograph (TIMAS) for the Polar mission, in *The Global Geospace Mission*, edited by C. T. Russell, p. 497, Kluwer Academic Publishers, Dordrecht, The Netherlands,1995.

Siscoe, G. L., W. Lotko, P. H. Reiff, and B. U. Ö. Sonnerup, A high-latitude, low-latitude boundary layer model of the convection current system, *J. Geophys. Res.*, *96*, 3487, 1991.

Smiddy, M., W. J. Burke, M. C. Kelley, N. A. Saflekos, M. S. Gussenhoven, D. A. Hardy, and F. J. Rich, Effects of high-latitude conductivity on observed convection electric fields and Birkeland currents, *J. Geophys. Res.*, *85*, 6811, 1980.

Tsyganenko, N. A., Effects of the solar wind conditions on the global magnetospheric configuration as deduced from data based field models, *Third International Conference on Substorms (ICS-3)*, ESA SP-389, p. 181, ESA Pub. Div., Noordwijk, The Netherlands, 1996.

Weimer, D. R., A flexible, IMF dependent model of high latitude electric potentials having "space weather" applications, *Geophys. Res. Lett.*, *23*, 2549, 1996.

Weimer, D. R., Models of high-latitude electric potentials derived with a least square error fit of spherical harmonic coefficients, *J. Geophys. Res.*, *100*, 19,595, 1995.

Winningham, J. D., J. L. Burch, N. Baker, V. A. Blevins and R. A. Hoffman, The low altitude plasma instrument (LAPI), *Space Sci. Instru.*, *5*, 465, 1981.

---

W. J. Burke, Phillips Laboratory, 29 Randolph Road, Hanscom AFB, MA 01731. (email: burke@plh.af.mil)

N. C. Maynard and D. R. Weimer. Mission Research Corporation, One Tara Blvd., Suite 302, Nashua, NH 03062. (email: nmaynard@mrcnh.com; dweimer@mrcnh.com)

F. S. Mozer, Space Science Laboratory, Grizzly Peak Drive, University of California, Berkeley, CA 94720. (email: fmozer@sunspot.ssl.berkeley.edu)

W. K. Peterson, Lockheed Martin Palo Alto Research Laboratory, 3251 Hanover St., Palo Alto, CA 94304. (email: pete@space.lockheed.com)

C. T. Russell, Institute for Geophysics and Planetary Physics, University of California at Los Angeles, Los Angeles, CA 90049. (email: ctrussell@igpp.ucla.edu)

J. D. Scudder, Department of Space Physics and Astronomy University of Iowa, Iowa City, IA, 52240. (email: jds@space-theory.physics.uiowa.edu)

# Interball Tail Probe Measurements in Outer Cusp and Boundary Layers

S. P. Savin,[1] N. L. Borodkova,[1] E. Yu. Budnik,[1] A. O. Fedorov,[1] S. I. Klimov,[1] M. N. Nozdrachev,[1] I. E. Morozova,[1] N. S. Nikolaeva,[1] A. A. Petrukovich,[1] N. F. Pissarenko,[1] V. I. Prokhorenko,[1] S. A. Romanov,[1] A. A. Skalsky,[1] Yu. I. Yermolaev,[1] G. N. Zastenker,[1] L. M. Zelenyi,[1] P. Triska,[2] E. Amata,[3] J. Blecki,[4] J. Juchniewicz,[4] J. Buechner,[5] M. Ciobanu,[6] R. Grard,[7] G. Haerendel,[8] V. E. Korepanov,[9] R. Lundin,[10] I. Sandahl,[10] U. Eklund,[10] Z. Nemecek,[11] J. Safrankova,[11] J. A. Sauvaud,[12] J. Rustenbach,[13] and J. L. Rauch[14]

Near cusp the magnetopause (MP) coexists with magnetosheath (MSH) plasma stagnation sites (PSS). The magnetic field inside the PSS drops by an order of magnitude, the thermal energy density excess is close to that of the MSH kinetic energy, which can be transformed into thermal energy via multiscale cascades: reflection of the MSH plasma flows from the outer throat (OT) magnetic walls (large scales of 1–3 Re), interaction with nonlinear vortex waves of medium scales and, finally, at the microscales (from the ion gyroradius to the electron inertial length, i.e. down to few km). The energetic electrons inside the PSS in OT indicate that the magnetic field lines had been reconnected with those of the Earth. Inside the MP, the ion distributions from the PSS evolve to quasi-perpendicular ones after the field aligned ions escape. The outer cusp population consist of these quasi-perpendicular hot (500–600 eV) ions and the newly injected and/or reflected ones. Just outside and at the MP, the turbulent boundary layers (TBLs) exist, the wave amplitudes provide nonlinear ion trapping. The TBL characteristic scales are 300 km to 5000 km. Most PSSs seem to be created in the TBL. The cusp field lines that cross the TBL can be partially disconnected from the OT having different particle distributions just inside and outside the TBL. The lines should be treated as connected in the statistical sense since they conserve only the average direction inside and outside the MP/TBL. The MSH ions are heated in the TBL up to the temperature of 1 keV; the TBL AC magnetic field energy densities reach 20% of the ion ones. The statistical OT depth is 1–2 Re. The cusp has X spread in SM frame from –2 to 7 Re, and ±7 Re in the Y direction The MSH-origin ion flows are permanently seen inside the MP and in the LLBL and at high latitude mantle field lines.

[1] Space Research Institute, Russian Academy of Sciences, Profsoyuznaya 84/32, G, Moscow, 117810, Russia
[2] Institute of Atmospheric Physics, Praha, Czechia
[3] Interplanetary Space Physics Institute, CNR, Frascati, Italy
[4] Space Research Center, Polish Academy of Sciences, Warsaw, Poland
[5] Max-Planck-Institut fur Aeronomy, Lindau, Germany
[6] Institute of Gravitation and Space Sciences, Bucharest, Romania
[7] Space Science Department, ESA, Noordwijk, The Netherlands
[8] Max-Planck-Institut fur Extraterrestrische Physik, Garsching, Germany
[9] Space Research Institute, Division, Ukrainian Academy of Sciences, Lviv, Ukraine
[10] Swedish Institute of Space Physics, Kiruna, Sweden
[11] Faculty of Mathematics and Physics of Charles University, Praha, Czechia
[12] Centre d'Etude Spatiale des Rayonnements, Toulouse, France
[13] Max-Planck-Institut fur Extraterrestrische Physik, Aussenstelle Berlin, Germany
[14] Laboratory of Physics and Chemistry of the Environment, CNRS, Orleans, France

Geospace Mass and Energy Flow: Results From the International Solar-Terrestrial Physics Program
Geophysical Monograph 104
Copyright 1998 by the American Geophysical Union

## INTRODUCTION

We present here preliminary INTERBALL Tail Probe pair results on the boundary layer (BL) and cusp studies during the first year of the spacecraft operation.

INTERBALL-1 satellite and MAGION-4 subsatellite [*Galeev et al., 1995*] have been operating in orbit since August 3, 1995. The observations reported here were made by the experiments on board the mother/daughter satellites [see *INTERBALL Mission and Payload, Galeev et al., 1995* for details):

- ASPI: Measurements of fields and waves aboard the mother satellite
- PROMICS-3: Measurements of the three-dimensional ion distribution function and masses in the energy range 4 eV–70 keV (INTERBALL-1)
- CORALL: Measurements of the three-dimensional ion distribution function in the energy range 50 eV–25 keV (INTERBALL-1)
- VDP (INTERBALL-1), VDP-S (MAGION-4): Measurements by Faraday cups of ion/electron fluxes and integral energy spectra (VDP only)
- ELECTRON: Measurements of three-dimensional electron distribution 10 eV–26 keV (INTERBALL-1)
- SPS/MPS: Measurements of two-dimensional electron and ion distribution 40 eV–25 keV (MAGION-4)
- SKA-2: Energetic particle spectrometer (INTERBALL-1)
- SGR-8: Magnetometer aboard Magion-4.

The detailed experiment descriptions with the first data examples have appeared in *Annales Geophysicae*, number 5 (e.g. [*Yermolaev, 1997*]) as well.

This article deals with one selected MP crossing in detail (April 21, 1996), characteristic scale studies, and preliminary 1-yr statistics of the BL and cusp crossings. We use the INTERBALL-1 published data for discussions as well.

## MAGNETOSPHERIC BOUNDARY CROSSINGS IN THE CUSP VICINITY ON APRIL 21, 1996

We would like to start with one characteristic example of an outbound orbit in which the MP is encountered just from the outer (or exterior) cusp. We use the term outer cusp (OC) for magnetosheath-like (MSH-like) plasma encounters on the Earth's magnetic field lines which are earthward of the MP and connected to the polar ionosphere. The OC appears to be controlled by the Earth's magnetic field. It differs from the stagnation region of *Paschmann et al.,* [1976] or the exterior cusp of *Lundin et al.* [1991A] which included both MSH plasma regions inside and outside the MP. In this case, the MP is understood in the sense introduced by *Haerendel and Paschmann* [1975]; i.e., the MP is the separatrix between the Earth's magnetic field and the MSH field. Similarly, the outer cusp throat (OT) is used here for the highly disturbed and/or stagnant

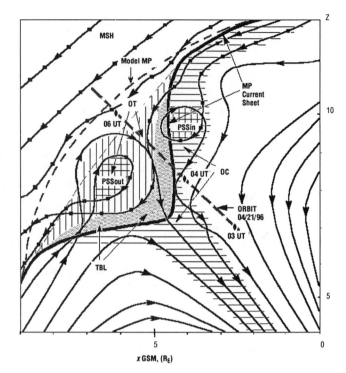

**Figure 1.** The sketch of the different regions and magnetic field lines (full lines with arrows) in the cusp vicinity in the GSM $xz$ plane. The MSH field lines are marked with full squares, the orbit is shown by a thick dashed line, the *Sibeck et al.* [1991] model MP by a dashed line, and the MP inferred from the Interball-1 magnetic field and the MP normal data on April 21, 1996 with usage of the statistical MP crossing by a thick full line. Outbound MP was crossed at 04:28 UT, spacecraft position was (4:63; –2:41; 8:76) $R_E$. In the 04–06 UT time interval, the $y$ changed from –2.7 to –2 $R_E$. See text for further details.

MSH plasma, which we believe is located outside the MP. However, when it is not possible to distinguish clearly if plasma is inside or outside the MP, we return to the "stagnation region" term usage. Figure 1 shows the relative location of the regions.

During the magnetospheric outbound pass on April 21, 1996, INTERBALL-1 and the subsatellite Magion-4 moved from the eveningside of the magnetosphere towards the Sun and noon meridian and from low to high Geomagnetic Solar Magnetospheric (GSM) $z$. The spacecraft orbit is shown in Fig. 1 with tick marks at 03, 04, and 06 UT. In the 04–06 UT time interval, Interball-1 and Magion-4 moved in $y$ direction from –2.7 to –2 $R_E$. The spacecraft were in the northern lobe prior to leaving the magnetosphere. Fig. 2 shows the magnetic field and ion parameters in the regions of interest. The regions are labeled in the same manner as in Fig. 1. In Fig. 2a, the magnetic field

The main MP current sheet is encountered at ~04:28 UT, as one can conclude from the change in the *Bl* component (see panel (c) in Fig. 2a) from positive to negative values. It's also possible that the spacecraft has returned quickly to the magnetospheric field lines at ~04:36:30, 04:38:00, and 05:06 UT. The first two events are seen inside the extremely turbulent zone (04:25–04:38 UT) in which the total magnetic field experiences changes up to 90% of its magnitude. After this mostly outer turbulent boundary layer (TBL) (shadowed by dots in Fig. 1), the spacecraft encounters a rather regular OT (core OT, 04:45–04:55 UT) with slightly deflected *Bn* and *Bm* as compared with the MSH at 05:20–05:30 UT. In the TBL, the irregularities in the ion velocity are bigger than in the regular OT, the $V_z$ of ~100 km/s being the dominant average velocity component (see panels (a)–(c) in Fig 2b). The general TBL feature that differentiates it from both the OT and the MSH is the much higher ion temperatures (excluding the PSS). In the TBL center, the ion temperature even exceeds that seen in the OC (see Fig. 2b).

Comparison with the *Sibeck et al.* [1991] MP model for a solar wind dynamic pressure of 1.8 nP (from WIND measurements) shows that the model MP should be crossed at ~06 UT, i.e., about 2 hr after the actual crossing. The spacecraft position moved 2.5 $R_E$ away from the Earth during that time (see Fig. 1).

At 05:02–05:16 UT in Fig. 2a one can see the highly disturbed magnetic field again. Here, a quite interesting feature is seen; namely, the magnetic field site with extremely

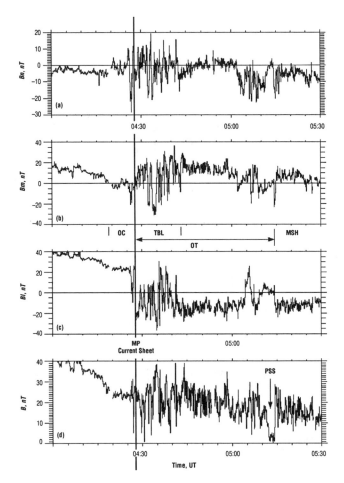

**Figure 2a.** The MIF-M magnetic field on April 21, 1996 (DD day=111). From the top to bottom: *Bn*, *Bm*, and *Bl* in the MP boundary coordinates frame; *B* is the total magnetic field.

from the MIF-M experiment on the main spacecraft is shown as it makes the transition from the field being close to the Tsyganenko model (04:00–04:15 UT) to a more or less undisturbed MSH field at 05:20–05:30 UT. The coordinate system is the local MP system introduced by *Russell and Elphic* [1978]; the direction of the MP normal is *N*=(0.92; –0.38; 0) in the GSM frame and was calculated as a vector product of internal and external magnetic fields (at 04:20–04:25 and 04:47–04:55 UT, respectively). The minimum variance analysis confirms the normal direction. Note that the local normal, being close to *x*, differs substantially from the model. In our case, the MP looks like a vertical magnetic wall for the MSH plasma. In Fig. 2b the ion parameters from the Corall experiment are displayed with the time resolution once per spin [see *Yermolaev et al.*, 1997 for details]. The spikes in *Vy* and *Vz* at 04:00–04:20 UT are partially due to ion distribution changes that take place in times less than the spin period (see Fig. 3(a)).

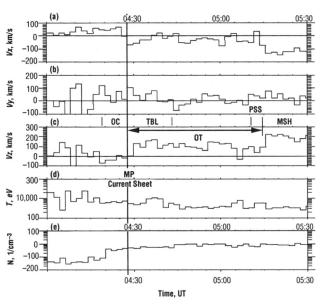

**Figure 2b.** The ion parameters (Corall experiment) on April 21, 1996 (DD day=111). From top to bottom: *Vx*, *Vy*, and *Vz* in the GSM frame; *T*=ion temperature; *N*=ion density; and time resolution=118 s.

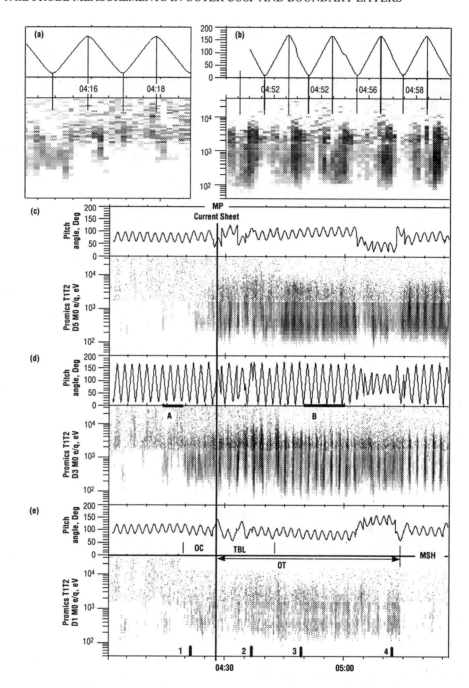

**Figure 3.** Panels (a) and (b) show details of the central panels in marked intervals; the vertical lines correspond to the maximum (ions flow antiparallel to the magnetic field) and minimun (ions flow parallel to the magnetic field, i.e., poleward for the field lines connected to the North Pole) pitch angles. Panel (c): Tailward-flowing ions and their pitch angles. The MP encounter by INTERBALL-1 was at 04:28 UT (see Fig. 2). Panel (d): Data from rotating channel, which is perpendicular to the Earth-Sun axis. Panel (e): The proton data of Promics-3, sunward-flowing ions (with data from 2 energy in keV for channels TRICS 1 and 2; the 1.5 keV value corresponds to the boundary between the channels) and pitch angle of the channels (bottom curve, in degrees; 0° corresponds to ion flows parallel to magnetic field) on April 21, 1996 (DD day=111). Numbers 1–4 in panel (e) correspond to the ion distribution cuts in Fig. 5.

low magnetic field magnitude (at 05:12–05:14 UT). The magnetic field here is rather stable compared to that of the transient (short duration) field in the turbulent zone just outside the MP. Note that the magnetic field of 1–5 nT is 4–7 times less than the surrounding MSH, including the normal field at the right edge of Fig. 2.

In the external MSH, the GSM velocity vector is (−120, 52, 195) km/s. That seems to agree with MSH plasma flow around the magnetosphere. The dominant velocity in the OT is GSM $V_z$=100 km/s, $V_x$ and $V_y$ are irregular, with $V_x$ 2–3 times less than $V_{yz}$.

In Fig. 3 the proton data of the Promics-3 mass spectrometer are presented for the same time interval as Fig. 2 [see Promics-3 description in *Sandahl et al.*, 1997] along with the pitch angles of the three channels. Panel (c) in Fig. 2 shows the tailward-flowing protons, panel (d) is looking perpendicular to the spacecraft spin axis (the axis is pointed to the Sun), and panel (e) shows the sunward-flowing protons and spin-modulated signals with a period of ~118 s, seen in both panel (c) and (d); panel (c) has a 0–18° field of view (fov). The two turbulent zones discussed above are evident in the irregular behavior of the pitch angles. Note that the spectrum changes at 1.5 keV are due to combining data from two different Promics-3 channels with different energy ranges and sensitivities. The SPS/MPS data from the subsatellite (not shown) demonstrate behavior that is very similar to that in Fig. 3. The time delays correspond to the magnetic field delays. We use the Promics-3 data because it has better time resolution.

One can see two regions inside the MP. In the first region (04:00–04:19 UT), low density, higher energy plasma co-exists with an MSH-like plasma (0.1–2 keV). This region corresponds to the boundary cusp or cleft at low altitudes [*Lundin et al.*, 1991A]. The 1–20 keV ions tend to flow along field lines antiparallel to the magnetic field (i.e., from the Earth), being likely of plasma sheet (PS) origin. The lowest energy ions at 04:04 UT are flowing tailward as well, while at 04:09 and 04:15 UT the ions flow mostly poleward. The latter case is better seen in Fig. 3(a). The MSH ion injection has typical V-shaped energy/pitch angle dependence [*Lundin et al.*, 1991A]. A new feature that one can see in Fig. 3(a) is the simultaneous presence of three separate populations: (a) injected (minimum pitch angles); (b) reflected (maximum pitch angles); and (c) quasi-perpendicular ions (QPIs), which are seen distinctly at 04:16:15 UT. One could argue that the QPI population can be due to an effect similar to a velocity filter. But comparison with the longer OC encounters (e.g., on March 14, 1996 and May 29, 1996 cases, etc.) shows that QPIs are a characteristic feature of the OC, as well as the OT (see Figs. 3 and 5(b) and related discussions below). Another strange thing is the coexistence of the low- and high-energy injections (at 04:15 and 04:16:50 UT in Fig. 3(a), respectively). The absence of lower energy ions suggests the presence of a retarding potential of a few kilovolts between the TBL and the boundary cusp (or cleft). Similar ion behavior has been reported also from the Prognoz-8 data [*Savin et al.*, 1994].

In the second inner BL region at 04:19–04:26 UT (OC, shadowed by horizontal lines in Fig. 1), intensive ion flows, which are both injected earthward and reflected from the convergent magnetic field at low altitudes, are seen simultaneously. This is a feature of the cusp proper [*Lundin et al.*, 1991A]. Note that the magnetic field magnitude drops and the direction changes in the OC region, the dominant component becomes GSM $B_y$ (Fig. 2a).

To prove that the OC region is located inside the MP, we have examined the electron distribution function at 04:18:47–04:24:48 UT [ELECTRON device data, *Sauvaud et al.*, 1997]. The result is presented in Fig. 4. It is evident that electrons in 68.2–811.3 eV energy range (and partially at 1331.3 eV) flow in both directions along the magnetic field line. The tailward-flowing (antiparallel to the magnetic field) electrons can only be the MSH field-aligned (injected) electrons that have reflected from the ionosphere. It is similar to the LLBL case [see *Sauvaud et al.*, 1997; *Savin et al.*, 1997]. The electron energy spread over 1 keV in Fig. 4 indicates that the electrons of MSH origin have been accelerated/heated at some point or possibly along their path into the MP. The electron heating is characterized by the average temperature, $T_e$. In the MSH and core OT, it is 36 eV; in the TBL, 68 eV; and in the OC, 92 eV. The 1331 keV electron beams in Fig. 4 and those with higher energies (not shown) are due to the suprathermal tails [cf. *Savin et al.*, 1997].

In Fig 5 the characteristic ion distribution cuts are presented in the frame which has the vertical axis along the spin-averaged magnetic field and the horizontal axis perpendicular to the spacecraft spin axis The plot numbers 1–4 in Fig. 5 correspond to those in panel (e) of Fig. 3. As with the electrons, one can also see in Fig. 5(a) two field-aligned ion populations in the OC, the colder and denser population flows along the magnetic field lines, being injected, we believe, from the OT. The reflected (downward in Fig. 5) ions have a wider spread in the velocity space. The bulk velocity along the magnetic field is 44 km/s; the average temperature is 594 eV, which is close to the average ion temperature in the TBL (see Fig. 2b). Both populations display the fine structure, which is believed to be due to the nonstationary source. The latter can be the case if the TBL is assumed to be the source region. In Fig. 5(a) the weak isotropic high-energy background is filtered out to outline the MSH-like ion distribution (see Fig. 3(a)).

In the TBL (04:28–04:43 UT) the ion energies reach 20 keV (Fig. 3). The ion distribution fine structure seems to correlate more with the magnetic field magnitudes than with pitch angles. The tailward-looking channel (Fig. 3(e)) registers ions in the OT until 05:15 UT, an effect of the stagnant plasma. The characteristic TBL distribution is shown in Fig. 5(b); the spin-averaged ion temperature is

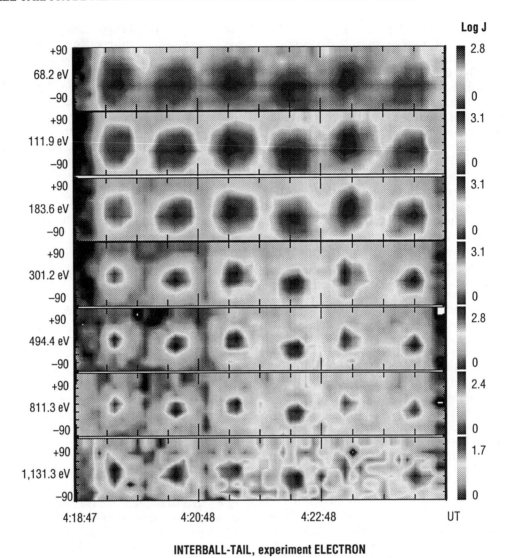

**Figure 4.** Angle distribution of electrons for 68.2 (top) 111.9, 183.6, 301.2, 494.4, 811.3, and 1331.3 eV in cusp on April 21, 1996, ELECTRON device. Polar angle, shown at the left side, is measured by eight channels, rotating perpendicular to the spin axis (pointed to the Sun). A three-dimensional distribution function is measured for full spin (about 118 s). The electrons are flowing mostly parallel and antiparallel to the magnetic field (dark spots twice per spin).

1100 eV (i.e., maximum in the TBL), the convection velocity perpendicular to the magnetic field is ~150 km/s. One could get the injected OC distribution (the upper part of Fig. 5(a)) by superimposing several random TBL distributions; the higher energy ions should escape along the field lines (deeper into the OC) or across them (into the boundary cusp (see Fig. 3(a)) if their gyroradius is bigger than the characteristic length of the magnetic waves. The alternative source region for the OC injections can be the core OT (04:45–04:55 UT; see Figs. 3(b) and 5(c), and discussion below).

In the TBL the ion thermal energy (1.5 nkT) dominates over the kinetic energy ($0.5 nmV^2$). Their average sum is ~0.8 of the values in the external MSH. Therefore, the TBL average ion temperature rise could be explained by the MSH plasma kinetic energy transformations into the OT/TBL. At 04:37 UT the maximum TBL plasma energy density (including small electron heating) exceeds the MSH density by ~20%, i.e., equivalent to the 24 nT amplitude turbulence (about 1500 eV/cm$^3$). Here the turbulence constitutes a substantial part of the cusp energy balance that suggests the magnetic field annihilation role in the MSH

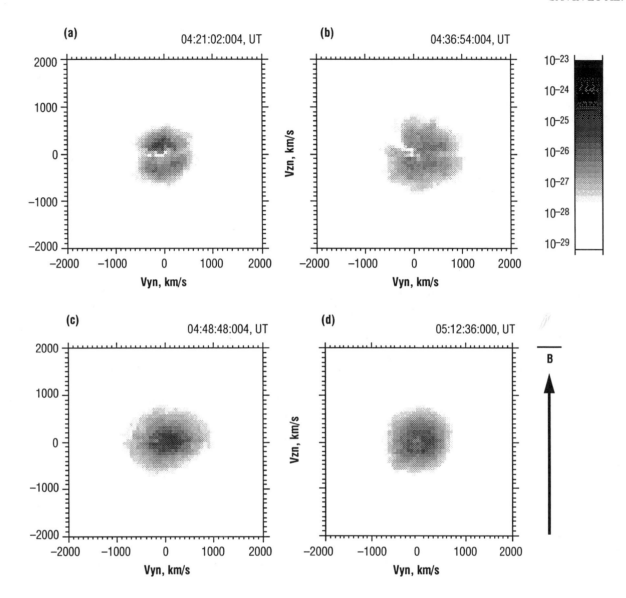

**Figure 5.** CORALL device ion distribution function cuts in the frames with the vertical axis along the spin-averaged magnetic field vector and the horizontal axis perpendicular to the spacecraft spin axis. The plot numbers 1–4 correspond to those in Fig. 3(e). The central times of the distribution cut intervals are given at the top of each plot.

plasma heating inside the BLs/regions. One can see that the magnetic field amplitude in Fig. 2a at this time is close to 25 nT. It is an open question how the magnetic turbulence energy can be transformed into plasma thermal energy. We discuss this point in the following sections.

The TBL with the heated plasma at 04:34–04:40 UT (electrons in the same region have a temperature of about a factor of 2 higher, similar to ions) does not correlate with MSH $Bz$ in the GSM frame (not shown). That might be concluded from the change of the average $Bz$ sign at 04:36 UT, i.e., almost in the middle of the TBL. The turbulence is replaced by the core OT plasma within 2 min after the GSM $By$ becomes dominant (04:41–05:00 UT). Because $By$ is dominant in the OC and in the OT, perhaps the OC field lines are reconnected (via the TBL) with the oppositely directed ones in the core OT. A strong argument for this is presented by the minimal ion count rate along the field lines which are marked by the vertical lines in Fig. 3(b). The general ion count rate maximums once per spin are due to the OT plasma motion perpendicular to the field lines with a bulk velocity of 63 km/s. It is also seen in the ion distribution cut of Fig. 5(c). In Fig. 3(b), one more valuable detail is seen, which is hardly possible to extract from Fig. 5(c). This is the tailward- (with maximum pitch angle) flowing ions that are seen best at 04:58:50 UT. Their appearance seems to have no correlation with the rest of the ion populations. We believe that these ions have been reflected from the ionosphere and transported back to the OT through the MP. The reflected ion intensities are ~10 times weaker than those in the OC. The ions with the maximum pitch angles (downward in Fig. 5(d)) could escape from the stagnated OT along the reconnected field lines which are bent around the magnetosphere (see PSSin in Fig. 1).

The ion source for the OC (Fig. 5(a)) could be the field-aligned ions which have escaped from the core OT into the magnetosphere (see Figs. 3(b) and 5(c)). If this is the case, (as an alternative for the source in the TBL) then the high ion temperature in the OC in Fig. 2b (~600 eV versus 300 eV in the core OT) should be due to the presence of the higher energy ions which originate in the TBL (see Fig. 3). One can imagine also the situation in which the spacecraft did not cross the inner counterpart of the core OT lines (04:45–04:55 UT). In that case, these lines should enter the MP at a different place compared to the spacecraft OC crossings shown in Fig. 1.

Another difference between the OC and core OT data is that in the core OT, the electrons are isotropic aboard both the satellite and subsatellite (not shown) with intensities that are ~5 times higher than those in the OC. Such electrons are seen in the MSH at 05:20–05:30 UT in contrast to those of the cusp lower density anisotropic. The latter, along with the opposite magnetic field directions, allows us to state the general difference between OC and OT as being inside or outside the MP, respectively.

The difference in the electron and ion behaviors in the core OT could be explained by the presence of electrostatic potential step(s) or barrier(s) of 100–200 V at the MP. This potential can stop the core MSH electrons but will not affect the field-aligned ions with higher energies. Another reason for the absence of the link for the MSH-origin electrons along the field line can be intensive, small-scale (of the order of the electron inertial length) waves in the TBL that mix the pitch angles and thermalize the field-aligned electrons (see Fig. 4). We demonstrate an example of such waves in the next section.

Now we present details of the striking part of the OT, the region in which the magnetic field almost vanishes. In Fig. 6 the total magnetic field from the satellite (solid line) and subsatellite (dashed line) are presented, the latter being shifted by 10 nT to distinguish the curves. The spacecraft were about 750 km apart along the orbit (405 km along the MP normal). There is a 9-s shift in the arrival times at each spacecraft for magnetic features in this region and a 9-s shift was also seen in the TBL at 04:28–04:43 UT. The structure speed is 45 km/s along the MP normal. For an outbound orbit and with the subsatellite being ahead, the similar time shifts imply that the site moves tailward as

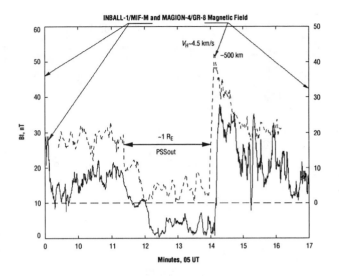

**Figure 6.** The total MIF-M magnetic field (full curve) at 05:09–05:17 UT on April 21, 1996. Dashed curve=SGR-8 total magnetic field from Magion-4 subsatellite; the curve is shifted by 10 nT to be distinguished from the mother spacecraft curve.

the MP and TBL do. The total depletion site spread along the MP local normal is ~1 $R_E$ (7 thermal ion gyroradii), the BL current sheet (at 05:14:15 UT) width being ~500 km. The latter equals 5 ion thermal gyroradii and 7 inertial ion gyroradii in the maximum barrier field (35 nT); the former calculated for the inner site plasma, the latter for the MSH. If one takes the average magnetic fields in each region, then the border current sheet will be 0.5 and 3.5 gyroradii thick, respectively. The total pressure is nearly constant across the PSS boundary. A field line configuration, compatible with the measured fields and assuming that the reconnection occurred equatorward of the spacecraft (for the southward IMF case), is shown in Fig. 1. The described structure is labeled as "PSSout." In Figs. 2 and 7, this region is labeled as "PSS," in Fig. 3 it is labeled by "4."

From Fig. 2a one can get average $B_n$ at 05:00–05:30 UT of 5–10 nT; that means a different local normal direction compared with the measured MP direction. Inside the vanishing field site (05:12–05:14 UT), $B_n$ is close to zero; the magnetic field fluctuations reach 90%, and the fluctuations are low in absolute value. Low spin modulation of ion flow inside the site (see Fig. 3) implies a low average velocity. The irregular flow fine structure is seen in the channel perpendicular to the Sun. The SPS/MPS data from the subsatellite (not shown) also observe ions that are very similar to those seen in Fig. 3(e). Spin-averaged $V_x$=40 km/s (sunward) is close to the error bar [see *Yermolaev et al.*, 1997 for details of the Corall instrument]. The ion distribution cut in the magnetic field depletion region is

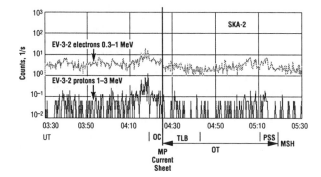

**Figure 7.** The SKA-2 energetic particle data for April 21, 1996. Top curve: Electrons 0.3–1 MeV; bottom curve: protons 1–3 MeV. Maximum signal is seen in the boundary cusp (04:20 UT). Note the energetic electron intensity rising in the PSS at about 05:03 and 05:12 UT.

shown in Fig. 5(d). The distribution is rather symmetric, having minor fine structure and low velocity $Vyz<60$ km/s. That is why we call such events "plasma stagnation sites" (PSSs).

A natural question arises: What can be the energy source for the plasma heating inside PSSs? The most probable candidate is the kinetic energy of the external MSH flow. It is indicated by the fact that the temperature excess inside the PSS in Fig. 5(d) of ~260 eV is very close to the ion kinetic energy in the external MSH (being 90% of the kinetic energy density). Another energy source for the PSS plasma heating might be found in the annihilation of wavy magnetic field with an amplitude of 27 nT. However, the wave turbulence could be only an intermediate state of the PSS ion thermal energy on the way from the streaming to the heated state. Further studies are necessary for problem clarification.

Inside the near-cusp MP, the ions of MSH origin are often seen with maximum intensities in the directions perpendicular to the magnetic field. The ion distributions resemble those in Fig. 5(c); e.g., at 05:09:25 UT on March 14, 1996, etc. The perpendicular QPI temperature in the OC (e.g., at 04:16:15 UT in Fig. 3(a)) is close to that of Fig. 5(d). The distribution in Fig. 5(c) itself looks like the double loss-cone distribution evolved from the isotropic distribution shown in Fig. 5(d). Our current understanding is that the MSH ions can gain the extra thermal energy in the PSS-like structures. The small-scale PSSs can be recognized in the TBL as well (see the total field drops in the lower panel of the Fig. 2a); then the field-aligned ions escape tailward or poleward. The poleward traveling ions are reflected by the convergent magnetic fields and are seen both together and separately from those perpendicular (see Figs. 3(a) and 5(a)). The QPI distributions represent a specific OC phenomenon that cannot be seen much deeper into the cusp due to the high perpendicular ion temperature. Their regular presence in the OC means that the PSS and/or TBL ion heating is a permanent process. We have not seen any difference in the QPI occurrence for the northward and southward IMF (for ~20 cases).

Comparison of symmetric ion distributions of MSH origin with symmetric ion distributions in the OC shows that the latter have few times less intensity. The PSS ion temperature in the OC is close to the ion temperature in the PSS in the OT. For example, on April 2, 1996 at 02:47:07 UT, a PSS with ion distribution similar to that in Fig. 5(d) was encountered deep inside the OC. The ion spin-averaged parameters were: $T=541.7$ eV and $Vyz=13.4$ km/s (in the spacecraft frame). The PSS was encountered 1.5 hr prior to the spacecraft entering the OT, i.e., ~2 $R_E$ from final MP crossing. On April 2, 1996, PSSs were encountered both inside and outside the MP. The PSS outside the MP is shown in Fig. 8. The situation with simultaneous PSS both inside and outside the MP is shown in Fig. 1 for the southward IMF case. Both PSSs are believed to originate as a result of the reconnection site moving tailward with the MSH plasma flow and becoming caught in the OT cavity. More details for such a process are given in the next section. Such distributions like the one on April 2, 1996 are rarely seen inside the MP; the QPI-like distributions are common instead. The PSSs in Fig. 1 are compatible with the Heos-2 stagnation region location both inside and outside the MP [*Haerendel and Paschmann*, 1975]. The difference between the *Haerendel and Paschman* [1975] stagnation regions and the PSS is the inclusion in the former of the regions with low temperatures (see Fig. 2b at 04:45–05:15 UT).

Returning to the PSS on April 21 (see Figs. 5(d) and 6), we would like to point out that electrons with energies 0.01–1.5 keV (not shown) are those of the isotropic MSH, i.e., they show no evidence for field-aligned distributions (see Fig. 4 and related discussion). The magnetospheric energetic electrons of 0.3–1 MeV (measured by SKA-2 device; see Fig. 7) have the same peak level in the PSS as in the OC. In Fig. 7, the energetic electron level drops by a factor of 2 and that of protons by a factor of 3 at the MP crossing at ~04:28 UT. The particle levels in the TBL at 04:28–04:43 UT are lower than in the PSS under study, but are slightly higher than those in the core OT and external MSH.

The higher level of energetic particles in the PSS is strong evidence for the occurrence of reconnection during the PSS formations, but neither the MSH ions nor the electrons display features of a free link with the magnetosphere at the moment of measurements (cf. Figs. 3(b), 4, 5(a), and 5(c)). An explanation could be the closed (or quasi-closed) field line configuration of the PSS (say, as the result of double reconnection). This could be the case after southward MSH field at 04:36:30–05:03:00 UT when the reconnection is believed to occur upstream of the cusp (15 min is enough for a reconnection site to reach the dawn-to-dusk meridian from the equator). The April 21

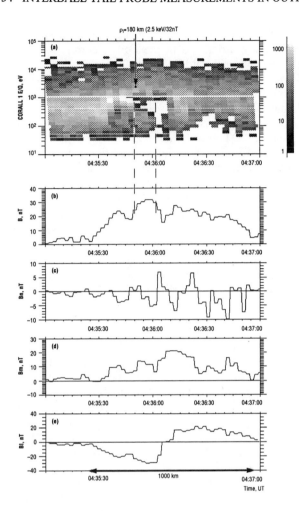

**Figure 8.** PSS and external MSH boundary on April 2, 1996 (DD day=092). (a) shows counts/sample, CORALL device ion distribution in the tailward direction. Panels (b)–(e) show the magnetic field in the local minimum variance frame. $B$=total magnetic field and $Bn$, $Bm$, $Bl$=minimum, medium, and maximum variance magnetic field components, respectively; the respective GSM eigenvectors are: (0.56; 0.58; 0.6), (–0.53; 0.8; –0.27) and (–0.64; –0.17; 0.75). The eigenvalues are 21, 32, and 152.

PSSout (along with that of April 2, 1996; see Fig. 8) could be the trailing part of the reconnection site; i.e., decelerated as compared with the original MSH flow (see Fig. 1). Strong turbulent mixing and heating of the MSH-origin particles in the TBL could be effective mechanisms for breaking the field line links. In the April 21 case the "reconnected" line could cross the TBL on the path from the PSSout to the OC. The energetic particles can feel no turbulence because their gyroradius is larger than the TBL width.

Similar to the PSSout, the April 2, 1996 internal PSSin could be attributed to the leading part of a reconnection site, being caught in the OT after impact with the OT tailward "magnetic wall." In that case, one side of the reconnected field line goes to the north magnetic pole, the other end is bent around the north MP. The latter becomes the newly closed-tail field line while drifting inside the magnetosphere, which in Fig. 1 corresponds to the upper magnetic field line loop (PSSin), which is mostly inside the vertical MP. The PSSin cannot move deep into the cusp towards the Earth because of its high thermal pressure and momentum perpendicular to the magnetic field. In most cases, the field-aligned ions should escape, resulting in the QPI distributions. These interpretations of the data are very preliminary, illustrating one possible opportunity. Certainly, further comparison of satellite/subsatellite data is foreseen for a number of cases as well as comparison of measurements with representative models.

## THE PHENOMENON SCALE STUDIES

After the general exterior cusp example description, let's briefly discuss different scale findings and open problems. We propose a rather usual division of the exterior cusp phenomena of large scale (of the order of 1 $R_E$ and larger), medium scale (from several hundred to several thousand km), and micro scale (of the order of 100 km and less). The large scales correspond to structures of the order of the magnetic field curvative and outer throat dimensions. The medium scales include MHD waves in time and space domains. The micro scales spread from the ion gyroradius to electron inertial length. In this paper we can only show several of the most interesting examples, from our point of view.

We have presented the most general large-scale phenomena for the April 21, 1996 exterior cusp in the previous paragraph: cusp proper, TBL and core OT, and, finally, PSS with vanishing magnetic field. We would like to add here the description of the large-scale ion flow structure in the second disturbed OT zone at 05:00–05:15 UT (see Figs. 2 and 3).

Three-dimensional plasma parameters are routinely measured once per spin (about 2 min). This time resolution limits the studies to the large-scale irregularities in the OT and OC. Nevertheless, these first vector plasma flow measurements, even with such low time resolution, have shown that the average flow pattern in turbulent zones can be presented as a slow drift of a vortex. In Fig. 9 we show the velocity hodograph at 05:00:42–05:14:35 UT in the minimum variance frame calculated for the 04:29–05:19 UT interval. The vertical axis is the maximum variance axis; it is 45° from $x$, 55° from GSE $y$, and 93° from the MP $l$ axis.

The horizontal axis is the medium one. The axis corresponding to the minimum variance axis is 34° from the MP normal. Eigenvalues of the variance matrix (1.94, 3.38, 10.5) show that only maximum variance direction is fairly well defined. In the plane, shown in Fig. 9, one can

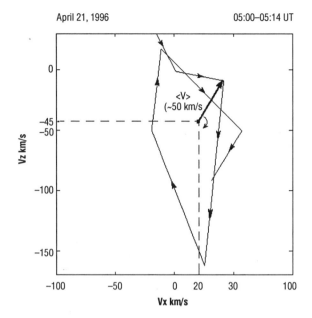

**Figure 9.** The CORALL device velocity hodograph at 05:00:42–05:14:35 UT in minimum variance frame calculated for 04:29–05:19 UT interval on April 21, 1996. The vertical axis is that of maximum variance (–0.42; 0.57; 0.71 in GSE), horizontal axis is the medium variance (0.47; –0.54; 0.7).

approximate the site as a vortex with bulk velocity (20, –45) km/s and an average radius of ~50 km/s.

An example of medium-scale sites is presented in Fig. 6: it is the magnetic barrier between the OT (PSS) and the external MSH at 05:14:30 UT. The barrier itself and the external MSH are modulated by waves from 0.05 to 0.5 Hz (proton cyclotron frequency being 0.3–0.5 Hz).

We would like also to point out the wavelet at 05:15:00–05:15:45 UT (Fig. 6) with total length of about 2000 km and wavelength of 500 km along the MP normal. Such high amplitude magnetic structures inside the turbulent zones modulate the ion distribution function. The modulation of the ion distribution function is a feature of nonlinear processes. This is the only example of this type that we have found that permits us to estimate the characteristic scales by two independent methods. This is also the OT/MSH interface magnetic barrier which separates the PSS with its strongly depleted magnetic field from the streaming MSH (Fig. 8). The spacecraft orbit in the OT was similar to that on April 21. In this case the fortunate Corall sensor orientations have been combined with long enough duration as compared with one Corall ion energy scan (about 3 s). The magnetic barrier is shown in Fig. 8. One can see again a very low (1–5 nT) magnetic field in the PSS (04:35:00–04:35:30 UT) in Fig. 8(b). In Fig. 8(c) the magnetic field magnitude exceeds 30 nT and the core MSH ions of 400–800 eV are depleted down to the Corall zero count rate (see Fig. 8(a)). The ions, which don't feel the barrier, have energies that start at 2.5 keV and a gyroradius in the maximum barrier magnetic field of ~170 km. The most striking feature of this magnetic field site is represented by the ions with energies of 40–300 eV at 04:35:50–04:36:00 UT, which are trapped inside the barrier! The trapped particles represent a very certain feature of nonlinear waves (as a result of interactions of the ions with nonlinear waves). The magnetic field magnitude itself, being 3 times higher as compared to the average MSH field, is another nonlinear feature as its magnetic pressure is comparable with the external MSH dynamic pressure. This "wavelet" represents a bipolar signal in the maximum variance magnetic field component $Bl$ (see Fig. 8(e)). The angle between the local $L$ component (maximum variance) and the model MP normal is 60°. The angle between the model MP normal and the local minimum variance axis is 73°. While eigenvalues for minimum and medium variance directions are close to one another, it is seen that the $Bn$ fluctuations are of higher frequency as compared to average $Bl$ and $Bm$ behavior.

We conclude the different scale phenomena examples with the TBL example of April 21, 1996. For the magnetic field turbulence study we apply the vector field correlation method [*Romanov*, 1998], which provides power distributions in the wave vector space, polarization, and vortex-like fluctuation powers. The result is presented in Fig. 10 for a 4-min interval, starting from 04:34:02 UT. In the interval, the peak-to-peak amplitude of the $Bl$ component reaches 65 nT (see Fig. 2a). In the bottom panel of Fig. 10, one can see the total wave power (full line) and the vortex power in the elliptically polarized wave (dashed line). The frequency band is 0.01–1 Hz. In the upper panel, the angular distribution of normals to the vortex planes (i.e., k-vectors) in the GSM frame is shown; the numbers are the median frequencies in milli-Hertz. In most of the spectral peaks (excluding 20–30 and 50–70 mHz frequencies), the vortex-like waves dominate. Their power exceeds half of the total power in the waves. These waves are well ordered, having wave vectors close to the $x$ GSM direction (that corresponds to the small TETA and FI angles on the upper panel of Fig. 10). In the 80–160 mHz band, almost all the wave power is in the circular polarized waves. This band contains the gyrofrequency of alpha particles; the 200–400 mHz band includes the proton gyrofrequency, but in such a highly variable field, average characteristic frequencies can have no explicit physical meaning (the proton cyclotron frequency band is shown in Fig. 10 for the total field range in the studied interval). The average magnetic field direction on the upper panel is shown by a cross; the inclined cross shows the antiparallel direction. The major vortice axes are close to the average magnetic field direction.

In the TBL (see Fig. 10), the average ion gyroradius is 150 km; therefore, according to our classification, the vortex-like structures of 10–20 and 100 mHz (2000–3000 and 450 km, respectively) are those of medium scale. The

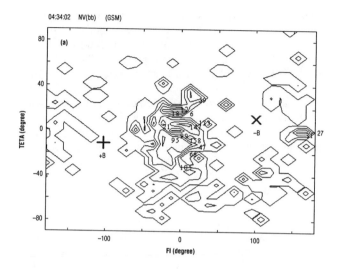

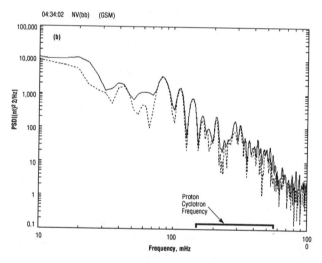

**Figure 10.** The turbulent TBL on April 21, 1996 at 04:34:02–04:38:04 UT. Power distributions in the wave vector space [*Romanov*, 1997]. (a) The angular distribution of normals to the vortex planes (i.e., k-vectors) in the GSM frame, the numbers are median frequencies in mHz. (b) Total wave power (full line) and vortex (elliptically polarized), (dashed line) are shown; the frequency band is 0.01–1 Hz. The proton cyclotron frequency band is shown for the total field range in the studied interval. The average magnetic field direction on the upper panel is shown by a cross. The inclined cross shows the antiparallel direction.

100 mHz waves could be attributed to micro scales as well if one takes into account the tremendous variability of the magnetic field magnitude. The processes at 200–400 mHz and higher should be the micro scale processes (i.e., with scales comparable or less than ion gyroradius) in which kinetic effects play a substantial role. The scale classification is done provided that the Doppler shift of the measured frequency is comparable with the frequencies in the plasma rest frame. If the Doppler shift is much less then the measured frequency, all wavelengths should be attributed to MHD waves.

In the MP frame, the vortices can either move tailward or sunward (assuming that the group velocity directions are close to $x$, i.e., to the wave vector directions, shown on the upper panel of Fig. 10). We don't think that the vortices are moving tailward, as in that case, they should be generated in the region of interaction of MSH flow with the dayside MP and then convected downstream into the OT. The interactions via the reconnections and/or plasma pulse penetrations are known to occur upstream of the cusp only for negative IMF $Bz$ [*Lundin et al.*, 1991A; *Russell*, 1995], while the average $Bz$ in the MSH was positive until 04:37 UT. Another opportunity is generation of the vortices at the tailward OT wall; in this case, the waves should propagate sunward. For positive IMF $Bz$, reconnection and/or plasma pulse penetration into the MP should occur, namely on the tailward (in respect to the cusp) field lines. The waves can propagate upstream of the OT and deflect the MSH flow at the outer boundary of the OT. This is a mechanism for the formation of an effectively collisionless boundary, i.e., the OT outer boundary in Fig. 1.

For higher frequency, large-amplitude magnetic waves (micro scales), only ion spectra with limited angle coverage, having 2–3 s time resolution, are available. The spectra are measured in five directions in two rotating planes (Promics-3, Corall instruments) or in four sectors along sunward- and tailward-looking cones, the cone angle switching in four steps (SCA-1). The combination of these factors means that features associated with higher frequency, large-amplitude magnetic waves can be seen by the particle devices only in those rare cases in which the sensors fortuitously match the orientation needed. The same is true for the electron measurements (Electron device). The integral particle flows are measured also by four Faraday cups with a sampling rate up to 16 Hz (VDP and VDP-S), but no proper information on the particle energies is available for these time scales.

A "fortunate" example of micro scale structure is the trapped plasma inside the magnetic barrier in Fig. 8. It is seen in two successive Corall spectra only, due to the low speed of the plasma relative to the spacecraft (12 km/s, deduced from the SPS/MPS and Corall/Promics-3 ion data comparison).

Let's now look at some of the minimum-scale structures we are able to detect. Fig. 11a shows VDP Faraday cup electron currents with a sampling rate of 16 Hz (the retarding potential is +2.4 keV, i.e. the "electron mode" of VDP). The axes of VDP 2 and 5 are at ±135° to the spacecraft $z$ axis, VDP 3 axis is at 45°; VDP 3 and 5 are looking in opposite directions. One can see signals at 04:34:04 UT in channels 3 and 5, with VDP 3 strongly dominant. The difference of the VDP 3 and 5 signals gives a net current of

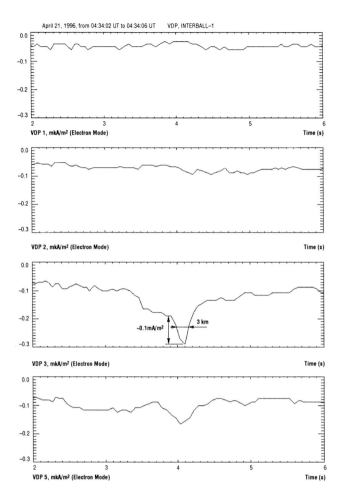

Figure 11a. Electron currents in TBL (VDP Faraday cups, looking to the Sun (upper curve) and perpendicular to the Sun); retarding potential was +2.4 kV.

scales is correct for the interspacecraft distance along MP normal of 405 km (i.e., 100 times the micro scale!). So, the direct application of the usual time-of-flight method between two spacecraft for the spatial scales determination seems to be internally inconsistent for the micro scales.

In order to validate this scale determination, we apply an independent approach of so-called "combined wave diagnostics" (CWD) [see *Klimov et al.*, 1986; *Romanov et al.*, 1990; and references therein]. The CWD uses the Maxwell equation:

$$\text{rot } \mathbf{B} = \mathbf{j} \; .$$

To get an order of magnitude estimate, we will replace **rot** by the wave vector $k=2\pi/L$, where $L$ is the characteristic length. In order to use the CWD method, it is essential that the current and magnetic signals are filtered and amplified in a consistent manner. This is accomplished in MIF-M device using the ac split probe current ($Iz$) and the search coil ($Bz$) sensors [see *Klimov et al.*, 1995, 1997].

In Fig. 11b the current ($Iz$) and search coil magnetic field fluctuations ($Bz$) are presented in the spacecraft frame with the sampling rate of 64 Hz. Their transfer gains are linearly decreasing with frequencies starting from 20 Hz. The $Iz$ fov

~0.1 µA/m² for the shortest pulse in the middle of Fig 11a with characteristic time of the order of 0.1 s. The average dc magnetic field unit vector in the spacecraft frame in Fig. 11 time interval is (–0.09, –0.45, –0.89), i.e., the magnetic field is 19° apart from the VDP 3–5 axis. Taking into account the VDP fov of ±50° and the very small signals perpendicular to the VDP 3–5 axis (i.e., VDP 1 and 2 sensors in Fig. 11a), one could conclude that a field-aligned electron current is seen with the electrons flowing parallel to the magnetic field.

To get an estimate of the current pulse scale (Fig. 11a), we assume that the variations seen are spatial variations rather than changes over time and apply the medium-scale characteristic velocities (obtained from the medium-scale feature time arrival delays on satellite and subsatellite) to transform the time scale into a space scale (i.e., distance). Getting the characteristic scale of 4 km, one might be in doubt if the medium-scale velocity application to micro

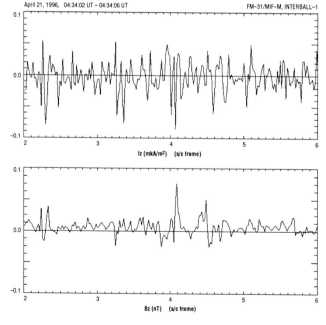

Figure 11b. (Upper panel): The split probe ac current $Iz$ in spacecraft frame; (Lower panel): Search coil ac magnetic field $Bz$ waveforms (spacecraft frame, MIF-M, and FGM-I devices) for the same interval (04:34:02–04:34:06 UT) as the VDP signals. See text for details.

overlaps the VDP 3 and 5 fov by 10°. The average dc magnetic field (for Fig. 11 time interval) is 27° from the spacecraft z (and Iz) axis and 19° from the VDP3–5 axis. The peak current disturbance amplitudes are close for dc (four Faraday cups in VDP experiment) and ac (split probe in ASPI experiment) current measurements. It indicates that Iz peak-to-peak amplitude at about 04:34:04 UT in Fig. 11b might be used as representative current amplitude and characteristic scale length could be estimated as $Lx \sim By/Iz$. We have checked that the ac $By$ amplitude (spacecraft frame) is comparable with the $Bz$ amplitude, while $Bx$ is several times less (MIF-M fluxgate sensor ac data, not shown). So, for the order of magnitude estimate of the scale length, we put $L \sim Bz/Iz$. For the $Iz$ and $Bz$ disturbance amplitudes in the middle of Fig. 11b we get $L = (1-2)$ km using the CWD approach. Multiplying the current burst characteristic width by the medium-scale disturbances velocity (45 km/s) we get the value of ~4 km. Combining these two independent estimates of the scales, we can conclude that, within a factor of 2 accuracy, the minimum electron current micro scale is ~2 km.

Our medium-scale and micro-scale studies show that the high latitude BL disturbances can have time scales comparable with or less than the time needed for the INTERBALL-1 plasma instruments to determine the three-dimensional particle distribution function. The energy of the magnetic field fluctuations is comparable to the external MSH dynamic pressure, i.e., at least medium-scale waves are of primary importance.

The presence of intense current microfilaments at scales of the electron inertial length (2 km) and electron gyroradius (1.8–6 km) can have a substantial impact on the dynamics of energy conversion of the thin currents, and on the plasma transport across the current layers in the vicinity of the MP. For example, such small-scale waves in the magnetic field associated with the thin currents can "break" the link of the MSH-origin electrons along the reconnected field lines: (a) the waves at electron gyroradius scale should provide isotropization of the electron distribution function via diffusion in the velocity space; (b) the waves at electron inertial length might even reflect MSH electrons back to MSH. Both of them should provide the collisionless mechanism of the thin current energy dissipation into the suprathermal electron distribution function tails [see *Savin et al.*, 1997 and references therein]. The minimum scale filament/wave studies are limited to the times when the Faraday cups and split probe on INTERBALL-1 are properly orientated. Neither INTERBALL-1/Magion-4 distance measurement accuracy nor the future Cluster measurements permit reliable determination of these minimum scales from the interspacecraft data comparison only.

### STATISTICAL REVIEW

Now that we have demonstrated the high latitude boundary features of the data, mainly one outbound MP crossing,

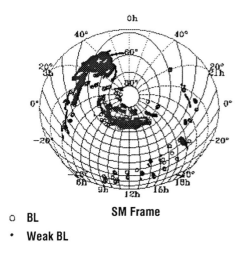

**Figure 12.** The projection on the MP surface of the MSH-origin ion encounters inside the MP in SM frame. View at 30° from the Sun-Earth line. Open symbols=BL with densities of one particle per cubic centimeter and more; Full symbols=weaker BLs. Magnetic latitude and MLT grids are also shown.

we would like to present preliminary statistical data for the first year of INTERBALL-1 operation. We are concentrating on the MSH-origin (that we call "MSH" below) plasma penetration inside the MP.

In Fig. 12 the MSH-like ion flows inside the MP (i.e., mostly BLs and cusp crossings) are shown in the projection on the model MP [*Sibeck*, 1991] for the first year of INTERBALL-1 operation along with the magnetic latitudes versus the MLT grid [see *Haerendel and Paschmann*, 1975 for details]. The projections for the BLs/cusps were done along the spacecraft radius vector. Cusp crossings at the altitudes less than 4 $R_E$ were excluded (mostly southern cusps). For the nightside crossings, the projections on the MP were done along the projection of the spacecraft radius vector on the plane with fixed $x$ (i.e., in the $yz$ plane containing the spacecraft). The strong BLs (densities of the order of one particle per cubic centimeter and more) are shown by open circles, the weaker ones by dots. The SM frame is used in Fig. 12 as the cusp and near-cusp BLs are known to be controlled by the geomagnetic field. The high latitude dayside MSH plasma encounters correspond well with the entry layer and exterior cusp/cleft positions introduced from HEOS-2 and low altitude satellite data [*Haerendel and Paschmann*, 1975; *Lundin et al.*, 1991B]. The poleward and nightside high latitude BL correspond to the mantle.

The lack of BLs in the 21–01 MLT frame is due to the INTERBALL-1 orbit which was inside the magnetosphere (see Fig. 13). We would like to note that the dayside gap between BLs near the equator and at higher latitudes is

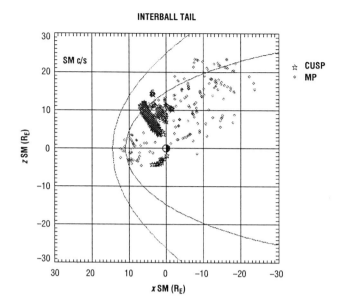

**Figure 13a.** First-year Interball-1 cusp (asterisks) and MP (rombs) crossings in SM frame; *xz* projection.

mainly due to the orbit as high latitude BLs were encountered by INTERBALL-1 on the outbound orbit legs, while on inbound orbit legs, the spacecraft was close to the equator, with poor coverage in between.

There is a substantial asymmetry between the frequent morningside BL occurrence and the less often eveningside BLs, partially due to: (a) different solar wind conditions (we plan to study this point further, using both second-year INTERBALL-1 and WIND data); (b) sliding of the INTERBALL-1 orbit along BLs in the 01–03 MLT sector; and (c) inclusion of the plasma sheet boundary layer (PSBL) encounters at SM *z* close to 10 $R_E$ and higher in the MSH-like plasma sites shown in Fig. 12. The PSBL's were mixed with the tailward-streaming mantle and were seen mainly during northward IMF at high *z* (A. Fedorov, personal communication). In any case, the majority of the MPs (25) in the 18–21 MLT sector and 30°–50° latitude were not accompanied by BLs while the dayside MPs near the equator and in the 01–05 MLT sector, as a rule, were (for MP coverage by INTERBALL-1, see Fig. 13). This fact is consistent with the dawn-dusk BL asymmetry in the near tail.

We think that the BL dawn-to-dusk asymmetry could be associated with the average IMF field spiral component which is close to the MP normal in the morning hours. We don't mean that it will induce an average magnetic field normal component at the MP (which would imply plasma flow across the MP). Different indirect reasons of the solar wind plasma penetration inside the MP might exist. For example, in the quasi-parallel bow shock (BS) morning sector (with the average IMF being close to the BS normal), the MSH is more variable and the respective plasma flow pulses can penetrate into the MP sunward or tailward of the cusp, depending on the IMF $B_z$ [see *Lundin*, 1991B and *Lemaire*, 1977]. In the quasi-perpendicular BS evening sector, the MSH flow seems to be stable as a consequence of sharp and stable bow shocks and absence of the foreshock. The foreshock was proposed as a reason for the generation of plasma pressure pulses and penetration of the MSH plasma inside the MP [*Sibeck*, 1995 and references therein].

For evaluation of integral mass flows in the BL, one needs simultaneous subsatellite plasma data. Such a statistical study is in progress. What can be concluded now is the rather permanent presence of MSH ion flows inside the MP at high latitudes (at least near the cusp and on the morning side) and substantially less regular internal MSH flows near the equator. The latter can be regarded as an argument for the dominance of the high latitude source of MSH plasma in the magnetosphere.

Figure 13 displays the measured MP and cusp projections onto the SM *xy* and *yz* planes. The *x* spread of the OC is from –2 to 7 $R_E$ (Fig. 13a), and the *y* spread is ±7 $R_E$ (Fig. 13b). The low-altitude cusps are shifted 1 to 2 $R_E$ in the negative *y* direction in the northern hemisphere and 1 to 2 $R_E$ in the positive *y* direction in the southern hemisphere in the SM frame (Fig. 13b). The MP distribution in Fig. 13 is defined by the INTERBALL-1 orbits and, as a result, show gaps in dayside midlatitudes and close to the midnight meridian (northern hemisphere). In summary, the southern and northern cusps cover all altitude ranges from INTERBALL-1 pericenter to the MP. Most cusps were encountered in the winter-spring periods in 1996. We have not separated them into different groups

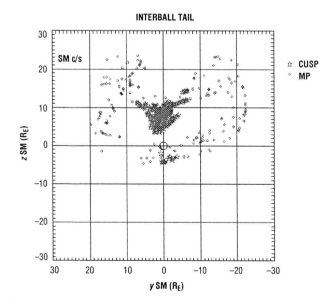

**Figure 13b.** First-year Interball-1 cusp (asterisks) and MP (rombs) crossings in SM frame; *zy* projection.

for different IMF $B_z$ and for other SW parameters because of poor statistics for different SW conditions. One can see that the cusps are seen at any altitude inside a rather narrow region in the $xz$ plane. The high altitude encounters are often seen continuously for a few hours, covering several $R_E$. This is evidence of the permanent entry of the MSH plasma into the magnetosphere through the cusps, independent of external SW parameters. *Sandahl et al.* [1996] draw similar conclusions from four successive orbits on January 17–28, 1996 only, during which both northward and southward IMF cases are encountered.

Figure 13 gives not only the full altitude cusp coverage but also shows the interrelation between cusp and MP statistics. One can see that in the vicinity of the exterior cusp, the MPs tend to be inside the model MP, while upstream and downstream MP crossings occur rather symmetrically with respect to the model MP. In the April 21 case study, we have already outlined that the MP crossing occurred about 2 hr earlier than expected. Based on the location of the model MP, and approximately between the real and model MPs, the stagnation sites and nonlinear turbulence have been seen (i.e., "OT"). To clarify how common the April 21, 1996 case is and if the MPs are closer to the Earth in the OT on average, we have examined the difference between the actual MP crossings and the model MP position as calculated for the measured SW dynamic pressure [see *Sibeck et al.*, 1991]. The results are presented in Fig. 14a where the GSM $x$ dependence of the MP deflection from model MP is shown for two GSM $z$ groups; namely, for GSM $z$ less or more than 7 $R_E$. This $z$ approximately separates the low and high latitude MPs in the first half of 1996 (the time interval used for producing Fig. 14).

One can see that for $-2<x<7$ $R_E$ the high latitude MPs (full circles) are deflected strongly inside the model MP. This interval corresponds to the exterior cusp/OT (Fig. 13). The average deflection for 31 MP crossings is 1.4 (±0.7) $R_E$. Based on this we conclude that the "Outer Cusp Throat" (being outside the MP is inferred from magnetic field direction changes; see also Fig. 1) is of 1–2 $R_E$ depth towards the Earth from the regular MSH flow. To illustrate that this indentation is rather substantial in terms of dynamic pressure, we calculated what dynamic pressure would be needed for the model MP to be located at the measured positions. The calculated dynamic pressures are shown by open circles (predicted) in Fig. 14b versus the maximum, average, and minimum SW dynamic pressure (upper dots, full circles, and lower dots in Fig. 14b, respectively) measured during 1-hr time intervals on WIND (the intervals are shifted by SW time of flight to INTERBALL-1). The measured and predicted dynamic pressures can differ by factors as large as 5, giving the average ratio of 3 (±2.7). Our data seem to fit the MP current layer shape better in the TSY87 and TSY89 [*Tsyganenko*, 1987] magnetic field models than in those of the modern MP [e.g., *Sibeck et al.*, 1991]. These are

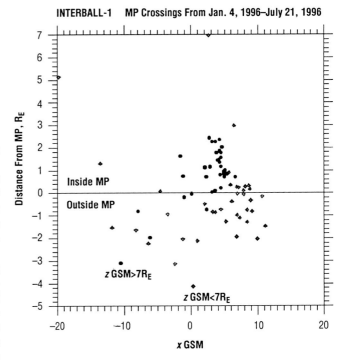

**Figure 14a.** Comparison of the real and model MP crossings for the period Jan. 4, 1996–July 21, 1996. $x$ (GSM) dependence of the MP crossing deflections from the model *Sibeck* [1991]; full circles=$z$ GSM > 7 $R_E$ and open rombs=$z$ GSM < 7 $R_E$.

arguments for the need of a new high latitude MP model. Such work is being planned after 2–3 yr of INTERBALL-1 high latitude measurements for better statistics (*Tsyganenko*, personal communication).

As for the low $z$ GSM MPs, they tend to occur about 1 $R_E$ out of the model MP (see Fig. 14a, open rhombs). We should note that the two low latitude MPs at $x=-20$ could be MSH plasma injection into the PSBL, due to SW disturbances. It is difficult to distinguish them from the MP proper. The bigger horizontal magnetosphere width as compared to the vertical width is in agreement with the findings of *Sibeck et al.* [1991].

## DISCUSSION AND CONCLUSIONS

In this section we summarize results presented above and also those published by other INTERBALL Tail Probe teams. The first high time resolution, two-point study of the near-cusp BL revealed multiscale nonlinear structures therein. A rather well defined (especially in terms of the magnetic field) MP coexists with MSH plasma vortices and stagnation sites ("PSS") caught in the exterior cusp on both sides of the main MP current sheet. The medium-scale (hydromagnetic) PSS are bounded by magnetic barriers

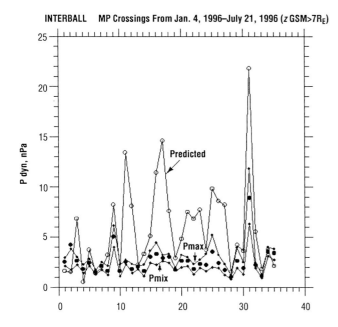

**Figure 14b.** Measured (full symbols) and calculated (using *Sibeck* [1991] model (open circles) represent dynamic pressures for the detected MP positions for $z$ GSM >7 $R_E$. See text for details.

with pressures of the order of the MSH plasma kinetic energy density (0.5 nmV$^2$). The magnetic field inside the PSS is an order of magnitude less than that outside the PSS; the thermal energy excess inside the PSS is close to the MSH kinetic energy density (being ~0.9 of that).

We don't think that the transformation of the MSH kinetic energy to the PSS plasma thermal energy is a one-step process. While Alfvenic and sound Mach numbers in the external MSH flow were 1.9 and 1.3, respectively, even a shock upstream of the OT shouldn't transform all kinetic energy into thermal energy. The PSSout (see Fig. 1) encounters the tailward boundary ("magnetic wall") of the OT (with $Bx$ close to zero, see Fig. 2a) and reflects, probably, several times. The MP indentation, proposed originally by *Choe et al.* [1973] and *Haerendel and Paschmann* [1975], can play a crucial role here. The ULF medium-scale waves, growing in the PSS/MP interaction region, start the chaotization of the flow energy. Their phase velocity directions are close to the MP normal at the tailward OT wall (Fig. 10). This chaotization finishes at micro-scale current sheets, i.e., at scales of the order of the electron inertial length and/or electron gyroradius scales (about 2 km; see Fig. 11). The heated plasma inside the PSS pushes the internal magnetic field out to the borders of the PSS. A part of the magnetic turbulence energy could be transformed into the ion thermal energy also. The picture proposed here is rather simplified and speculative; a comparison with representative models is needed in the future.

The turbulent BLs, being mostly outside the MP, contain a deflected magnetic field, high ULF turbulence, and accelerated/heated MSH particles. The TBL represents a characteristic feature of the high latitude MP, having maximum turbulence amplitudes near the cusps [*Klimov et al.*, 1986, 1997; *Savin et al.*, 1997]. One could regard the TBL as a part of the thick MP. In some cases the MP current sheet cannot be separated from the TBL [cf. *Haerendel and Paschmann*, 1975]. The decreases of the total field in the TBL resemble the PSS but with much smaller scale.

The interior cusp is believed to be on the reconnected field lines which come out of the magnetosphere through the OT [see *Lundin et al.*, 1991A]. These reconnected lines should cross the TBL where the extremely high level turbulence mix them and heat the MSH ions (see Fig. 2). This picture differs dramatically from the picture presented by laminar plane reconnection at low latitudes characterized by the presence of the accelerated Alfvenic flows in the deHoffman-Teller frame [see *Cowley*, 1995]. One could argue that in the TBL the sporadic burst reconnection of the small scale (compared with Fig. 2b time resolution) could operate. But the trapped ions inside the magnetic barrier (Fig. 8) show that nonlinear processes in the TBL cannot be properly described by the isolated burst reconnections (that again, accelerate ions do not heat). Individual field lines can hardly be traced from the OC through the TBL to the OT. The lines should be treated as connected in the statistical sense; they conserve the average direction inside and outside the MP. So, the cusp field lines (at least the portion that crosses the TBL) can be partially disconnected from the OT, having very different particle distributions just inside and outside the MP/TBL (see Figs. 3 and 5).

The energy balance for the ions, heated in the TBL, shows that the ac and/or dc magnetic field annihilation could provide up to 20% of the ion thermal energy (e.g., at 04:37 UT; see Fig. 2a). But the main energy source for the heating is believed to be the kinetic energy of the external streaming MSH plasma. The extremely high amplitude waves serve as the intermediate process in the cascade of the transformation of the SW energy. This process should be included in future models for a proper description of the exterior cusp processes.

Isolated FTE-like bipolar signatures of $Bn$ cannot be easily extracted in the vicinity of the exterior cusp as: (a) no unique normal exists in the complicated OT geometry with PSS magnetic loops; (b) nonlinear turbulence even in $Bn$ reaches amplitudes comparable with the total magnetic field magnitudes; (c) reconnection features can be seen almost continuously in the TBL (e.g., counterstreaming field-aligned ions, energization of the MSH particles in the TBL, escaping and reflected ions in the core OT (see Figs. 3, 5(a), and 5(c)) and energetic particle penetrations in the OT outside the MP). This is unlike tail and low-latitude MPs where classical and multiple FTEs are seen slightly more than 50% of the time [*Klimov et al.*, 1997; *Savin et al.*, 1997].

The reconnection features in the INTERBALL-1 data have also been reported at low latitudes and on the high latitude tail field lines at $x=-20\ R_E$ [*Klimov et al.*, 1997; *Vaisberg et al.*, 1997; and *Savin et al.*, 1997]. The latter case seems to be connected with the magnetic cloud interaction with the magnetosphere on October 18, 1995; the dominating sunward flow inside the MP shows that plasma could be locally heated to energies of the cool PS during injection inside the magnetosphere.

Filaments of a few kilometer thickness (Fig. 11) have been predicted by a number of thin MP current sheet simulations [see *Drake et al.*, 1995 and references therein] but to our knowledge they never have been measured directly in the near-Earth space. In *Savin et al.* [1997] the shortest scales of a few kilometers at the MP and in FTE for the deep tail, high-latitude MP crossings by INTERBALL-1 on October 18, 1995 have been obtained by applying the medium-scale disturbance velocities to the micro scales. The ion VDP mode and low sampling rate of the split probe haven't permitted the proper application of the CWD approach. Similar single Faraday cup and magnetic fluctuation data in the OC regions aboard Prognoz-8 have been used by *Savin et al.* [1994] for indirect estimations of the current fine-structure scales. The Prognoz-8 estimates are close to the INTERBALL-1 measurements, while for the Prognoz-8 case, the average MP speed has been used and the currents couldn't be distinguished from anisotropic field-aligned bidirectional electrons (see Fig. 4). From the magnetic field and plasma data similarities, one can estimate that the characteristic scales at low latitude BLs are on the same order as the high latitude BLs [*Savin et al.*, 1997]. Namely, the scales of the MP current sheets are of the order of 100–500 km; 50–200 km are characteristic for medium-scale sheets (TBL turbulence, FTE, etc.) and of a few kilometers for the turbulent microstructures.

Statistically, the OT is of 1–2 $R_E$ depth (part of stagnation region outside MP). The OC inside the MP have ±7 $R_E$ spread in SM $y$ near the MP, it is shifted from 7 to 2 $R_E$ in SM $x$. The INTERBALL-1 orbit permitted us to see the cusp throughout all the dayside magnetosphere from 1000 km altitude to the MP (see Fig. 13). The MSH ion flows are also repeatedly seen also in the LLBL and mantle (Fig. 12), the dawn mantle being much more thick and frequent. Comparison with the MP *Sibeck et al.* [1991] model shows that in the cusp vicinity, a new MP model should be developed.

Our BL studies concentrated mostly on the ion flows, having a kinetic energy excess which could drive ionospheric current systems and play a major role in the development of substorms [*Lundin et al.*, 1991B]. MSH electrons in the BL are dense and heated; the ULF turbulence can heat the electrons mostly along magnetic field lines [see *Savin et al.*, 1997]. Inside the cusp and BL these electrons are bidirectional, proving that these lines could be closed. In the MSH, such electrons could escape from the MP [see *Sauvaud et al.*, 1997], while in the core OT the electrons seem to be isolated from the magnetosphere. The potential barrier of 100–200 V at the MP can account for the difference in the ion and electron behaviors [cf. *Vaisberg et al.*, 1983]. The correlation of heated electrons with ULF waves and small-scale currents at high latitudes have been also reported by *Blecki et al.* [1987]. The microturbulence can be another reason for the electron "disclosures" that both support differences in the electron distributions just inside and outside the main MP current sheets and might locally heat the electrons in the current sheets and TBL.

The characteristic medium and micro time scales could be less than the time resolution of the INTERBALL three-dimensional particle energy spectra measurements. This should be taken into account for the interpretation of phenomena such as ion beams in the LLBL [*Vaisberg et al.*, 1997], where the ion distribution could quickly oscillate as well as decay into separate beams (see Fig. 8 and related discussions).

Quasi-periodic disturbances in the inner BL with characteristic time scales of a few minutes (e.g., at 04:02, 04:08, and 04:14 UT in Fig. 3) are observed in many cases [*Sandahl et al.*, 1997; *Sauvaud et al.*, 1997; *Safrankova et al.*, 1997; *Vaisberg et al.*, 1997]. At the high latitude tail BL, simultaneously taken two-point particle data are compatible with the quasi-periodic movements of the plasma sites or flows with amplitudes of 500 to 3000 km [*Savin et al.*, 1997]. The MP crossing itself seems to have periodicity of 5–10 min. It is an open question if the latter periodicity is driven by SW disturbances or by MP surface waves. *Safrankova et al.* [1997] have shown that simultaneous wavy motion of the inner and outer boundaries of the MP BL could be attributed to the Kelvin-Helmholtz instability. It has been proposed that in the cusp, such disturbances are driven by standing Alfvén waves [*Sandahl et al.*, 1997].

During the revision of this paper, the Hawkeye-1 cusp crossing example was published [*Chen et al.*, 1997] in which many features similar to those of INTERBALL-1 are shown: deflected and heated MSH plasma, depressed magnetic field with intermediate (between MSH and magnetosphere) direction, the MP crossing with highly deflected normal deep inside the model MP, etc. Their exterior cusp corresponds to our OT. The interior cusps resemble the PSS in the depressed field and heated MSH plasma with one substantial difference: loss cone-like gaps in the ion distributions along the magnetic field (see their Fig. 7). The latter should mean merging with the Earth's magnetic field at the moment of the measurement or just prior to it (cf. QPI in Fig. 5(c) and related discussions). Relying on INTERBALL-1 data, we think that the plasma heating has taken place after reconnection in the multistep manner we have described in this paper. Therefore, we believe that the interior cusps in *Chen et al.* [1997] can be the OT substructures (PSS) or reconnected field line filaments rather than multiple MP crossings. We should also mention that the interior cusp ion distributions, labeled 1–4 in Fig. 7 in *Chen et al.* [1997], are inconsistent with placing the distribution on the inner field lines (connected to the north magnetic pole; see their Fig. 9). In this case, reconnection models predict plasma flow components towards the pole [see Fig. 10 of *Chen et al.*, 1997], while the ion distributions have minimums in the field direction, meaning that the plasma is flowing antiparallel to the magnetic field. The only ion distribution consistent with the reconnection field-aligned flow direction in the OC is the "5" in Fig. 7 of *Chen et al.* [1997]. The latter looks like our core OT data (see Fig. 3, 04:45–04:55 UT and related discussions). Our speculative considerations, of course, couldn't prove which interpretation is correct. We do need to find similar cases in the INTERBALL-1/Magion-4 data to make progress on this quite interesting problem. The three-dimensional electron distributions can be critical for the problem, as we have shown here.

It is worth mentioning one more manuscript with Hawkeye-1 data; namely, *Fung et al.* [1997] which was sent to press just after our paper. Their statistical exterior cusp distributions are very similar to ours, even though they have many more crossings for 4 yr of operation. Their wider cusp spread (out of average bow shock, see their Fig. 4) is, we believe, not only due to the higher number of crossings and different SW parameters, but also partially due to differences in the identification of their exterior cusp, which contains our OT and OC. We would like to do detailed comparison of our statistical data on the OT, OC, TBL, and PSS with the Hawkeye-1 data in the future.

Our preliminary cusp and BL INTERBALL-1 studies show that the multidevice and multipoint data are very promising. Comparison with the Polar and Geotail data should substantially enrich our understanding of BL processes.

*Acknowledgments.* We are grateful to Drs. C. T. Russell and N. A. Tsyganenko for helpful discussions. We thank Dr. K. W. Ogilvie and the SWE team for providing WIND SW dynamic pressure and velocity data. We are grateful to the first referee for valuable suggestions on the paper improvement and to I. Dobrovolsky for help in the paper preparation. The work was partially supported by INTAS through grant INTAS-93-2031.

## REFERENCES

Blecki, J., K. Kossacki, S. I. Klimov, M. N. Nozdrachev, A. N. Omelchenko, S. P. Savin, and A. Yu. Sokolov, ELF/ULF plasma waves observed on Prognoz-8 near the Earth magnetopause, *Artificial Satellites*, Space Physics, No. 7, vol. 22, No. 4, 1987.

Chen, S.-H., S. A. Boardsen, S. F. Fung, J. L. Green, R. L. Kessel, L. C. Tan, T. E. Eastman, and J. D. Craven, Exterior and interior polar cusps: Observations from Hawkeye, *J. Geophys. Res., 102*(A6), 11335–11347, 1997.

Choe, J. Y., D. B. Bearb, and E. C. Sullivan, Precise calculation of the magnetosphere surface for a titled dipole, *Planet. Space Sci., 21,* 485–498, 1973.

Cowley, S. W. H., Theoretical perspectives of the magnetopause: A tutorial review, in *Physics of the Magnetopause*, edited by P. Song, B. U. O. Sonnerup, and M. F. Thomsen, American Geophysical Union, Washington, DC, pp. 29–43, 1995.

Drake, J. F., Magnetic Reconnection: A Kinetic Treatment, in *Physics of the Magnetopause,* edited by P. Song, B. U. O. Sonnerup, and M. F. Thomsen, American Geophysical Union, Washington, DC, pp. 155–165, 1995.

Fung, S. F., T. E. Eastman, S. A. Boardsen, and S.-H. Chen, High-Altitude Cusp Positions Sampled by the Hawkeye Satellite, *Physics and Chemistry of the Earth,* accepted, 1997.

Galeev, A. A., Yu. I. Galperin, and L. M. Zelenyi, The INTERBALL Project to study solar-terrestrial physics, in *INTERBALL Mission and Payload*, CNES-IKI-RSA, pp. 11–42, 1995.

Haerendel, G., and G. Paschmann, Entry of solar wind plasma into the magnetosphere, in *Physics of the Hot Plasma in the Magnetosphere,* edited by B. Hultqvist and L. Stenflo, p. 23, Plenum, NY, 1975.

Klimov, S. I., M. N. Nozdrachev, P. Triska, Ya. Vojta, A. A. Galeev, Ya. N. Aleksevich, Yu. V. Afanasiev, V. E. Baskakov, Yu. N. Bobkov, R. B. Dunetz, A. M. Zhdanov, V. E. Korepanov, S. A. Romanov, S. P. Savin, A. Yu. Sokolov, and V. S. Shmelev, Investigation of plasma waves by combined wave diagnostic device BUDWAR (PROGNOZ-10-INTERCOSMOS), *Cosmic Research* (Transl. from Russian), *24,* 177, 1986.

Klimov, S., S. Romanov, E. Amata, J. Blecki, J. Buechner, J. Juchniewicz, J. Rustenbach, P. Triska, L. J. C.Woolliscroft, S. Savin, Yu. Afanas'yev, U. de Angelis, U. Auster, G. Bellucci, A. Best, F. Farnik, V. Formisano, P. Gough, R. Grard, V. Grushin, G. Haerendel, V. Ivchenko, V. Korepanov, H. Lehmann, B. Nikutowski, M. Nozdrachev, S. Orsini, M. Parrot, A. Petrukovich, J. L. Rauch, K. Sauer, A. Skalsky, J. Slominski, J. G. Trotignon, J. Vojta, and R. Wronowski, ASPI experiment: Measurements of fields and waves onboard of the INTERBALL-TAIL mission, in *INTERBALL mission and payload*, IKI-CNES, p. 128, 1995.

Klimov, S., S. Romanov, E. Amata, J. Blecki, J. Buechner, J. Juchniewicz, J. Rustenbach, P. Triska, L. J. C. Woolliscroft, S. Savin, Yu. Afanas'yev, U. de Angelis, U. Auster, G. Bellucci, A. Best, F. Farnik, V. Formisano, P. Gough, R. Grard, V. Grushin, G. Haerendel, V. Ivchenko, V. Korepanov, H. Lehmann, B. Nikutowski, M. Nozdrachev, S. Orsini, M. Parrot, A. Petrukovich, J. L. Rauch, K. Sauer, A. Skalsky, J. Slominski, J. G. Trotignon, J. Vojta, and R. Wronowski, ASPI Experiment: Measurements of Fields and Waves Onboard the INTERBALL-1 Spacecraft, *Ann. Geophys., 15,* 514, 1997.

Lemaire, J., Impulsive penetration of filamentary plasma elements into magnetosphere of the Earth and Jupiter., *Planet. Space Sci., 25,* 887, 1977.

Lundin, R., J. Woch, and M. Yamauchi, The present understanding of the cusp, in Proceedings of the Cusp Workshop, *Eur. Space Agency Spec. Publ., ESA SP-330,* p. 83–95, 1991A.

Lundin, R., I. Sandahl, J. Woch, and R. Elphinstone, The contribution of the boundary layer EMF to magnetospheric substorms, in *Magnetospheric Substorms, Geophysical Monograph 64,* Amercian Geophysical Union, Washington, DC, p. 335, 1991B.

Paschmann, G., G. Haerendel, N. Sckopke, H. Rosenbauer, and P. C. Hedgecock, Plasma and magnetic field characteristics of the distant polar cusp near local noon: The entry layer, *J. Geophys. Res., 81*, 2883, 1976.

Romanov, S. A., Vector field correlation analysis for the space plasma measurements, *Cosmic Research* (Transl. from Russian), in press, 1998.

Romanov, S. A., S. I. Klimov, P. A. Mironenko, Space characteristics and dispersion relations of ULF waves on the Earth bow shock from the Prognoz-10 satellite data, *Cosmic Research* (Transl. from Russian), *28*, No. 5, 750–759, 1990.

Russell, C. T., and R. C. Elphic, Initial ISEE magnetometer results: magnetopause observations, *Space Sci. Rev., 22*, 681, 1978.

Russell, C. T., The structure of the magnetopause, in *Physics of the Magnetopause*, edited by P. Song, B. U. O. Sonnerup, and M. F. Thomsen, American Geophysical Union, Washington, DC, pp. 81–98, 1995.

Sandahl, I., R. Lundin, M. Yamauchi, U. Eklund, J. Safránková, Z. Nemecek, K. Kudela, R. P. Lepping, R. P. Lin, V. N. Lutsenko, and J-A. Sauvaud, Cusp and boundary layers observations by INTERBALL, *Adv. Space Res., 20*, No. 4-5, pp. 823-832, 1997.

Sandahl, I., R. Lundin, M. Yamauchi, U. Eklund, J. Safrankova, Z. Nemecek, R. P. Lepping, R. P. Lin, K. Kudela, V. N. Lutsenko, and J. A. Sauvaud, Cusp and Boundary Layers Observations by INTERBALL, *Adv. Space Res., 15*, 542, 1997.

Safrankova, J., Z. Nemecek, L. Prech, G. Zastenker, A. Fedorov, S. Romanov, D. Sibeck, and J. Simunek, Two-point observations of magnetopause motion, *Adv. Space Res., 19*, in press, 1997.

Sauvaud, J. A., P. Koperski, T. Beutier, H. Barthe, C. Aoustin, J. J. Thocaven, J. Rouzaud, O. Vaisberg, and N. Borodkova, The INTERBALL-Tail electron experiment, Initial results on the low latitude boundary layer of the dawn magnetosphere, *Ann. Geophys., 15*, 587, 1997.

Savin, S. P., ELF waves near the high latitude magnetopause, in *Abstracts of AGU Chapman Conference on Physics of the Magnetopause*, p. 41, March 14–18, 1994.

Savin, S. P., O. Balan, N. Borodkova, E. Budnik, N. Nikolaeva, V. Prokhorenko, T. Pulkkinen, N. Rybjeva, J. Safrankova, I. Sandahl, E. Amata, U. Auster, G. Bellucci, A. Blagau, J. Blecki, J. Buechner, M. Ciobanu, E. Dubinin, Yu. Yermolaev, M. Echim, A. Fedorov, Formisano, R. Grard, V. Ivchenko, F. Jiricek, J. Juchniewicz, S. Klimov, V. Korepanov, H. Koskinen, K. Kudela, R. Lundin, V. Lutsenko, O. Marghitu, Z. Nemecek, B. Nikutowski, M. Nozdrachev, S. Orsini, M. Parrot, A. Petrukovich, N. Pissarenko, S. Romanov, J. Rauch, J. Rustenbach, J. A. Sauvaud, E. T. Sarris, A. Skalsky, J. Smilauer, P. Triska, J. G. Trotignon, J. Vojta, G. Zastenker, L. Zelenyi, Yu. Agafonov, V. Grushin, V. Khrapchenkov, L. Prech, and O. Santolik, INTERBALL Magnetotail Boundary Case Studies, *Adv. Space Res., 19*, 993, 1997.

Sibeck, D. G., R. E. Lopez, and E. C. Roelof. Solar wind control of the magnetopause shape, location, and motion. *J. Geophys. Res., 96*, 5489, 1991.

Sibeck, D. G., The magnetospheric response to Foreshock Pressure pulses, in *Physics of the Magnetopause*, edited by P. Song, B. U. O. Sonnerup, and M. F. Thomsen, American Geophysical Union, Washington, DC, pp. 293–302, 1995.

Tsyganenko, N. A., Global quantitative models of the geomagnetic field in the cislunar magnetosphere for different disturbance levels, *Planet. Space Sci., 35*, 1347, 1987.

Vaisberg, O. L., A. A. Galeev, L. M. Zeleny, G. N. Zastenker, A. N. Omelchenko, S. I. Klimov, S. P. Savin, Yu. I. Yermolaev, V. N. Smirnov, and M. N. Nozdrachev, Fine structure of the magnetopause from the measurements of the satellites PROGNOZ-7 and PROGNOZ-8, *Cosmic Research* (Transl. from Russian), *21*, 49, 1983.

Vaisberg, O. L., L. A. Avanov, V. N. Smirnov, J. L. Burch, J. H. Waite, A. A. Petrukovich, and A. A. Skalsky, Interball observations of the dayside magnetopause, *Adv. Space Res., 19*, 779, 1997.

Yermolaev, Yu. I., A. O. Fedorov, O. L. Vaisberg, V. M Balebanov, Yu. A. Obod, R. Jimenez, J. Fleites, L. Llera, and A. N. Omelchenko, Ion distribution dynamics near the Earth's bow shock: first measurements with the two-dimensional ion energy spectrometer CORALL on the INTERBALL/Tail-probe satellite, *Ann. Geophys., 15*, 533, 1997.

---

S. P. Savin, Space Research Institute, Russian Academy of Sciences, Profsoyuznaya 84/32, GSP-7, Moscow, 117810, Russia. (e-mail: SMTP:ssavin@mx.iki.rssi.ru)

# Convection of Plasmaspheric Plasma into the Outer Magnetosphere and Boundary Layer Region: Initial Results

Daniel M. Ober,[1,2] J. L. Horwitz

*CSPAAR, University of Alabama in Huntsville, Huntsville, Alabama*

D. L. Gallagher

*Space Science Laboratory, NASA Marshall Space Flight Center, Huntsville, Alabama*

We present initial results on the modeling of the circulation of plasmaspheric-origin plasma into the outer magnetosphere and low-latitude boundary layer (LLBL), using a dynamic global core plasma model (DGCPM). The DGCPM includes the influences of spatially and temporally varying convection and refilling processes to calculate the equatorial core plasma density distribution throughout the magnetosphere. We have developed an initial description of the electric and magnetic field structures in the outer magnetosphere region. The purpose of this paper is to examine both the losses of plasmaspheric-origin plasma into the magnetopause boundary layer and the convection of this plasma that remains trapped on closed magnetic field lines. For the LLBL electric and magnetic structures we have adopted here, the plasmaspheric plasma reaching the outer magnetosphere is diverted anti-sunward primarily along the dusk flank. These plasmas reach $X = -15$ $R_E$ in the LLBL approximately 3.2 hours after the initial enhancement of convection and continues to populate the LLBL for 12 hours as the convection electric field diminishes.

## 1. INTRODUCTION

Convection of cold plasma through the inner magnetosphere can be broadly divided into two different flow regimes by the separatrix between open and closed drift paths. Flux tubes on closed drift paths convect eastward with the earth filling to high (~$10^3$ cm$^{-3}$) densities [e.g., Nishida, 1966]. Flux tubes on open drift paths convect sunward from the tail region, filling to comparatively lower densities (~1-10 cm$^{-3}$), before being lost to the outer magnetosphere. During large-scale convection enhancements, the separatrix between open and closed drift paths penetrates closer to the earth. Plasmaspheric flux tubes on previously closed drift paths may now be located outside the separatrix and convect sunward on open drift paths into the outer magnetosphere creating a plasmaspheric tail [Chen and Wolf, 1972] or possibly a 'detached' plasma region [Chappell, 1974]. Freeman et al. [1977] were probably the firsts to suggest that detached plasmaspheric clouds [e.g., Chappell, 1974] could circulate from the dusk bulge region during large-

---
[1]Current address is Space Science Laboratory, NASA Marshall Space Flight Center, Huntsville, Alabama.
[2]NAS/NRC Research Associate.

Geospace Mass and Energy Flow: Results From the International Solar-Terrestrial Physics Program
Geophysical Monograph 104
Copyright 1998 by the American Geophysical Union

scale storm-time convection gust into the outer magnetosphere, magnetosheath, and low-latitude boundary layer (LLBL) regions. Their scenario further proposed that such plasmas would be energized, and convect into the tail region, so that the plasmasphere could thus be a viable source of energetic magnetospheric plasmas.

There have been several observations for which spacecraft crossed the magnetopause and enhanced density regions were observed next to and just inside the initial magnetopause crossing and in between successive multiple crossings for a single pass [Chappell, 1974]. Adjacent observations of plasmaspheric plasmas and magnetopause crossings have also been reported from geosynchronous orbit during times of high magnetospheric activity [Elphic et al., 1996]. The presence of He$^+$ in the magnetospheric boundary layers can perhaps be used to identify the loss of plasmaspheric plasma out of the magnetosphere, since He$^+$ is a significant component of plasmaspheric plasma [Horwitz et al., 1984]. Observations from ISEE 1 have shown He$^+$ densities near the subsolar magnetopause of 0.2 cm$^{-3}$, He$^+$ densities inside the magnetopause of 0.8 cm$^{-3}$, and significant He$^+$ concentrations for two cases in the magnetosheath layer outside the magnetopause [Peterson et al., 1982]. Observations from AMPTE showed cold He$^+$ ions at densities in the range 0.1-0.8 cm$^{-3}$ convecting into the subsolar low latitude boundary layer where it was evidently heated and accelerated to energies of a few keV [Fuselier et al., 1989]. Observations on PROGNOZ-7 have shown cases where the cold plasma component was frequently found to dominate the local magnetospheric plasma density in the dayside boundary layer indicated by a high percentage of He$^+$ ions [Lundin et al., 1985]. This cold plasma component may have the largest influence on the flow of momentum across the boundary layer [Lundin et al., 1985; Lundin et al., 1984].

Here we seek to explore quantitatively the circulation of plasmaspheric plasma into the outer magnetosphere and boundary layer regions using a dynamic global core plasma model (DGCPM) [Ober et al., 1997; Ober et al., 1995]. We present the results for a sequence in which the large-scale convection increases following a magnetically quiet period. First, we will discuss the formulation of the model magnetic and electric field structures in the outer magnetosphere regions used in the DGCPM, and then the implications of our model results will be discussed.

## 2. MODEL DESCRIPTION

The electric field structure used here for the DGCPM derives from a time-dependent ionospheric two-cell electric potential model [Sojka et al., 1986] which is mapped along assumed equipotential magnetic field lines given by the Tsyganenko-89 model (Kp=0) [Tsyganenko, 1989]. These magnetospheric electric and magnetic field structures are used to calculate the E×B drift of approximately 140,000 convecting cold plasma flux tubes. The plasma density is assumed uniform along a flux tube and the content of a tube is controlled by ionospheric refilling and tube volume variations during convection. The total ion content of a magnetic flux tube evolves in time as

$$\frac{D_\perp N}{Dt} = \frac{F_N + F_S}{B_i}$$

where D/Dt is the convective derivative in the moving frame of the flux tube, N is the total ion content per unit magnetic flux, $F_N$ and $F_S$ are the ionospheric fluxes in or out of the tube at the northern and southern ionospheres, and $B_i$ is the magnetic field in the ionosphere at the foot point of the flux tube [Rasmussen, 1993; Chen and Wolf, 1972]. The net flux of particles into the flux tube on the dayside is

$$F_D = \frac{n_{sat} - n}{n_{sat}} F_{max}$$

where $F_D$ (=$F_N$+$F_S$) is the dayside ionospheric flux, $n_{sat}$ is the saturation density of plasma in the flux tube, n is the density of plasma in the flux tube, and $F_{max}$ (=2.9·10$^8$ particles/cm$^2$/sec) is the limiting flux of particles from the ionospheres [Rasmussen, 1992; Chen and Wolf, 1972]. The saturation density is approximated from the empirical model of Carpenter and Anderson [1992]. The nightside flux is approximated as exponential drainage of the flux tube content into the nightside ionosphere.

Observations of the LLBL from IMP 6 indicate that the bulk flow always has an anti-sunward component, that the LLBL is at times on closed field lines, and that the thickness appears to increase with increasing longitudinal distance from the subsolar point [Eastman et al., 1979]. Flow speeds are largest close to the magnetopause [Sckopke et al., 1981]. We can achieve these characteristics for our model LLBL by allowing closed magnetic field lines in the outer magnetospheric flanks and subsolar region to map into the ionosphere around the polar cap boundary and flow reversal region. Magnetic field lines in the LLBL are expected to have significant field aligned potential drops. Here the electric potential is mapped assuming equipotential field lines. The effects of field aligned potential drops on the structure of the electric field in the LLBL will be considered in future work. Figure 1 shows the model electric potential pattern in the ionosphere for a total polar cap potential drop of 72 kV. The shaded region in Figure 1 shows approximately the area in our model ionosphere that maps to the outer magnetosphere and LLBL regions. The LLBL flow velocities in our model are on the order of 0-300 km/sec while the magnetospheric flow velocities in the outer magnetosphere tend to be in the range of 0-60 km/sec [e.g., Eastman et al., 1979]. Figure 2 shows the electric potential pattern in the equatorial magnetosphere mapped from the ionosphere for a total polar cap potential drop of 72 kV. The dashed line shows the inner boundary of the LLBL. Within the model LLBL, the cold plasma flow is away from the subsolar point and anti-sunward along the flanks. The total potential drop across our model LLBL shown in Figure 2, at MLT = 15, is 26 kV which is slightly larger than what has been observed [e.g., Hapgood and Lockwood, 1993; Mozer, 1984]. The width of the model LLBL increases with increasing anti-sunward distance.

For the preliminary investigation here, the ionospheric electric potential changes in time but not the magnetic field structure. The location of the inner boundary of the LLBL is determined by the mapping of the ionospheric electric potential pattern and changes as the electric potential evolves in time. The location of the magnetopause is determined by the transition from closed to open magnetic field lines in the magnetic field model and does not change in time. Therefore, the width of the LLBL may become unrealistically large (2.6 $R_E$ at 15 MLT) at times in the simulation but we anticipate that the physics of the model still remains useful to beginning to understand the flow of plasmaspheric plasma into the outer magnetosphere and boundary layer regions.

For the simulation presented here we consider a case in which the total polar cap potential drop increases suddenly and then decays. Figure 3 shows the total polar cap

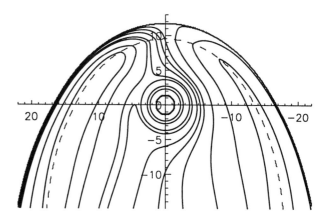

**Figure 2.** Electric potential pattern in the equatorial magnetosphere for a total polar cap potential of 72 kV. The dashed line shows the inner boundary of the LLBL.

potential drop, the total potential drop across the LLBL and the width of the LLBL at MLT=15, all as a function of time. The polar cap potential shown in Figure 3 rises suddenly during the first 6 hours of the simulation.

## 3. RESULTS

The simulation shown here will illustrate the convection of bulge region plasmaspheric plasma into the dayside magnetopause boundary layer. Initially, the simulation is ran with a constant potential pattern until a self-consistent steady state density distribution is established that is consistent with observations of a quiet time plasmasphere.

Plate 1 shows the initial (0 hours) density distribution in the magnetosphere prior to the sudden rise of the total polar cap potential. At 3 hours in the simulation plasmaspheric flux tubes convecting sunward into the outer magnetosphere reach the dayside magnetopause and are diverted towards the dusk flank of the magnetosphere. At 3.2 hours into the simulation, anti-sunward streaming flux tubes of plasmaspheric origin in the LLBL reach x = -15 $R_E$. Shown here is the continued transport of plasmaspheric origin plasma in the LLBL at 5 hours into the simulation. At 15 hours the sunward transport of plasmaspheric flux tubes has diminished and at 21 hours the plasmasphere has returned to a near quiet time configuration except that the plasmapause has shifted earthward.

In our model simulation the cold plasma densities in the LLBL were in the range of 1-10 $cm^{-3}$. Using a $He^+/H^+$ ratio of 0.2 that is typical of the plasmasphere [Horwitz et al., 1984], our model would predict a $He^+$ density in the LLBL of 0.2-2 $cm^{-3}$. This is in the range of observed $He^+$ densities in the LLBL [e.g. Peterson et al., 1982; Fuselier et al., 1989].

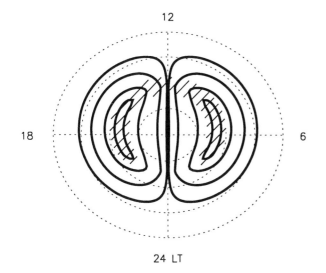

**Figure 1.** Electric potential pattern in the ionosphere for a total polar cap potential of 72 kV. The shaded region shows the approximate area that maps to the outer magnetosphere and LLBL regions.

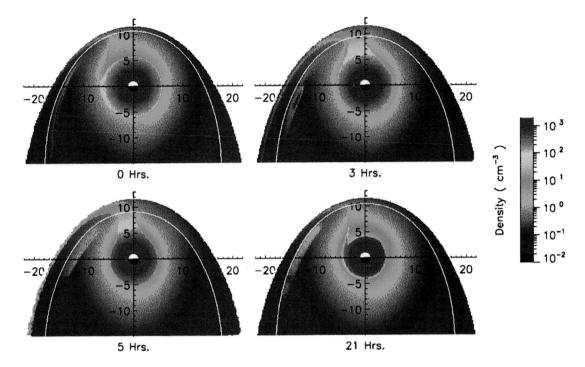

**Plate 1.** Modeled cold plasma density distribution in the equatorial magnetosphere at 0, 3, 5, and 21 hours during the simulation. The white line marks the location of the inner edge of the low-latitude boundary layer. The axis are geocentric in units of $R_E$. Color bar shows density color scale used.

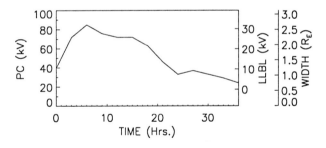

**Figure 3.** Plotted are the total polar cap potential drop, the total potential drop across the LLBL and the width of the LLBL, both at MLT=15, used in the model simulation.

modeled He$^+$ densities in the LLBL are around 0.2-2 cm$^{-3}$ that is in the range of observed values of total densities appropriate to He$^+$ in the LLBL. In our simulation we have only considered the convection of plasma that remains on closed field lines. Flux tubes in the outer plasmasphere may also convect out to the magnetopause, reconnect with the IMF, and convect over the polar cap [Elphic et al., 1997; Freeman et al., 1977].

*Acknowledgments.* This research was supported by NASA grant NGT-51299 and NSF grant ATM-9301024 to the University of Alabama in Huntsville.

## 4. CONCLUSION

In our model simulation it was observed that plasmaspheric flux tubes convect sunward from the dusk bulge region during a large-scale convection enhancement and are diverted along the dusk flank in the LLBL when reaching the outer magnetosphere. Plasmaspheric flux tubes convecting in the LLBL region require about 20 minutes to reach X= -15 $R_E$. Transport of plasmaspheric origin flux tubes into the dusk flank LLBL continues for about 12 hours until the outer plasmasphere has become depleted. The

## REFERENCES

Carpenter, D. L., and R. R. Anderson, An ISEE/whistler model of equatorial electron density in the magnetosphere, J. Geophy. Res., 97, 1157, 1992.

Chappell, C. R., Detached plasma regions in the magnetosphere, J. Geophy. Res., 79, 1861, 1974.

Chen, A. J., and R. A. Wolf, Effects on the plasmasphere of a time-varying convection electric field, Planet. Space Sci., 20, 483, 1972.

Elphic, R. C., M. F. Thomsen, and J. E. Borovsky, The fate of the outer plasmasphere, Geophys. Res. Lett., 24, 365, 1997.

Elphic, R. C., L. A. Weiss, M. F. Thomsen, D. J. McComas, and M. B. Moldwin, Evolution of plasmaspheric ions at geosynchronous orbit during times of high geomagnetic activity, Geophys. Res. Lett., 23, 2189, 1996.

Eastman, T. E., and E. W. Hones, Jr., Characteristics of the magnetospheric boundary layer and magnetopause layer as observed by Imp 6, J. Geophys. Res., 84, 2019, 1979.

Freeman, J. W., H. K. Hills, T. W. Hill, and P. H. Reiff, Heavy ion circulation in the earth's magnetosphere, Geophys. Res. Lett., 4, 195, 1977.

Fuselier, S. A., W. K. Peterson, D. M. Klumpar, and E. G. Shelley, Entry and acceleration of He+ in the low latitude boundary layer, Geophys. Res. Lett., 16, 751, 1989.

Hapgood, M. and M. Lockwood, On the voltage and distance across the low latitude boundary layer, Geophys. Res. Lett., 20, 145, 1993.

Horwitz, J. L., R. H. Comfort, and C. R. Chappell, Thermal ion composition measurements of the formation of the new outer plasmasphere and double plasmapause during the storm recovery phase, Geophys. Res. Lett., 11, 704, 1984.

Lundin, R., E. M. Dubinin, Solar wind energy transfer regions inside the dayside magnetopause: Accelerated heavy ions as tracers for MHD-processes in the dayside boundary layer, Planet. Space Sci., 33, 891, 1985.

Lundin, R., E. M. Dubinin, Solar wind energy transfer regions inside the dayside magnetopause-1. Evidence for magnetospheric plasma penetration, Planet. Space Sci., 32, 745, 1984.

Mozer, F. S. Electric field evidence on the viscous interaction at the magnetopause, Geophys. Res. Let., 11, 135, 1984.

Ober, D. M., J. L. Horwitz and D. L. Gallagher, Formation of density troughs embedded in the outer plasmasphere by subauroral ion drifts (SAID), J. Geophys. Res., 14595, 1997.

Ober, D. M., J. L. Horwitz and D. L. Gallagher, Global plasmasphere evolution during a sub-auroral ion drift (SAID) event, in Physics of Space Plasmas (1995). Number 14, T. Chang and J. R. Jasperse, eds. (MIT Center for Theoretical Geo/Cosmo Plasma Physics, Cambridge, MA, 1996), 691.

Nishida, A., Formation of plasmapause, or magnetospheric knee, by the combined action of magnetospheric convection and plasma escape from the tail, J. Geophys. Res., 71, 5669, 1966.

Peterson, W. K., E. G. Shelley, G. Haerendel, and G. Paschmann, Energetic ion composition in the subsolar magnetopause and boundary layer, J. Geophy. Res., 87, 2139, 1982.

Rasmussen, C. E., S. M. Guitar, and S. G. Thomas, A two-dimensional model of the plasmasphere: Refilling time constants, Planet. Space Sci., 41, 35, 1993.

Rasmussen, C. E., The plasmasphere, In Physics of space plasmas, number12 in SPI Conference proceedings and reprint series, page 279, Scientific publishers, 1992.

Sckopke, N., G. Paschmann, G. Haerendel, B. U. O. Sonnerup, S. J. Bame, T. G. Forbes, E. W. Hones, Jr., and C. T. Russell, Structure of the low-latitude boundary layer, J. Geophy. Res., 86, 2099, 1981.

Sojka, J. J., C. E. Rasmussen, and R. W. Schunk, An interplanetary magnetic field dependent model of the ionospheric convection electric field, J. Geophy. Res., 91, 11281, 1986.

Tsyganenko, N. A., A magnetospheric magnetic field model with a warped tail current sheet, Planet. Space Sci., 37, 5, 1989.

---

D. L. Gallagher, and D. M. Ober, ES83, Space Science Laboratory, NASA Marshall Space Flight Center, Huntsville, Al 35812.

J. L. Horwitz, CSPAAR, The University of Alabama in Huntsville, Huntsville, Al 35899.

# AMPTE-IRM Observations of Particles and Fields at the Dayside Low-latitude Magnetopause

T. M. Bauer, G. Paschmann, N. Sckopke, W. Baumjohann, and R. A. Treumann

*Max-Planck-Institut für extraterrestrische Physik, Garching, Germany*

T.-D. Phan

*Space Sciences Laboratory, University of California, Berkeley*

Using AMPTE-IRM data, we study the structure of the low-latitude magnetopause (MP) and its boundary layer on the dayside (08:00–16:00 LT). We analyze three MP passes in detail and present an overview of the entire data set. In about half of the passes the boundary layer is divided into an outer part (OBL) dominated by solar wind plasma and an inner part (IBL) where the densities of solar wind and magnetospheric particles are comparable. The distinctive feature of the IBL are "warm" counterstreaming electrons originating mainly from the magnetosheath. There is evidence that the IBL is on closed field lines: The bulk flow changes from tailward in the OBL to stagnant in the IBL, and the density of ring current electrons increases sharply at the interface between OBL and IBL. For two of the three passes analyzed in detail, we find reconnection signatures in the OBL and magnetosheath: bulk flows in reasonable agreement with the stress balance of a rotational discontinuity, reflection and "D-shaped" distributions of solar wind ions, and counterstreaming of solar wind and magnetospheric particles. During the third pass reconnection signatures are absent, and a plasma depletion layer is observed in the magnetosheath. There is a clear tendency for reconnection signatures to be more frequent when the magnetic shear across the magnetopause is high. Estimating diffusion coefficients, we find that cross-field diffusion cannot transport solar wind plasma at a rate that can account for the formation of the OBL or IBL.

## 1. INTRODUCTION

For many years it has been known that immediately earthward of the magnetopause (MP) at low latitudes is a boundary layer (BL) commonly populated by shocked solar wind plasma from the magnetosheath and magnetospheric plasma. Explaining the formation of the low latitude boundary layer, i.e., determining the mode of plasma transport across it, is one of the outstanding problems of magnetospheric physics. Magnetic reconnection with subsequent mixing along open field lines and diffusion across closed magnetic field lines have been invoked to provide the transport. With the aid of spacecraft data the problem of boundary layer formation can be approached in several ways:

1) Important information can be deduced from measured time profiles of the partial densities of the various plasma populations, i.e., solar wind, ring current, and ionospheric particles. Whereas sharp steps in the profiles may mark topological boundaries or discontinuities associated with reconnection [*Sonnerup et al.*,

1981; *Gosling et al.*, 1990b; *Song et al.*, 1990; *Walthour et al.*, 1994], gradual transitions are expected for a BL formed by diffusion [*Thorne and Tsurutani*, 1991].

2) If the MP is open it should have the properties of a rotational discontinuity. Therefore the problem of BL formation can be approached by comparing the measured time series of plasma moments with predictions for a rotational discontinuity. Predictions used for experimental tests are the existence of a de Hoffmann-Teller (HT) frame [*Sonnerup et al.*, 1990; *Paschmann et al.*, 1990], the tangential stress balance [*Sonnerup et al.*, 1981, 1990; *Paschmann et al.*, 1986, 1990; *Gosling et al.*, 1990a; *Phan et al.*, 1996], the energy balance [*Paschmann et al.*, 1986], and the mass balance [*Fuselier et al.*, 1993].

3) In addition to applying the fluid concept of a rotational discontinuity, one can also check whether the measured distribution functions show the single particle signatures expected on open field lines. These signatures are a consequence of the interpenetration of solar wind and magnetospheric particles and of their reflection at a thin current layer [*Cowley*, 1982; *Gosling et al.*, 1990b; *Fuselier et al.*, 1991; *Bauer et al.*, 1997].

4) If magnetic reconnection does not occur no magnetic flux can be transferred across the MP. This may lead to pile-up of the magnetic field and simultaneous depletion of the magnetosheath plasma adjacent to the MP [*Zwan and Wolf*, 1976; *Denton and Lyon*, 1996; *Phan et al.*, 1994]. Therefore the observation of a plasma depletion layer (PDL) in the magnetosheath adjacent to the MP can be regarded as indirect evidence for reconnection being absent or inefficient in the vicinity of the spacecraft.

5) The amplitudes of electric and magnetic fluctuations observed in the BL can be used to compute diffusion coefficients derived from quasilinear theory [*Thorne and Tsurutani*, 1991; *Treumann et al.*, 1995]. These semi-empirical diffusion coefficients can then be used to elucidate whether or not cross-field diffusion can transport solar wind plasma into the BL at a rate that can account for the thickness of the BL.

In the present paper, we will apply the aforementioned methods 1 to 5 to data of the AMPTE-IRM spacecraft obtained at the dayside MP. Section 2 describes methods 2 and 3, i.e., how we test for the fluid and particle properties of a rotational discontinuity. In section 3 three MP passes are analyzed in detail. In order to discuss the question of how typical these individual events are we compare them with other IRM events in section 4. In section 5 quasilinear diffusion coefficients are estimated. Finally, we draw our conclusions in section 6.

We use measurements of the triaxial flux gate magnetometer [*Lühr et al.*, 1985], the plasma instrument, and the wave instrument on board the IRM spacecraft. The plasma instrument [*Paschmann et al.*, 1985] consisted of two electrostatic analyzers of the top hat type, one each for ions and electrons. Three-dimensional distributions with 128 angle and 30 energy channels in the energy-per-charge range from 15 V to 30 kV for electrons and 20 V to 40 kV for ions were obtained every satellite rotation period, i.e., every 4.4 s. From each distribution, microcomputers within the instruments computed moments of the distribution functions of ions and electrons: densities in three contiguous energy bands, bulk velocity vector, pressure tensor, and heat flux vector. In these computations it was assumed that all the ions are protons. Whereas the moments were transmitted to the ground at the full time resolution, the distributions themselves were transmitted less frequently, because the allocated telemetry was limited. The ELF/VLF spectrum analyzer of the IRM wave experiment package [*Häusler et al.*, 1985] used the signal from the 47-m tip-to-tip dipole antenna to provide a relatively coarse frequency resolution, yet rapid temporal resolution with essentially continuous coverage, of electric wave signals from 25 Hz to 250 kHz.

## 2. PREDICTIONS FOR A ROTATIONAL DISCONTINUITY

If the MP is time stationary and tangential gradients are small compared to the normal gradients then it can be modelled as a magnetohydrodynamic discontinuity [*Sonnerup et al.*, 1990]. A closed MP, which has a normal component $B_n = 0$ of the magnetic field, can be modelled as a tangential discontinuity. In the case of an open MP, reconnection produces a finite $B_n$ and plasma can cross the MP along field lines. Since the rotational discontinuity is the only discontinuity with $B_n \neq 0$ that can provide an arbitrary change of the direction of **B**, an open MP should have the properties of a rotational discontinuity. In this paper we use the measured time series of plasma moments to check the existence of a HT frame and the tangential stress balance.

The HT frame is a frame moving at velocity $\mathbf{V}_{HT}$ in which the transformed plasma bulk velocity, $\mathbf{V}' = \mathbf{V} - \mathbf{V}_{HT}$, is purely field aligned and therefore the convection electric field, $\mathbf{E}'_c = -\mathbf{V}' \times \mathbf{B}$, vanishes. Whereas a rotational discontinuity must have a HT frame, a tangential discontinuity does, in general, not have such a frame [*Sonnerup et al.*, 1990]. $\mathbf{E}'_c = 0$ can be used to estimate the de Hoffmann-Teller velocity, $\mathbf{V}_{HT}$, of an observed MP from the measured time series of the proton bulk velocity, $\mathbf{V}_p$, and the magnetic field, **B**. Hereby $\mathbf{V}_{HT}$ is obtained as the vector that minimizes the quadratic form

$$D = \langle |(\mathbf{V}_p - \mathbf{V}_{HT}) \times \mathbf{B}|^2 \rangle \quad (1)$$

which is approximately the square of $\mathbf{E}'_c$ averaged over measurements taken in the vicinity of the MP [*Sonnerup et al.*, 1990]. If the minimum of $D$ is well defined and the estimated convection electric field, $\mathbf{E}_c = -\mathbf{V}_p \times \mathbf{B}$, is approximately equal to the transformation electric field, $\mathbf{E}_{HT} = -\mathbf{V}_{HT} \times \mathbf{B}$, we can conclude that a HT frame exists for the MP crossing under consideration.

The tangential stress balance (Walén relation) of a rotational discontinuity states that the plasma bulk velocity in the HT frame is Alfvénic. Replacing the plasma bulk velocities, $\mathbf{V}$ and $\mathbf{V}'$, by the proton bulk velocities, $\mathbf{V}_p \approx \mathbf{V}$ and $\mathbf{V}'_p \approx \mathbf{V}'$, this reads

$$\mathbf{V}'_p = \mathbf{V}_p - \mathbf{V}_{HT} = \pm \mathbf{c}_A = \pm \frac{\mathbf{B}(1-\alpha)^{1/2}}{(\mu_0 N m_p)^{1/2}} \quad (2)$$

where $\mathbf{c}_A$ is the Alfvén velocity in a plasma with number density $N$ and pressure anisotropy $\alpha = (P_\parallel - P_\perp)\mu_0/B^2$. The + sign (− sign) is valid when the normal component $V_{pn}$ of the proton bulk flow has the same (opposite) direction as $B_n$. *Sonnerup et al.* [1990] and *Paschmann et al.* [1990] quantified the fit between the data and the prediction of Eq. (2) by computing correlation and regression coefficients for the time series of $\mathbf{V}'_p$ and $\mathbf{c}_A$. For the MP crossings analyzed in section 3 we compute in addition the quantities $C_{V,c_A}$ and $V'_{p\parallel}/c_A$. The ratio $V'_{p\parallel}/c_A$ is evaluated for each measurement of the field-aligned component of $\mathbf{V}'_p$ and the Alfvén speed. $C_{V,c_A}$ is the cross correlation of the components of $\mathbf{V}_p$ and $\mathbf{c}_A$ along the maximum variance direction of $\mathbf{B}$. The maximum variance direction of $\mathbf{B}$ [*Sonnerup et al.*, 1990] is tangential to the MP and is chosen, because it is approximately the direction along which the variation of $\mathbf{c}_A$ has the highest dynamic range. If $C_{V,c_A}$ and $V'_{p\parallel}/c_A$ are both close to $\pm 1$ the data agree with the prediction for a rotational discontinuity. Across a tangential discontinuity the variation of $\mathbf{V}$ does not depend on the variation of $\mathbf{c}_A$. Therefore $C_{V,c_A}$ can assume arbitrary values in the case of a closed MP and $V'_{p\parallel}$ cannot be defined if no HT frame exists.

Particle distribution functions expected at an open MP have been described by *Cowley* [1982]. After reconnection has produced a finite $B_n$, ring current and ionospheric particles can move outward, i.e., toward the solar wind end of an open field line and solar wind particles can move inward, i.e., toward its terrestrial end. If the magnetopause current layer is sufficiently thin the ion motion in the current layer becomes non-adiabatic. Then an ion component incident on the current layer is only partly transmitted, the other part is reflected. For reflection at a thin current layer the field-aligned flow velocities in the HT frame of the reflected ($V'_{r\parallel}$) and incident ($V'_{i\parallel}$) component fulfill $V'_{r\parallel} = -V'_{i\parallel}$. In the HT frame the particle velocities $\mathbf{v}'$ of inward moving particles fulfill $v'_\parallel > 0$ when $B_n$ points inward and $v'_\parallel < 0$ when $B_n$ points outward. For outward moving particles it is the other way round. Hence, each component of the incident, reflected, and transmitted plasma populations should have a velocity cutoff at $v'_\parallel = 0$. Distribution functions with such a velocity cutoff are called "D-shaped" distributions and were observed by *Gosling et al.* [1990b], *Smith and Rodgers* [1991] and *Fuselier et al.* [1991]. Ion reflection off the MP was reported by *Sonnerup et al.* [1981], *Gosling et al.* [1990a], and *Fuselier et al.* [1991].

## 3. CASE STUDY

In this section three MP passes are analyzed in detail. $\mathbf{B}$ and $\mathbf{V}_p$ are given in $LMN$ boundary normal coordinates [*Russell and Elphic*, 1979]. This coordinate system is defined such that the $N$ axis points outward along the MP normal and the $LN$ plane contains the $Z$ axis of the GSM system. The MP normal is taken from the model of *Fairfield* [1971]. For the MP crossings examined in sections 3.1 and 3.2, the shear angle, $\Delta \varphi_B$, between the magnetic fields in the magnetosheath and in the BL is high ($\Delta \varphi_B > 45°$). The crossing examined in section 3.3 is a low shear crossing ($\Delta \varphi_B < 30°$). Figures 1 to 3 give the time series of the plasma moments measured during the three passes. In Plate 1 electron and ion distributions are shown in a planar cut through velocity space that is nearly tangential to the MP.

### 3.1. Observations on 84/09/21

Figure 1 presents an overview of the outbound MP crossing on 84/09/21, which occurs at 13° northern GSM latitude and 11:10 LT. The MP at 13:01:11 UT can be identified as a change of the angle $\varphi_B$, i.e., a rotation of the magnetic field tangential to the MP. After 13:01:11, IRM is located in the magnetosheath. Earthward of the MP three different regions can be distinguished. From ~12:57 to 12:58:51, IRM is in the magnetosphere proper and from 13:01:02 to 13:01:11 it is located in the outer boundary layer (OBL), a region of dense magnetosheath-like plasma. Before ~12:57 and during the intervals 12:58:51–13:00:01 and 13:00:18–13:01:02, the total density is somewhat higher than in the magnetosphere proper and the contribution of solar wind particles to the density is comparable to the contribution of magnetospheric particles. We call this region the inner boundary layer (IBL). In the plasma moments of Fig. 1, the difference between the IBL and the magnetosphere is hard to see, but it will become clearly visible in the distributions. The division of the BL into an outer and inner part was already reported by *Sckopke et al.* [1981] for the flanks as well as by *Song et al.* [1990] and *Hall et al.* [1991] for the dayside MP.

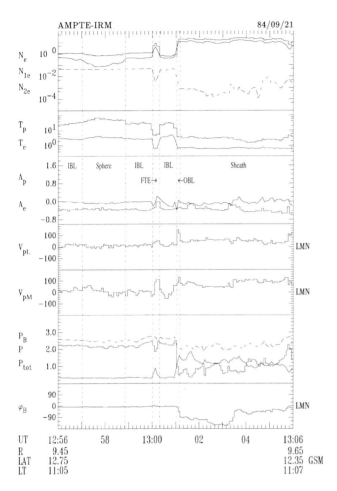

**Figure 1.** Magnetopause pass on 84/09/21. The upper panel shows the total (15 eV–30 keV) electron density, $N_e$ (histogram line), in cm$^{-3}$ and the partial densities, $N_{1e}$ (solid line) and $N_{2e}$ (dashed line), of electrons in the energy ranges 60 eV–1.8 keV and 1.8 keV–30 keV, respectively. In the next two panels the proton and electron temperatures, $T_p$ (histogram line) and $T_e$ (solid line), in $10^6$ K and the respective anisotropies, $A_p = T_{p\parallel}/T_{p\perp} - 1$ (histogram line) and $A_e = T_{e\parallel}/T_{e\perp} - 1$ (solid line), are given. The next two panels present the components $V_{pL}$ and $V_{pM}$ of the proton bulk velocity in km/s. $V_{pL}$ and $V_{pM}$ refer to the boundary normal coordinate system. In the sixth panel the magnetic pressure, $P_B$ (histogram line), plasma pressure, $P = N_p K T_p + N_e K T_e$ (solid line), and total pressure, $P_{tot} = P_B + P$ (dashed line), in nPa are shown. The last panel gives the angle $\varphi_B$ the magnetic field makes with the $L$ axis in the $LM$ plane of the boundary normal coordinate system. Vertical dashed lines indicate boundaries separating different plasma regions.

The enhancement of $N_e$ and depression of $T_p$, $T_e$ around 13:00:10 correspond to a flux transfer event (FTE). It shows the +− bipolar signature of $B_n$ (not shown) expected for open magnetic flux tubes moving northward [e.g., *Cowley*, 1982].

In the panel of $V_{pL}$ we recognize a northward directed reconnection flow in the OBL. The interval 13:01:02–13:01:28 around the MP has indeed a good HT frame and the time series of $\mathbf{V}_p$ and $\mathbf{B}$ are well correlated. For example, the cross correlation $C_{V,c_A}$ of the components along the maximum variance direction of $\mathbf{B}$ equals +0.9. The sign of $C_{V,c_A}$ indicates $B_n < 0$, i.e., open field lines connected to the northern hemisphere.

Panel a of Plate 1 presents a series of electron distributions measured on 84/09/21 in the magnetosphere (12:58:32), the IBL (13:00:21), the OBL (13:01:05), and the magnetosheath (13:01:48). In the sheath, IRM detects solar wind electrons with temperature $KT \approx 50$ eV. The distribution taken in the magnetosphere proper at 12:58:32 shows hot ($KT \approx 5$ keV) ring current electrons at velocities $v > 10000$ km/s and cold ($KT \sim 10$ eV) electrons presumably of ionospheric origin at velocities $v < 4000$ km/s. From the sign of $C_{V,c_A}$ we inferred that the local MP has an inward directed normal magnetic field, $B_n < 0$. This result is strongly supported by the electron distribution taken in the OBL at 13:01:05. We see solar wind electrons streaming parallel to $\mathbf{B}$ (inward along open field lines) and simultaneously hot ring current electrons streaming antiparallel to $\mathbf{B}$ (outward). In the IBL at 13:00:21, IRM detects hot ring current electrons and another population at field-aligned velocities $v_\parallel \approx 8000$ km/s. This population was already observed by the ISEE satellites [*Ogilvie et al.*, 1984] and by AMPTE-UKS [*Hall et al.*, 1991] and was called "counterstreaming" electrons, because the flux parallel to $\mathbf{B}$ equals the flux antiparallel to $\mathbf{B}$. Since it is not a priori clear that this electron population does always exhibit the balance between the flux parallel and antiparallel to $\mathbf{B}$, we prefer to call it "warm" electrons. The term "warm" shall indicate that the field-aligned temperature of this population is mostly higher than that of solar wind electrons in the magnetosheath and in the OBL. The origin of the warm electrons will be discussed in section 3.3.

Let us turn to the ion distributions (Plate 1d) taken in the magnetosphere (12:58:18), the IBL (13:00:34), the OBL (13:01:05), and the magnetosheath (13:01:39). As expected on open field lines with $B_n < 0$, the distributions in the sheath and in the OBL show solar wind ion plasma with the flow velocity $\mathbf{V}'$ in the HT frame parallel to $\mathbf{B}$. The distribution in the OBL has the characteristic "D shape" predicted by *Cowley* [1982]. Its cutoff velocity is consistent with $\mathbf{V}_{HT}$: There are only few ions with field-aligned particle velocities $v'_\parallel < 0$. Checking the ratio $V'_{p\parallel}/c_A$ of the field-aligned proton bulk velocity in the HT frame and the Alfvén speed, we find that it is +0.2 in the magnetosheath and +0.5 in the OBL, which differs considerably from the value +1 predicted by Eq. (2). Nevertheless, the ion and electron

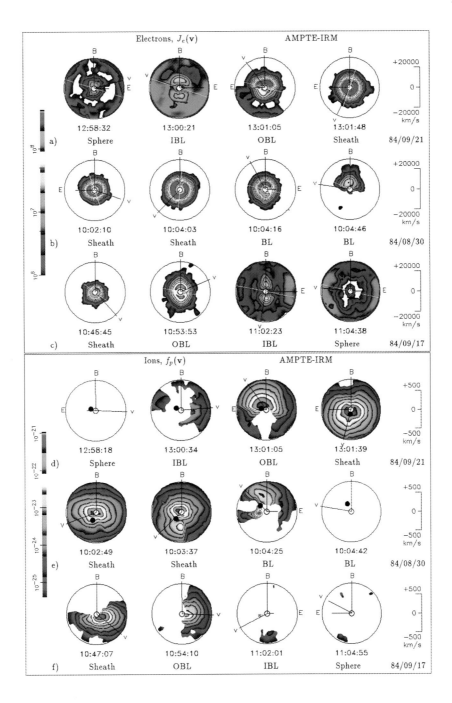

**Plate 1.** Ion and electron distributions across the dayside magnetopause. The distribution functions are shown in a two-dimensional cut through velocity space that contains the magnetic field direction, **B** (upward), and $\mathbf{n} \times \mathbf{B}$ (to the left), where **n** is the magnetopause normal. Moreover, the projection of the direction of the proton bulk flow, $\mathbf{V}_p$, is given. The energy range shown is 15 eV–3 keV for the electrons and 20 eV–4 keV for the ions. The black dot in the ion distributions gives the projection of the de Hoffmann-Teller velocity, $\mathbf{V}_{HT}$, onto the cut. $\mathbf{V}_{HT}$ is determined by minimization of $D$ (Eq. (1)) and is the origin of the $\mathbf{v}'$ system used in the text. The yellow line in the electron distributions gives the projection of the IRM spin axis. Due to an instrumental defect some distributions exhibit a reduced electron flux along the spin axis at low energies. Panels a to c show the differential directional energy flux density $J_e$ (in eV/s cm$^2$ eV sr) of electrons measured during the three magnetopause passes analyzed in section 3. Panels d to f show the phase space density $f_p$ (in cm$^{-6}$ s$^3$) of ions measured during these passes.

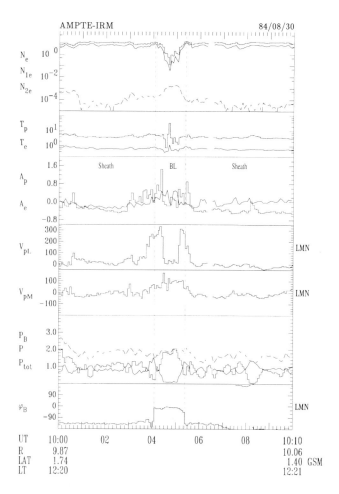

**Figure 2.** Magnetopause pass on 84/08/30. The format is the same as in Fig. 1.

distributions observed in the OBL provide evidence for the OBL being on open field lines with $B_n < 0$.

In the limited energy range shown in Plate 1, no ions are measured in the magnetosphere proper. However, IRM detects hot ring current ions with temperature $KT \approx 10\,\text{keV}$ (not shown). This ring current population is also observed in the IBL. Moreover, the distributions taken in the IBL after the FTE (e.g., the one given in Plate 1d) show solar wind ions ($KT \sim 1\,\text{keV}$), whereas before the FTE cold ($KT \sim 10\,\text{eV}$) ions of ionospheric origin are detected instead. The electron distributions measured before and after the FTE are similar to one another. For many of the distributions taken in the IBL, e.g., for the one given in Plate 1d, the proton bulk velocity $\mathbf{V}'_p$ in the HT frame has a substantial component perpendicular to $\mathbf{B}$. This indicates that the IBL is not located on open field lines crossing the OBL. Important information about the IBL can also be deduced from the time series of $N_{2e}$ and $V_{pM}$. In the IBL

the partial density $N_{2e}$ of electrons above $1.8\,\text{keV}$ has about the same value as in the magnetosphere proper, but it drops at the interface between IBL and OBL. Such a drop is expected at the interface between closed and open field lines. In the OBL, $V_{pM}$ is directed dawnward as expected for plasma on tailward moving open field lines on the dawnside (11:10 LT). In contrast, $V_{pM}$ is highly variable in the IBL before the FTE and even directed duskward after the FTE. These features taken together suggest that the IBL is on closed field lines.

No ion and only one electron distribution was transmitted to the ground during the FTE. Like the electron distribution taken in the OBL, the distribution during the FTE shows solar wind electrons streaming parallel to $\mathbf{B}$. This indicates that the field lines of the FTE are also connected to the northern hemisphere and is consistent with the $+-$ signature of $B_n$ during the FTE.

### 3.2. Observations on 84/08/30

Figure 2 as well as panels b and e of Plate 1 present a close pair of MP crossings on 84/08/30 at 2° northern GSM latitude and 12:20 LT. Both crossings can be identified as a sudden change of the angle $\varphi_B$. The inbound crossing occurs at 10:04:05 UT and the outbound crossing at 10:05:23. Between the two crossings IRM encounters the BL. For this event it is not possible to distinguish two separate parts of the BL. While the electron distributions change gradually, the ion distributions are highly variable. Note the rather smooth transition of the total density $N_e$, and the partial densities $N_{1e}$, $N_{2e}$ on the one hand, and the large variation of $T_p$ and $T_{p\parallel}/T_{p\perp}$ on the other hand. As we will see, the high values of $T_{p\parallel}/T_{p\perp}$ in the vicinity of the MP are due to counterstreaming of different ion components.

In the panel of $V_{pL}$ we recognize northward directed reconnection flows. The existence of a HT frame and the agreement with the Walén relation (2) was already tested for these flows by *Paschmann et al.* [1986] and *Sonnerup et al.* [1990]. They found a good HT frame and a fairly good correlation of the time series of $\mathbf{V}_p$ and $\mathbf{B}$. For the interval 10:03:48–10:04:27 around the inbound crossing and for the interval 10:05:06–10:05:45 around the outbound crossing, the cross correlation $C_{V,c_A}$ of the components along the maximum variance direction of $\mathbf{B}$ equals $+0.6$ and $+0.8$, respectively, indicating $B_n < 0$. The existence of a normal magnetic field $B_n$ directed inward is confirmed by the electron distributions taken in the BL at 10:04:16 and 10:04:46 (Plate 1b), which show solar wind electrons streaming parallel to $\mathbf{B}$, i.e., inward along open field lines. In the magnetosheath (10:02:10 and 10:04:03) the solar wind electrons exhibit a reduced flux along the spin axis which is due to an instrumental defect, described in Appendix 1 of *Paschmann et al.* [1986].

In Plate 1e we see a series of ion distributions measured in the magnetosheath well before the inbound crossing (10:02:49), in the magnetosheath close to the inbound MP crossing (10:03:37), in the dense part of the BL (10:04:25), and finally in its dilute part (10:04:42). The two magnetosheath distributions show an incident solar wind component flowing parallel to **B** with $V'_{i\parallel} \approx +0.6 c_A$ in the HT frame. Close to the MP, reflected solar wind ions appear. As expected for reflection at a thin current layer, the field-aligned flow velocities in the HT frame of the reflected ($V'_{r\parallel}$) and incident ($V'_{i\parallel}$) component fulfill $V'_{r\parallel} = -V'_{i\parallel}$ (section 2). $V'_{i\parallel} > 0$ and $V'_{r\parallel} < 0$ is consistent with $B_n < 0$, as deduced from the test of the Walén relation and the electron distributions in the BL. The appearance of the reflected ions leads to the detected increase of $T_{p\parallel}/T_{p\perp}$ after ~10:03.

In the BL at 10:04:25, we recognize a maximum of the temperature anisotropy, $T_{p\parallel}/T_{p\perp} \approx 2.5$. As can be seen in Plate 1e, this field-aligned anisotropy is also due to counterstreaming of two components. One component are solar wind ions that have been transmitted across the MP: They have $v'_\parallel > 0$, which is again consistent with $B_n < 0$. The other component is much colder and presumably of ionospheric origin: These ions have $v'_\parallel < 0$ and thus stream outward along open field lines with $B_n < 0$. Because of the presence of the ionospheric ions the field-aligned bulk velocity $V'_{p\parallel}$ in the HT frame is only $+0.05 c_A$ in the BL. As noted earlier, $\mathbf{V}_p$ was computed under the assumption that all the ions are protons. If the ionospheric component contained many heavy ions the actual $V'_{p\parallel}$ might even be negative. Although $V'_{p\parallel}/c_A$ is significantly different from $+1$, the reflected ions in the sheath, the counterstreaming ions in the BL, and the electron distributions in the BL provide evidence for open field lines. At 10:04:42, in the dilute part of the BL, no ions are visible within the energy range of Plate 1. However, IRM detects hot ring current ions with temperature $kT \approx 5$ keV at that time.

### 3.3. Observations on 84/09/17

Figure 3 presents an overview of the inbound MP crossing on 84/09/17, which occurs at 22° southern GSM latitude and 14:10 LT. The rotation of the magnetic field across the MP is low ($\Delta\varphi_B < 30°$). *Phan et al.* [1994] performed a statistical survey of the magnetosheath adjacent to the MP and found that all AMPTE-IRM low shear crossings have a plasma depletion layer, whereas they did not find a PDL for the high shear ($\Delta\varphi_B > 45°$) crossings. This suggests that for low magnetic shear less magnetic flux is transferred across the MP, consistent with the expectation that magnetic reconnection is absent or less efficient between magnetic fields that are nearly parallel. The low shear MP crossing on 84/09/17 was included in the data set of *Phan et al.* [1994] and it was also studied by *Paschmann et al.* [1993]. In this section we will show that the absence of magnetic reconnection, as inferred from the existence of a PDL, is confirmed by tests of the predictions of section 2, which will reveal that the local MP is closed.

The MP crossing occurs at 10:47:58 UT. We can see that it has indeed a clear PDL: In Fig. 3 the plasma pressure decreases before 10:47:58 and the magnetic pressure increases. Furthermore, the existence of a PDL is reflected in the strong perpendicular anisotropy, $T_{p\parallel}/T_{p\perp} \approx 0.2$, of the proton temperature in the sheath adjacent to the MP. This anisotropy can be understood both in the fluid and particle picture. In the particle picture it arises because ions with high field-aligned velocities leave the region where the plasma is depleted earlier than ions with low field-aligned velocities [*Paschmann et al.*, 1993]. In the fluid picture it is due to stretching of the field lines [*Denton et al.*, 1994].

Since $\Delta\varphi_B$ is small, it is not possible to identify the MP with the magnetic field data. But it is clearly visible in the plasma moments. Most striking is the sharp

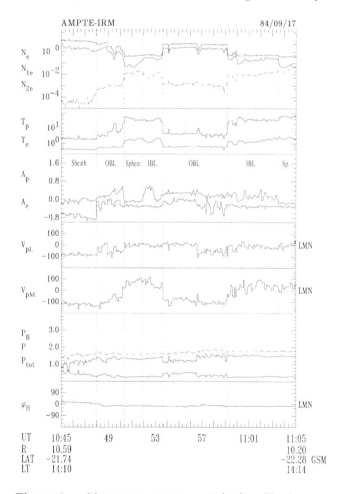

**Figure 3.** Magnetopause pass on 84/09/17. The format is the same as in Fig. 1.

increase of $T_{p\parallel}/T_{p\perp}$ from its low value of about 0.2 in the PDL to values of almost 1 after 10:47:58.

Like for the high shear crossing on 84/09/21, three different regions can be distinguished earthward of the MP. From 10:47:58 to ~10:50:20 and from 10:53:35 to 10:59:09, IRM encounters the dense plasma of the OBL. Between ~10:50:20 and ~10:51:50 and after ~11:03:30, it is located in the magnetosphere proper. An IBL with properties similar to those of the IBL observed on 84/09/21 is encountered from ~10:51:50 to 10:53:35 and from 10:59:09 to ~11:03:30. In Fig. 3 the difference between the IBL and the magnetosphere is visible in the traces of $N_{1e}$ and $A_e$. It is not possible to find a well-defined HT frame for the MP at 10:47:58. Moreover, the time series of $\mathbf{V}_p$ and $\mathbf{B}$ are not correlated with one another, confirming that the local MP is closed.

Plate 1f presents a series of ion distributions measured on 84/09/17 in the magnetosheath (10:47:07), the OBL (10:54:10), the IBL (11:02:01), and the magnetosphere (11:04:55). We recognize that the solar wind population has a strong perpendicular anisotropy in the PDL and is nearly isotropic in the OBL. A few solar wind ions are also detected in the IBL: Note the narrow green patch at $v \approx 200$ km/s in the distribution taken at 11:02:01. Furthermore, hot ring current ions are observed in the IBL and magnetosphere proper. Having $KT \approx 10$ keV, they lie outside the energy range selected for Plate 1. None of the ion distributions shows single particle signatures predicted for open field lines.

Plate 1c presents electron distributions measured in the four regions. At 10:46:45 in the sheath, IRM detects solar wind electrons with temperature $KT \approx 30$ eV. Across the MP the field-aligned temperature of the solar wind electrons increases by a factor of 2, while their perpendicular temperature increases only slightly. Unlike the distribution taken in the OBL on 84/09/21, the electron distributions observed in the OBL on 84/09/17, e.g., at 10:53:53, do not show any evidence for open field lines. The distributions taken in the IBL (11:02:23) and magnetosphere proper (11:04:38) are similar to those observed on 84/09/21. In the magnetosphere proper we find cold ($KT \sim 10$ eV) electrons presumably of ionospheric origin at velocities $v < 4000$ km/s and hot ($KT \approx 1$ keV) ring current electrons at velocities $v > 10000$ km/s. Outside the energy range shown in Plate 1, a second ring current component with temperature $KT \sim 10$ keV is detected. In the IBL, e.g., at 11:02:23, we recognize again the warm electrons at field-aligned velocities $v_{\parallel} \approx 8000$ km/s. What is the origin of this electron population that is observed in the dilute part of the BL for almost all high shear and low shear MP passes? Whereas *Hall et al.* [1991] concluded that the warm counterstreaming electrons are solar wind electrons on closed field lines, *Ogilvie et al.* [1984] suggested that they are beams from the ionosphere that reach the low-latitude BL along $\mathbf{B}$.

We want to address the question of the origin of the warm electrons with the aid of Fig. 4, which presents one-dimensional cuts through the distributions of Plate 1c. Let us first have a look at the phase space density, $f_e$, in a cut parallel to $\mathbf{B}$ (left diagram). Typical field-aligned velocities, $v_{\parallel}$, of the warm electrons are 6000–15000 km/s. In this range, the value of $f_e(v_{\parallel}, v_{\perp} = 0)$ in the IBL is comparable to its value in the OBL. On the other hand, the phase space density of the cold ionospheric electrons in the magnetosphere falls below the detection threshold for $v_{\parallel} > 6000$ km/s and is thus at least 1 order of magnitude lower than the phase space density of the warm electrons. If the warm electrons originated also from the ionosphere the high phase space density for $v_{\parallel} > 6000$ km/s could only be explained as the consequence of field-aligned acceleration on field lines mapping to the IBL. Field-aligned beams of ionospheric electrons accelerated upward are indeed observed at low altitudes on auroral field lines [e.g., *Lundin et al.*, 1987]. Those upgoing electrons may form part of the population of warm electrons. However, the relatively sharp boundary between IBL and magnetosphere argues against the possibility that the ionosphere is the main source of the warm electrons: It is hard to understand why the process accelerating ionospheric electrons upward at low altitudes should work only in the limited region of field lines mapping to the IBL, but not on field lines mapping to the magnetosphere proper. The fact that the curves of $f_e(v_{\parallel}, v_{\perp} = 0)$ in the IBL and OBL are pretty close to one another in the range 6000–15000 km/s rather suggests that the electrons in the IBL and OBL have a common source: the shocked solar wind plasma of the magnetosheath. For $v_{\parallel} < 6000$ km/s the phase space density in the IBL is considerably less than that in the OBL. This indicates that the mechanism which transports solar wind electrons into the IBL does not work effectively for electrons with small field-aligned velocities. This is also reflected in the diagram for $f_e(v_{\parallel} = 0, v_{\perp})$. For $v_{\parallel} \approx 0$ the phase space density in the IBL is not very different from that in the magnetosphere proper.

Like for the crossing on 84/09/21, important information about the IBL can be deduced from the time series of $N_{2e}$ and $V_{pM}$. In the IBL the partial density $N_{2e}$ of electrons above 1.8 keV is again comparable to $N_{2e}$ in the magnetosphere proper, but it drops in the OBL. Of course, this drop is also visible in Plate 1c. The trace of $V_{pM}$ indicates again a flow reversal at the interface between OBL and IBL. Since IRM is located at 14:00 LT, the sheath flow has a duskward component, $V_{pM} \approx -100$ km/s. Whereas the flow in the OBL shares this duskward motion, the flow in the IBL and magnetosphere is rather directed dawnward. This reveals that the plasma in the IBL is not magnetically or viscously coupled to the sheath plasma. Rather, the dawnward motion is consistent with the return flow of closed magnetic flux from the tail back to the dayside.

While the time series of $N_{2e}$ and $V_{pM}$ provide evidence that the IBL is on closed field lines, it is difficult to decide on the state of the OBL. On the one hand, the existence of a PDL and tests of the predictions of section 2 imply that the MP is locally closed. On the other hand, cross-field diffusion should not be able to form an OBL whose density and temperature profiles show a plateau (10:53:35–10:59:09) with a sharp step at its inner edge. A possible explanation for the OBL on 84/09/17 would be that it is on open field lines that cross the MP at a location farther away from the spacecraft. In this case the solar wind plasma detected in the OBL may have entered along open field lines. If these field lines do not cross the MP locally but farther away from the spacecraft, there is no reason why the observed local MP should have the properties of a rotational discontinuity.

## 4. COMPARISON WITH OTHER MAGNETOPAUSE PASSES

Now we want to compare the MP crossings of section 3 with other crossings of the IRM spacecraft in order to discuss the question: How typical are the events of section 3? A detailed statistics of AMPTE-IRM magnetopause crossings is given by *Bauer* [1997].

Looking at the time series of plasma densities (e.g., $N_e$, $N_{1e}$, $N_{2e}$), and field-aligned and perpendicular temperatures, we find that only ∼20% of the BL crossings show gradual profiles, whereas ∼80% of the crossings show steplike profiles with two plateaus (OBL and IBL) or one plateau (OBL or IBL). In the discussion of the present paper we focus on BL crossings with steplike profiles, since we hope that the distinction between OBL and IBL provides additional insight into the physics of the BL.

Deconvolving the time series with the measured proton bulk velocity, $V_{pn}$, normal to the MP, *Phan and Paschmann* [1996] performed a statistical survey of the thickness of the dayside low-latitude boundary layer and found an average thickness of ∼1000 km. They defined the inner edge of the BL as the point where the total proton density $N_p$ drops below 5% of its magnetosheath value. In the case of most steplike profiles this drop below 5% occurs at the interface between OBL and IBL. Inspection of many IRM events shows that the durations of OBL and IBL crossings are both very variable. On average the durations of OBL and IBL crossings are comparable and the normal motion of IRM relative to the sublayers has about the same speed in the OBL and IBL [*Bauer*, 1997]. So we arrive at the result that both sublayers are on average roughly 1000 km thick.

The characteristic feature of the IBL are the warm electrons (Plate 1 and Fig. 4). The flux of warm electrons parallel to **B** is almost always balanced by the flux antiparallel to **B**. This balance was interpreted as evidence for closed field lines [*Hall et al.*, 1991]. However, the balance of field-aligned fluxes may also be observed for electrons on open field lines that are mirrored at low altitudes [*Fuselier et al.*, 1995]. As to the ions, there are always hot ring current ions detected in the IBL. For many IBL crossings we see in addition solar wind ions like on 84/09/17 and on 84/09/21 after the FTE, or ionospheric ions like on 84/09/21 before the FTE. Only on rare occasions all three ion populations are detected at the same time.

On 84/09/17 and 84/09/21 the density $N_{2e}$ of electrons above 1.8 keV has roughly the same value in the IBL and magnetosphere proper, but starts to decrease at the interface between IBL and OBL. This feature is observed for most steplike boundary layers and provides evidence for the IBL being on closed field lines. Further evidence is provided by the time series of the proton bulk flow, which changes both its magnitude and direction at the interface between OBL and IBL. For most steplike boundary layers the magnitude $V_p$ of the bulk flow in the IBL is distinctly lower than in the OBL. On 84/09/21 we found a good HT frame and transformed the measurements of the proton bulk velocity into this frame. The result was that $\mathbf{V}'_p$ is field-aligned in the magnetosheath and in the OBL, but it is not in the IBL. From that it can be concluded that the plasma in the IBL is not on open field lines crossing the OBL.

The finding of *Phan et al.* [1994] that there is a PDL adjacent to the MP for all low shear crossings of IRM suggests that for local magnetic shear $\Delta \varphi_B \leq 30°$ the transfer of magnetic flux across the MP occurs only at a low rate. On the other hand, there are low shear crossings of IRM for which the plasma moments measured across the MP agree with the Walén relation (2) [*Paschmann et al.*, 1990]. Furthermore, *Fuselier et al.* [1995] reported on AMPTE-CCE observations of an electron heat flux in a magnetosheath layer close to the MP and inferred that this layer is on reconnected field lines. Finding this electron signature in ∼90% of the low shear MP crossings of AMPTE-CCE, they concluded that even for low shear the MP is open most of the time. We find such an electron heat flux also for low shear crossings of IRM. However, our occurrence frequency is considerably less (∼30%).

*Phan et al.* [1996] performed a statistical survey of the fit between the observed change $\Delta \mathbf{V}_p$ of the proton bulk velocity across the MP and the change $\pm \Delta \mathbf{c}_A$ of

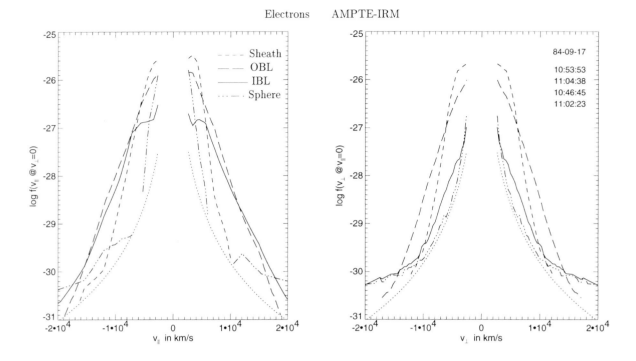

**Figure 4.** One-dimensional cuts through the electron distributions measured on 84/09/17. The measurements are taken at the same times as in Plate 1c in the magnetosheath (short dashed line), in the outer BL (long dashed line), in the inner BL (solid line), and in the magnetosphere proper (dash-dotted line). Whereas the left diagram shows the phase space density $f_e$ (in cm$^{-6}$ s$^3$) as a function of $v_\parallel$ for $v_\perp = 0$, the right diagram shows $f_e$ as a function of $v_\perp$ for $v_\parallel = 0$. The dotted line gives the detection threshold.

the Alfvén velocity for IRM high shear crossings. $\Delta \mathbf{V}_p$ and $\Delta \mathbf{c}_A$ were both computed for each measurement in the BL as the difference between the respective measurement in the BL and the average of a reference interval in the sheath. For each MP crossing the agreement with the prediction of Eq. (2) was then quantified by computing the index

$$\Delta V^* = \frac{\Delta \mathbf{V}_p \cdot \Delta \mathbf{c}_A}{|\Delta \mathbf{c}_A|^2} \quad (3)$$

Figure 5 shows $|\Delta V^*|$ evaluated at the time of the maximum observed velocity change $\Delta \mathbf{V}_p$ for each IRM high shear crossing with stable magnetosheath conditions plotted versus the plasma $\beta$ of the reference interval in the magnetosheath. Case studies of *Paschmann et al.* [1986] and *Sonnerup et al.* [1990] revealed that there are events where the observation agrees perfectly with the prediction for a rotational discontinuity. In Fig. 5 those events lie close to $|\Delta V^*| = 1$. On the other hand, there are also events that lie far below $|\Delta V^*| = 1$. For these events the angle enclosed by $\Delta \mathbf{V}_p$ and $\pm \Delta \mathbf{c}_A$ is large or $|\Delta \mathbf{V}_p|$ is much less than $|\Delta \mathbf{c}_A|$. Thus they have rather the properties of a tangential discontinuity than

those of a rotational discontinuity. Furthermore, it can be seen that for low $\beta$ the fit between observation and prediction is better on average than for high $\beta$.

The high shear crossing on 84/09/21 and the two crossings on 84/08/30 have $|\Delta V^*| \approx 0.6$, which reflects the fact that $|V'_{p\parallel}/c_A|$ is less than 1 during these crossings. Although $|V'_{p\parallel}/c_A|$ is less than predicted by Eq. (2), the single particle signatures observed during these crossings provide alone clear evidence that the local MP is open. Why is $|V'_{p\parallel}|$ less than the Alfvén speed measured simultaneously for these crossings and also for many other crossings [*Paschmann et al.*, 1986; *Sonnerup et al.*, 1990] that show a good correlation between $\mathbf{V}_p$ and $\mathbf{B}$? One reason may be the presence of heavy ions. If heavy ions are present the actual $c_A$ is less than the computed $c_A$, which we compute assuming that all the ions are protons. A second reason may be that the rotational discontinuity is not well separated from the slow expansion fan [*Levy*, 1964] located earthward of the rotational discontinuity. The slow expansion fan is associated with an increase of $c_A$ in the BL. On 84/08/30, $c_A$ has indeed increased by a factor of 2 between 10:03:37 and 10:04:25. If we correct for this factor of 2 the very low value of $|V'_{p\parallel}/c_A|$ at 10:04:25

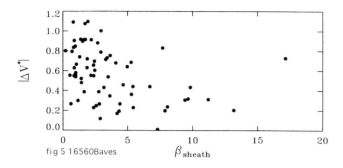

**Figure 5.** Scatterplot of the agreement index $|\Delta V^*|$ (Eq. (3)) versus the plasma $\beta$ in the magnetosheath. Each dot corresponds to an AMPTE-IRM high shear crossing with stable conditions in the sheath [*Phan et al.*, 1996].

becomes 0.1. The measurement in the BL on 84/09/21 at 13:01:05 is taken sheathward of the increase of $c_A$. Another reason for $|V'_{p\parallel}/c_A| < 1$ may be gradients of the plasma pressure tangential to the MP. In a plasma with curved field lines, the force associated with plasma pressure gradients tends to oppose the tension force due to the field line curvature. At an open MP tangential gradients of the plasma pressure will thus act to reduce the plasma acceleration by the magnetic tension force.

For the two events described in sections 3.1 and 3.2, the particle properties predicted by *Cowley* [1982] seem to be a more reliable indicator of open field lines than the fluid properties. However, it must be noted that there are also MP crossings where the particle signatures are absent although the fluid properties provide evidence of open field lines ($|\Delta V^*| \geq 0.5$). In Fig. 5 we can see that 60% of the high shear crossings have $|\Delta V^*| > 0.5$. How many of the high shear crossings show the particle signatures expected on open field lines? In sections 3.1 and 3.2 we found the following signatures: field-aligned streaming of solar wind and ring current electrons, "D-shaped" distribution of solar wind ions with cutoff at $v'_\parallel = 0$, solar wind ion component reflected off the MP having $V'_{r\parallel} = -V'_{i\parallel}$, counter-streaming of solar wind and ionospheric ions. If we look for one particular of those signatures we find it only for the minority of MP crossings. But ~80% of the high shear crossings of IRM show at least one of the single particle signatures [*Bauer*, 1997].

When we observe the particle signatures of open field lines, we can infer the sign of the normal magnetic field $B_n$. We find that this sign of $B_n$ is almost always consistent with the sign of $B_n$ as deduced from the sign of $\Delta V^*$ or $C_{V,c_A}$. Because of this consistency, both the various types of particle signatures and a qualitative agreement with the Walén relation, $|\Delta V^*| \geq 0.5$, can in a statistical sense be considered as a useful indicator of open field lines.

## 5. ESTIMATES OF DIFFUSION COEFFICIENTS

Diffusion in the BL has been extensively discussed in the literature [e.g., *Thorne and Tsurutani*, 1991; *Treumann et al.*, 1995]. Here we focus on the differences that exist between diffusion in the outer and inner boundary layer. If the OBL is formed by inward diffusion then its thickness, $h_{OBL}$, at a certain location is given by the diffusion coefficient, $D_{OBL}$, and the time, $\tau_{OBL}$, that has passed since the flux tubes first came into contact with the MP: $h_{OBL} \sim (D_{OBL}\tau_{OBL})^{1/2}$. At the dayside MP the contact time $\tau_{OBL}$ should be of the order of 10 min, which corresponds to a flow length of $15 R_E$ travelled at a flow speed of 150 km/s typical for the OBL. Combining this with a typical thickness, $h_{OBL} \sim 1000$ km, of the OBL, one arrives at the often quoted result [*Sckopke et al.*, 1981] that $D_{OBL} \sim 10^9 \text{m}^2/\text{s}$ is needed to populate the OBL by means of cross-field diffusion. For the diffusion of solar wind particles from the OBL into the IBL we must consider the time, $\tau_{IBL}$, that has passed since the flux tubes in the IBL first came into contact with the interface between the two sublayers. Since the flow in the IBL is distinctly slower than in the OBL and often directed sunward rather than tailward, $\tau_{IBL}$ may be considerably longer than $\tau_{OBL}$.

According to *Thorne and Tsurutani* [1991] average power spectra of electric and magnetic fluctuations measured in the BL close to the MP on board the ISEE satellites can be approximated by the expressions

$$S_E = \left(\frac{f}{1\,\text{Hz}}\right)^{-2.8} \cdot 30\,\frac{\text{mV}^2}{\text{m}^2\text{Hz}} \quad (4)$$

$$S_B = \left(\frac{f}{1\,\text{Hz}}\right)^{-3.9} \cdot 10\,\frac{\text{nT}^2}{\text{Hz}} \quad (5)$$

for frequencies $f > 10$ Hz. Let us compare these results with AMPTE-IRM measurements. Using the data of the wave instrument, we computed average spectra of the electric field in the frequency range 25 Hz–250 kHz [*Bauer*, 1997]. At 1 kHz the average power observed by IRM in the OBL and IBL is comparable to the expression (4). However, since our average spectra are flatter, $S_E \propto f^{-2.3}$, they are almost 1 order of magnitude lower than expression (4) at 30 Hz. This difference is presently not understood. As to the magnetic fluctuations, we performed a Fourier analysis of the high resolution magnetometer measurements and computed average power spectra from 0.008 Hz up to the Nyquist frequency of 16 Hz [*Bauer*, 1997]. The power obtained in the OBL at 10 Hz is consistent with Eq. (5). In the IBL the power in the three field components is almost

Table 1. Diffusion coefficients for solar wind particles in the OBL and IBL.

| Parameter | Outer boundary layer | | Inner boundary layer | |
|---|---|---|---|---|
| | $s = p$ | $s = e$ | $s = p$ | $s = e$ |
| $B$ in nT | 30 | | 50 | |
| $N$ in cm$^{-3}$ | 10 | | 1 | |
| $KT_{s\parallel}$ in eV | 500 | 60 | 500 | 90 |
| $KT_{s\perp}$ in eV | 500 | 60 | 500 | 60 |
| $\delta E = \left(\int_{0.1f_{LH}}^{10f_{LH}} S_E\,df\right)^{1/2}$ in mV/m | 2 | | 1.5 | |
| $D_{\text{LHDI}}$ in m$^2$/s | $10^7$ | | $10^6$ | |
| $\delta E = \left(\int_{0.9f_{gs}}^{1.1f_{gs}} S_E\,df\right)^{1/2}$ in mV/m | 5 | $5 \cdot 10^{-3}$ | 3 | $3 \cdot 10^{-3}$ |
| $D_{s,\text{ESPAS}}$ in m$^2$/s | $10^{10}$ | $10^1$ | $10^9$ | $3 \cdot 10^{-1}$ |
| $\delta B = \left(\int_{0.9f_{gs}}^{1.1f_{gs}} S_B\,df\right)^{1/2}$ in nT | 4 | $0.1 \cdot 10^{-3}$ | 2 | $0.04 \cdot 10^{-3}$ |
| $D_{s,\text{EMPAS}}$ in m$^2$/s | $3 \cdot 10^8$ | $3 \cdot 10^{-2}$ | $3 \cdot 10^7$ | $10^{-3}$ |
| $\delta B_n = \left(\int_{0.5c_A/\lambda_\parallel}^{2c_A/\lambda_\parallel} S_{B_n}\,df\right)^{1/2}$ in nT | 8 | | 1.5 | |
| $D_{s,\text{KAW}}$ in m$^2$/s | $10^8$ | $10^7$ | $10^2$ | $3 \cdot 10^6$ |

The upper four rows give averages for the magnetic field magnitude, $B$, the density, $N$, and the field-aligned and perpendicular thermal energies, $KT_{s\parallel}$ and $KT_{s\perp}$, of solar wind protons ($s = p$) and electrons ($s = e$). These quantities are combined with typical amplitudes of electric and magnetic fluctuations in order to estimate proton and electron diffusion coefficients caused by lower hybrid drift instability ($D_{\text{LHDI}}$), gyroresonant pitch-angle scattering due to electrostatic ($D_{s,\text{ESPAS}}$) and electromagnetic ($D_{s,\text{EMPAS}}$) waves, and by kinetic Alfvén wave turbulence ($D_{s,\text{KAW}}$). The relevant wave amplitudes are given one row above the respective diffusion coefficients. Whereas $\delta E$ and $\delta B$ are derived from antenna measurements, $\delta B_n$ is derived from flux gate magnetometer measurements normal to the magnetopause.

1 order of magnitude lower than in the OBL. For an estimate of the diffusion by kinetic Alfvén waves, we use average spectra, $S_{B_n}$, of the normal magnetic field $B_n$.

In Table 1 we estimate diffusion coefficients caused by lower hybrid drift instability, gyroresonant pitch angle scattering, and kinetic Alfvén wave turbulence. The lower hybrid drift instability may be driven by the cross-field current in the MP [*Winske et al.*, 1995; *Treumann et al.*, 1995]. The anomalous resistivity associated with this instability leads to ambipolar diffusion with a common diffusion coefficient for protons and electrons

$$D_{\text{LHDI}} \approx 0.6 r_{ge}^2 \omega_{\text{LH}} \frac{m_p \omega_{pe}^2}{m_e \omega_{ge}^2}\left(1 + \frac{T_{p\perp}}{T_{e\perp}}\right)\frac{\epsilon_0 \delta E^2}{2NKT_{p\perp}} \quad (6)$$

$\omega_{ps}$ and $\omega_{gs}$ are the plasma frequency and gyrofrequency of species $s$ ($s = p$ for protons, $s = e$ for electrons). The thermal speed and gyroradius of species $s$ are defined as $v_{Ts} = (KT_s/m_s)^{1/2}$ and $r_{gs} = v_{Ts}/\omega_{gs}$. $\omega_{\text{LH}}$ is the lower hybrid frequency and $\delta E$ is the electric field amplitude of the waves. For the estimate of $D_{\text{LHDI}}$ in Table 1 we assume that the lower hybrid drift instability is responsible for the electric power observed in the frequency range $0.1$–$10 f_{\text{LH}}$ and compute $\delta E$ by integrating Eq. (4) over that frequency range. Since the anomalous resistivities associated with other cross-field current-driven instabilities are lower than that of the lower hybrid drift instability [*Winske et al.*, 1995; *Treumann et al.*, 1995], also their diffusion coefficients are lower than that given in Eq. (6).

Postulating that the collision frequency of the particles equals the coefficient of pitch angle diffusion by gyroresonant waves, *Thorne and Tsurutani* [1991] derive expressions for the spatial proton ($s = p$) and electron ($s = e$) diffusion coefficients

$$D_{s,\text{ESPAS}} \sim r_{gs}^2 \omega_{gs}\left(\frac{\delta E}{v_{Ts} B}\right)^2 \quad (7)$$

$$D_{s,\text{EMPAS}} \sim r_{gs}^2 \omega_{gs}\left(\frac{\delta B}{B}\right)^2 \quad (8)$$

associated with pitch angle scattering (PAS) by electrostatic (ES) and electromagnetic (EM) waves, respectively. In Table 1 we estimate $\delta E$ and $\delta B$ by integrating Eqs. (4) and (5) from $0.9 f_{gs}$ to $1.1 f_{gs}$.

Kinetic Alfvén waves in the BL may be generated because of the coupling to the turbulence in the magnetosheath [Lee et al., 1994]. The diffusion coefficients

$$D_{s,\text{KAW}} \approx \frac{0.6 c_A^2}{k_\| v_{Ts\|}} \exp\left(-\frac{c_A^2}{2 v_{Ts\|}^2}\right) \left(\frac{T_{e\perp}}{T_{p\perp}} k_\perp^2 r_{gp}^2 \frac{\delta B_n}{B}\right)^2 \quad (9)$$

associated with these waves depend on the amplitude $\delta B_n$ of magnetic fluctuations normal to the MP and on the components $k_\perp$ and $k_\|$ of the wave vector $\mathbf{k}$ perpendicular and parallel to $\mathbf{B}$. For our estimate of $D_{s,\text{KAW}}$ in Table 1 we assume $k_\perp r_{gp} = 0.5$ and $\lambda_\| = 2\pi/k_\| = 5 R_E$. $\delta B_n$ is computed by integrating $S_{B_n}$ from $0.5 c_A/\lambda_\|$ to $2 c_A/\lambda_\|$. So our estimate of $D_{s,\text{KAW}}$ is based on the assumption that the power of $B_n$ observed in this frequency range is due to kinetic Alfvén waves.

Looking at the results of Table 1, we notice that cross-field diffusion by lower hybrid waves cannot explain the formation of the BL. The value found for $D_{p,\text{ESPAS}}$ in the OBL is 1 order of magnitude higher than required for the transport of solar wind protons into that sublayer, but pitch angle scattering by gyroresonant waves cannot explain the presence of solar wind electrons in the BL. Note that an ambipolar electric field generated by the fast ion diffusion could enhance $D_e$ only by a factor of $1 + T_{p\perp}/T_{e\perp}$. In the IBL, $D_{\text{LHDI}}$, $D_{s,\text{ESPAS}}$, and $D_{s,\text{EMPAS}}$ are all about 1 order of magnitude lower than in the OBL. Since $D_{p,\text{KAW}}$ is very sensitive to the ratio $c_A/v_{Ts\|}$, its value in the IBL is 6 orders of magnitude lower than in the OBL, whereas the values of $D_{e,\text{KAW}}$ in the IBL and OBL are not very different. Among the diffusion coefficients considered, $D_{e,\text{KAW}}$ is the best candidate for explaining the entry of the warm electrons from the magnetosheath onto closed field lines in the IBL. However, with $D_{\text{IBL}} \sim D_{e,\text{KAW}} \sim 3 \cdot 10^6 \text{m}^2/\text{s}$, a contact time $\tau_{\text{IBL}}$ of the order of 100 hours is needed to form an IBL of thickness $\delta_{\text{IBL}} \sim 1000$ km, which is much too long even in view of the stagnant flow observed in the IBL. In summary, none of the diffusion mechanisms considered can explain the presence of both solar wind protons and electrons in the OBL or IBL.

## 6. CONCLUSIONS

Confirming previous studies of ISEE and AMPTE-UKS data, the AMPTE-IRM data investigated in this paper show that the low-latitude boundary layer is often divided into two distinct parts. Whereas the outer BL is clearly dominated by solar wind plasma, the partial densities of solar wind and magnetospheric particles are comparable to one another in the inner BL.

The characteristic feature of the IBL are warm counterstreaming electrons that originate mainly from the magnetosheath and have a field-aligned temperature that is a factor of 1–5 higher than the electron temperature in the sheath. Profiles of the proton bulk flow and of the density of hot ring current electrons provide evidence that the IBL is on closed field lines. According to Table 1 cross-field diffusion can probably not account for the formation of the IBL, because it cannot transport solar wind electrons into this sublayer at a sufficiently high rate. Hence, the entry of the warm electrons into the IBL remains an open question.

In the OBL and the magnetosheath close to the MP, IRM frequently observes fluid and particle signatures of magnetic reconnection: Plasma moments show at least a qualitative agreement with the Walén relation of a rotational discontinuity and the measured distribution functions exhibit signatures expected on open field lines. When a qualitative agreement with the Walén relation and the single particle signatures are seen at the same time, like for the two high shear crossings examined in sections 3.1 and 3.2, the sign of the normal magnetic field $B_n$ inferred from the particle signatures is mostly consistent with the sign of $B_n$ as deduced from the test of the Walén relation. The occurrence frequencies of particle signatures of open field lines and of $|\Delta V^*| > 0.5$ indicate that for high magnetic shear the observed MP is open at least half of the time.

The existence of a PDL for all IRM low shear crossings is an indirect evidence for reconnection being absent or inefficient in the vicinity of the spacecraft. However, searching for the fluid and single particle properties of a rotational discontinuity, one finds that there are without doubt cases where the low shear MP is open.

In section 3.3 we examined a low shear crossing for which the existence of a PDL and the simultaneous absence of fluid or particle signatures expected on open field lines imply that the MP is closed in the vicinity of the spacecraft. How is an OBL formed that looks like the one described in section 3.3? According to Table 1 cross-field diffusion cannot account for the formation of the OBL, because it cannot transport solar wind electrons into this sublayer at a sufficiently high rate. A possible explanation for observations of an OBL earthward of a closed MP is magnetic reconnection at a location farther away from the spacecraft.

*Acknowledgments.* We are grateful to H. Lühr and B. Häusler for making available the magnetic field and wave data. We thank W. Lotko for useful discussions and A. Czaykowska for helpful comments on the manuscript.

# REFERENCES

Bauer, T. M., Particles and Fields at the Dayside Low-latitude Magnetopause, Dissertation, Ludwig-Maximilians-Universität München, Germany, 1997.

Bauer, T. M., G. Paschmann, R. A. Treumann, W. Baumjohann, and N. Sckopke, Ion signatures of reconnection at the magnetopause, *Adv. Space Res.*, 19(12), 1947–1950, 1997.

Cowley, S. W. H., The causes of convection in the Earth's magnetosphere: A review of developments during the IMS, *Rev. Geophys. Space Phys.*, 20, 531–565, 1982.

Denton, R. E., B. J. Anderson, S. P. Gary, and S. A. Fuselier, Bounded anisotropy fluid model for ion temperatures, *J. Geophys. Res.*, 99, 11,225–11,241, 1994.

Denton, R. E., and J. G. Lyon, Density depletion in an anisotropic magnetosheath, *Geophys. Res. Lett.*, 23, 2891–2894, 1996.

Fairfield, D. H., Average and unusual locations of the Earth's magnetopause and bow shock, *J. Geophys. Res.*, 76, 6700–6716, 1971.

Fuselier, S. A., D. M. Klumpar, and E. G. Shelley, Ion reflection and transmission during reconnection at the Earth's subsolar magnetopause, *Geophys. Res. Lett.*, 18, 139–142, 1991.

Fuselier, S. A., D. M. Klumpar, and E. G. Shelley, Mass density and pressure changes across the dayside magnetopause, *J. Geophys. Res.*, 98, 3935–3942, 1993.

Fuselier, S. A., B. J. Anderson, and T. G. Onsager, Particle signatures of magnetic topology at the magnetopause: AMPTE/CCE observations, *J. Geophys. Res.*, 100, 11,805–11,821, 1995.

Gosling, T. J., M. F. Thomsen, S. J. Bame, R. C. Elphic, and C. T. Russell, Plasma flow reversals at the dayside magnetopause and the origin of asymmetric polar cap convection, *J. Geophys. Res.*, 95, 8073–8084, 1990a.

Gosling, J. T., M. F. Thomsen, S. J. Bame, T. G. Onsager, and C. T. Russell, The electron edge of the low latitude boundary layer during accelerated flow events, *Geophys. Res. Lett.*, 17, 1833–1836, 1990b.

Häusler, B., R. R. Anderson, D. A. Gurnett, H. C. Koons, R. H. Holzworth, O. H. Bauer, R. Treumann, K. Gnaiger, D. Odem, W. B. Harbridge, and F. Eberl, The plasma wave instrument on board the IRM satellite, *IEEE Trans. Geosci. Remote Sens.*, GE-23, 267–273, 1985.

Hall, D. S., C. P. Chaloner, D. A. Bryant, D. A. Lepine, and V. P. Tritakis, Electrons in the boundary layers near the dayside magnetopause, *J. Geophys. Res.*, 96, 7869–7891, 1991.

Lee, L. C., J. R. Johnson, and Z. W. Ma, Kinetic Alfvén waves as a source of plasma transport at the dayside magnetopause, *J. Geophys. Res.*, 100, 17,405–17,411, 1994.

Levy, R. H., H. E. Petschek, and G. L. Siscoe, Aerodynamic aspects of the magnetospheric flow, *AAIA J.*, 2, 2065–2076, 1964.

Lühr, H., N. Klöcker, W. Oelschlägel, B. Häusler, and M. Acuña, The IRM fluxgate magnetometer, *IEEE Trans. Geosci. Remote Sens.*, GE-23, 259–261, 1985.

Lundin, R., L. Eliasson, B. Hultqvist, and K. Stasiewicz, Plasma energization on auroral field lines as observed by the VIKING spacecraft, *Geophys. Res. Lett.*, 14, 443–446, 1987.

Ogilvie, K. W., R. J. Fitzenreiter, and J. D. Scudder, Observations of electron beams in the low latitude boundary layer, *J. Geophys. Res.*, 89, 10,723–10,732, 1984.

Paschmann, G., H. Loidl, P. Obermayer, M. Ertl, R. Laborenz, N. Sckopke, W. Baumjohann, C. W. Carlson, and D. W. Curtis, The plasma instrument for AMPTE/IRM, *IEEE Trans. Geosci. Remote Sens.*, GE-23, 262–266, 1985.

Paschmann, G., I. Papamastorakis, W. Baumjohann, N. Sckopke, C. W. Carlson, B. U. Ö. Sonnerup, and H. Lühr, The magnetopause for large magnetic shear: AMPTE/IRM observations, *J. Geophys. Res.*, 91, 11,099–11,115, 1986.

Paschmann, G., B. Sonnerup, I. Papamastorakis, W. Baumjohann, N. Sckopke, and H. Lühr, The magnetopause and boundary layer for small magnetic shear: Convection electric fields and reconnection, *Geophys. Res. Lett.*, 17, 1829–1832, 1990.

Paschmann, G., W. Baumjohann, N. Sckopke, T.-D. Phan, and H. Lühr, Structure of the dayside magnetopause for low magnetic shear, *J. Geophys. Res.*, 98, 13,409–13,422, 1993.

Phan, T.-D., and G. Paschmann, Low-latitude dayside magnetopause and boundary layer for high magnetic shear: 1. Structure and motion, *J. Geophys. Res.*, 101, 7801–7815, 1996.

Phan, T.-D., G. Paschmann, W. Baumjohann, N. Sckopke, and H. Lühr, The magnetosheath region adjacent to the dayside magnetopause: AMPTE/IRM observations, *J. Geophys. Res.*, 99, 121–141, 1994.

Phan, T.-D., G. Paschmann, and B. U. Ö. Sonnerup, Low-latitude dayside magnetopause and boundary layer for high magnetic shear: 2. Occurrence of magnetic reconnection, *J. Geophys. Res.*, 101, 7817–7828, 1996.

Russell, C. T., and R. C. Elphic, ISEE observations of flux transfer events at the dayside magnetopause, *Geophys. Res. Lett.*, 6, 33–36, 1979.

Sckopke, N., G. Paschmann, G. Haerendel, B. U. Ö. Sonnerup, S. J. Bame, T. G. Forbes, E. W. Hones, Jr., and C. T. Russell, Structure of the low latitude boundary layer, *J. Geophys. Res.*, 86, 2099–2110, 1981.

Smith M. F., and D. J. Rodgers, Ion distributions at the dayside magnetopause, *J. Geophys. Res.*, 96, 11,617–11,624, 1991.

Song, P., R. C. Elphic, C. T. Russell, J. T. Gosling, and C. A. Cattell, Structure and properties of the subsolar magnetopause for northward IMF: ISEE observations, *J. Geophys. Res.*, 95, 6375–6387, 1990.

Sonnerup, B. U. Ö., G. Paschmann, I. Papamastorakis, N. Sckopke, G. Haerendel, S. J. Bame, J. R. Asbridge, J. T. Gosling, and C. T. Russell, Evidence for magnetic field reconnection at the Earth's magnetopause, *J. Geophys. Res.*, 86, 10049–10067, 1981.

Sonnerup, B. U. Ö., I. Papamastorakis, G. Paschmann, and H. Lühr, The magnetopause for large magnetic shear: Analysis of convection electric fields from AMPTE/IRM, *J. Geophys. Res.*, 95, 10,541–10,557, 1990.

Thorne, R. M., and B. T. Tsurutani, Wave-particle interactions in the magnetopause boundary layer, in *Physics of Space Plasmas (1990), SPI Conf. Proc. Rep. Ser.*, vol. 10, ed. by T. S. Chang et al., pp. 119–150, Scientific Publishers, Cambridge, Mass., 1991.

Treumann, R. A., J. LaBelle, and T. M. Bauer, Diffusion processes: An observational perspective, in *Physics of the Magnetopause, Geophys. Monogr. Ser.*, vol. 90, ed. by B. U. Ö. Sonnerup et al., pp. 331–341, AGU, Washington, D. C., 1995.

Walthour, D. W., J. T. Gosling, B. U. Ö. Sonnerup, and C.

T. Russell, Observation of anomalous slow-mode shock and reconnection layer in the dayside magnetopause, *J. Geophys. Res.*, *99*, 23,705–23,722, 1994.

Winske, D., V. A. Thomas, and N. Omidi, Diffusion at the magnetopause: A theoretical perspective, in *Physics of the Magnetopause, Geophys. Monogr. Ser.*, vol. 90, ed. by B. U. Ö. Sonnerup et al., pp. 321–330, AGU, Washington, D. C., 1995.

Zwan, B. J., and R. A. Wolf, Depletion of solar wind plasma near a planetary boundary, *J. Geophys. Res.*, *81*, 1636–1648, 1976.

___

T. M. Bauer (e-mail: thb@mpe-garching.mpg.de), W. Baumjohann, G. Paschmann, N. Sckopke, R. A. Treumann, Max-Planck-Institut für extraterrestrische Physik, Postfach 1603, D-85740 Garching, Germany, T.-D. Phan, Space Sciences Laboratory, University of California, Berkeley, CA 94720-7450

# The Effect of the Interplanetary Magnetic Field on the Dayside Magnetosphere

Daniel W. Swift

*Geophysical Institute, University of Alaska, Fairbanks, Alaska*

The results of a two-dimensional, global-scale simulation of the dayside magnetosphere are presented. The major conclusions are that (1) field-aligned currents linking the cusp and ionosphere are an ever-present feature, independent of the magnitude and orientation of the IMF. This suggests that aurora should be an ever-present feature of the cusp ionosphere. (2) The appearance of accelerated ions in the cusp occurs primarily when the IMF is southward. No evidence for accelerated ions is seen under conditions of zero IMF. (3) Clumps of plasma can become trapped between the compressed IMF and the magnetopause. These might be interpreted as impulsive injection events sometimes seen at the magnetopause.

## 1. INTRODUCTION

The effect of the interplanetary magnetic field (IMF) has been well documented, and has been investigated in numerous MHD simulations of the magnetosphere. Most of these simulations have used a rather dissipative code and a course mesh in order to simulate a large spatial volume. The simulations to be presented here have a much finer scale resolution and include ion kinetic effects, which make it possible to see new details of the interaction between the IMF and magnetosphere. Moreover, the simulation domain extends all the way down to the Earth's ionosphere, making it possible to observe the effect of the ionosphere.

The simulations presented here use the hybrid code algorithm originally developed by Harned [1981]. Basically, in the hybrid code the ions are fully kinetic, while the electrons are treated as a zero-mass fluid. The magnetic field is advanced through Faraday's law and is fully self-consistent with the evolution of the plasma. Quasi-neutrality is assumed, so there is no electrostatic interaction. The displacement current is also neglected, so the fastest wave allowed is the whistler mode.

Geospace Mass and Energy Flow: Results From the International Solar-Terrestrial Physics Program
Geophysical Monograph 104
Copyright 1998 by the American Geophysical Union

The simulations reported here use the code described by Swift [1996], which includes several important new features. One is the use of a curvilinear coordinate system, the grid points of which are shown in Figure 1. The simulation domain is the noon meridian plane and extends from the Earth's surface at 1 $R_E$ to 18 $R_E$ in the undisturbed solar wind. The boundary at the bottom is the polar axis. Note the high density of grid points near the magnetopause region at 10 $R_E$. There is also a high grid point density near the Earth. The selective placement of grid points permits high spatial resolution in critical regions.

The solar wind carrying an IMF is incident at the outer boundary. The solar wind ions are fully kinetic. Inside the magnetopause the Earth is surrounded by a cold dense plasma of ionospheric origin. This plasma sill serve to conduct current and transmit Alfven waves between the magnetopause and ionosphere. Another innovative feature of the code is the treatment of this cold dense component as an MHD fluid. We are justified in treating this ionospheric plasma as a fluid since the plasma is strongly magnetized and the Alfven velocities are always much larger than and thermal or bulk flow velocity.

The basic simulation will be that as described in Swift [1995], except that that a comparison will be made between runs with northward, southward and zero IMF. At $t=0$, the solar wind is assumed uniform and there is no IMF inside 18 $R_E$ The geomagnetic field is compressed so there is an approximate balance between the solar wind ram

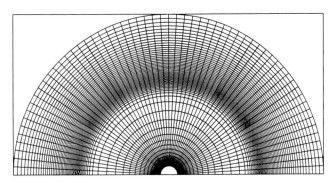

**Figure 1.** The 121×123 coordinate mesh used in this simulation

pressure and the confined geomagnetic field. At the Earth's surface, the Earth's dipole field reaches a value of 500 s$^{-1}$, or 8×10$^4$ nT. The magnetic field is expressed in terms of the O$^+$ gyrofrequency. Beginning at $t = 0$, a an IMF ( for the cases of northward or southward IMF) is convected inward with the solar wind, ramping up to a constant value over a distance of 1 R$_E$. The solar wind speed is about 1 R$_E$ /s or 6000 km/s. The reason for this high value is that the two-dimensional dipole field falls off to $\frac{1}{100}$ of its near-Earth value, instead of $\frac{1}{1000}$ for a three-dimensional dipole field. The reason for the two-dimensional model is that a three-dimensional model in two dimensions has a non-zero curl, which would imply a non-existent current.

In units used here, the IMF has a value of 3s$^{-1}$, or 500 nT, which gives a nominal ion gyroradius of about 2000 km. The solar wind ion density was chosen such that the ion inertial length is about 1000 km. This compares with an actual value of 100 km in a solar wind density of 5 H$^+$ ions/cm$^{-3}$. The nominal Alfven velocity is 0.5 R$_E$ /s, giving a nominal Alfven Mach number in the undisturbed IMF of 2. The thermal velocity assumed was 0.1 R$_E$/s. The particle time step is 0.15 s, and the fields are updated with a 0.015 s time step.

## 2. SIMULATION RESULTS

The results of running this code for the case of a southward turning magnetic field have been described in Swift [1995]. Among the significant features of running the code were the generation of fast and slow-mode shocks. The slow mode shock appeared as a result of the southward turning of the IMF, and was swept downstream by the plasma flow around the Earth. Another feature was the appearance of field-aligned currents linking the polar cusp and polar ionosphere. Finally, the presence of accelerated ions in the cusp was noted. In this section, we shall examine the dependence of these features on the presence and direction of the IMF.

### 2.1 Field-Aligned Currents

Figure 2 shows contour plots of the y-component of the magnetic field. The top panel shows the case for a southward IMF, the middle panel for a northward IMF of the same strength, while the bottom panel is for zero IMF. Each of these cases is for relatively early in the simulation. The panels each have three distinct regions. The crescent region of field turbulence in the top two panels is the region between the slow and fast mode shocks. There are no shocks for the case of zero IMF in the bottom panel. The

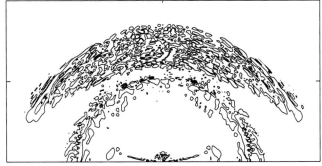

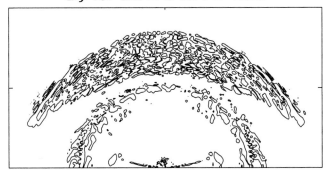

**Figure 2.** Contour plots of the y-component of the magnetic field, indicating presence of field-aligned currents linking the cusp and polar ionosphere. The top panel is for negative IMF, the middle is for positive IMF, while the bottom panel is for zero IMF.

next semicircular region of enhanced field fluctuations closer to the Earth is the magnetopause. Later in the simulations, the disturbed magnetosheath region merged with the fluctuations associated with the magnetopause in the in the two cases of finite IMF.

The main feature of note are the $B_y$ contours extending from the cusp to the polar ionosphere. These appear early in the simulation, and appear independent of the presence of the IMF. The polarity of the $B_y$-fields on opposite sides of the Earth is reversed, indicating up-down orientation of the currents is the same. These currents are generated by the partial demagnetization of the ions. The fields in the cusp are weak, so ions of solar wind energy move across field lines, while electrons are still constrained to follow the magnetic field. This constitutes a current generator, and the current apparently flows along the magnetic field carried by the cold "ionospheric" plasma which is assumed present inside the magnetopause. The cusp is a region where $\mathbf{J} \cdot \mathbf{E} < 0$, due to the fact that ions are giving up kinetic energy doing work against the electric field.

## 2.2 Ion Acceleration

Figure 3 is another set of contour plots showing the particle flow velocity. The top two panels with the finite IMF show the presence of the bow shock. The flow velocity is constant in the undisturbed solar wind upstream of the shock. The flow structure behind the shock does not seem to depend on the polarity of the IMF. There is no shock in the zero IMF case in the bottom panel.

Of most interest here are the velocity contours in the cusp. In the case of the southward IMF in the top panel, the contours, and more detailed analysis, indicate that some particles have been accelerated to velocities three times the undisturbed solar wind flow velocity, which indicates a factor of 9 energy gain for some particles. The middle panel, corresponding to a northward IMF indicates acceleration a few particles to velocities of about 1.5 times the undisturbed solar wind speed. The bottom panel corresponding to zero IMF indicates no particle acceleration in the cusp.

## 2.3 Magnetopause Structure

Figures 4 and 5 respectively show contour plots of $log|\mathbf{B}|$ and particle density. The bow shock is characterized by jumps in B and density for the cases of northward and southward IMF. The nearly circular contours inside the magnetopause in Figure 4 is mainly due to the strong dipole field near the Earth. The main features to be noted in figure 4 are the apparent islands of low field intensity appearing near the magnetopause. These nearly field-free

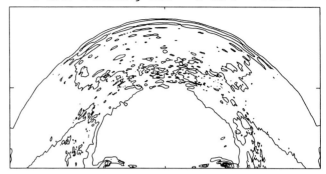
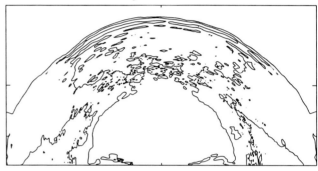
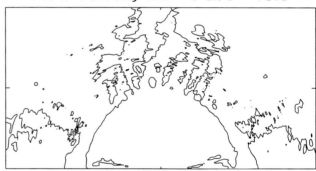

**Figure 3**. Same as Figure 2, except showing contour plots of the particle flow velocity.

islands correlate with blobs of enhance plasma density as seen in Figure 4.

It is tempting to attribute this structure to reconnection, but it appears with both northward and southward IMF. If it were reconnection, it would be seen only with southward IMF. Instead it appears that it is due to plasma that becomes trapped between the magnetopause and the and advancing front of enhanced IMF.

## 3. SUMMARY AND CONCLUSIONS

We have seen that field-aligned currents linking magnetopause cusp region and high-latitude ionosphere appear

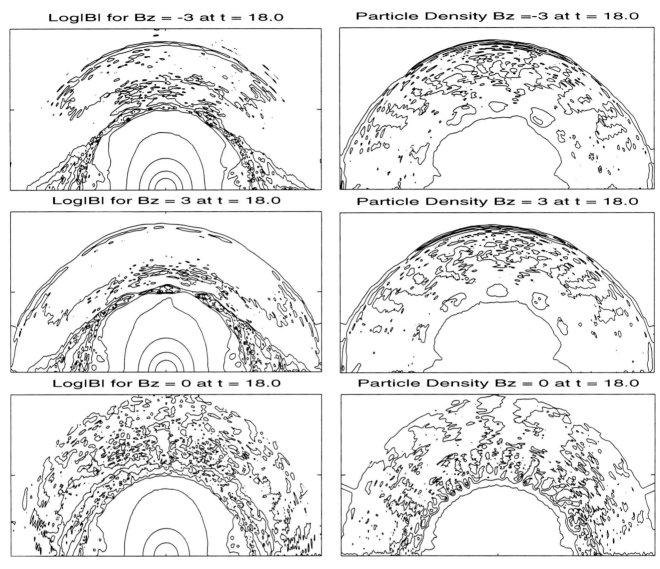

**Figure 4.** Same as Figure 2, except showing contours plots of $log|\mathbf{B}|$.

**Figure 5.** Same as Figure 2, except showing contour plots of particle density.

occur independently of the of the strength and polarity of the IMF. The currents appear to be driven by ion demagnetization in the weak field region of the cusp. Aurora is often associated with field-aligned currents. Given this association, one might expect that aurora to always be present on cusp field lines. A systematic analysis of DMSP satellite images of dayside aurora by Meng and Lundin [1986] do not reveal such a simple picture. Auroral arcs seem to be an ever-present feature of the dayside oval, but arcs do not seem to appear exactly in the cusp. There is a diffuse glow on the ionospheric footprint of the cusp, while discrete arcs appear to radiate away from the cusp footprint. The pattern of arcs surrounding the cusp region does seem to depend on the IMF, but is independent of auroral substorm activity on the nightside of the auroral oval. The observed alignment of the arcs may depend on convection driven by the IMF and solar wind.

Perhaps the best evidence of reconnection seen in the simulations is afforded by the observation of accelerated particles in the cusp. The strongest acceleration appears in the example of the southward IMF, which would be expected because of the large interface between oppositely directed magnetic fields draped across the magnetopause. There is also some evidence of particle acceleration in the northward IMF. This could occur as a result of magnetic merging poleward of the cusp. No evidence of acceleration is seen in the zero IMF example.

Finally, an approaching front of increased IMF can trap

blobs of plasma at the magntopause These blobs appear even though the solar wind velocity, density and magnetic field are all uniform. This may be an explanation for the apparent intrusions of magnetosheath plasma into the magnetosphere observed by Eastman and Hones [1979].

## REFERENCES

Eastman, T. E. and E. W. Hones, Characteristics of the magnetospheric boundary layer and magnetopause layer as observed by Imp 6, *J. Geophys. Res., 84,* 2019, 1979

Harned, D. S., Quasineutral hybrid simulation of macroscopic plasma phenomena, *J. Comput. Phys., 47,* 452, 1981

Meng, C.-I. and R. Lundin, Auroral morphology of the midday oval, *J. Geophys. Res., 91,* 1572, 1986

Swift, D. W., Use of a hybrid code to model the Earth's magnetosphere, *Geophys. Res. Lett., 22,* 311, 1995

Swift, D. W., Use of a hybrid code for global-scale plasma simulation, *J. Comp. Phys., 126,* 1996

---

Daniel W. Swift, Geophysical Institute, University of Alaska, Fairbanks, AK 99775-7320

# Entry of Solar Wind Plasma into the Magnetosphere: Observations Encounter Simulation

Patrick T. Newell and Simon Wing

*The Johns Hopkins University Applied Physics Laboratory, Laurel, Maryland*

The entry of magnetosheath plasma into the magnetosphere can now be modeled with unprecedented accuracy, at least on the global scale. A new class of models makes predictions about spectral details which have been subsequently verified by observations. We introduce here new refinements to magnetosheath entry modeling, including the use of the latest Tsyganenko magnetic field model (which correctly geolocates the cusp for the first time), to the use of kappa distributions for the magnetosheath ion spectra, and more realistic ionospheric convection velocities. These improvements do indeed upgrade the agreement between simulation and DMSP observations at low altitude. However some problems remained outstanding, including the existence of very-high-energy ions (~≥20–30 keV) equatorward of the cusp, and the tendency in the model for the potential retarding electron entry at the magnetosheath to rise until polar rain is excluded altogether. We also argue that the inability of this type of model to produce any precipitation region which resembles what we have previously considered to be "closed-field line low-lattitude boundary layer" demonstrates that the latter is appropriately named.

## 1. INTRODUCTION

The basic mode of solar wind–magnetosphere coupling is through merging of the interplanetary magnetic field with the Earth's magnetic field, which therefore is a basic means of shocked solar wind entry into the magnetosphere. Recently, powerful new tools have placed the investigation of the access of magnetosheath plasma to low altitude into a new era of quantitative study. It is now possible to investigate many aspects of this rich phenomenology with a precision that is rare in magnetospheric physics. In the present article we first review these recent advances in modeling the entry of solar wind plasma deep into the Earth's magnetosphere. We then introduce several improvements to existing work. Finally, the geophysical consequences of the model's successes and deficiencies are discussed, along with possible future directions.

Because several excellent reviews on aspects of the cusp have been published recently, we will elide most historical aspects as well as specialized topics (e.g., the temporal continuity of the cusp) and concentrate on the quantitative representation of large-scale morphology of shocked solar wind entry. Although interesting quantitative work was done as long as two decades ago [*Hill and Reiff*, 1977], global modeling (e.g., magnetic field models) was not yet advanced enough for many firm conclusions to be drawn. As these models became more advanced, so did efforts to track particle entry from the magnetosheath into the magnetosphere [*Richard et al.*, 1994].

*Sergeev et al.*, 1997]. However we suggest an alternative: the expansion of a plasma into a vacuum is known to create some ions with a speed about equal to the electron thermal velocity [*Samir et al.*, 1983]. Since the connection of a magnetospheric field line up to the solar wind must resemble expansion into a vacuum in the initial stages, this is another possibility. Finally stochastic processes have been known to produce higher than expected ion acceleration in the magnetosheath; it is likely this happens also in the magnetopause current layer, as indeed the work of *Richard et al.* [1994] suggests.

A second disappointment is that the κ distribution in the ions did not allow enough additional ions to enter downstream to avoid the necessity of artificially capping the retarding potential at the magnetosheath at −250 V. It is worth discussing in detail the behavior of the potential needed to maintain charge quasi-neutrality, which we now do.

### 3.3 *Latitudinal Profile of the Retarding Potential*

Figure 1 illustrates the potential inferred as a function of latitude (recall that the retarding potential is calculated by requiring equal numbers of electrons and ions after reflecting all electrons below the potential energy. Let us consider this figure moving from lower latitude (left) to higher (right). Just equatorward of the main cusp lie the most recently opened field lines, wherein the bulk of the ions have not had enough elapsed time since merging to reach the ionosphere. This is the open LLBL. The retarding potential needed is large, reaching the model-imposed cap value of −250 V. As the bulk of the ions arrive, the charge drops rapidly to zero. At first thought, it might appear that some residual potential should still be retarding electrons, because in the cusp, some ions (a few percent to more than 50%) have still not yet reached low altitude. However another factor compensates for this: the ions in the main cusp originate from close to the merging site, while the electrons originate from somewhat higher latitude. Since the sheath density on the frontside drops rapidly away from the nose, the electrons in the cusp arrive from a less dense portion of the magnetosheath than do the ions. The net result is that no retarding potential is required in the cusp.

As one moves to higher latitudes, this rapid variation of density with latitude begins to fade, and at about the same time ion entry becomes constrained by the increasing ratio of downstream bulk flow velocity to ion thermal velocity. These effects cause a sharp rise in the potential over the region of about 0.1° to 0.2° MLAT, thus creating the observed large negative gradient in electron density at the poleward edge of the cusp. Thereafter the potential plateaus, as relatively few ions are entering, but the long magnetic field lines do not empty out until the slower-moving ions bounce near the ionosphere and return to the magnetosheath (the model does not track these ions after the bounce). Finally (coinciding with the poleward edge of the cusp plume or mantle region) the potential rises again as at least the low-altitude end of the field line empties out. The final plateau is artificial, since we have capped the voltage at 250 V in order to allow the strong polar rain entry shown in Plate 1b. Incidentally, most polar cap passes show much weaker polar rain than in this instance.

## 4. IMPLICATIONS FOR MAGNETOSPHERIC STRUCTURE

The quantitative and detailed success of the model allows us to draw two types of conclusions. First, the open regions (open LLBL, cusp, mantle, and polar rain) are difficult to ascribe to any mechanism other than magnetosheath entry across a current layer in association with merging. Second, the closed LLBL type of precipitation cannot possibly arise in this manner, and in fact has yet to be successfully modeled in quantitative detail.

### 4.1 *The Open Regions Are Well Modeled*

To appreciate the success of this type of modeling effort, one must consider some of the quantitative details. Note also that the model is deterministic once solar wind conditions and the ionospheric convection have been specified.

A prime example is the well-known cusp dispersion. Although this effect was actually predicted based on the merging model prior to observations [*Rosenbauer et al.*, 1975], numerous alternative explanations have been proposed. It is outside the scope of this paper to discuss these, except to point out that none of these alternatives explain: (i) that the equatorward portion of the cusp has ions accelerated above magnetosheath energies with a narrow spectral distribution, indicating it is not a high thermal temperature but accelerated bulk flow [*Newell and Meng*, 1991]; (ii) that low-energy ion cutoffs are observed in the cusp [*Newell and Meng*, 1995], as was previously predicted from theory [*Lockwood and Smith*, 1992]; (iii) that the poleward portion of the cusp is de-energized compared with magnetosheath energies, as follows from the standard merging geometry; or (iv) that the highest-energy (above 10 keV) ions, apparently of magnetospheric origin moving poleward into the cusp,

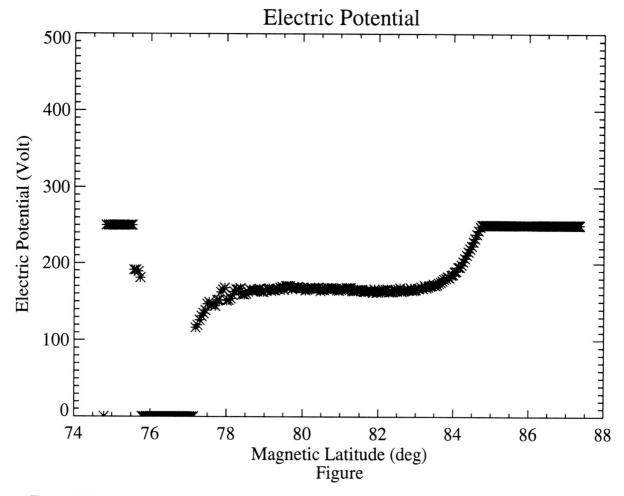

**Figure 1.** The potential between the Earth and the magnetosheath needed to retard electron entry enough to satisfy charge quasi-neutrality as a function of magnetic latitude. The region labels correspond to those shown in Plate 2.

typically drop out. It is not simply a question of declining intensity moving poleward (as some alternatives can explain) or even of declining average energy (as a few alternatives explain). Items (i) and (ii), predictions of the merging model later verified by observations, are particularly hard on the alternatives.

To our knowledge there is no other instance in magnetospheric physics of a model reproducing a complex set of observations in such quantitative detail, especially in view of the lack of free parameters in the model. We believe that the Onsager-class models of magnetosheath plasma entry are a major step forward in quantitative magnetospheric understanding.

*4.2 The Closed-LLBL Precipitation Is Not Well Modeled*

In Plate 3 we show two sample closed LLBL crossings by the DMSP F7 satellite from June 1984, chosen pretty much at random. Recently there has been an effort to understand the LLBL as open field lines equatorward of the cusp, wherein most of the plasma has not yet had time to reach low altitude [e.g., *Lockwood et al.*, 1993; *Lyons et al.*, 1994]. This explanation certainly seems to work for near-noon cases equatorward of the cusp; the simulation in Plate 2 reproduces such a region nicely. Moreover the observed ions in these near-noon cases equatorward of the cusp have low-energy ion cutoffs, highly indicative of their open (and recently merged) characteristics.

However, no field lines in the model at any angle of simulated satellite crossing reproduce anything like the examples we show in Plate 3. Consider the following circumstances: (i) LLBL crossings are observed for both northward and southward IMF, apparently about equally often [*Mitchell et al.*, 1987; *Phan and Paschmann*, 1996] (although there may be some difference in the ratio of

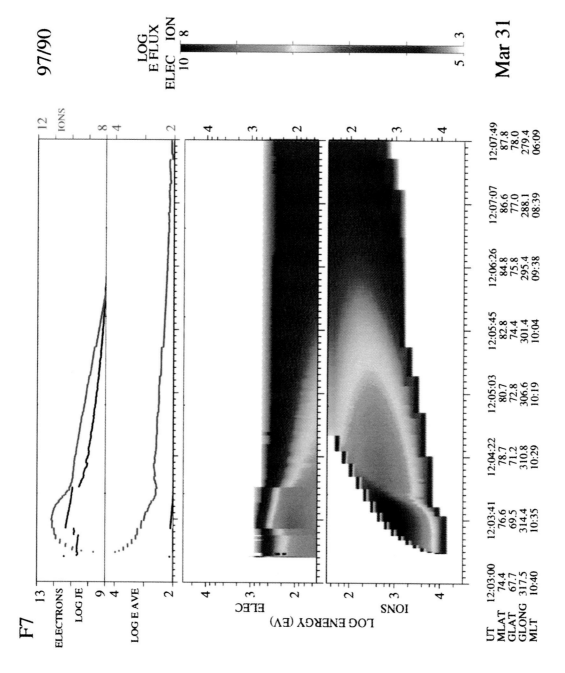

**Plate 2.** The improved simulation introduced in this paper. The cusp is now correctly geolocated, realistic convection velocities have been used, and a κ distribution for magnetosheath ions makes the cusp and mantle spectra more realistic.

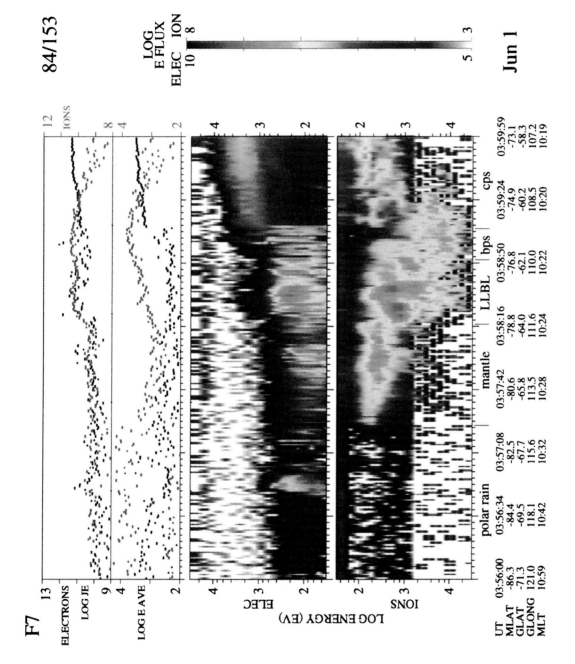

Plate 3. Two examples in the DMSP F7 data set of closed LLBL type precipitation that cannot be modeled by this type of simulation. (a) On June 1, 1984, around 0400 UT. (b) On June 2, 1984, around 0600 UT. Such signatures are most common in the DMSP data set 2–3 hours away from local noon, as in these examples.

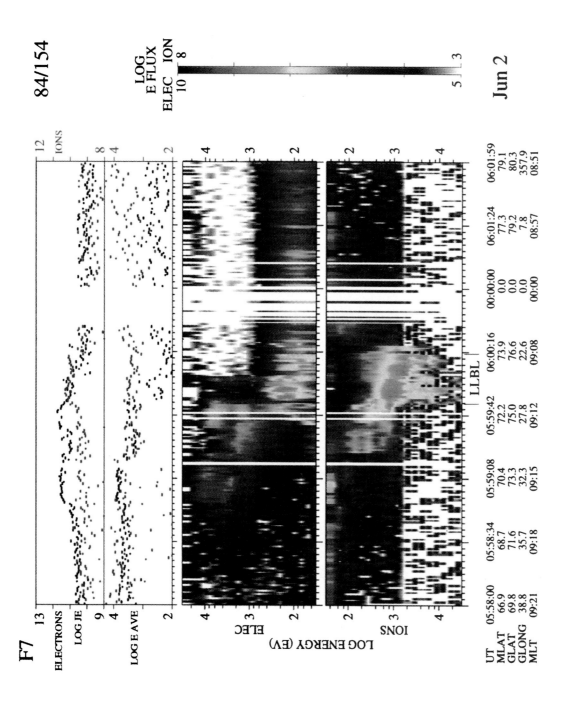

Plate 3 (continued)

closed to open type cases). (ii) The closed LLBL signatures shown above do *not* have low-energy ions cutoffs. (iii) These closed LLBL signatures have a broader spectral peak than in the cusp, indicating more of the energy is thermal and less bulk flow velocity. (iv) The electrons, while softer than the central plasma sheet or even the BPS, are not at the very low magnetosheath energies seen in the cusp [*Newell et al.*, 1991]. (v) Ions in the 10–30 keV range and presumably of magnetospheric origin are often present throughout. (vi) Finally, there is no evidence of a velocity filter–type dispersion in these cases. (Although there can often be a gradual softening of the spectra moving to higher latitudes, it lacks the details of the velocity filter discussed in the previous section.)

Consider: if the field lines in Plate 3 are open, where could they map? If they mapped to the frontside, we should get a cusp-like behavior; if down the tail, mantle or polar rain. What is the mechanism that allows these cases to have ions of all energies, for northward and southward IMF, and with a mixture of magnetosheath plasma and high-energy magnetospheric plasma? Such plasma on open field lines would escape after one bounce.

Since the LLBL signature shown in Plate 3 cannot be successfully modeled as open field lines, and since no modeling has reproduced the observations in any quantitative manner, there is clearly work to be done here. It may be that multipoint merging can explain the closed LLBL signature, as advocated by *Nishida* [1989], or it may well be that some type of impulsive penetration explanation is required [*Lemaire*, 1977]. By far the simplest explanation is the diffusion which appears in detailed (nonadiabatic) single-particle trajectory calculations [*Richard et al.*, 1994; *Omidi and Winske*, 1995]. Indeed, there are many reports of magnetosheathlike precipitation found embedded deep within the closed field line region [*Lundin*, 1987; *Lundin et al.*, 1987; *Yamauchi et al.*, 1993]. However, to achieve satisfactory closure requires that these processes be modeled in such detail that they reproduce observational spectra; indeed, ideally the models should make predictions about the observations which can then be verified (as has happened with the merging model and the cusp).

## 5. SUMMARY, CONCLUSIONS, AND FUTURE DIRECTIONS

Enormous progress has been made over the last two decades in understanding the entry of shocked solar wind plasma into the Earth's magnetosphere. The most successful approach has proven to lie in the assembly of many submodules that individually deal with certain physical aspects of the problem. This submodule approach provides better results than do self-contained simulations. The variability of magnetosheath plasma with entry position, the plasma energization or de-energization in crossing the magnetopause current layer, and the requirement of charge quasi-neutrality have all become understood. In this paper we have additionally used the *Tsyganenko and Stern* [1996] magnetic field model, which finally incorporates realistic Birkeland currents, to properly geolocate the cusp, allowing in turn realistic ionospheric convection velocities. The use of $\kappa$ distributions in the magnetosheath allows cusp and mantle ions to be more realistic, while the use of superthermal electron populations in the solar wind allows the polar rain to be accurately represented.

Remaining problems include the presence of high-energy ions equatorward of the main cusp at fluxes far above those predictable from the model, and the need to artificially cap the potential at the magnetosheath which retards electron entry from rising to very high values and eliminating the polar rain altogether. This latter may be related to the presence of polar wind outflows, which produce an oppositely directed electric field above the Earth's atmosphere to reduce electron escape from the ionosphere. The escaping polar wind may partially balance the entry of polar rain electrons (and vice versa).

The most glaring deficiency in modeling dayside precipitation patterns, however, is easily reproducing the closed LLBL type signatures shown in Plate 3. This discrepancy leads to the conclusion that these LLBL signatures originate not in the merging process but in the diffusion of magnetosheath plasma onto closed field lines.

Exciting possibilities which appear within grasp include investigating the coupling between polar rain and ionospheric outflow and using detailed SuperDARN ionospheric convection patterns to investigate specific satellite cusp crossings to determine such features as the apparent magnetopause merging site.

*Acknowledgments.* The DMSP SSJ/4 detectors were designed and built by D. Hardy and colleagues. T. Onsager was gracious enough to supply us with the original version of his model. This research was funded by NASA grant NAG5-4010 to The Johns Hopkins University Applied Physics Laboratory.

## REFERENCES

Alem, F., and D. C. Delcourt, Nonadiabatic precipitation of ions at the cusp equatorward edge, *J. Geophys. Res., 100*, 19,321, 1995.

Burch, J. L., Rate of erosion of dayside magnetic flux based on a quantitative study of the dependence of polar cusp latitude on interplanetary magnetic field, *Radio Sci., 8,* 955, 1973.

Christon, S. P., D. J. Williams, D. G. Mitchell, L. A. Frank, and C. Y. Huang, Spectral characteristics of plasma sheet ion and electron populations during undisturbed geomagnetic conditions, *J. Geophys. Res., 94,* 13,409, 1989.

Cowley, S. W. H., and C. J. Owen, A simple illustrative model of open flux tube motion over the dayside magnetopause, *Planet. Space Sci., 37,* 1461, 1989.

Hill, T. W., and P. H. Reiff, Evidence of magnetospheric cusp proton acceleration by magnetic merging at the dayside magnetopause, *J. Geophys. Res., 82,* 3623, 1977.

Lemaire, J., Impulsive penetration of filamentary plasma elements into the magnetospheres of the Earth and Jupiter, *Planet. Space Sci., 25,* 887, 1977.

Lockwood, M., and M. F. Smith, The variation of reconnection rate at the dayside magnetopause and cusp ion precipitation, *J. Geophys. Res., 97,* 14,841, 1992.

Lockwood, M., and C. J. Davis, An analysis of the accuracy of magnetopause reconnection rate variations deduced from cusp ion dispersion characteristics, *Annales. Geophys. 14,* 149, 1996.

Lockwood, M., W. F. Denig, A. D. Farmer, V. N. Davda, S. W. H. Cowley, and H. Lühr, Ionospheric signatures of pulsed reconnection at the Earth's magnetopause, *Nature, 361,* 424, 1993.

Lundin, R., Processes in the magnetospheric boundary layer, *Physica Scripta, T18,* 85, 1987.

Lundin, R., K. Stasiewicz, and B. Hultqvist, On the interpretation of different ion flow vectors of different ion species in the magnetospheric boundary layer, *J. Geophys. Res., 92,* 3214, 1987.

Lyons, L. R., M. Schultz, D. C. Pridmore-Brown, and J. L. Roeder, Low-latitude boundary layer near noon: An open field line model, *J. Geophys. Res., 99,* 17,367, 1994.

Mitchell, D. G., F. Kutchko, D. J. Williams, T. E. Eastman, L. A. Frank, and C. T. Russell, An extended study of the low-latitude boundary layer on the dawn/dusk flanks of the magnetosphere, *J. Geophys. Res., 92,* 7394, 1987.

Newell, P. T., and C.-I. Meng, Ion acceleration at the equatorward edge of the cusp: Low altitude observations of patchy merging, *Geophys. Res. Lett., 18,* 1829, 1991.

Newell, P. T., and C.-I. Meng, Cusp low-energy ion cutoffs: A survey and implications for merging, *J. Geophys. Res., 100,* 21,943, 1995.

Newell, P. T., C.-I. Meng, D. G. Sibeck, and R. Lepping, Some low-altitude cusp dependencies on the interplanetary magnetic field, *J. Geophys. Res., 94,* 8921, 1989.

Newell, P. T., W. J. Burke, E. R. Sánchez, C.-I. Meng, M. E. Greenspan, and C. R. Clauer, The low-latitude boundary layer and the boundary plasma sheet at low-altitude: Prenoon precipitation regions and convection reversal boundaries, *J. Geophys. Res., 96,* 21,013, 1991.

Newell, P. T., D. G. Sibeck, and C.-I. Meng, Penetration of the interplanetary magnetic field $B_y$ and magnetosheath plasma into the magnetosphere: Implications for the predominant magnetopause merging site, *J. Geophys. Res., 100,* 234, 1995.

Nishida, A., Can random reconnection on the magnetopause produce the low latitude boundary layer?, *Geophys. Res. Lett., 16,* 227–230, 1989.

Omidi, N., and D. Winske, Structure of the magnetopause inferred from one-dimensional hybrid simulations, *J. Geophys. Res., 100,* 11,935, 1995.

Onsager, T. G., C. A. Kletzing, J. B. Austin, and H. MacKiernan, Model of magnetosheath plasma in the magnetosphere: Cusp and mantle particles at low-altitudes, *Geophys. Res. Lett., 20,* 479, 1993.

Phan, T. D., and G. Paschmann, Low-latitude dayside magnetopause and boundary layer for high magnetic shear 1. Structure and motion, *J. Geophys. Res., 101,* 7801, 1996.

Richard, R. L., R. J. Walker, and M. Ashour-Abdalla, The population of the magnetosphere by solar wind ions when the interplanetary magnetic field is northward, *Geophys. Res. Lett., 21,* 2455, 1994.

Rosenbauer, H., H. Grunwaldt, M. D. Montgomery, G. Paschmann, and N. Sckopke, Heos 2 Plasma observations in the distant polar magnetosphere: The plasma mantle, *J. Geophys. Res., 80,* 2723, 1975.

Samir, U., K. H. Wright, and N. H. Stone, The expansion of a plasma into a vacuum: basic phenomena and processes and applications to space plasma physics, *Rev. Geophys. Space Phys., 21,* 1631, 1983.

Sergeev, V. A., G. R. Bikkuzina, and P. T. Newell, Dayside isotropic precipitation of energetic protons, *Annales Geophys.,* in press, 1997.

Sibeck, D. G., R. W. McEntire, A. T. Y. Lui, R. E. Lopez, S. M. Krimigis, R. B. Decker, L. J. Zanetti, and T. A. Potemra, Energetic magnetospheric ions at the dayside magnetopause: Leakage or merging?, *J. Geophys. Res., 92,* 12,097, 1987.

Spreiter, J. R., and S. S. Stahara, Magnetohydrodynamic and gasdynamic theories for planetary bow waves, in *Collisionless Shocks in the Heliosphere: Reviews of Current Research,* Geophys. Monogr. Ser. vol. 35, edited by B. T. Tsurutani and R. G. Stone, p. 85, AGU, Washington, D. C., 1985.

Stern, D. P., Parabolic harmonics in magnetospheric modeling: The main dipole and the ring current, *J. Geophys. Res., 90,* 10,851, 1985.

Tsyganenko, N. A., and D. P. Stern, Modeling the global magnetic field of the large-scale Birkeland current systems, *J. Geophys. Res., 101,* 27,187, 1996.

Vasyliunas, V. M., A survey of low-energy electrons in the evening sector of the magnetosphere with OGO 1 and OGO 3, *J. Geophys. Res., 73,* 2839, 1968.

Wing, S., P. T. Newell, and T. G. Onsager, Modeling the entry of magnetosheath electrons into the dayside ionosphere, *J. Geophys. Res., 101,* 13,155, 1996.

Yamauchi, M., J. Woch, R. Lundin, M. Shapshak, and R. Elphinstone, A new type of ion injection event observed by Viking, *Geophys. Res. Let., 20,* 795, 1993.

---

P. T. Newell and S. Wing, The Johns Hopkins University Applied Physics Laboratory, Johns Hopkins Road, Laurel, MD 21023-6099. E-mail: patrick.newell@jhuapl.edu; simon.wing@jhuapl.edu.

# Under What Conditions Will Ionospheric Molecular Ion Outflow Occur?

G. R. Wilson[1] and P. D. Craven

*Space Science Laboratory, ES83, NASA/MSFC, MSFC, AL 35812*

The presence of molecular ions at high altitudes in the polar cap and in more distant portions of the magnetosphere poses a theoretical problem. Typically these ions are abundant in the *E* region ionosphere but have very small scale heights (due to large mass and short chemical life times). They are a negligible constituent in the topside ionosphere where most energization processes, that produce escaping heavy ions, occur. One possible solution to this dilemma is that, prior to the commencement of molecular ion outflow, the ionosphere must be significantly modified so that molecular ions can reach the topside in large numbers where they would be accelerated to escape energies. Convection driven ion-neutral heating is one process that could lead to the needed ionospheric modification. This process occurs, to the required degree, in regions where the convection electric field is large (> 50 mV/m). We present here evidence from Dynamics Explorer (DE) 1 and 2 data that support this idea, specifically, that large quantities of molecular ions are present in and above the topside ionosphere after the underlying ionosphere has been subjected to strong convection.

## 1. INTRODUCTION

Molecular ions ($NO^+$, $N_2^+$, $O_2^+$) are a normal constituent of the lower ionosphere in the E region. In the F region and the topside ionosphere they constitute a minor species. This is due to their larger mass compared to $O^+$ and to their short chemical life times which, under normal conditions in the F region, can be a few minutes. One would not expect to see these ions in significant quantities above the topside ionosphere or in more distant parts of the magnetosphere. That such is the case has been demonstrated by many observations over the last 20 years.

*Taylor* [1974] reported, using OGO-6 data, seeing spatially localized regions in the topside ionosphere (~1000 km) where molecular ion densities were greatly enhanced over their normal levels during magnetically active intervals. *Hoffman et al.* [1974] presented a case where molecular ions with densities as high as 1000 $cm^{-3}$ were seen at 1400 km altitude during a large magnetic storm. Molecular ions have been seen at much higher altitudes over the polar cap by the RIMS (Retarding Ion Mass Spectrometer) instrument on DE 1 [*Craven*, 1985] and the SMS (Superthermal Mass Spectrometer) instrument on AKEBONO [*Yau et al.*, 1993], at high energies in the ring current by the SULEICA (Suprathermal Energy Ionic Charge Analyzer) instrument on the AMPTE/IRM spacecraft [*Klecker et al.*, 1986], and in the distant magneto-tail by the EPIC (Energetic Particles and Ion Composition) instrument on GEOTAIL [*Christon et al.*, 1994].

---

[1] On leave from the Department of Physics, University of Alabama in Huntsville. Now at Mission Research Corp., Nashua, NH

Geospace Mass and Energy Flow: Results From the International Solar-Terrestrial Physics Program
Geophysical Monograph 104
Copyright 1998 by the American Geophysical Union

Some preliminary conclusions about outflowing molecular ions can be drawn from the above listed observations. 1) Molecular ion outflow occurs during, or immediately following intervals of elevated magnetic activity. For the two cases in Taylor [1974] Kp = 4 to 5; for the *Hoffman et al.* [1974] and *Craven et al.* [1985] cases Kp = 9; and in the total of four events reported by *Klecker at al.* and *Christon et al.* molecular ions were observed during or shortly after intervals when Kp = 7⁻ to 8⁻. In the majority of the AKEBONO cases Kp was 4 or greater at the time interval of the observations and in the preceding hours [*Yau et al.*, 1993]. In 19 of 25 cases from the RIMS data set Kp was 4 or greater in the 3 hour Kp interval containing the observation (more on the RIMS data will be presented below). 2) The ion composition observed in high altitude molecular ion events does not match that typically observed in the ionosphere. The $N^+/O^+$ density or flux ratio is often near 1.0 instead of the more typical value of about 0.1 [*Hoffman et al.*, 1974; *Yau et al.*, 1993]. Among the molecular ions the dominant species is often $N_2^+$ [*Hoffman et al.*, 1974; *Craven et al.*, 1985; *Yau et al.* 1993] which is frequently the scarcer of the three molecular ions in the *F* region [*Brinton et al.*, 1978]. The least abundant molecular ion is, in many cases, $O_2^+$ [*Hoffman et al.*, 1974; *Craven et al.*, 1985; *Yau et al.* 1993] which is often the most abundant molecular ion at 300 km altitude [*Brinton et al.*, 1978].

For a molecular ion to reach an altitude of 1 $R_E$, starting from 300 km, it must acquire an energy of about 10 eV. The energization time and the transit time to 1 $R_E$ must together, be less than the average dissociative recombination time. In the *F* region with an electron density of $10^5$ cm⁻³ and an electron temperature of 3000 K the dissociative recombination time for $NO^+$ is 230 s, for $O_2^+$ is 180 s, and for $N_2^+$ is 110 s using recombination coefficients from *Torr et al.* [1990]. These short times have lead several to conclude that outflowing molecular ions must be energized, at low altitudes, to a minimum of about 10 eV in just a few minutes [*Yau et al.*, 1993; *Peterson et al.*, 1994]. However, if the outflow of molecular ions were preceded by a significant and appropriate modification of the ionosphere/thermosphere in the source region, this requirement could be greatly reduced. Such a modification could involve some or all of the following: 1) increased production rate of molecular ions, 2) a reduction in the column integrated total electron content of the *F* region, 3) an increase in the electron temperature, and 4) changes in the composition of the neutral atmosphere. (The dissociative recombination rates are inversely proportional to the electron temperature to the power *a*, where *a* = 0.4 - 1.0.) Auroral zone processes which are know to cause these kinds of modifications are ion-neutral frictional heating driven by convection or neutral winds [*Schunk et al.*, 1975; *Häggström and Collis*, 1990; *Anderson et al.*, 1991], and energetic electron precipitation [*Wahlund et al.*, 1992].

If the outflow of large numbers of molecular ions from the ionosphere requires that a modification of the ionosphere precede the event, then it should be possible to trace backwards the trajectory of these ions to a point and time in the ionosphere where conditions for the elevated production of, or longer life times for molecular ions exist. The purpose of this paper is to demonstrate this by showing such examples using DE 1 and DE 2 data. In particular we show cases where the evidence suggests that convection-driven ion-neutral frictional heating is the responsible process since large convection electric fields are seen in the vicinity of the outflowing molecular ions.

## 2. SUMMARY OF DE 1 RIMS MOLECULAR ION OBSERVATIONS

After a careful review of DE 1 RIMS data we have identified 25 separate orbits where molecular ions were observed at altitudes below 3000 km. This altitude was chosen as a cutoff simply as a means of identifying events where the RIMS instrument observed molecular ions near their source location. Molecular ions were seen by DE 1 at altitudes much higher than 3000 km [*Craven et al.*, 1985]. Table 1 contains the list of these events along with Kp and AE for the time interval of observation and the preceding one, and the peak number of counts seen in 30 seconds along with the time interval (in minutes) along the orbital track over which molecular ions were seen. These last two quantities give a rough measure of how strong the event was. They must be viewed with caution since the peak count rate and the time interval over which molecular ions will be seen will depend on the position of the orbit relative to the event as much as on the strength of the event.

Some conclusions which can be drawn from this table include: 1) The average Kp during the time interval of observation, for all 25 events, was 5. The average Kp for the preceding 3 hour interval was 5-. 2) The number of events for which the Kp was 4 or larger during the interval of observation was 19. 3) The average value of the AE index during the hour of observation was 607 and for the preceding hour was 631. 4) Generally speaking the higher the level of magnetic activity the stronger the event. The weakest 6 events (peak counts less than 500) had an average concurrent Kp of 4- while the events stronger than that had an average concurrent Kp of 5+.

Table 1. Low Altitude Molecular Ion Events Seen by DE 1

| Date Year/Day | Time (UT) | Kp | Previous Kp | AE | Previous AE | Peak Counts | Time Interval |
|---|---|---|---|---|---|---|---|
| 82 021 | 15-16 | 4 | 3 | 70 | 92 | 5000 | 9 |
| 82 028 | 18-19 | 1+ | 3- | 58 | 58 | 200 | 3 |
| 82 035 | 21-22 | 6- | 5- | 604 | 699 | 3000 | 8.5 |
| 82 041 | 14-15 | 5+ | 5+ | 989 | 716 | 4000 | 9.5 |
| 82 060 | 17-18 | 7 | 6+ | 314 | 851 | 3000 | 15.5 |
| 82 079 | 19.5-20.5 | 2+ | 3- | 256 | 307 | 2000 | 14.5 |
| 82 083 | 12.5-13.5 | 2+ | 3 | 69 | 116 | 90 | 1.5 |
| 82 092 | 15.5-16.5 | 6- | 5- | 1214 | 1083 | 10000 | 2 |
| 82 093 | 12-13 | 5- | 4+ | 657 | 693 | 7000 | 11 |
| 82 115 | 10.5-11.5 | 6+ | 7- | 930 | 1124 | 10000 | 15 |
| 82 120 | 14.5-15.5 | 4- | 4- | 578 | 264 | 400 | 3.5 |
| 82 163 | 16-17 | 7- | 6 | 1165 | 1464 | 1500 | 15 |
| 82 176 | 19-20 | 1- | 1+ | 132 | 137 | 8000 | 4 |
| 82 181 | 8 - 9 | 5 | 7- | 455 | 456 | 500 | 12 |
| 82 193 | 15-16 | 7- | 7- | 1195 | 1190 | 600 | 15.5 |
| 82 194 | 4 - 5 | 6- | 5- | 978 | 878 | 900 | 4.5 |
| 82 195 | 14-15 | 6 | 7 | 613 | 494 | 1300 | 18.5 |
| 82 196 | 4 - 5 | 4 | 5- | 717 | 841 | 3000 | 9 |
| 82 249 | 5 - 6 | 8+ | 8 | 959 | 1371 | 10000 | 7.5 |
| 82 249 | 12-13 | 9 | 8+ | 847 | 490 | >10000 | 25.5 |
| 82 266 | 7.5-8.5 | 4+ | 5 | 336 | 483 | 10000 | 2 |
| 82 302 | 13-14 | 5- | 4+ | 577 | 401 | 3000 | 4 |
| 82 305 | 9 -10 | 4- | 4- | 573 | 264 | >10000 | 5.5 |
| 83 314 | 19.5-20.5 | 6- | 4 | 672 | 828 | 200 | 10 |
| 83 319 | 16-17 | 4 | 4 | 211 | 480 | 100 | 10.5 |

Table of molecular ion events seen by the RIMS instrument on DE 1. The time is the hour interval containing the event. The first Kp value is for the 3 hour interval containing the event while the second Kp value is for the three hour interval preceding that. The first AE value is for the hour containing the event while the second is for the hour preceding that. Peak counts are the maximum number of molecular ions (either $NO^+$, $N_2^+$, or $O_2^+$) seen in a 30 s interval during the event. The time interval is the length of time in minutes during which molecular ions were seen.

3. CASE 1: FEBRUARY 4, 1982.

Figures 1 to 4 show the details of the 82/035 event from Table 1. With the exception of the magnetic indices and the IMP 8 IMF (Interplanetary Magnetic Field) data all data are from instruments on either DE 1 or DE 2. Figure 1 gives the AE, Kp and IMF history for about 10 hours preceding the event. Around 21:00 UT on February 4th IMP 8 was at a GSM position of x = -10.5 $R_E$, y = -25.9 $R_E$ and z = -26.5 $R_E$, which means that the IMF values in Figure 1 were observed about 5 minutes after they reached the nose of the magnetosphere assuming a solar wind speed of 400 km/s. In each panel of Figure 1 the filled diamond indicates the time when the molecular ions were seen by DE 1. The IMF was southward for at least 1.5 hours prior to the molecular ion interval.

Figure 2 a) shows the DE 1 and DE 2 orbits in latitude and altitude along with representative magnetic field lines projected into the orbit plane. Figures 2 b) and c) show respectively the same DE 2 and DE 1 orbit segments in invariant latitude and magnetic local time coordinates. Those points marked by the open boxes indicate where the strong perpendicular electric fields were seen and the points marked by the smaller filled squares indicate where molecular ions were seen. The DE 2 spacecraft saw the strong electric field as it climbed from 320 to 330 km altitude. DE 1 saw a strong DC electric field between 830 and 1150 km altitude and molecular ions between 1270 and 2676 km altitude.

88 MOLECULAR ION OUTFLOW

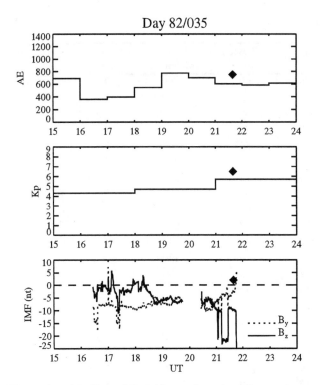

**Figure 1.** AE, Kp and IMF history for February 4, 1982 (day 82/035). The filled diamonds indicate the time when RIMS instrument on DE 1 saw molecular ions.

Figure 3 panel a) shows the perpendicular (to **B**) component of the DC electric field as seen by DE 1. This field is measured by the PWI (Plasma Wave Instrument) on DE 1. It gives, however, only the component of the true perpendicular electric field which lies in the spacecraft spin plane. The dashed line is this same electric field mapped down to 200 km altitude assuming no field-aligned potential drop. Figure 3 panel b) gives the total number of counts during a 30 second interval for the peak of the mass scan curves corresponding to $N_2^+$ and $NO^+$. No value is given for $O_2^+$ because at this time the instrument was operated in such a way that the $O_2^+$ peak was not fully scanned. Between 21:30 and 21:35 the instrument was shut down due to high ion count rates at low altitude.

The three panels in Figure 4 give the spacecraft x component of the DC electric field as measured by VEFI (Vector Electric Field Instrument), and the electron temperature and density as measured by LANG (Langmuir Probe Instrument), on DE 2. The full perpendicular DC electric field could not be obtained because one component (spacecraft z) of the electric field was missing due to the failure of the z antenna to deploy. The data are all sampled on a half second time interval. During this time the DE 2 spacecraft was climbing in altitude from 323 to 341 km.

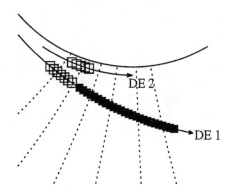

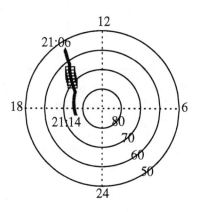

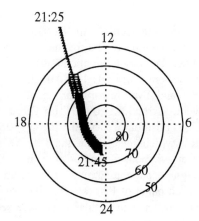

**Figure 2.** For case 1: a) DE 1 and DE 2 orbit segments in geographic latitude and altitude with dipolar magnetic field lines as projected into the orbit plane. This event occurred in the southern hemisphere. b) DE 2 orbit segment between 21:06 and 21:14 UT in magnetic local time and invariant latitude coordinates. c) DE 1 orbit between 21:25 and 21:45 UT in magnetic local time and invariant latitude coordinates. Points marked by open boxes indicate where a strong convection electric field was seen. Points marked by the filled boxes indicate where molecular ions were seen.

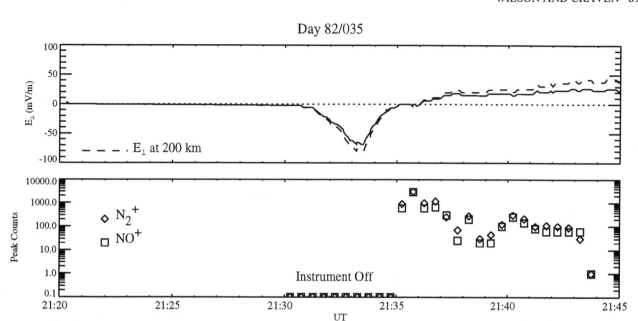

**Figure 3.** DE 1 data for the February 4, 1982 molecular ion event. a) Perpendicular (to B) component of the DC electric field measured by the PWI instrument. The dashed curve is this electric field mapped down to 200 km altitude. b) Number of molecular ions seen in a 30 s interval by the RIMS instrument.

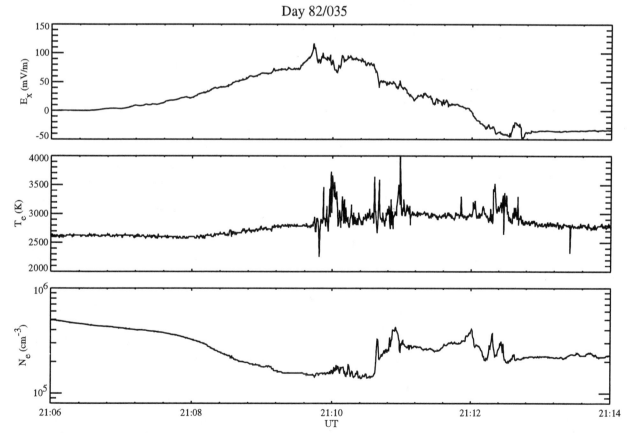

**Figure 4.** DE 2 data for the February 4, 1982 molecular ion event. a) X component (in spacecraft reference frame) of the electric field measured by the VEFI instrument. b) Electron temperature, and c) electron density measured by the LANG instrument.

Given the similarity in the profile of the electric field peaks and the nearly identical location in which each was observed, it is very likely that the two spacecraft saw the same electric field structure despite the approximately 23 minutes separating each observation. (It is not possible to directly compare the two electric fields since neither spacecraft was able to obtain a full three component measurement so that the full perpendicular electric field component for each could be found.) This suggests that the polar cap convection pattern was fairly stable, having lasted for at least 23 minutes. The existence of southward IMF for 1.5 hours preceding the event suggests that a well developed two cell convection pattern existed at this time. The strong electric field seen at low latitude by both spacecraft, in this view, represents sunward flow. As each spacecraft moved to higher latitude the electric field reversed sign as they moved into a region of antisunward flow.

One possible interpretation of this data is that the ionosphere was significantly modified in the strong sunward flow channel, primarily by ion-neutral frictional heating. This resulted in a drop in the overall electron density as $O^+$ was converted to the shorter lived ions $NO^+$ and $O_2^+$. The frictional heating of $O^+$ and the molecular ions increased their scale heights. An increase in the electron temperature also aided the molecular ions by increasing their life times since the ion dissociation reaction rates are inversely proportional to the electron temperature to the power 0.4-1 [*Torr et al.*, 1990]. The molecular ion upwellings produced in the high speed sunward flow channel convect through the dayside convection throat and then spread out over the polar cap. The convection driven frictional heating discussed here is unlikely to produce molecular ions with energies needed to escape; that would need to be done by another process, possibly at higher altitude. Those molecular ions which do not get energized to escape energies in the cusp region would then move over the polar cap and fall back to the ionosphere in the region where they are seen by DE 1.

## 4. CASE 2: NOVEMBER 1, 1982.

The data for this case are presented in Figures 5-8 which are, with one exception, in the same format as the corresponding figures for case 1. Around 8:30 UT on November 1st IMP 8 was at a GSM position of x = 30 $R_E$, y = 0.4 $R_E$ and z = -21 $R_E$, which means that the IMF values in Figure 5 were observed about 5 minutes before they reached the nose of the magnetosphere assuming a solar wind speed of 400 km/s. Figure 6 gives the DE 1 and DE 2 orbit segments marked to show where the large

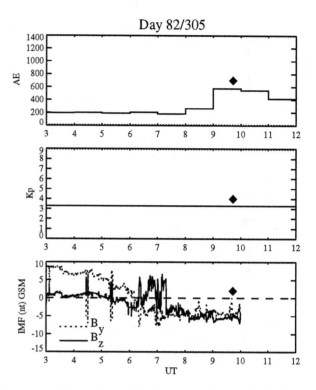

**Figure 5.** AE, Kp and IMF history for November 1, 1982 (day 82/305). The filled diamonds indicate the time when RIMS instrument on DE 1 saw molecular ions.

convection electric field and molecular ions were seen. DE 2 sees the strong electric field near 520 km altitude while DE 1 sees it near 2420 km altitude. The DE 1 observations of molecular ions start near 2300 km altitude and continue down to 1400 km altitude where the RIMS instrument shuts down for perigee passage. Figure 7 shows the DE 1 electric field and molecular ion data while Figure 8 shows the DE 2 data, now with the ion temperature added.

The DE 2 data in this case further supports the interpretation of these events discussed above. The electric field shown in panel a) of Figure 8 would imply a convection speed of about 3 km/s. The full DC electric field is not available from the VEFI because its z axis antenna pair failed to deploy, however, the RPA and IDM (Ion Drift Meter) instruments together found a perpendicular (to **B**) convection speed of about 4.5 km/s. Such large convection speeds would lead to a significant amount of ion heating which can be seen in panel c) of Figure 8. (Note: The RMS error for the high $T_i$ interval was near 100%, however, the sharp increase in $T_i$ on the low latitude side is well resolved.) Panel b) shows that the electron temperature is also elevated in the region, possibly due to energy transfer from the ions via Coulomb

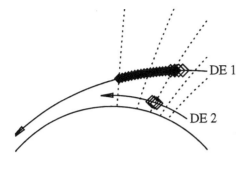
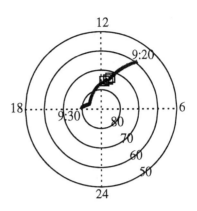
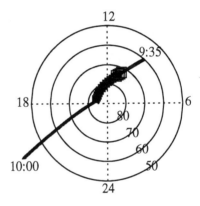

**Figure 6.** For case 2: a) DE 1 and DE 2 orbit segments in geographic latitude and altitude with dipolar magnetic field lines as projected into the orbit plane. This event occurred in the northern hemisphere. b) DE 2 orbit segment between 9:20 and 9:30 UT in magnetic local time and invariant latitude coordinates. c) DE 1 orbit between 9:35 and 10:00 UT in magnetic local time and invariant latitude coordinates. Points marked by open boxes indicate where a strong convection electric field was seen. Points marked by the filled boxes indicate where molecular ions were seen.

collisions. Alternately $T_e$ may be elevated in this region due to energetic electron precipitation. Enhanced fluxes of downgoing electrons with energies up to about 1 keV where seen by the LAPI (Low Altitude Plasma Instrument) near 9:23:30 and 9:23:50 UT. The electron density in the strong **E** field region drops, very likely due to the conversion of $O^+$ to molecular ions. The fact that this density remains low into the polar cap (between 9:24:30 and 9:27:00) beyond the region of strong convection heating may be a consequence of depleted flux tubes convecting into the polar cap where the lack of sunlight delays the recovery of preheated densities.

The DE 2 RPA (Retarding Potential Analyzer) data also suggest that the relative abundance of $O^+$ (and $N^+$) in the strong **E** field region drops to less than 80% while the abundance of $H^+$ and $He^+$ remains below 1%. Outside of the region of large Ex the $O^+$ fraction gradually returns to near 100% by 9:27:00 UT. Such compositional analysis is obtained from inflection points seen in the collector current versus retarding potential voltage curve and can only separate different ion species whose masses differ by a large amount. Thus $O^+$ cannot be distinguished from $N^+$ but these two can be separated from $He^+$ and from the molecular ions [see *Hanson et al.*, 1973 and *Hoffman et al.*, 1973]. If these relative abundances of $O^+$, $H^+$ and $He^+$ are accurate this would indicate an increase in the amount of molecular ions present.

In high speed convection, resonant charge exchange collisions between the $O^+$ and atomic oxygen can produce a toroidal $O^+$ velocity distribution [*St.-Maurice and Schunk*, 1979]. This occurs when the $\mathbf{E} \times \mathbf{B}$ drift speed is large compared to the original thermal speeds of either the ions or the neutrals and the ions are transferred (in velocity space) by the collisions to the rest frame of the neutrals. Above the region where the ion-neutral collision rate is high and Coulomb self collisions become important, the $O^+$ will tend to relax toward Maxwellian, passing through intermediate, nonMaxwellian, distributions in the process [*Wilson*, 1994]. Such nonMaxwellian velocity distributions could account for the large RMS errors associated with the ion temperatures in the high electric field region [*Heelis* 1997, private communication]. This occurs because the

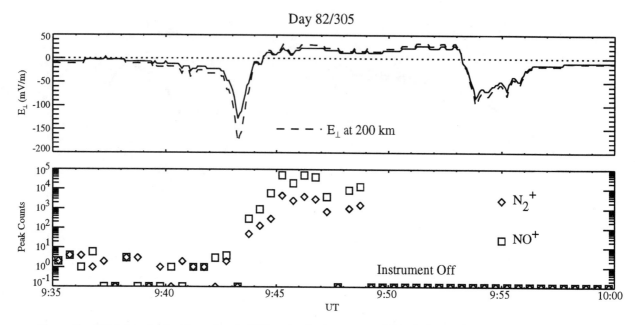

**Figure 7.** DE 1 data for the November 1, 1982 molecular ion event. a) Perpendicular (to B) component of the DC electric field measured by the PWI instrument. The dashed curve is this electric field mapped down to 200 km altitude. b) Number of molecular ions seen in a 30 s interval by the RIMS instrument.

automatic fitting routine used to process the data assumed a Maxwellian velocity distribution which will produce a well defined RPA (Retarding Potential Analyzer) curve that will differ in functional form from that produced by a toroidal distribution. What is needed is to go back to the RPA curves and analyze them using a toroidal velocity distribution. Outside of the region of large Ex the RMS errors are small, only a few percent.

This event is much stronger than the case 1 event. The number of molecular ions seen in a 30 s interval at the peak of the event is about an order of magnitude higher than the peak counts seen in case 1. The largest count rate (~2000/s) is seen between 9:45 and 9:47 UT when DE 1 is descending from about 2040 to 1660 km altitude. The peak count rate (~200/s) for case 1 occurs at 1280 km altitude. The difference between the count rates seen in these two events cannot be attributed to differences in sampling altitudes. The peak convection electric field, as measured by DE 1, when projected to 200 km altitude was about 2.3 times larger in case 2 than case 1. This would lead to about a factor of 5 increase in the ion temperatures in the altitude region where ion-neutral collisions are frequent. As a result the production rate of molecular ions could be greater and the depth of the density decrease in the $F$ region would be greater.

## 5. DISCUSSION

Figure 9 shows DE 1 orbit segments for nine cases where molecular ions were seen. The format of each individual plot within this figure is the same as the invariant latitude-magnetic local time orbit segment plots in figure 2. The two cases discussed above are included in this figure. The common feature for each of these nine events is that the IMF was southward during the time interval when molecular ions were seen and for some time preceding that event. The interval of southward IMF ranged from a low of about 20 minutes (82/266) to a high of about 6 hours (82/193). The maximum perpendicular electric field (at 200 km) seen among these nine events was 180 mV/m (82/305) and the minimum was 67 mV/m (82/196). The average was 105 mV/m.

In most of these cases one can see that the molecular ions are seen either where the strong electric field is or poleward of it. This can be explained in terms of the two cell convection pattern with the upflowing molecular ions being produced on the low latitude side where the convection is sunward and then spread across the polar cap when the flow turns antisunward in the convection throat region. A molecular ion with an energy of 2 eV would convect about 1000 km in the time it takes for the ion to

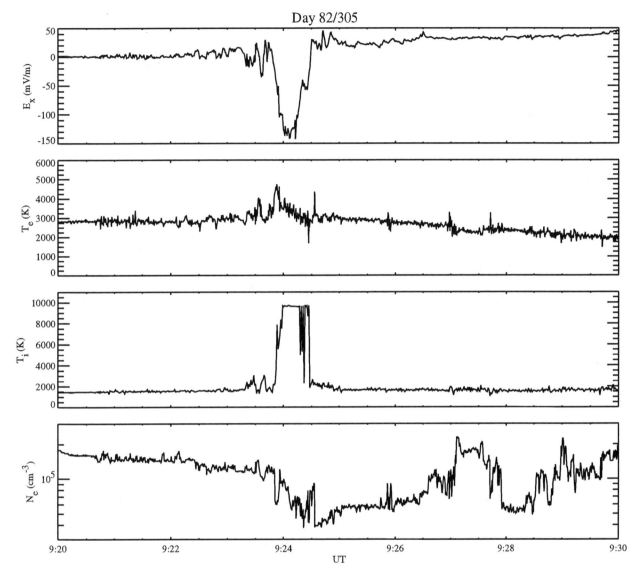

**Figure 8.** DE 2 data for the November 1, 1982 molecular ion event. a) X component (in spacecraft reference frame) of the electric field measured by the VEFI instrument. b) Electron temperature measured by the LANG instrument. c) Ion temperature measured by the RPA instrument. d) Electron density measured by the LANG instrument.

travel upward from the ionosphere and fall back to it, assuming an average convection electric field of 50 mV/m at 200 km altitude.

The basic mechanism proposed here is that molecular ion upflows originate from regions of the ionosphere where strong convection electric fields exist. These strong electric fields cause the plasma to flow at up to several kilometers per second relative to the neutrals. This high speed flow can significantly energize the ions at low altitude where their collision rate with the neutral atmosphere is high. The cross section for the reaction of $O^+$ with $N_2$ and $O_2$ is a very sensitive function of energy and, for the reaction of $O^+$ with $N_2$, can increase by two orders of magnitude when the collision energy increases from 0.1 to 1 eV [*Albritton et al.*, 1977]. The heated $O^+$ is then rapidly converted to $NO^+$ and $O_2^+$. Because these ions are shorter lived than $O^+$ the overall plasma density drops as a new equilibrium is established. In this new equilibrium the concentrations of molecular ions in the *F* region are greatly elevated in an environment where the ions are hotter. Upflowing, but not dramatically energized molecular ions then result. The signatures of this process at low altitude are high speed ion

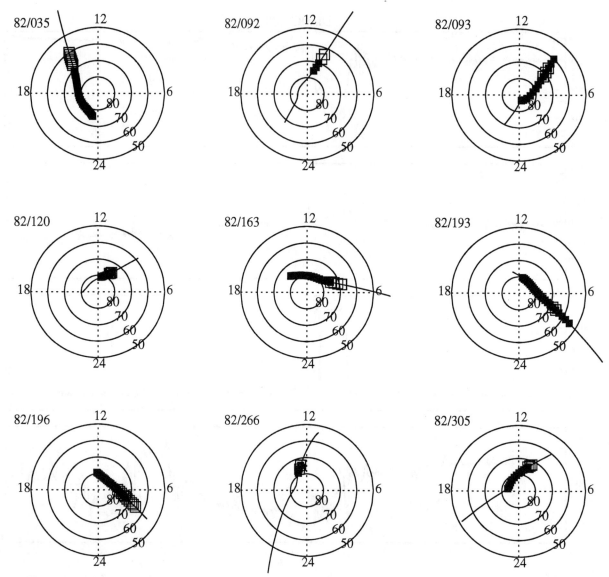

**Figure 9.** Nine DE 1 orbit segments for time intervals when molecular ions were seen. The dates for each plot are indicated to the left and above each. The format of each individual dial plot is the same as given in Figure 2.

flows, a drop in electron density, an increase in ion temperature and a decrease in the relative abundance of $O^+$. All of these have been seen in the DE 2 data for the 82/305 event with the caveats noted above.

The mechanism described above would predict that $NO^+$ and possibly $O_2^+$ would be the dominant molecular ion in ion upflows. As was discussed in the introduction, $N_2^+$ is often a dominant molecular ion in these upflows. In case 1 the $N_2^+$ and $NO^+$ fluxes are about equal while in case 2 the $NO^+$ flux clearly dominates $N_2^+$. One possible way to account for the larger relative abundance of $N_2^+$ is that much of the $O^+$ being converted to molecular ions is excited $O^+$ ($O^+(^2D)$) which when it reacts with $N_2$ produces $N_2^+$ instead of $NO^+$. The rate for this reaction is larger than the rate for the reaction converting ground state $O^+$ to $NO^+$, even over the whole energy range from 0.01 to 1.0 eV [*Torr et al.*, 1990]. $O^+(^2D)$ is produced from atomic oxygen when it is ionized by either an EUV photon or an energetic electron. The presence of large fluxes of $N_2^+$ relative to the other molecular ions in these upflows may indicate that the ionospheric source region has been subjected to a significant amount of precipitation of energetic electrons.

There are a number of things left to be done with the data from the two cases reported above and some of the other ones listed in the table. One is to use the RPA and spin curves from the RIMS data to constrain the energy and flow speed of the molecular ions so as to determine if they have been energized beyond what could be accounted for by frictional heating. If additional heating is apparent then one could ask whether or not the additional heating process is directly associated with the ionospheric modification process or incidentally present. A second thing is to examine the abundance of $N^+$ relative to $O^+$ to see if this ratio remains near typical ionospheric values or approaches 1 [*Yau et al.*, 1993]. Also the relative abundance of the three molecular ions needs to be cataloged to see how it varies from event to event. Such compositional studies, when coupled with modeling efforts, could reveal the degree and means by which the ionosphere is perturbed. It may also indicate how the deposited energy is partition among the various ion species.

*Acknowledgments.* The work of G. R. Wilson was supported by the National Research Council. We thank D. Menietti from the University of Iowa for providing the DE 1 electric field data and the NSSDC for providing the DE 2 VEFI, LANG and RPA data, and the IMP8 magnetic field data through SPyCAT. We also acknowledge useful discussions with W. K. Peterson and A. W. Yau.

## REFERENCES

Albritton, D. L., I. Dotan, W. Lindinger, M. McFarland, J. Tellinghuisen, and F. C. Fehsenfeld, Effect of ion speed distributions in flow-drift tube studies of ion-neutral reactions, *J. Chem. Phys.*, 66, 410, 1977.

Anderson, P. C., R. A. Heelis, and W. B. Hanson, The ionospheric signatures of rapid subauroral ion drifts, *J. Geophys. Res.*, 96, 5785, 1991.

Brinton, H. C., J. M. Grebowsky, and L. H. Brace, The high-latitude winter *F* region at 300 km: Thermal plasma observations from AE-C, *J. Geophys. Res.*, 83, 4767, 1978.

Christon, S. P., G. Gloeckler, D. J. Williams, T. Mukai, R. W. McEntire, C. Jacquey, V. Angelopoulos, A. T. Y. Lui, S. Kokubun, D. H. Fairfield, M. Hirahara, and T. Yamamoto, Energetic atomic and molecular ions of ionospheric origin observed in distant magnetotail flow-reversal events, *Geophys. Res. Lett.*, 21, 3023, 1994.

Craven, P. D., R. C. Olsen, C. R. Chappell, and L. Kakani, Observations of Molecular Ions in the Earth's Magnetosphere, *J. Geophys. Res.*, 90, 7599, 1985.

Häggström, I., and P. N. Collis, Ion composition changes during *F*-region density depletions in the presence of electric fields at auroral latitudes, *J. Atmos. Terr. Phys.*, 52, 519, 1990.

Hanson, W. B., D. R. Zuccaro, C. R. Lippincott, and S. Sanatani, The retarding-potential analyzer on Atmospheric Explorer, *Radio Science*, 8, 333, 1973.

Hoffman, J. H., W. B. Hanson, C. R. Lippincott, and E. E. Ferguson, The magnetic ion-mass spectrometer on Atmospheric Explorer, *Radio Science*, 8, 315, 1973.

Hoffman, J. H., W. H. Dodson, C. R. Lippincott, and H. D. Hammack, Initial ion composition results from the Isis 2 satellite, *J. Geophys. Res.*, 79, 4246, 1974.

Klecker, B., E. Möbius, D. Hovestadt, M. Scholer, G. Gloeckler, and F. M. Ipavitch, Discovery of energetic molecular ions ($NO^+$ and $O_2^+$) in the storm-time ring current, *Geophys. Res. Lett.*, 13, 632, 1986.

Peterson, W. K., T. Abe, H. Fukunishi, M. J. Greffen, H. Hayakawa, Y. Kasahara, I. Kimura, A. Matsuoka, T. Mukai, T. Nagatsuma, K. Tsuruda, B. A. Whalen, and A. W. Yau, On the sources of energization of molecular ions at ionospheric altitudes, *J. Geophys. Res.*, 99, 23257, 1994.

Schunk, R. W., W. J. Raitt, and P. M. Banks, Effects of electric fields on the daytime high-latitude *E* and *F* regions, *J. Geophys. Res.*, 80, 3121, 1975.

St.-Maurice, J.-P., and R. W. Schunk, Ion velocity distributions in the high-latitude ionosphere, *Rev. Geophys.*, 17, 99, 1979.

Taylor, H. A., Jr., High latitude ion enhancements: A clue for studies of magnetosphere-atmosphere coupling, *J. Atmos. Terr. Phys.*, 36, 1815, 1974.

Torr, M. R., D. G. Torr, P. G. Richards, and S. P. Yung, Mid- and low-latitude model of thermospheric emissions 1. $O^+(^2P)$ 7320 Å and $N_2$ ($^2P$) 3371 Å, *J. Geophys. Res.*, 95, 21147, 1990.

Wahlund, J.-E., H. J. Opgenoorth, I. Häggström, K. J. Winser, and G. O. Jones, EISCAT observations of topside ionospheric ion outflows during auroral activity: Revisited, *J. Geophys. Res.*, 97, 3019, 1992.

Wilson, G. R., Kinetic modeling of $O^+$ upflows resulting from $\mathbf{E} \times \mathbf{B}$ convection heating in the high-latitude *F* region ionosphere, *J. Geophys. Res.*, 99, 17453, 1994.

Yau, A. W., B. A. Whalen, C. Goodenough, E. Sagawa, and T. Mukai, EXOS D (Akebono) observations of molecular $NO^+$ and $N_2^+$ upflowing ions in the high-altitude auroral ionosphere, *J. Geophys. Res.*, 98, 11205, 1993.

---

P. D. Craven, Space Science Laboratory, ES83, NASA/MSFC, MSFC, AL 35812.

G. R. Wilson, Mission Research Corp., One Tara Blvd., Suite 302, Nashua, NH 03062.

# Day-Night Asymmetry of Polar Outflow Due to the Kinetic Effects of Anisotropic Photoelectrons

Sunny W. Y. Tam, Fareed Yasseen[1] and Tom Chang

*Center for Space Research, Massachusetts Institute of Technology, Cambridge, MA*

Recent polar wind measurements between 5000 and 9000 km altitude by the Akebono satellite indicate that both $H^+$ and $O^+$ ions can have remarkably higher outflow velocities in the sunlit region than on the nightside. In addition, electrons also display an asymmetric behavior: the dayside difference in energy spread, greater for upward-moving than downward-moving electrons, is absent on the nightside. We use a self-consistent hybrid model [*Tam et al.*, 1995b] that was developed for the polar wind outflow to address these observed day-night asymmetric features. The model takes into account the evolution of the polar wind self-consistently by properly recognizing the global, kinetic, collisional effects of the sunlit photoelectrons. By studying the effects of the presence and absence of photoelectrons on the polar outflow, we compare the daytime and night-time polar wind results, and demonstrate the asymmetries observed by the Akebono satellite.

## 1. MOTIVATION

The existence of the polar wind, an outflow of plasma along the open magnetic field lines emanating from the polar region of the ionosphere, was first proposed by *Axford* [1968] and *Banks and Holzer* [1968]. These early studies recognized the ambipolar electric field as one of the mechanisms governing the plasma outflow. Because the polar cap, in general, is a relatively quiescent region, the ambipolar effect is a major contribution to the electric field in the "classical" polar wind, which is steady-state, quasi-neutral, and current-free outflow.

The ambipolar field and the background plasma exist self-consistently. The ambipolar field is therefore influenced by the mechanisms that determine the dynamics of the particles. For example, the geomagnetic field, which decreases with altitude, gives rise to the mirror force that changes the particles' pitch angles. Coulomb interactions among all the species lead to energy exchange and pitch angle diffusion. These effects are essential to the dynamics of the particles, and therefore, can affect the polar wind electric field.

Recent observations have suggested that another effect — the generation of photoelectron populations in the sunlit ionosphere — can alter the polar wind significantly. Our study on the photoelectron-driven polar wind is motivated by these increasingly convincing experimental indications.

Early polar cap measurements obtained by the ISIS-1 satellite showed evidence of "anomalous" field-aligned photoelectron fluxes in both upward and downward directions, where the downgoing (return) fluxes were considerably smaller than the outgoing fluxes above a certain energy [*Winningham and Heikkila*, 1974]. Such non-thermal features were confirmed by the DE-1 and -2 satellites [*Winningham and Gurgiolo*, 1982]: outgoing field-aligned electron fluxes in the photoelectron energy range were observed by the HAPI (High Altitude Plasma Instrument) on DE-1 and the LAPI (Low Altitude Plasma Instrument) on DE-2; evidence of downstreaming electron fluxes was also found in the low-altitude distribution measured by the LAPI. These

---

[1]Now at Climate Change Secretariat, Bonn, Germany

Geospace Mass and Energy Flow: Results From the International Solar-Terrestrial Physics Program
Geophysical Monograph 104
Copyright 1998 by the American Geophysical Union

fluxes are considered anomalous because their existence cannot be related to the idea of thermal conductivity and temperature gradient in classical fluid theories. Similar to the ISIS-1 measurements, the return fluxes observed by DE-2 were comparable to the outgoing fluxes below some truncation energy, but considerably smaller above that. As suggested by *Winningham and Gurgiolo* [1982], the existence of such downstreaming fluxes may be due to reflection of electrons by the ambipolar electric field along the geomagnetic field line above the satellite. The truncation energy, obtained by comparing the outgoing and the return electron fluxes, would thus provide an estimate for the potential drop due to the electric field. These authors observed that this truncation energy ranged from 5 to 60 eV, and thus were able to deduce the magnitude of the potential drop above the altitude of the satellite ($\sim$ 500 km). Unfortunately, existing classical polar wind theories can only account for a much smaller potential drop [*Ganguli*, 1996, and references therein]. *Winningham and Gurgiolo* [1982] also pointed out that variation of the truncation energy was due to changes in the solar zenith angle at the production layer below the satellite. The solar zenith angle is related to the photoionization rate, which itself is related to the local ionospheric photoelectron density [*Jasperse*, 1981]. These observations therefore imply a relationship between the local photoelectron density below the satellite and the potential drop along the field line above it, and are consistent with the idea that the photoelectrons may significantly affect the ambipolar electric field.

While the observations discussed above have implied that photoelectrons may contribute to the dynamics of the polar wind, more recent evidence indicates that the polar wind characteristics themselves are affected by the photoelectrons. *In situ* measurements by the Akebono satellite have revealed novel features in the polar wind: day-night asymmetries in the ion and electron features. The most dramatic are asymmetries in the ion outflow velocities [*Abe et al.*, 1993b]: satellite data between 5000 and 9000 km altitude have indicated remarkably higher outflow velocities for the major ion species, $H^+$ and $O^+$, in the sunlit region than on the nightside. For example, the $H^+$ velocity ($u_h$) was found to be about 12 km/s on the dayside, but only about 5 km/s on the nightside. Similarly, the $O^+$ velocity ($u_o$) in the sunlit region ($\sim$ 7 km/s) is about twice that in the midnight sector ($\sim$ 3 km/s). A day-night asymmetry was also observed in the electron behavior. Electrons were distinguished according to their velocities along the geomagnetic field line. On the dayside, it was found that the temperature of the upstreaming population is greater than that of the downstreaming population, *i.e.*, $T_{e,up} > T_{e,down}$, indicative of an upwardly directed heat flux [*Yau et al.*, 1995]. On the nightside, in contrast, no such up-down anisotropy was observed [*Abe et al.*, 1996].

Besides the day-night asymmetries, Akebono measurements between 5000 and 9000 km altitude have also revealed other sometimes unexpected ion transport properties in the polar region [*Abe et al.*, 1993b]. For example, $O^+$ was most often found to be dominant over $H^+$ as the major ion species, a situation noted earlier by several authors [*Shelley et al.*, 1982; *Yau et al.*, 1984; *Waite et al.*, 1985]. These observations were contrary to the traditional belief that very few $O^+$ ions are able to overcome the gravitational force and escape to such high altitudes due to their heavier mass. The measured outflow velocities for both the $H^+$ and $O^+$ ions in general increase monotonically with altitude, and the flows for both species are supersonic at high altitudes. In fact, the measured $O^+$ outflow velocities (see above) are much larger than the values expected by classical polar wind models [*Schunk and Watkins*, 1981, 1982; *Blelly and Schunk*, 1993, e.g.]. All these ion outflow characteristics, particularly the enhanced ion outflow velocities, suggest a higher ambipolar electric field than that predicted by classical polar wind models [*Ganguli*, 1996, and references therein], and are consistent with the values of the field-aligned potential drop deduced by *Winningham and Gurgiolo* [1982] based on the DE-2 measurements.

Because of the marked day-night asymmetries observed in several characteristics of the polar wind, and the fact that photoelectrons exist primarily in the sunlit ionosphere, they are *the* natural candidate to account for the day-night asymmetries. Indeed, collisionless kinetic calculations by *Lemaire* [1972] showed that escaping photoelectrons may enhance the electric field and increase the ion outflow velocities in the polar wind. Photoelectrons, therefore, may provide a possible explanation for the observations of both sets of satellites — the magnitude of the ambipolar electric field deduced from the DE-2 measurements, and the day-night asymmetries and enhanced ion outflow velocities observed by the Akebono satellite. Because Coulomb collisions may also influence the dynamics of the photoelectrons, for example, by transferring their energy to other particle components in the polar wind, and thereby reducing the escaping photoelectron flux, collisional effects should also be taken into account in determining the impact of photoelectrons on the electric field. Moreover, collisions should be included because the upward acceleration that accounts for the higher escape velocities of the heavier ions should begin below the transition for collisional to collisionless outflow. Our goal, therefore, is to address these observations by incorporating the complete photoelectron physics into a self-consistent, global description of the polar wind.

We should note that other effects may also contribute to a day-night difference in high-latitude plasmas. For example, the strong plasma heating observed in the cusp/cleft region has been proposed to produce a day-night asymmetry, particularly in the $O^+$, which has been observed to be persistently downward moving on the nightside around 1000 km altitude [*Heelis et al.*, 1993], and slowly moving or downward moving, up to 4000 km [*Chandler*, 1995].

Let us first specify some criteria that will enable us to define the problem more precisely. First, we will consider the polar wind only at altitudes above 500 km (which corresponds roughly to the polar orbits of DE-2). At such altitudes, neutral densities are low enough to neglect the "chemical" reactions such as photoionization, recombination, *etc.* Second, the magnitude of the geomagnetic field is such that the gyration period and Larmor radius, for all particle species, are much smaller than any relevant time or length scales. We can therefore use the guiding center approximation. Third, the gradients of the geomagnetic field are such that transport *along* the magnetic field line dominates. We shall, for the time being, neglect transport perpendicular to the magnetic field. The time-dependent distribution function $f(t, s, v_\parallel, v_\perp)$ for a given particle species is therefore governed by the following collisional gyrokinetic equation:

$$\left[\frac{\partial}{\partial t} + v_\parallel \frac{\partial}{\partial s} - \left(g - \frac{q}{m}E_\parallel\right)\frac{\partial}{\partial v_\parallel} - v_\perp^2 \frac{B'}{2B}\left(\frac{\partial}{\partial v_\parallel} - \frac{v_\parallel}{v_\perp}\frac{\partial}{\partial v_\perp}\right)\right]f = \frac{\delta f}{\delta t} = Cf, \quad (1)$$

where $s$ is the distance along the magnetic field line $B$, $q$ and $m$ are the algebraic electric charge and mass of the species respectively, $E_\parallel$ is the field-aligned electric field, $g$ is the gravitational acceleration, $B' \equiv dB/ds$, $\delta f/\delta t$ represents the rate of change of the distribution function due to collisions, and $C$ is a collisional operator for Coulomb interaction, which is the dominant type of collision above 300 km altitude. Equation (1) includes the major forces a particle experiences as it travels along the geomagnetic field line: gravitational force, field-aligned electric force, mirror force, and forces that are due to Coulomb collisions.

In this paper, we will study the role of photoelectrons in the day-night transition of the polar wind, and their impact on the outflow based on a self-consistent hybrid theory recently developed by *Tam et al.* [1995b].

## 2. PHOTOELECTRONS, ENERGY FLUXES AND ELECTRIC FIELD

The global kinetic collisional physics of suprathermal electrons in a steady-state space plasma outflow was first considered by *Scudder and Olbert* [1979] in their study of the solar wind halo electrons. These authors related the anomalous field-aligned electron heat fluxes observed in the solar wind to the non-local nature of the electron distributions, and demonstrated the formation of such non-thermal features using a simplified collisional operator. They also suggested that these suprathermal electrons, through their anomalous contribution to the energy flux, may significantly increase the ambipolar electric field along the magnetic field lines, thereby "driving" the solar wind [*Olbert*, 1982].

An analogous situation exists for the dayside, photoelectron-driven polar wind. It has been shown by *Yasseen et al.* [1989] that the polar wind photoelectrons can give rise to the non-thermal distributions observed by the DE satellites [*Winningham and Gurgiolo*, 1982]. The effect on the polar wind due to the energy fluxes associated with these photoelectrons has been examined by *Tam et al.* [1995a], who concluded that such anomalous electron energy fluxes may significantly increase the ambipolar electric field. Because photoelectrons exist primarily in the sunlit ionosphere, they enhance the dayside ambipolar electric field, thereby increasing the ion outflow velocities on the dayside. Photoelectrons, with their associated energy fluxes, can therefore provide not only a mechanism for the enhanced ion outflow velocities observed by the Akebono satellite [*Abe et al.*, 1993a, 1993b], but also the explanation for the observed day-night asymmetric ion and electron features in the polar wind [*Abe et al.*, 1993b, 1996; *Yau et al.*, 1995].

Energetic suprathermal electrons in the polar wind or in other ionospheric/magnetospheric settings have been considered by various authors. For example, kinetic collisional calculations by *Khazanov et al.* [1993] have examined the role of photoelectrons on plasmaspheric refilling. Collisionless kinetic calculations by *Lemaire* [1972] have shown that escaping photoelectrons may increase the ion outflow velocities in the polar wind. Collisionless kinetic calculations by *Barakat and Schunk* [1984] have examined the impact of hot magnetospheric electrons, and concluded that such particles may also increase the ion outflow velocities. Generalized semikinetic (GSK) calculations by *Ho et al.* [1992], however, have found that downward heat conduction from hot magnetospheric electrons would decelerate the polar wind outflow and lessen its flux.

We should add that other mechanisms besides suprathermal electron effects may also be proposed as alternative explanations for the enhanced ion velocities. Parallel ion acceleration driven by $\mathbf{E} \times \mathbf{B}$ convection was considered by *Cladis* [1986], and shown to significantly energize $O^+$ ions escaping to the polar magnetosphere. This force can also be seen as a centrifugal force in the convecting frame of reference, and was included in

**Table 1.** Dayside polar wind observations by the Akebono satellite. Ion measurements were made at 5000 – 9000 km altitude [Abe et al., 1993b]. Electron observations were made at about 1700 km altitude [Yau et al., 1995].

| Qualitative Features | Quantitative Estimates |
|---|---|
| $O^+$ dominance over $H^+$ ions | $O^+$ density : lower limit $\sim 10^8$ m$^{-3}$ |
| monotonically increasing ion outflow velocities | $H^+$ density : lower limit $\sim 10^7$ m$^{-3}$ |
| supersonic flows for both $H^+$ and $O^+$ | Ion temperatures : upper limit $\approx 10^4$ °K |
| anisotropy between upwardly and downwardly moving electron population | $H^+$ outflow velocity about 12 – 13 km/s |
| upwardly directed total electron heat flux | $O^+$ outflow velocity about 6 – 7 km/s |

this form in the time-dependent, GSK model developed by *Horwitz et al.* [1994]. These mechanisms, including the suprathermal electron effects, have recently been reviewed by *Ganguli* [1996].

Recently, a self-consistent hybrid model has been developed by *Tam et al.* [1995b] to take into account the global kinetic collisional nature of the polar wind physics introduced by the photoelectrons. Our results in this paper (which we shall present in Section 3) are based on this model. The model represents two breakthroughs in polar wind theoretical modeling. One, it has enabled us to incorporate the global kinetic collisional photoelectron effects into a self-consistent polar wind description. Two, due to its treatments of the ions, the solutions it generates span the simulation range continuously from the collisional subsonic regime at low altitudes to the collisionless supersonic regime at high altitudes.

The model is hybrid in that it consists of a kinetic and a fluid component. Photoelectrons (treated as test particles because of their low relative density) and both the $H^+$ and $O^+$ ions are described using a global kinetic collisional approach while thermal electron properties (density, drift velocity, and temperature) are determined from a simpler, fluid approach that also calculates the self-consistent ambipolar electric field. Because of its treatment of the thermal electrons, the model should be distinguished from traditional hybrid approaches where electrons are treated as a massless neutralizing fluid. The model is based on an iterative scheme combining the kinetic and fluid calculations, that should converge to physically meaningful solutions.

### 3. APPLICATION TO THE PHOTOELECTRON-DRIVEN POLAR WIND

#### 3.1. Comparison of Theoretical Predictions With Observations

The goal of this paper is to demonstrate the observed day-night asymmetry of the polar outflow based on our study of the anisotropic kinetic photoelectron effects. We shall present polar outflow results for the daytime situation, as well as for the nightside where photoelectrons are practically absent. Application of the self-consistent hybrid model to the sunlit polar region has generated results [*Tam et al.*, 1995b] that are consistent with various qualitative features observed in the polar wind (see Table 1) [*Abe et al.*, 1993b; *Yau et al.*, 1995]. However, in order for our results to be more convincing, our dayside solution should also be consistent with the observed polar wind quantities at satellite altitudes, which we summarize in Table 1.

We have generated such an admissible solution for the dayside polar outflow. The local electric field in this solution has converged to within 2% throughout the simulation range (500 km – 2 $R_E$, where $R_E$ is the radius of the earth). In this calculation, initial distributions for the kinetic part of the model are in the form of an upper-half Maxwellian, and are applied at the lower boundary. The ions are represented by Maxwellians (with no drift velocity) with temperatures ($T_*$) of 1 eV for $H^+$ and 2.2 eV for $O^+$. In the simulation, we will consider only the upward-moving ions in the boundary conditions. The initial photoelectron distribution is of 21.6 eV energy spread, with energy ranging from 2 to 62 eV (based on the photoelectron spectra measured by [*Lee et al.*, 1980]). For the fluid component of the model, boundary conditions are also imposed at the lower end: the $H^+$ and $O^+$ densities are $1.5 \times 10^9$ and $4 \times 10^{10}$ m$^{-3}$ respectively; a photoelectron to thermal electron density ratio of $10^{-3}$ is assumed, and the thermal electron temperature is taken to be 3000 °K.

Note that our boundary conditions are in fact functions. These boundary functions should be viewed as approximations to the solution (a choice based on an educated guess). The difference between these approximations and the solution may lead to some spurious manifestations (*e.g.*, an exaggerated increase in ion temperature). We are trying to develop a systematic procedure that will allow us to reduce these effects, and improve our comparisons with observations.

The self-consistent electric potential profile for this set of initial and boundary conditions is shown in Fig 1. A potential drop of about 5 V is obtained across our simulation range. Note that this potential drop is smaller than that in *Tam et al.* [1995b] ($\sim$ 12 V), but is still considerably larger than most classical polar wind models predicted. This smaller potential drop compared with our previous results [*Tam et al.*, 1995b] is mainly due to the larger $O^+$ temperature at the boundary. The effect of the ionospheric $O^+$ temperature on the polar wind, however, is beyond the scope of this paper, and will be discussed in detail in a forthcoming paper.

The density profiles for all the species in our calculation are shown in Fig. 2. First of all, the photoelectron density $n_s$ is small compared with the thermal electron density. Our test-particle approach for photoelectrons is therefore justified. The results also show that $O^+$, being driven by the self-consistent ambipolar field, maintains its dominance over $H^+$ even at high altitudes, in agreement with the Akebono observations. From 5000 to 9000 km, the $O^+$ and $H^+$ densities are to the orders of $10^9$ and a few times $10^7$ m$^{-3}$, consistent with the respective lower limit estimates — $10^8$ and $10^7$ m$^{-3}$ — obtained from the Akebono measurements [*Abe et al.*, 1993b]. Because of the quasi-neutrality condition, the two major particle species, $O^+$ and thermal electron, have very comparable densities.

Figure 3 shows the outflow velocities of the ions. Note that in general the outflow velocities for both ion species increase with altitude. Not only are these ion velocity profiles consistent with the Akebono observations qualitatively, but, moreover, their high-altitude (5000 – 9000 km) values of about 20 km/s for the $H^+$, and 5 km/s for the $O^+$ are also within the measurement limits of the dayside polar wind [*Abe et al.*,

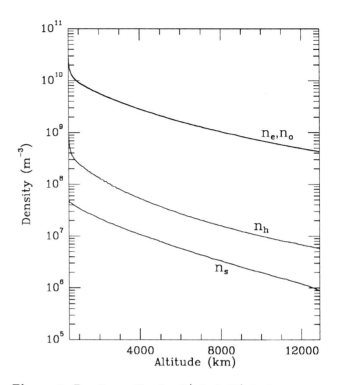

**Figure 2.** Density profiles for $O^+$ ($n_o$), $H^+$ ($n_h$), thermal electron ($n_e$), and photoelectron ($n_s$) in the dayside polar wind solution. Note that $n_e$ and $n_o$ are almost equal because of quasi-neutrality. Their lines virtually overlap in the plot.

1993b]. Of course, when there are special magnetospheric/ionospheric conditions other than the photoelectrons affecting the plasma flow in the polar cap, these results may be altered [*Chandler*, 1995]. Such effects include cross-field drifts, and magnetospheric sources. While the inclusion of magnetospheric sources may be accommodated by this model in its present form without major modification, that of cross-field drifts would require more effort. Hence, such situations are, at present, beyond the scope of this model.

The ion parallel and perpendicular temperatures obtained from the kinetic calculations are shown in Fig. 4. The order of temperatures between 5000 and 9000 km altitude are no higher than $10^4$ °K, which is the upper limit estimated from the Akebono measurements [*Abe et al.*, 1993b]. Note that these temperatures at the lower boundary may be different from their respective $T_*$ values, which characterize the energy spread of the initial distributions. In the calculations, the initial distributions only consist of upgoing ions. These ions may be reflected in the simulations due to Coulomb collisions and the gravitational force. Because the ion distributions in the solution comprise both the upgoing and reflected populations, the ion temperatures at the lower boundary are generally not the same as the initial conditions $T_*$.

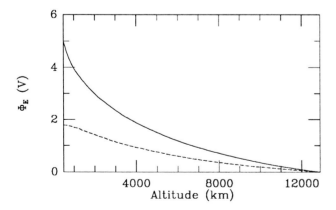

**Figure 1.** Electric potential profiles in the dayside (solid) and nightside (dashed) polar wind solutions. The magnitude of the potential drop on the dayside is comparable to values deduced from observations.

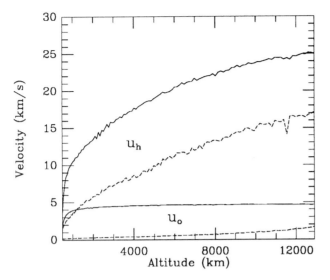

**Figure 3.** Profiles of ion outflow velocities in the dayside (solid) and nightside (dashed) polar wind solutions.

The Mach number, defined for each ion species $j$ as:

$$\mathcal{M}_j = \sqrt{\frac{m_j u_j^2}{2 T_j}}, \qquad (2)$$

increases in general with altitude. In the case of the H$^+$, the increase is mainly due to the upward gradient of the outflow velocity; while in the case of the O$^+$, the downward gradient of the temperatures is primarily responsible for the Mach number increase (see Fig. 3 and 4). From Fig. 5, it is clear that the continuous solution spans the subsonic and supersonic regimes of both ion species. In particular, both ion species attain supersonic flows at high altitudes, in agreement with the Akebono observations [Abe et al., 1993b].

In order to compare our results with Akebono's electron measurements [Yau et al., 1995], we combine all the electrons (thermal and photoelectrons) in the calculation into a single population. The thermal electrons in this total population are assumed to be in a drifting Maxwellian distribution whose density and velocity are obtained from the local quasi-neutrality conditions, and whose temperature is determined from the fluid equations. The resulting Maxwellian and the kinetic results for the photoelectrons combine to give the total distribution of the electron population. This total distribution is then separated into two components: $f_{e,up}$ and $f_{e,down}$, corresponding to upwardly and downwardly moving electrons, respectively. The parallel temperature profiles obtained from these components are shown in Fig. 6. The solution reveals a temperature anisotropy between the upwardly and downwardly moving electrons, i.e. $T_{e,up} > T_{e,down}$ (the $\|$ subscript is omitted for simplicity). In the calculation, this anisotropy is entirely due to the photoelectrons. Because such a temperature anisotropy was observed in the dayside polar wind [Yau et al., 1995] but seems absent on the nightside [Abe et al., 1996], the role of photoelectrons in the model is consistent with the observed polar wind scenario. Despite such a temperature anisotropy introduced by the photoelectrons, the overall temperature profile still seems to be dominated by the thermal electrons. Because temperature is directly related to the collisional frequencies of Coulomb interaction, the temperature dominance by the thermal population justifies our test-particle approach for photoelectrons.

However, if we consider the next higher order of velocity moment, namely the heat flux, the contribution by the photoelectrons is significant. To show that, we shall compare the heat fluxes carried by the thermal and photoelectrons in the total electron distribution. Such heat flux moment should be taken about the mean velocity of the total electron distribution, $u_{e,total}$. Note that the thermal electrons, though assumed to be distributed in a drifting Maxwellian, have a finite heat flux contribution. This thermal electron heat flux arises from the finite difference between $u_{e,total}$ and the drift velocity of the Maxwellian itself ($u_e$). Because the heat flux is proportional to the number density, one would expect that the major contribution to this heat flux comes from the relatively dense portion of the Maxwellian, i.e. the bulk of the thermal electron distribution. In reality, of course, the thermal electron distribution may deviate slightly from a drifting Maxwellian, giving rise to an additional contribution to the heat flux. However, such a heat flux would mainly come from the tail portion of the distribution. We found that this contribution is negligibly small compared with the heat flux carried by the drifting Maxwellian due to the shift in the mean velocity. By comparing the heat flux contribution due to the thermal electron Maxwellian with that carried by

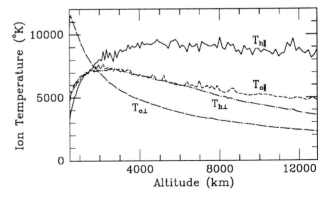

**Figure 4.** Profiles of ion temperatures in the dayside polar wind solution.

the photoelectrons, we find that the heat fluxes carried by the two electron components are in opposite directions — downward for the thermal but upward for the suprathermal, as shown in the top panel of Fig. 7. We also find that the photoelectron heat flux is much larger in magnitude than its thermal counterpart, as shown in the bottom panel of Fig. 7. The two heat flux components combine to give the total electron heat flux, which is shown in Fig. 8 and whose direction is dictated by the photoelectron contribution, *i.e.* upward. The upwardly directed total electron heat flux is consistent with the data from *Yau et al.* [1995] and the calculated results by *Tam et al.* [1995a, 1995b]. The results here further agree with *Tam et al.* [1995a] regarding the directions and the relative magnitudes of the heat fluxes carried by the two electron components. These results, however, might seem different from the observations by *Yau et al.* [1995], who concluded that both the thermal and photoelectron heat fluxes are both upwardly directed, and that the thermal population dominates the overall contribution. The discrepancies between the observations and our results can be explained by the difference in the cutoff energies for the photoelectrons: *Yau et al.* [1995] only took into account photoelectrons of $> 10$ eV while recognizing electrons with energy below that as thermal; this study, however, uses a considerably lower energy cutoff (2 eV) for the photoelectron population. Our results would agree with the observations considering that photoelectrons with energy between 2 and 10 eV make a significant heat flux contribution in the upward direction.

### 3.2. Day-Night Asymmetry

Our polar wind results presented in the previous section compare well with observations by the DE-1, DE-

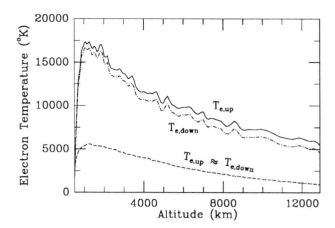

**Figure 6.** Parallel temperatures for upwardly and downwardly moving electrons, the $\parallel$ subscript has been omitted for simplicity. The dayside temperatures are represented by the solid and dot-dashed lines; the nightside temperatures, which are virtually equal for the two electron populations, are represented by the dashed line.

2 and Akebono satellites [*Winningham and Gurgiolo*, 1982; *Abe et al.*, 1993a, 1993b; *Yau et al.*, 1995]. These polar-orbiting satellites have measured or estimated various polar wind quantities under different conditions. Their measurements have indicated that the polar wind ion and electron characteristics, and the corresponding ambipolar electric potential difference, can be very different due to the variability of the polar wind settings.

Indeed, the dynamics of the polar wind depends on a number of conditions. For example, the correlation between the $H^+$ outflow velocity and the amount of magnetic activity, *i.e.* the $K_p$ index, calculated by *Abe et al.* [1993a] from the Akebono data, seems to indicate an increase of the velocity gradient with the $K_p$ index on a time scale of several days.

The polar wind dynamics, evidently, also depends on the magnetic local time, as implied by the day-night asymmetry that has been observed by the Akebono satellite [*Abe et al.*, 1993b, 1996; *Yau et al.*, 1995], and discussed in Section 1. The magnetic local time is related to the solar zenith angle. Indeed, *Winningham and Gurgiolo* [1982] have observed that the field-aligned electric potential difference of the polar wind varied as the solar zenith angle. Because the solar zenith angle is related to the photoionization rate, which directly contributes to the local ionospheric photoelectron density [*Jasperse*, 1981], we expect that variation of this density may be closely related to the day-night transition of the polar wind, despite the relatively small photoelectron population compared with the thermal electrons.

In this study, we characterize the night-time polar wind conditions with the absence of photoelectrons. We will examine the day-night asymmetry of the polar wind

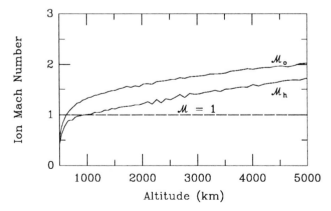

**Figure 5.** Ion Mach numbers in the dayside polar wind solution, shown for the region spanning the subsonic-to-supersonic transition. The Mach numbers increase monotonically at higher altitudes.

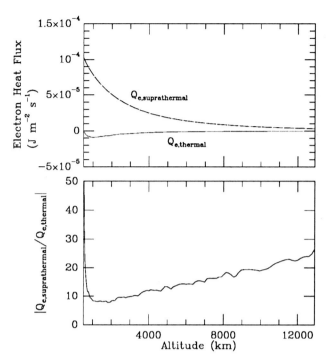

**Figure 7.** Top panel: heat fluxes carried by the thermal electrons ($Q_{e,thermal}$) and photoelectrons ($Q_{e,suprathermal}$). Note that the heat flux contribution by the thermal electrons is in the downward direction (negative sign) while that by the photoelectrons is upwardly directed (positive sign). Bottom panel: the ratio of the magnitudes of the two heat fluxes.

by comparing results of the dayside and nightside outflows. We have presented in the previous section a polar wind solution that agrees with the observed daytime outflow characteristics. To examine the night-time polar outflow, we have generated another admissible solution with the same boundary conditions, except without the presence of photoelectrons, i.e. $n_s = 0$.

The self-consistent electric potential profile for the night-time polar outflow is shown in Fig. 1. The potential drop on the nightside is only about 2 V, considerably smaller compared with the dayside. The larger potential drop in the daytime polar outflow is due to the anisotropic kinetic nature of photoelectrons. Because photoelectrons are suprathermal particles, they skew the electron distribution, giving rise to a larger upward electron energy flux (and heat flux), which in turn increases the ambipolar potential difference and the resulting electric field. For this reason, we expect the ambipolar electric field to be larger if the initial low-altitude photoelectron spectrum is more energetic or if photoelectrons have a higher density relative to the thermals. On the other hand, in the night-time polar outflow, where the heat flux contribution by the photoelectrons is absent, the overall electron heat flux is carried by the thermal population. In our calculations, we have assumed the thermal electron distribution to be a drifting Maxwellian. Thus, without the presence of photoelectrons, the thermal electron population by itself does not carry any heat flux. However, as discussed in Section 3.1 and shown in Fig. 7, the heat flux contribution by the thermal electrons in reality would be negligibly small compared with the corresponding suprathermal contribution. Therefore, for comparison purpose, the magnitude of the ambipolar potential difference in the night-time polar outflow is properly reflected by our results on the nightside.

The absence of photoelectrons in our nightside calculations also leads to the absence of a noticeable electron anisotropy, in contrast to the dayside situation. The temperatures for the upward- and downward-moving electron populations in our night-time polar outflow solution are shown in Fig. 6. The temperatures for the two populations only differ by about 0.2% at the altitudes where the satellite measurements were made (900 – 1700 km) [*Abe et al.*, 1996]. The dayside and nightside electron temperatures together thus demonstrate the consistency between our photoelectron scenario and the day-night asymmetry observed by the Akebono satellite [*Yau et al.*, 1995; *Abe et al.*, 1996].

We have also found that the overall electron temperatures are generally higher in the presence of photoelectrons. The higher temperature is due to collisional heating of the thermal electron population by the photoelectrons. Notice from Fig. 6 that the electron temperature has an upward gradient at low altitudes both on the dayside and nightside. The upward temperature gradients are related to collisional heating of the thermal electron population. On the nightside, the thermal electrons are heated through their collisions with the ions. However, on the dayside, energy is transferred to the thermal electrons also through collisions

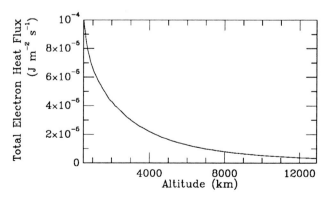

**Figure 8.** Total heat flux profile for the total electron population in the dayside polar wind solution.

with the photoelectrons. In fact, this heating is more efficient because energy from the photoelectrons is preferentially transferred to the more energetic part of the thermal electron population, thereby increasing the energy spread more effectively.

Because the absence of photoelectrons in the nightside polar outflow, as we have shown, leads to a significantly different ambipolar electric field from the dayside, we expect the ion dynamics to be also affected by this day-night difference. For example, as shown in Table 2, the ion number fluxes is much larger on the dayside than on the nightside. This correlation is mainly due to the magnitude of the self-consistent ambipolar electric field, which is larger in the presence of photoelectrons: the larger the ambipolar field, the greater the number of escaping ions contributing to the number fluxes at high altitudes, because of the acceleration by the field.

As one would expect, the magnitude of the self-consistent ambipolar electric field also has a significant impact on the ion outflow velocities. The $O^+$ and $H^+$ velocities in our two polar wind solutions are shown in Fig. 3. It is clear that both ion species have considerably smaller outflow velocities in the nightside results, in agreement with the observed day-night asymmetry in ion velocities [Abe et al., 1993b].

## 4. CONCLUSION

Our work on the polar wind is motivated by the increasing experimental evidence that photoelectrons may affect the dynamics of the polar wind. In particular, the day-night asymmetry in the ion and electron features suggests that photoelectrons may be the dominant effect in the polar outflow.

In order to examine the anisotropic kinetic effects of photoelectrons as a source of the day-night asymmetry in the polar outflow, we have used a hybrid kinetic-fluid code to generate two polar wind solutions, one corresponding to the daytime outflow, and the other for the nightside. The dayside solution is in quantitative agreement with observations in various aspects. For example, the electric potential drop is consistent with the experimentally deduced value. The ion properties such as densities, outflow velocities and temperatures are also within experimental limits. An anisotropy between upwardly and downwardly going electrons, and an upwardly directed electron heat flux, both observed in the sunlit polar wind, also appear in our calculations for the dayside.

By comparing our dayside and nightside results, we have demonstrated the consistency between our photoelectron scenario and the observations. In particular,

**Table 2.** The ion number fluxes for the dayside and nightside polar wind solutions.

| Number fluxes at 10000 km ($m^{-2}s^{-1}$) | | |
|---|---|---|
| | $H^+$ | $O^+$ |
| Night | $9.49 \times 10^{10}$ | $3.25 \times 10^{11}$ |
| Day | $2.35 \times 10^{11}$ | $3.20 \times 10^{12}$ |

we have shown that the presence (absence) of photoelectrons on the dayside (nightside) may explain the observed day-night asymmetry in the polar outflow. For example, the anisotropy between the upward- and downward-moving electrons, which is observed in the sunlit polar outflow but seems absent on the nightside, only appears in our dayside polar outflow results. The ion velocities in our results are remarkably larger on the dayside, in agreement with satellite measurements.

*Acknowledgments.* The authors would like to thank J. D. Winningham, Andrew Yau, and Takumi Abe for many useful discussions with regard to the experimental data, John Retterer for participating with them in the initial stage of the investigation, and Supriya Ganguli for constant collaboration. This work is supported by NSF Grant ATM-9634599, AFOSR Grant F49620-96-1-0340, NASA Grant NAG5-2255, and AF Contract F19628-91-K-0043.

## REFERENCES

Abe, T., B. A. Whalen, A. W. Yau, S. Watanabe, E. Sagawa, and K. I. Oyama, Altitude profile of the polar wind velocity and its relationship to ionospheric conditions, *Geophys. Res. Lett.*, **20**, 2825, 1993a.

Abe, T., B. A. Whalen, A. W. Yau, R. E. Horita, S. Watanabe, and E. Sagawa, EXOS D (Akebono) suprathermal mass spectrometer observations of the polar wind, *J. Geophys. Res.*, **98**, 11191, 1993b.

Abe, T., B. A. Whalen, A. W. Yau, E. Sagawa, and S. Watanabe, Akebono observations of thermal ion outflow and electron temperature in the polar wind region, in *Physics of Space Plasmas (1995)*, edited by T. Chang, and J. R. Jasperse, no. 14, p. 3, Cambridge, MA. MIT Center for Theoretical Geo/Cosmo Plasma Physics, 1996.

Axford, W. I., The polar wind and the terrestrial helium budget, *J. Geophys. Res.*, **73**, 6855, 1968.

Banks, P. M., and T. E. Holzer, The polar wind, *J. Geophys. Res.*, **73**, 6846, 1968.

Barakat, A. R., and R. W. Schunk, Effect of hot electrons on the polar wind, *J. Geophys. Res.*, **89**, 9771, 1984.

Blelly, P. L., and R. W. Schunk, A comparative study of the time-dependent standard 8-, 13- and 16-moment transport formulations of the polar wind, *Ann. Geophys.*, **11**, 443, 1993.

Chandler, M. O., Observations of downward moving $O^+$ in the polar topside ionosphere, *J. Geophys. Res.*, **100**, 5795, 1995.

Cladis, J. B., Parallel acceleration and transport of ions from polar ionosphere to plasma sheet, *Geophys. Res. Lett.*, **13**, 893, 1986.

Ganguli, S. B., The polar wind, *Rev. Geophys.*, **34**, 311, 1996.

Heelis, R. A., G. J. Bailey, R. Sellek, R. J. Moffett, and B. Jenkins, Field-aligned drifts in subauroral ion drift events, *J. Geophys. Res.*, **98**, 21493, 1993.

Ho, C. W., J. L. Horwitz, N. Singh, G. R. Wilson, and T. E. Moore, Effects of magnetospheric electrons on polar plasma outflow: a semikinetic model, *J. Geophys. Res.*, **97**, 8425, 1992.

Horwitz, J. L., C. W. Ho, H. D. Scarbro, G. R. Wilson, and T. E. Moore, Centrifugal acceleration of the polar wind, *J. Geophys. Res.*, **99**, 15051, 1994.

Jasperse, J. R., The photoelectron distribution function in the terrestrial ionosphere, in *Physics of Space Plasmas*, edited by T. S. Chang, B. Coppi, and J. R. Jasperse, no. 4 in SPI Conference Proceedings and Reprint Series, p. 53, Cambridge, MA. Scientific Publishers, Inc., 1981.

Khazanov, G. V., M. W. Liemohn, T. I. Gombosi, and A. F. Nagy, Non-steady-state transport of superthermal electrons in the plasmasphere, *Geophys. Res. Lett.*, **20**, 2821, 1993.

Lee, J. S., J. P. Doering, T. A. Potemra, and L. H. Brace, Measurements of the ambient photoelectron spectrum from Atmosphere Explorer: II. AE-E measurements from 300 to 1000 km during solar minimum conditions, *Planet. Space Sci.*, **28**, 973, 1980.

Lemaire, J., Effect of escaping photoelectrons in a polar exospheric model, *Space Res.*, **12**, 1413, 1972.

Olbert, S., Role of thermal conduction in the acceleration of the solar wind, *NASA Conf. Publ.*, p. 149, 1982.

Schunk, R. W., and D. S. Watkins, Electron temperature anisotropy in the polar wind, *J. Geophys. Res.*, **86**, 91, 1981.

Schunk, R. W., and D. S. Watkins, Proton temperature anisotropy in the polar wind, *J. Geophys. Res.*, **87**, 171, 1982.

Scudder, J. D., and S. Olbert, A theory of local and global processes which affect solar wind electrons: 1. the origin of typical 1 AU velocity distribution functions — steady state theory, *J. Geophys. Res.*, **84**, 2755, 1979.

Shelley, E. G., W. K. Peterson, A. G. Ghielmetti, and J. Geiss, The polar ionosphere as a source of energetic magnetospheric plasma, *Geophys. Res. Lett.*, **9**, 941, 1982.

Tam, S. W. Y., F. Yasseen, T. Chang, S. B. Ganguli, and J. M. Retterer, Anisotropic kinetic effects of photoelectrons on polar wind transport, in *Cross-Scale Coupling in Space Plasmas*, edited by J. L. Horwitz, N. Singh, and J. L. Burch, no. 93 in Geophysical Monograph, p. 133, Washington D.C. American Geophysical Union, 1995a.

Tam, S. W. Y., F. Yasseen, T. Chang, and S. B. Ganguli, Self-consistent kinetic photoelectron effects on the polar wind, *Geophys. Res. Lett.*, **22**, 2107, 1995b.

Waite, Jr., J. H., T. Nagai, J. F. E. Johnson, C. R. Chappell, J. L. Burch, T. L. Killeen, P. B. Hays, G. R. Carignan, W. K. Peterson, and E. G. Shelley, Escape of suprathermal $O^+$ ions in the polar cap, *J. Geophys. Res.*, **90**, 1619, 1985.

Winningham, J. D., and W. J. Heikkila, Polar cap auroral electron fluxes observed with Isis 1, *J. Geophys. Res.*, **79**, 949, 1974.

Winningham, J. D., and C. Gurgiolo, DE-2 photoelectron measurements consistent with a large scale parallel electric field over the polar cap, *Geophys. Res. Lett.*, **9**, 977, 1982.

Yasseen, F., J. M. Retterer, T. Chang, and J. D. Winningham, Monte-Carlo modeling of polar wind photoelectron distributions with anomalous heat flux, *Geophys. Res. Lett.*, **16**, 1023, 1989.

Yau, A. W., B. A. Whalen, W. K. Peterson, and E. G. Shelley, Distribution of upflowing ionospheric ions in the high-altitude polar cap and auroral ionosphere, *J. Geophys. Res.*, **89**, 5507, 1984.

Yau, A. W., B. A. Whalen, T. Abe, T. Mukai, K. I. Oyama, and T. Chang, Akebono observations of electron temperature anisotropy in the polar wind, *J. Geophys. Res.*, **100**, 17451, 1995.

---

Sunny W. Y. Tam and Tom Chang, Center for Space Research, Massachusetts Institute of Technology, Cambridge, MA 02139

Fareed Yasseen, Climate Change Secretariat, P.O. Box 260124, D-53153 Bonn, Germany

# POLAR Observations of Properties of H$^+$ and O$^+$ Conics in the Cusp Near ~5300 km Altitude

M. Hirahara,[1,2,3,4] J. L. Horwitz,[1] T. E. Moore,[2,5] M. O. Chandler,[2] B. L. Giles,[2]

P. D. Craven,[2] and C. L. Pollock,[6]

Observations by the thermal ion dynamics experiment (TIDE) on POLAR are used to explore properties of low-energy ionospheric ion conical distributions at ~5300 km altitude over the southern cusp under different interplanetary magnetic field (IMF) conditions with negative and positive $B_z$ components. The properties are summarized as follows: (1) At the edge upstream of the convection in the cusp, the energy of outflowing ions abruptly increased from a few eV to ~100 eV; (2) The angular distributions also abruptly changed from rammed $<\sim 5$ eV polar wind distributions to ~10-100 eV conics and the cone angles are wider for O$^+$ than for H$^+$; (3) The uppermost energy of the O$^+$ conics was larger than that of H$^+$, while the O$^+$ flux was lower than that of H$^+$; (4) The cone angles for both light and heavy ion conics were largest in the upstream region of the convection and gradually decreased in the convection direction as well as the conic energies; (5) The energization region could distribute over $>\sim 1°$ in the latitudinal direction; (6) These conic signatures gradually gave way again to polar wind components further downstream of the cusp; (7) The UFI beams and conics were sometimes observed alternating, particularly for H$^+$, and (8) The distinct ion conic bursts often occurred multiple times, especially downstream of the cusp.

---

[1] Center of Space Plasma, Aeronomy, and Astrophysics Research, University of Alabama in Huntsville

[2] Space Sciences Laboratory, NASA Marshall Space Flight Center, Huntsville, Alabama.

[3] Department of Earth and Planetary Physics, Faculty of Science, University of Tokyo, Bunkyo-ku, Tokyo, Japan.

[4] Now at Department of Physics, College of Science, Rikkyo University, Nishi-Ikebukuro, Toshima-ku, Tokyo, Japan.

[5] Now at Interplanetary Physics Branch, NASA Goddard Space Flight Center, Greenbelt, Maryland.

[6] Instrumentation and Space Research Division, Southwest Research Institute, San Antonio, Texas

Geospace Mass and Energy Flow: Results From the International Solar-Terrestrial Physics Program
Geophysical Monograph 104
Copyright 1998 by the American Geophysical Union

## 1. INTRODUCTION

The cusp/cleft region is thought the most important source of ionospheric ions transported into the polar magnetosphere and the lobe/mantle regions of the magnetotail [e.g., *Horwitz and Lockwood*, 1985; *Lockwood et al.*, 1985a, b]. Suprathermal ionospheric ions outflowing from the polar cap and the cusp/cleft regions were suggested as an important contributor to the plasma sheet content [*Chappell*, 1987, and references therein].

In the dayside auroral region, ion conics are a typical form of the upward flowing ions (UFI). The energization process acting on the thermal ionospheric ions to produce the suprathermal outflows was considered by *Moore et al.* [1985] based on Dynamics Explorer measurements, who concluded that the low-altitude transverse ion acceleration (TIA) is important for the ionospheric ion supply to the magnetosphere. *André et al.* [1990] used simulations and DE

1 observations to suggest that the heating region could be narrow in the latitudinal direction but extended in altitude. *Whalen et al.* [1991] also showed abrupt enhancement of the perpendicular heating in the cusp/cleft at 5000-7000 km altitudes. *Miyake et al.* [1993] proposed that the conics are continuously energized over a wide altitudinal range during the upward transport from the topside ionosphere.

*Knudsen et al.* [1994] recently reported Akebono observations of the $H^+$ and $O^+$ conics in the low-altitude cusp. They observed that the cone angles of $O^+$ are always larger than those of $H^+$. They concluded that a latitudinally narrow but altitudinally wide ("wall" like) region upstream of the cusp is the location of the predominant process energizing the ionospheric ions perpendicular to the magnetic field. They also proposed that the velocity filter effect due to the $\mathbf{E} \times \mathbf{B}$ drift after the heating could produce the overall cone angle variations in the convection direction and the mass dependence of the cone angles at a given position/time.

More recently, the suborbital mission, sounding of cleft ion fountain energization region (SCIFER [*Kintner et al.*, 1996]), provided fine-scale observations of the ionospheric ion heating. *Arnoldy et al.* [1996] reported that several narrow heating regions were repeatedly observed at altitudes of ~1000-1500 km near 10 MLT. The ionospheric ions were significantly accelerated to energies of 500 eV in the perpendicular direction.

*Woch and Lundin* [1992] presented typical energy dispersions of the solar wind proton component penetrating into the low-altitude cusp to investigate interplanetary magnetic field (IMF) effects on the solar wind injection into the ionospheric cusp under different IMF conditions. It was evident that the solar wind injection had a significant influence on the TIAs, which suggests the importance of the relative location of the upflowing ion signatures to the solar wind injection events.

The thermal ion dynamics experiment (TIDE) onboard the POLAR satellite is able to measure fully three-dimensional velocity distribution functions of thermal and suprathermal ion composition typical in the ionosphere and the magnetosphere with high time and angular resolutions. The fine energy resolution and wide energy range sampled by TIDE enable us to investigate precise distributions not only of the upflowing ionospheric ions but also of the solar wind component. We devote this paper to presenting recent observational results concerning energization processes and transport of the ionospheric ions observed in and near the cusp by TIDE onboard POLAR.

## 2. DATA SOURCES

We present the TIDE data obtained in the southern hemisphere near perigee (~5300 km). TIDE has seven identical sensors, each of which consists of two portions: an energy analyzer using an electrostatic mirror and a retarding potential analyzer, and a mass spectrometer using a time-of-flight technique [*Moore et al.*, 1995]. The combination of the wide field of view (157.5°) and the satellite spin motion realizes the measurement of fully three-dimensional velocity distributions of major ion species ($H^+$, $He^{2+}$, $He^+$, $O^+$, and some molecular ions) with high angular resolution (11.25°×22.5°). The energy range of the data shown here is 0.1-300 eV/q divided into 16 steps on a logarithmic scale. The time resolution is 6 s, the same as the spin period.

The key parameter data obtained by the magnetic field instrument (MFI [*Lepping et al.*, 1995]) and the solar wind experiment (SWE [*Ogilvie et al.*, 1995]) onboard the Wind satellite are briefly described to indicate the IMF conditions before and during the POLAR cusp observations.

## 3. OBSERVATIONS

We present two observations of the upflowing ions in cusp under different IMF conditions in Plate 1. The first observation was for southward IMF conditions, and the second for northward IMF conditions, according to observations by the Wind satellite. At southern perigee, POLAR always crossed the cusp from higher to lower latitudes during April 1996.

### 3.1. Event 1: Southward IMF

POLAR crossed the low-altitude cusp during 0429-0432 UT on April 12, 1996. If the approximately 20-minute time lag of the solar wind arrival between the Earth and Wind is taken into account, the $z$ component of IMF was southward (~−2.5 nT) on the average, although the $x$ and $y$ components were also significant (+6 and −5 nT, respectively).

Plate 1a shows PES (polar angle in abscissa, energy in ordinate, spin angle in color code) chromograms of $H^+$ and $O^+$. The abscissa of the overall plots is universal time (UT in hours and minutes), and the ordinate is energy (in electron volts: eV) on a logarithmic scale. Each bin separated by short tick marks shows a polar angle-energy spectrogram for a spin. The spin angles in which the largest ion fluxes were detected are indicated by a color code wheel, as shown at the middle top. The fluxes are expressed by brightness in the left-side portion of the color code wheel.

An intense solar wind proton injection was seen in the high-energy range (>50 eV/q), and the energy dispersion is consistent with the antisunward convection in the cusp under southward IMF condition [e.g., *Woch and Lundin*, 1992]. Namely, the proton energy was highest in the upstream region of the cusp because of a velocity filter effect due to $\mathbf{E} \times \mathbf{B}$ drift. The fluxes colored by white suggest no dependence on the polar angle, which means that the solar wind component was isotropic. A vertical purple line embedded in each bin indicates the loss cone position in the upward direction. Sudden decreases of fluxes seen at the middle-energy range of the energy dispersion were due to the instrumental sensitivity variations. Unfortunately, the TIDE data for a spin of 0431:41-0431:47 UT were not available, as displayed by gray areas.

At energy ranges lower than those of the solar wind component, $H^+$ and $O^+$ fluxes are seen. There are several important characteristics seen in these spectrograms: The energies of the $O^+$ conics were higher than those of the $H^+$ conics, and the energy variations for both ions were largest toward the injection region for the solar wind component, especially

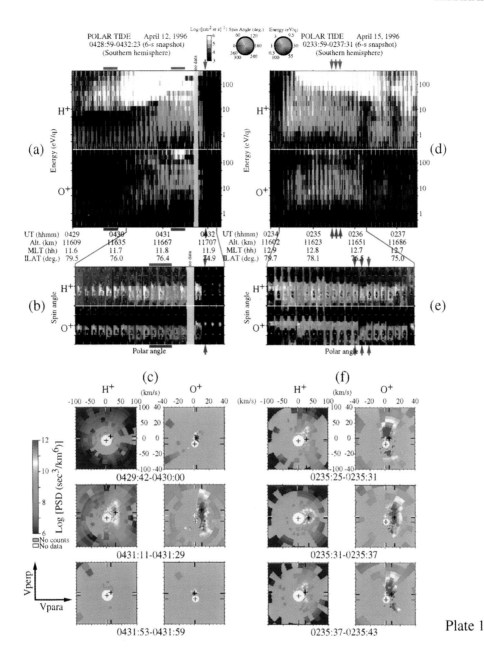

**Plate 1.** From the top, PES and PSE chromograms and velocity distribution functions for low-energy $H^+$ and $O^+$ ions observed by TIDE in the southern low-altitude cusp are shown for the interval of April 12 (a-c) and April 15 (d-f) in 1996. PES chromograms (a and d) are energy-time ($E$-$t$) diagrams consisting of energy-polar angle bins, and PSE chromograms (b and e) show variations of the angular distributions. The scales of the ion energies and fluxes are logarithmic, as indicated by ordinates, or color code and the brightness level at the middle top. In the two types of chromograms, universal time (UT, in hours and minutes), geocentric altitude (km, in kilometers), magnetic local time (MLT, in hours), and invariant latitude (ILAT, in degrees) of POLAR are shown. The red horizontal lines or vertical arrows along the abscissas indicate the intervals of the velocity distribution functions (c and f) displayed below the chromograms. In each of the distribution contour plots, the rightward and upward directions in each subpanel correspond to field-aligned upward direction and perpendicular direction closest to the satellite ram motion, respectively. The origins of the Earth and satellite frames are shown by white and thin black crosses and tick marks, respectively, while the bulk center of the low-energy upflowing ions from moment result is located at the thick cross and tick mark. We assume that the satellite potential was 1.5 volts throughout.

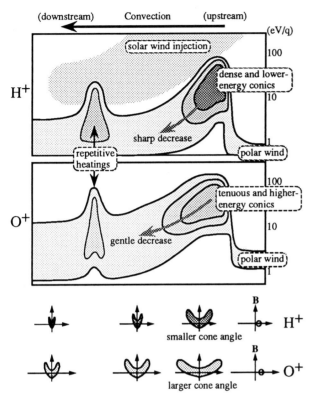

**Figure 1.** Schematic pictures summarizing the general properties of $H^+$ and $O^+$ seen in Events 1 and 2.

before 0431 UT. In the upstream portion of the cusp (that is, near the equatorward edge during 0431-0432 UT), the $O^+$ energy was particularly steady for about 30 s within the energy range 5-100 eV/q. On the other hand, the energies of the $H^+$ conics varied more rapidly, although this is somewhat indefinite since the $H^+$ fluxes partially overlapped the solar wind components at higher energies. Also, in the downstream portion of the cusp before 0431 UT, the energy variation of the $O^+$ conics was generally gradual as compared with that of $H^+$. Throughout the cusp observation, the flux of the $H^+$ conics was equal to or frequently larger than that of $O^+$. The important properties of the outflowing ions are schematically shown in Figure 1.

Plate 1b shows two PSE (polar angle in abscissa, spin angle in ordinate, energy in color code) chromograms of $H^+$ and $O^+$, displaying the angular distributions of the largest fluxes of ions among all energy steps. The energy can be roughly estimated from the color code wheel shown at the top of Plate 1; the lowest energy (<0.3 eV/q) is indicated by red, the highest (>100 eV/q) by blue, and the intermediate ($\sim 10$ eV/q) by green. During the interval when the energy of the $O^+$ conics did not change significantly, there were no significant variations in the pitch angle distributions either (see the interval marked by red underlines at 0431:20 UT). This may imply that the satellite crossed above a latitudinally-wide $O^+$ heating region. Because the satellite velocity perpendicular to the magnetic field direction was $\sim 7.4$ km/s, the roughly-estimated width of the $O^+$ heating region would be at least 133 km.

In the mantle downstream of the cusp during 0430-0430:30 UT, the $O^+$ energy was still enhanced, in contrast to that of $H^+$. It is likely that the $O^+$ conics observed in the mantle were not locally energized but were transported from the cusp proton injection site. On the other hand, the $H^+$ energy rapidly decreased with increasing latitude. Because the $H^+$ velocity was higher than that of $O^+$, the $H^+$ conics might escape to high altitudes along close-by field lines before being convected very far poleward from the energization region.

As seen in Plate 1a, poleward in the cusp, TIDE observed a large amount of the $H^+$ and $O^+$ polar wind, although they might contain some low-energy residual of the conics at higher latitudes due to the convection effect [*Gurgiolo and Burch*, 1982] or some small heating processes was still active in the polar cap near the cusp. The pitch angle distributions evolved from the complete conic signatures through asymmetric distributions enhanced in the satellite-ram direction as reported by *Giles et al.* [1994] to the polar wind components. The pitch angle evolution from the polar wind to the conics was more abrupt at the equatorward edge of the cusp than in the poleward region, as observed by other satellites and rockets [e.g., *André et al.*, 1990; *Knudsen et al.*, 1994; *Arnoldy et al.*, 1996].

In Plate 1c, we present velocity distributions of upflowing ions in the plane containing the local magnetic field direction during three intervals (from the top, mantle, pure cusp, and at lower latitudes than the cusp), as indicated by red underlines or arrows in Plates 1a-b. As seen in the middle plot of Plate 1c, the cone angle of $O^+$ conics was larger than that of $H^+$. Figure 2 shows an example of the pitch angle distributions of $H^+$ and $O^+$ during this interval. The $O^+$ distribution had a peak at $\sim 55°$, whereas the peak for $H^+$ was at $\sim 40°$. The wider cone angle for $O^+$ is also evident from Plate 1b. That is, the circular distributions in the PSE chromograms and the holes at the centers of the distributions were larger for $O^+$ than for $H^+$.

### 3.2. Event 2: Northward IMF

During the event 2 on April 15, 1996, the IMF was strongly northward ($B_z \sim +6$ nT). The energy dispersion of the solar wind protons observed in the energy range more than 50 eV/q was consistent with an equatorward convection for such northward IMF conditions, as seen in Plate 1d. The characteristic energy of the dispersion decreased gradually and monotonically with decreasing latitudes during 0234:30-0236:30 UT.

The characteristic $O^+$ energy was nearly constant during the interval 0234:40-0236:10 UT, and the pitch angle distributions of $O^+$ changed from nearly transverse features to the upflowing conic features at $\sim 0235$ UT. The energy variation of the $H^+$ conics was more abrupt than that of $O^+$. It is also noted that the characteristic $O^+$ energy was slightly higher that that of $H^+$ during 0235-0236 UT. These results imply that the energization rates and/or the influence of the transport at and below the satellite altitude are different between

$H^+$ and $O^+$. For example, it is likely that the $O^+$ ions could be energized more efficiently since the frequencies of plasma waves accelerating the ionospheric ions transversely would be closer to the gyro frequency of $O^+$ than for $H^+$. Inferring from the gradual energy variation of $O^+$, the energization region of $O^+$ may be wider in altitude and latitude than that of $H^+$.

At 0235:34 UT, the pitch angle distributions of the $H^+$ conics abruptly changed from conical to field-aligned, and after several seconds the distribution showed wide ($\sim 60°$) cone angle again, as displayed in Plate 1e (see the PSE chromograms indicated by three red arrows). Plate 1f shows the velocity distributions during the three satellite spins. As seen in the middle subpanel of $H^+$, the $H^+$ distribution consists of an almost field-aligned component. The characteristic energy was several eV/q, and there was no large difference between the energies for the conics and the beams. After one satellite spin (6 s) the field-aligned component evolved back to the conical form. *Liu et al.* [1994] reported that the alternating occurrence of low-energy $H^+$ beam and conic features on the nightside was related to upward field-aligned current density. Also in the cusp, the field-aligned current may have a significant influence on the pitch angle distribution of the upflowing $H^+$ ions. On the other hand, the pitch angle variation of the $O^+$ conics was not as clear as that in $H^+$, as seen in the successive $O^+$ plots.

Also observed in this event were the repeated TIAs at $\sim$0236:40 and $\sim$0237:15 UT, as shown schematically in Figure 1. The SCIFER results also showed that there were multiple heating walls in the cleft region [*Arnoldy et al.*, 1996]. These results imply that the structure of the heating sources and spatial distributions is more complicated than perhaps previously indicated [*Knudsen et al.*, 1994; *Watanabe et al.*, 1995].

## 4. DISCUSSION

The TIDE observations presented here suggest that the energization of the ionospheric ions occurs presumably along the field lines where highest-energy solar protons are injected. Moreover, the heating regions were observed to be quite structured. It is also noteworthy that the cone angles of the $O^+$ conics were wider than those of $H^+$ for most of the intervals. *Knudsen et al.* [1994] proposed a heating wall model to explain the pitch angle distributions and variations. On the other hand, the latitudinal energy variations imply that the heating rates for $H^+$ and $O^+$ ions could be different and dependent on altitude, which may be associated with the differences of the energies, pitch angle distributions, and their latitudinal variations. We consider that the heating region of the ionospheric ions may be distributed more widely in the altitudinal direction in the upstream region of the cusp than in the downstream region. It is also possible that the gravitational effect would be more significant for the heavy ion ($O^+$) conics.

With respect to the fluxes of UFI conics, the TIDE data showed tendencies that are not inconsistent with the data from DE 1 and Akebono. On the other hand, the energy

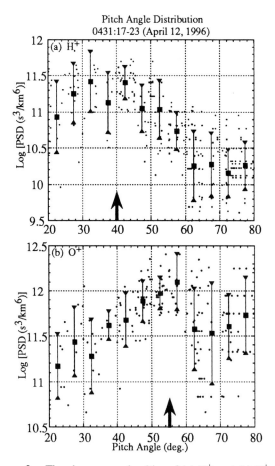

**Figure 2.** The phase space densities of (a) $H^+$ and (b) $O^+$ from multiple channels and spin sectors of TIDE versus the corresponding pitch angles calculated in the plasma convection frame moving perpendicular to the magnetic field.

of $O^+$ is larger than that of $H^+$ in the TIDE observations. In Freja results from lower altitudes (<2000 km), the peak energy of $O^+$ was comparable to or slightly smaller than that of $H^+$, while the flux of $O^+$ was larger than that of $H^+$ [e.g., *Norqvist et al.*, 1996].

The differences of the energies and fluxes between the TIDE data and these Freja results may imply that the $O^+$ ions could be continuously accelerated further at higher altitudes than for $H^+$. This scenario would lead to wider cone angles for the $O^+$ conics than for $H^+$. It is also possible that a cool component of $H^+$ ions at altitudes higher than the $O^+$-dominated region would continuously contribute to the $H^+$ conic population and that the heating rate for $H^+$ would be smaller than for $O^+$. This would explain the higher $H^+$ flux than that of $O^+$ above 5000 km. Some observations by Akebono and DE 1 also implied that the $H^+$ fluxes were higher than those for $O^+$ [*Peterson et al.*, 1993; *Watanabe et al.*, 1995].

Solar activity and/or seasonal variations may also be involved in the explaining the differences in these results, as discussed by *Yau et al.* [1984] and *Moore et al.* [1996]. The indices of sunspot number and $F10.7$ show that the solar activity decreased during the periods of the Freja observations in February and March 1994 and the POLAR observations.

In the mantle downstream of the direct cusp proton injection, the low-velocity $O^+$ ions could be transported from the active heating region upstream in the cusp, and so, the energy decrease for the $O^+$ fluxes would be more gradual than for $H^+$, because of the velocity filter effect. Within the upstream direct injection site, both ion species would be energized over $>\sim 1°$ of latitude because the energies and the pitch angle distributions of $H^+$ and $O^+$ showed no noticeable variations for several spins, as described in the observation section.

The heating regions were localized to the cusp region, but the latitudinal width estimated in our observations could be larger than those proposed by *Knudsen et al.* [1994] and *Watanabe et al.* [1995].

## 5. SUMMARY

The upflowing ionospheric ion signatures consisting of $H^+$ and $O^+$ observed by TIDE in the cusp at $\sim 5300$ km show several common features for the different convection patterns: Sunward convection under northward IMF conditions and antisunward under southward IMF. UFI conics under sunward convection conditions have been frequently reported on the basis of observations from DE 1, Viking, and Akebono [e.g., *André et al.*, 1988; *Watanabe et al.*, 1995]. While the two cases shown here were obtained under different IMF conditions, the variations of the conic signatures were both consistent with the directions of the actual plasma convection.

The general properties seen in the POLAR observations are as follows. Enhancements of the energies and cone angles of the conics were abruptly observed at the edge most upstream of the convection in the cusp where the highest-energy magnetosheath protons were precipitating into the ionosphere, and the conic energies were the highest there. In other words, the polar wind component was abruptly energized and transformed into the conics at the upstream edge of the cusp, and the widest cone angle distributions were observed. The cone angles of the $O^+$ conics were wider than those of $H^+$ for most of these intervals. Also, the cone angles for these conics gradually decreased in the convection direction. The energies of the conics decreased as well as the energy of the high-energy proton precipitation of magnetosheath origin. The latitudinal decrease of the $O^+$ characteristic energy was more gradual than that for $H^+$. The low-energy conic distributions in the downstream region gradually returned to the polar wind characteristics. The energy enhancements of the conics were observed repetitively during a cusp crossing. These results are consistent with those previously reported by, for instance, *Knudsen et al.* [1994], *Watanabe et al.* [1995], and *Arnoldy et al.* [1996].

Finally, we summarize our new results from the POLAR observations.

1. The $O^+$ conics were more energetic than the $H^+$ conics, while the flux of $O^+$ was lower than those of $H^+$.

2. The energies of the conics remained high for $>130$ km latitudinal distance.

3. The cone angles, particularly for the $H^+$ conics, sometimes suddenly decreased as the distribution became nearly field-aligned at the cusp injection site, and then soon recovered to conics downstream within $0.2°$ of latitude. This corresponding folding of the $O^+$ conics was more gradual than that of $H^+$.

*Acknowledgments.* R. P. Lepping and K. W. Ogilvie provided us with their Wind data to monitor the IMF and solar wind plasma conditions, respectively. One of the authors, M. H., thanks S. Watanabe and D. J. Knudsen for fruitful discussions on the heating wall of the conics. We acknowledge that the works regarding the instrument developments and data processing of TIDE were supported under NASA grants NAG8-114, NAG8-693, NASA contract NAS8-38189, and SwRI Internal Research Project 15-9455.

## REFERENCES

André, M., H. Koskinen, L. Matson, and R. Erlandson, Local transverse ion energization in and near the polar cusp, *Geophys. Res. Lett.*, *15*, 107-110, 1988.

André, M., G. B. Crew, W. K. Peterson, A. M. Persoon, C. J. Pollock, and M. J. Engebretson, Ion heating by broadband low-frequency waves in the cusp/cleft, *J. Geophys. Res.*, *95*, 20,809-20,823, 1990.

Arnoldy, R. L., K. A. Lynch, P. M. Kintner, J. Bonnell, T. Moore, and C. J. Pollock, SCIFER - Structure of the cleft ion fountain at 1400 km altitude, *Geophys. Res. Lett.*, *23*, 1869-1872, 1996.

Chappell, C. R., The terrestrial plasma source: A new perspective in solar-terrestrial processes from Dynamics Explorer, *Rev. Geophys.*, *26*, 229-248, 1988.

Giles, B. L., C. R. Chappell, T. E. Moore, R. H. Comfort, and J. H. Waite Jr., Statistical survey of pitch angle distributions in core (0-50 eV) ions from Dynamics Explorer 1: Outflow in the auroral zone, polar cap, and cusp, *J. Geophys. Res.*, *99*, 17,483-17,501, 1994.

Gurgiolo, C., and J. L. Burch, DE-1 observations of the polar wind - A heated and an unheated component, *Geophys. Res. Lett.*, *9*, 945-948, 1982.

Horwitz, J. L., and M. Lockwood, The cleft ion fountain: A two-dimensional kinetic model, *J. Geophys. Res.*, *90*, 9749-9762, 1985.

Kintner, P. M., et al., The SCIFER experiment, *Geophys. Res. Lett.*, *23*, 1865-1868, 1996.

Knudsen, D. J., B. A. Whalen, T. Abe, and A. Yau, Temporal evolution and spatial dispersion of ion conics: Evidence for a polar cusp heating wall, *Geophys. Monogra. Ser.*, vol. 84, pp. 163-169, AGU, Washington, D.C., 1994.

Lepping, R. P., et al., The WIND magnetic field investigation, *Space Sci. Rev.*, *71*, 207-229, 1995.

Liu, C., J. D. Perez, T. E. Moore, C. R. Chappell, and J. A. Slavin, Fine structure of low-energy $H^+$ in the nightside auroral region, *J. Geophys. Res.*, *99*, 4131-4141, 1994.

Lockwood, M., J. H. Waite Jr., J. F. E. Johnson, T. E. Moore, and C. R. Chappell, A new source of suprathermal $O^+$ ions near the dayside polar cap boundary, *J. Geophys. Res.*, *90*, 4099-4116, 1985a.

Lockwood, M., M. O. Chandler, J. L. Horwitz, J. H. Waite Jr., T. E. Moore, and C. R. Chappell, The cleft ion fountain, *J. Geophys. Res.*, *90*, 9736-9748, 1985b.

Miyake, W., T. Mukai, and N. Kaya, On the evolution of ion conics along the field line from EXOS D observations, *J. Geophys. Res., 98,* 11,127-11,134, 1993.

Moore, T. E., C. R. Chappell, M. Lockwood, and J. H. Waite Jr., Superthermal ion signatures of auroral acceleration processes, *J. Geophys. Res., 90,* 1611-1618, 1985.

Moore, T. E., C. J. Pollock, M. L. Adrian, P. M. Kintner, R. L. Arnoldy, K. A. Lynch, and J. A. Holtet, The cleft ion plasma environment at low solar activity, *Geophys. Res. Lett., 23,* 1877-1880, 1996.

Norqvist, P., M. André, L. Eliasson, A. I. Eriksson, L. Blomberg, H. Lühr, and J. H. Clemmons, Ion cyclotron heating in the dayside magnetosphere, *J. Geophys. Res., 101,* 13,179-13,193, 1996.

Ogilvie, K. W., et al., SWE, A comprehensive plasma instrument for the WIND spacecraft, *Space Sci. Rev., 71,* 55-77, 1995.

Peterson, W. K., A. W. Yau, and B. A. Whalen, Simultaneous observations of $H^+$ and $O^+$ ions at two altitudes by the Akebono and Dynamic Explorer 1 satellites, *J. Geophys. Res., 98,* 11,177-11,190, 1993.

Whalen, B. A., S. Watanabe, and A. W. Yau, Observations in the transverse ion energization region, *Geophys. Res. Lett., 18,* 725-728, 1991.

Woch, J., and R. Lundin, Magnetosheath plasma precipitation in the polar cusp and its control by the interplanetary magnetic field, *J. Geophys. Res., 97,* 1421-1430, 1992.

Watanabe, S., T. Abe, E. Sagawa, B. A. Whalen, A. W. Yau, T. Mukai, and H. Hayakawa, EXOS-D observations of thermal ion energy distributions in transverse ion energization regions, *J. Geomagn. Geoelectr., 47,* 1161-1169, 1995.

Yau, A. W., P. H. Beckwith, W. K. Peterson, and E. G. Shelley, Long-term (solar cycle) and seasonal variations of upflowing ionospheric ion events at DE 1 altitudes, *J. Geophys. Res., 90,* 6395-6407, 1985.

---

M. O. Chandler, P. D. Craven, and B. L. Giles, Space Sciences Laboratory, NASA Marshall Space Flight Center, Huntsville, AL 35812. (e-mail: michael.chandler@msfc.nasa.gov; paul.craven@msfc.nasa.gov; barbara.giles@msfc.nasa.gov)

M. Hirahara, Department of Physics, College of Science, Rikkyo University, 3-34-1 Nishi-Ikebukuro, Toshima-ku, Tokyo 171, Japan. (e-mail: hirahara@se.rikkyo.ac.jp)

J. L. Horwitz, Center of Space Plasma, Aeronomy, and Astrophysics Research, University of Alabama in Huntsville, AL 35899. (e-mail: horwitzj@cspar.uah.edu)

T. E. Moore, Interplanetary Physics Branch, NASA Goddard Space Flight Center, Greenbelt, MD 20771. (e-mail: T.E.moore@gsfc.nasa.gov)

C. L. Pollock, Instrumentation and Space Research Division, Southwest Research Institute, PO Drawer 28510, San Antonio, TX 78228. (e-mail: cpollock@swri.edu)

# Recent Developments of Ion Acceleration in the Auroral Zone

Tom Chang

*Center for Space Research, Massachusetts Institute of Technology, Cambridge, MA, and International Space Science Institute, Bern, Switzerland, and*

Mats André

*Swedish Institute of Space Physics, Umeå University, Umeå, Sweden*

A brief review of the recent developments in the knowledge of auroral ion acceleration is presented. During the past several years, sufficient experimental data on ion acceleration in the auroral zone have been collected by the Freja satellite, allowing for an in-depth examination of some of the proposed energization mechanisms. In addition, new results from the Akebono, POLAR, and FAST satellites are providing new insights into the details of the microphysics as well as the related mesoscale interactions of ion energization and evolution in the ionosphere and magnetosphere through high-precision and high-resolution measuring instruments. These results are further augmented by a number of equally innovative sounding rocket experiments such as SCIFER and AMICIST. It is not an exaggeration to state that, despite more than twenty years of intense interest and research efforts in ion acceleration, this exotic phenomenon continues to attract the attention and imagination of contemporary auroral physicists.

## 1. INTRODUCTION

Interest in understanding the fundamental mechanism(s) of ion heating perpendicular to the geomagnetic field in the auroral zone has not waned since the first observations of "ion conics" [Sharp et al., 1977; Klumpar et al., 1979] over two decades ago. The name "conic" refers to the fact that the observed ion distributions are peaked in pitch angle in velocity space. The generally accepted scenario for transverse ion acceleration is some sort of wave-particle interaction. The transversely accelerated ions are then folded into conical shape by the diverging geomagnetic field as they drift to higher altitudes. Candidates of plasma waves that could be responsible for this process include lower hybrid waves [Chang and Coppi, 1981], electrostatic ion cyclotron waves [Kindel and Kennel, 1971; Palmadesso et al., 1974; Lysak et al., 1980; Ashour-Abdalla and Okuda, 1984], electromagnetic ion cyclotron waves [Chang et al., 1986; Temerin and Roth, 1986], non-resonant static structures and low frequency waves [Borovsky, 1984; Hultqvist, 1988; Lundin and Hultqvist, 1989], and velocity-shear generated localized electrostatic eigenmodes [Ganguli et al., 1994]. Through the years, investigators have come to recognize that, depending on the physical situation, available free-energy, and boundary conditions, nearly all proposed mechanisms are viable processes for transverse ion acceleration. However, only a few mechanisms seem to be of major importance in the terrestrial auroral zone. Sometimes, a combination (or combinations) of these basic mechanisms is needed to provide a reasonable explanation of the observed ion conic distributions.

Since there already exists a number of review articles on this subject [e.g., Klumpar, 1986; Lysak, 1986; Chang, 1993; André and Chang ,1993; André and Yau, 1997; Yau and André, 1997], this paper will only briefly touch upon the description of several fundamental heating mechanisms. It is then supplemented by results that are relevant to the more recent observations of the Freja satellite, and the SCIFER and AMICIST sounding rockets. Because of the preliminary and limited nature of the reported data from POLAR and FAST, we shall refrain from making any definitive comments about these results in this review. In some respect, this paper may be considered as a supplement to the two previous review articles written by the authors [André and Chang, 1993; Chang. 1993].

## 2. RESONANT HEATING BY WAVES NEAR THE ION GYROFREQUENCY

A left circularly polarized wave with frequency equal to the ion gyrofrequency can efficiently accelerate a positive ion perpendicularly to a homogeneous magnetic field. Consider an ion in a homogeneous (infinite wavelength) monochromatic left-handed electric field with frequency $f$ equal to the ion gyrofrequency $f_c$. Since the electric field rotates in the same direction and with the same angular velocity as the ion, it applies a constant force on the ion along its orbit, providing a velocity increase proportional to the time $t$, and hence an energy increase proportional to $t^2$. However, a very narrow-banded coherent wave with frequency around $f_c$ is hard to obtain even in a laboratory, and usually does not occur in the magnetosphere. Rather, random phased waves covering a fairly broad frequency band including $f_c$ are often observed. The left-handed component of these waves still interacts efficiently with the ion. However, the ion motion in velocity space must now be regarded as a random walk. The velocity increase is then proportional to $\sqrt{t}$, and the energy increase is thus proportional to $t$ [Chang et al., 1986.] Broadband waves at frequencies around and below the ion gyrofrequency are often observed in the magnetosphere [e.g., Gurnett et al., 1984; Chang et al., 1986; André et al., 1988]. These waves are often associated with transverse ion heating. Thus, both theory and observations seem to indicate that broadband waves around $f_c$ are important for ion heating.

The theory of ion cyclotron resonance heating by broadband waves is based on a diffusion operator [Retterer et al., 1987]. This operator gives the diffusion rate corresponding to the random walk of ions in velocity space. Using the long wavelength approximation, the operator can be written in a very simple form. This approximation should be valid, for example, for broadband Alfvén waves in most regions of the magnetosphere. The appropriate diffusion coefficient is then simply proportional to the electric field spectral density ($S$) at the local gyrofrequency of the ion species of interest [Chang et al., 1986; Retterer et al., 1987]. The resulting average heating rate per ion ($Q$) can also be shown to be proportional to the spectral density (and thus to the mean square of the electric field fluctuations) at the local ion gyrofrequency [Chang et al., 1986]:

$$Q = (q^2/2m)S_L \quad (1)$$

where $q$ and $m$ are the charge and mass of the ion, respectively. The spectral density, $S_L$, in equation (1) is the fraction of the spectral density due to left-hand polarized waves. Actually the resonance condition $2\pi|f - Nf_c| = k_\| v_\|$ should be used to find out at which frequencies ($f$) ions are in resonance with a wave. Here $k_\|$ and $v_\|$ are the parallel wavevector and the parallel component of the particle velocity, while $N$ is an integer. This relation takes into account, e.g., the finite wavelength, and the possibility of interactions at higher harmonics. For emissions with long wavelengths such as most Alfvén waves, the effect of the term $k_\| v_\|$ becomes small. Furthermore, interactions at higher harmonics often require a nonzero perpendicular component of the wavevector.

Retterer et al. [1987] applied the cyclotron resonance heating mechanism to observations in the central plasma sheet. In this study, an observed wave spectrum (Fig. 1) together with a Monte Carlo simulation were used to produce an ion distribution quantitatively similar to that observed by Winningham and Burch [1984] at a geocentric distance of about 2 earth radii ($R_E$), Fig. 2. (The theory predicts that the ions were oxygen ions. This prediction was confirmed by measurements from the mass spectrometer EICS on DE-1 [Peterson, Klumpar and Shelley, private communications.]) Other detailed and successful tests of the cyclotron resonance mechanism have been performed using satellite data from the cusp-cleft region [André et al., 1990; Norqvist et al., 1996] and from the nightside auroral region [Crew et al., 1990].

In the central plasma sheet where the ion conics studied by Retterer et al. [1987] were observed, there were no obvious local energy sources that could power the broadband electromagnetic waves. This led Johnson et al. [1989] to suggest that the waves were generated by anisotropic ion distributions in the equatorial plane and then

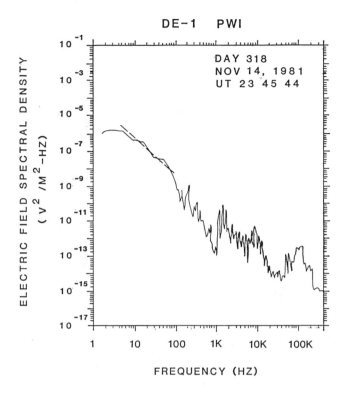

**Figure 1.** Typical electric field spectral density in the central plasma sheet. Geocentric distance ≈ 2 $R_E$, invariant latitude ≈ 60°. (M. Mellott and D. Gurnett, private communication, 1986).

propagated to the ion heating region. Such a scenario has been demonstrated to be plausible using ray-tracing, wave distribution function, and mode-conversion techniques [Rönnmark and André, 1991; Johnson et al., 1995].

Statistical studies of ion mass spectrometer data from both DE-1 [Peterson et al., 1992] and Akebono [Miyake et al., 1996] indicate that ions are not suddenly heated at one altitude and then adiabatically move upward. Rather, the pitch angles of the ion distributions suggest height-integrated transverse acceleration of ions over a wide altitude range. This is consistent with gradual heating, e.g., by waves near the ion gyrofrequency as the ions move upward.

The successful use of relation (1), and our use of the term "cyclotron resonance", does not imply that long wavelength Alfvén waves always dominate during the ion heating events. The assumed fraction of left-hand-polarized waves may be regarded as a parameter describing how large fraction of the waves is in resonance with the ions. For various reasons, less than 100% of the observed emissions, which may include so-called electrostatic waves, are in resonance. Assuming that some fraction of the observed electric field spectral density near the ion gyrofrequency is due to left-hand Alfvén waves is, as a first approximation, the same as assuming that the same fraction of the waves is in resonance with the ions, regardless of the wave mode.

## 3. RESONANT HEATING BY WAVES NEAR THE LOWER HYBRID FREQUENCY

In 1981, Chang and Coppi suggested that under favorable conditions, lower hybrid waves may be responsible for the transverse energization of ionospheric ions in the auroral zone. The original paper assumed a random, broadband wave spectrum. Subsequent theoretical calculations and computer simulations suggest that the broad band turbulence generally is intermittent and inhomogeneous [Retterer et al., 1986; Chang, 1993; and references contained therein.] The turbulence may in fact be viewed as an assemble of slowly propagating (i.e., nearly stationary) density cavities filled with emissions near the

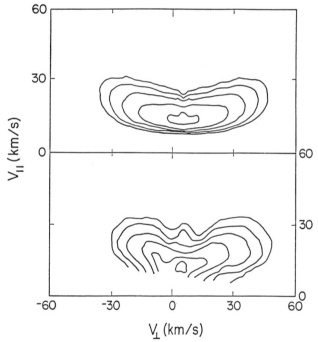

**Figure 2.** Bottom panel is a contour diagram of the observed ion conic distribution function, measured by the HAPI instrument on the DE-1 satellite [Winningham and Burch, 1984]; the fact that the ions were oxygen ions were confirmed by the EICS instrument on DE-1, (Peterson, Klumpar and Shelley, private communications). Top panel is the calculated ion-velocity distribution (Retterer et al., 1987) using the electromagnetic ion cyclotron resonance theory (Chang et al., 1986) and the wave spectrum in Figure 1, plotted in the same way as the observed distribution.

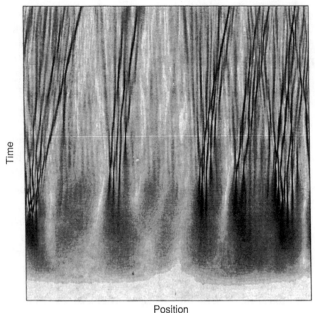

**Figure 3.** Evolution of lower hybrid caviton turbulence in real space and time based on the theory of Retterer, Chang and Jasperse [1986]. Shading indicates the strength of the electric field amplitude. The horizontal scale is approximately 100 Debye lengths and the vertical scale is approximately 100 ion plasma periods. Notice how the initial nearly uniform wave amplitude intensity evolves into intense localized solitary structures.

lower hybrid frequency (lower hybrid cavities.) [See Figure 3.] The theory of Retterer et al. was based on a self-consistent nonlinear theory. Subsequently, Bell and Ngo [1990] suggested that such a wave spectrum could also be the results of linear conversion of the often present VLF hiss due to the pre-existing irregularities in the suprauroal region. Generally, the wave-particle interaction process may be relegated to a model diffusion operator [Crew and Chang, 1985]. Using such an operator, mesoscale evolution of ionsopheric ions in the magnetosphere due to lower hybrid heating may be evaluated [Retterer et al., 1994; André et al. 1994; and references contained therein]. Figures 4 and 5 show simulations and observations of ion distributions heated by waves near the lower hybrid frequency.

Recently, data have been obtained by sounding rockets at altitudes up to 1400 km. Observations by MARIE, sounding rockets in the TOPAZ series, SCIFER and AMICIST indicated that lower hybrid cavities could provide the initial energization of ionospheric ions at low altitudes [Yau et al., 1987; Kintner et al., 1986, Kintner et al., 1992; Kintner et al., 1996; Arnoldy et al., 1996; Lynch et al., 1996]. One possible explanation for the occurrence of these isolated density cavities, in fact, may also be provided by the self-consistent nonlinear picture [Chang, 1993; Shapiro et al., 1993; Retterer et al., 1994]. A self-consistent two-dimensional simulation including wave-particle interactions performed by Retterer [1997] seems to support these conjectures. Actually, for purely two-dimensional propagations of lower hybrid waves perpendicular to the magnetic field, nonlinear condensation of lower hybrid waves (or lower hybrid collapse) cannot happen. This is proven rigorously by Tam et al. [1995]. However, Tam et al. have also demonstrated analytically that pseudo two-dimensional collapse can proceed for oblique lower hybrid propagations provided certain propagation characteristics and physical conditions are satisfied.

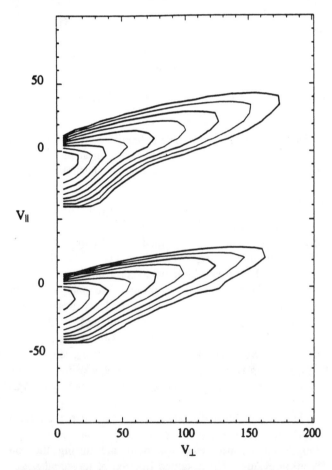

**Figure 4.** Simulation of the MARIE conic event (Kintner et al., 1986; Yau et al., 1987). Contour plots of the ion distribution as a function of perpendicular and parallel velocities (in km/sec) at 620 km (bottom) and 740 km (top) altitudes. The contours are a half decade apart.

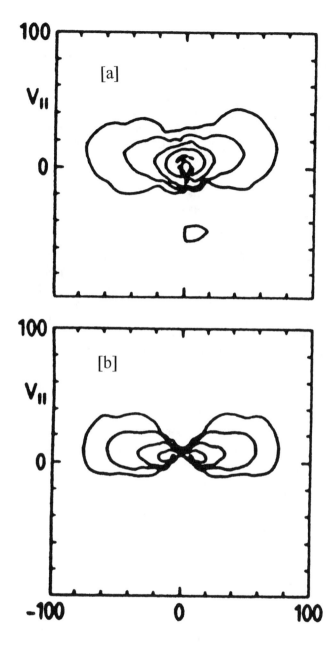

**Figure 5.** Top panel gives the $O^+$ distribution function with mean energy of 31 eV obtained by the Freja satellite on orbit 790, December 5, 02.35.20-02.35.32 UT at 1760 km., MLT=18.3 and CGLAT=72.8. Bottem panel corresponds to the $O^+$ distribution obtained by a Monte Carlo simulation plotted in the same way as the observed distribution.

Additionally, as demonstrated by Retterer et al. [1994] and Retterer [1997], the nonlinear collapse process will generally be arrested by wave-particle interactions.

On the other hand, individual lower hybrid cavities may also result from the trapping of VLF waves in pre-existing density cavities and the subsequent linear conversion into localized lower hybrid eigenmodes [Seyler, 1994]. Recently Pinçon et al. [1997] and Schuck et al. [1997] have shown that the observed oscillations extracted from the TOPAZ data do indeed support the possibility of such an interpretation.

It has been suggested by Melatos and Robinson [1996] that for coherent lower hybrid cavities, the ion heating may not be well described by a diffusion operator and transition-time heating theories may need to be employed. On the other hand, the relative time scales in the auroral zone (where the lower hybrid emissions generally are not truly coherent) seem to indicate that diffusion is the probably the more appropriate wave-particle interaction process to be considered [Retterer et al., 1994].

At higher altitudes, individual lower hybrid cavities have been observed by satellites [Potellette et al., 1992; Eriksson et al., 1994; Dovner et al., 1997]. These lower hybrid cavities are not associated with intense ion heating, but from observations it is hard to exclude the possibility that they could cause ion heating to energies of a few eV.

Although lower hybrid waves are more efficient in heating lighter ions, it has been suggested by Chang et al. [1988] and demonstrated by André et al [1994] that oxygen ions may also be heated by broadband lower hybrid waves if these ions were pre-heated by the co-existing broadband waves near the ion gyrofrequency. [See Fig. 5].

## 4. OTHER HEATING MECHANISMS

### 4.1 Multiple Cyclotron and Subharmonic Resonant Heating

It has been suggested by Temerin and Roth [1986] that nonlinear resonance could also produce transverse ion heating. One process involves two waves with finite wavenumbers, where the sum of the frequencies matches the gyrofrequency. This mechanism may be called "double cyclotron absorption" (or "double cyclotron resonance"). Assuming low amplitude electric field fluctuations, the average heating rate per ion for this process can be shown to be proportional to the product of the electric field spectral densities at the two frequencies. Although this mechanism is less efficient than the ion cyclotron resonance heating (since it is of higher order in spectral densities), it can involve the entire frequency range (below the local gyrofrequency) of the spectrum, thereby producing a substantial amount of transverse ion acceleration.

A detailed comparison of double-cyclotron absorption with cyclotron resonance heating in the cusp/cleft region of the magnetosphere was performed by Ball and André [1991]. This study used the DE-1 data and the heating rates presented by Ball [1989a, b]. It was concluded that double-cyclotron absorption might give an appreciable contribution to $O^+$ energization, especially when the heating was accomplished locally (in altitude). However, the major conclusion is that cyclotron resonance heating is the more efficient mechanism for the oxygen conics observed in the cusp/cleft region.

From a theoretical point of view, there exist obviously other higher order resonance processes involving spectral densities at more than two frequencies ("multiple cyclotron resonance"). However, the heating rates under these circumstances will be of even higher order in wave amplitudes (or spectral densities). For reasonably low wave amplitudes, these higher order effects will be much less efficient and can usually be neglected.

From the discussion above, it follows that waves with frequencies $f$ around the $N$th order subharmonic ($f = f_c/N$, $N = 2,3,4...$) can resonantly heat ions for finite wavelength electric field fluctuations. The average heating rate is then proportional to the $N$th power of the electric field spectral density at the subharmonic. For sufficiently low amplitude fluctuating fields, this leads to the following conclusions: (1) Cyclotron resonance ($N = 1$) is, for typical broadband spectra, often more efficient than subharmonic resonances, (2) 2nd order subharmonic resonance heating (which is equivalent to the special case of double cyclotron resonance for $f_1 = f_2 = f_c/2$) is generally more efficient than those due to other higher order subharmonic resonances, and (3) For broadband electric field spectra, it is more important to include all the double cyclotron resonance effects (in addition to the $N = 2$ subharmonic effect) than to include other higher order ($N > 2$) subharmonic effects.

*4.2. Nonresonant Energization*

Waves with frequencies much below the ion gyrofrequency may energize ions. Since both the subharmonic heating and multiple cyclotron resonance heating will be of very high order in wave amplitude here, they can generally be neglected. It has been suggested that such waves instead might nonresonantly cause significant ion heating in the magnetosphere [Lundin et al., 1990; Lundin and Hultqvist, 1989; Hultqvist, 1996].

To test this idea of nonresonant energization, Ball and André [1991] performed a numerical calculation based on a model of an observed broadband wave spectrum. The model spectrum was subdivided into three parts: one resonant part around the proton gyrofrequency, one low frequency nonresonant part, and one high frequency nonresonant part. The effect of the different parts of the spectrum on ion energization was tested by simply solving the equation of motion for protons subject to the (known) time-varying electric field of the model in the presence of a static uniform magnetic field. Using a realistic wave spectrum, the difference between resonant heating by waves around the proton gyrofrequency and nonresonant energization by other parts of the spectrum is dramatic. It is found that the low frequency part of the spectrum causes the perpendicular proton energy to increase rapidly to some moderate energy (some eV), after which no further net energy transfer occurs. The energization process due to the high frequency part of the spectrum is similar. The resulting energy is lower, however, since the spectral density is typically lower there. After a few gyroperiods the mean proton energy due to resonant heating by waves around $f_c$ is higher than the total energy obtained from nonresonant energization processes. We note that no peak in the spectrum around the gyrofrequency is needed for efficient resonant heating, provided sufficient wave power is present near the gyrofrequency.

The numerically obtained resonant heating rate agrees well with the theoretical value from Eq. (1), as discussed by Ball and André [1991]. Such resonant heating can readily explain the observed keV protons reported by Hultqvist et al. [1988]. On the other hand, nonresonant energization can only account for a small fraction of the observed ion energy. It was concluded that fundamental ion cyclotron resonance heating can easily explain observed high energy proton conics (above 100 eV) which are hard to explain with any other mechanism.

For events involving heavier ions, the situation is less clear. As discussed above, detailed studies of $O^+$ heating events show that resonant heating by observed broadband waves can heat the ions to the observed energies. However, since the gyrofrequencies of heavy ions are rather low, it will be more difficult to differentiate the contributions of individual energization mechanisms. At least in some instances, double-cyclotron absorption may significantly contribute to $O^+$ heating [Ball and André, 1991]. Similarly, nonresonant interaction with waves below $f_c$ may produce significant $O^+$ heating. Furthermore, such nonresonant energization may well give high enough energies to explain some ion outflow from the ionosphere [Lundin and Hultqvist, 1989; Hultqvist, 1996]. Nevertheless, Ball and André [1991a; 1991b] concluded that energization by waves near the ion gyrofrequency provides the best explanation for the majority of observed ion conic events (that are generated either locally or over an extended region in space) at altitudes above a few thousand kilometers in the auroral and cusp/cleft regions. Recent observations by sounding rockets [Kintner et al., 1996; Lynch et al., 1996] and by Freja [André et al., 1997] indicate that the same type of energization dominates at lower altitudes.

### 4.3. Heating at Higher Harmonics

Resonant heating of ions may occur also at higher harmonics of the ion gyrofrequency ($f = Nf_c$, $N = 2, 3, 4,...$). As in the case of subharmonic heating, energization at higher harmonics requires finite wavelength electric field fluctuations. Heating at higher harmonics is important, e.g., in energization of Tokamak fusion plasmas with radio frequency waves. Here waves are artificially launched on the magnetosonic branch from the low magnetic field side of the plasma. In the magnetosphere, wave absorption at the second harmonic may be important, e.g., when waves are propagating down a magnetic field line, and thus into a stronger magnetic field [Horne and Thone, 1990.] In this situation, the frequency of a downgoing wave will first match the second harmonic before reaching the fundamental gyrofrequency. Indeed, in some cases the downgoing wave may be reflected, e.g., at the ion-ion hybrid frequency, before reaching the gyrofrequency.

For suitable conditions, wave absorption at higher harmonics might be significant. However, for the broadband spectra, which very often are associated with conics, the wave intensity is much higher at the fundamental ion gyrofrequency than at higher harmonics. Thus, resonant heating at the gyrofrequency usually is the more important mechanism. It should be noted that ion energization by waves near the lower hybrid frequency is a special case of heating at higher harmonics. In this case, reasonably high wave intensities occur near the lower hybrid frequency, i.e., usually at several times the ion gyrofrequency. In conclusion, we find that during ion heating events with broadband wave spectra, with a spectral density simply decreasing with frequency, energization by waves at the gyrofrequency dominates over heating at higher frequencies. When waves near the lower hybrid frequency are present, these may cause ion heating via resonant interaction at higher harmonics of the ion gyrofrequency.

### 4.4. Low Frequency Electrostatic Waves or Eigenmodes.

One mechanism that has received considerable attention is the heating of ionospheric ions by electrostatic ion cyclotron waves above multiples of the ion gyrofrequency. This possibility was suggested a number of years ago [Kindel and Kennel, 1971; Palmadesso et al., 1974] and studied by numerous researchers [Lysak et al., 1980; Ashour-Abdalla and Okuda, 1984; and references contained therein]. Electrostatic ion cyclotron waves may be generated by the simultaneously observed upgoing ion beams, or by drifting electrons (currents) in the auroral zone. More recently, it has been suggested by Ganguli [Ganguli et al., 1994; and references contained therein] that electrostatic eignmodes of localized fluctuations in the ion cyclotron [and lower hybrid] range of frequencies may be the prime candidates for the observed electrostatic modes. These fluctuations may be produced locally by velocity-shear and narrowly confined electric fields. Simulations have demonstrated that these electrostatic eigenmodes can accelerate the ions quite efficiently [Romero and Ganguli, 1992, and references contained therein]. Recently, Kintner et al. [1996] and Lynch et al. [1997] have identified a number of events in the low altitude auroral and cusp/cleft regions where low frequency electrostatic modes are observed simultaneously with the transverse energization of ionospheric ions. These results were based on observations of the SCIFER and AMICIST sounding rocket experiments.

## 5. FREJA OBSERVATIONS OF VARIOUS ION ENERGIZATION MECHANISMS

We conclude this brief review by presenting some recent statistical studies of ion energization events collected by the Freja satellite. The Freja satellite was launched October 6, 1992 into an orbit with 63° inclination, an apogee in the northern hemisphere of 1750 km, and a perigee of 600 km. Data used in the studies described below were obtained in the northern hemisphere at altitudes above about 1450 km. Freja has a set of high-resolution wave instruments [Marklund et al., 1994; Holback et al., 1994; Zanetti et al., 1994] and particle detectors [Eliasson et al., 1994b; Boehm et al., 1994] for studies of space plasma wave-particle interaction processes [André, 1993; Lundin et al., 1994a, b]. The spacecraft is Sun-pointing and is spin stabilized with a spin period of 6 seconds.

Several studies of ion energization events observed by Freja have been performed [André et al., 1994; Eliasson et al., 1994a; Eriksson et al., 1994; Norqvist et al., 1996; Knudsen et al., 1997; Wahlund et al., 1997]. In the following we are mainly interested in intense perpendicular ion energization to mean energies above 5-10 eV. During these events a large fraction of the ions has high enough velocities to eventually escape from the Earth. Studies of several Freja ion heating events show that the most common and important energization mechanism seems to be resonant heating by broadband low-frequency waves at frequencies of the order of the ion gyrofrequency. At Freja altitudes this mechanism causes the highest number flux of $O^+$ ions (a few times $10^{13}$ ions/$(m^2s)$ on the dayside) and the highest $O^+$ energies (average energies of up to hundreds of eV, at other local times). The energization occurs within the auroral oval at all local times, and on the nightside the broadband low-frequency waves are loosely associated with keV auroral electrons accelerated by a quasi-static potential drop. However, these waves and the ion energization do not occur on exactly the same field lines as the auroral electrons. Other, less common and less intense, ion energization mechanisms are directly associated with auroral electrons. These electrons may generate waves near the lower hybrid frequency $f_{LH}$, or electromagnetic ion cyclotron (EMIC) waves below the proton gyrofrequency, which then resonantly heats the ions. Yet another possibility is that precipitating $H^+$ or $O^+$ ions generate waves near $f_{LH}$, which then locally heat other ions. These latter mechanisms may require ``pre-heating'' by some broadband low-frequency waves for the ions to obtain high enough velocities to be in resonance with the other wave modes.

When performing a statistical study of perpendicular heating of $O^+$ ions, André et al. [1997] defined a few different types of ion heating regions. Types 1 and 2 are both associated with broadband low-frequency waves. However, Type 1 often occurs adjacent to regions with auroral electrons, while Type 2 is associated with precipitating protons and electrons typical of the cusp/cleft region. Type 3(LH) events occur in regions of auroral electrons and waves near the lower hybrid frequency, while Type 3(EMIC) also is directly associated with auroral electrons but with EMIC waves at about half the proton gyrofrequency. During Type 4 events precipitating ions ($H^+$ or $O^+$) with keV energies are correlated with ion heating, and the events are different from Type 2 events typical of the cusp/cleft region, e.g., since no precipitating electrons are observed. One objective of the study was to investigate the relation between wave intensity at various frequencies and the $O^+$ mean energy. The basic idea was to perform a test to see if the observed wave intensities at some resonant wave frequencies were high enough to cause the observed ion energies. Accordingly, a total of 20 events of intense ion energization were selected. During each event, twelve satellite spin periods of six seconds, each corresponding to about 40 km along the spacecraft orbit, were studied in detail. The periods were chosen

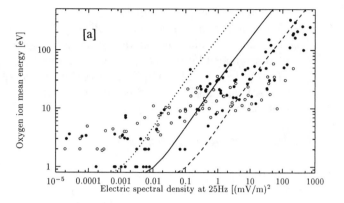

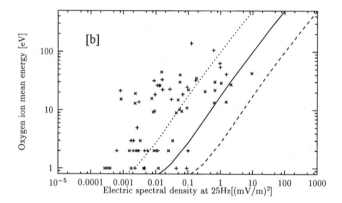

**Figure 6.** Mean perpendicular $O^+$ energy as a function of electric field spectral density at 25 Hz (approximately the oxygen gyrofrequency) for (a) Type 1 (filled circles) and 2 (circles) ion heating events, and for (b) Type 3(LH) (x) and 3(EMIC) (+). Each data point corresponds to observations from a six second period. The dashed, solid and dotted lines correspond to ion mean energies that can be obtained from cyclotron heating at 25 Hz when 1%, 10% and 100% (respectively) of the observed spectral density contributes to ion energization. The ions are launched at 300 km below the satellite. Essentially all points in (a) with energies above 5-10 eV can be explained by resonant cyclotron heating, while this is not the case for (b).

either within the regions of intense ion heating, or near but outside the heating region. The $O^+$ mean energy $W_{O^+}$ (obtained by integrating over perpendicular velocities) as a function of electric field spectral density at 25 Hz (close to the $O^+$ gyrofrequency at these altitudes), $S_E(25Hz)$, for the Type 1 and 2 events is given in Figure 6a. In Figure 6a it is obvious that high (low) wave intensities correspond to high (low) ion mean energies. Rather than just investigating the correlation between ion and wave observations, we also estimate the O$^+$ energies can be obtained from the waves. To do so, we use the ion cyclotron resonance heating mechanism [Chang et al., 1986]. A maximum heating rate corresponding to waves with frequencies covering the ion gyrofrequency can be obtained by assuming that the perpendicular wavevector is much smaller than the inverse of the ion gyroradius (as would be the case for some Alfvén waves), and that the left-hand polarized fraction of the waves is heating the ions. The ion cyclotron heating rate [Chang et al., 1986; Retterer et al., 1987] is then given by Eq. (1) of Section II. This relation is convenient to use when estimating the ion heating rate from the observed spectral densities. An exact calculation of the ion energies that can be caused by the observed waves requires detailed information concerning the waves, such as distribution in wavevector space, which usually is not available. The successful use of relation (1) in several studies is not necessarily an indication that the long wavelength Alfvén wave is the dominant wave mode in the observed broadband low-frequency spectra. For example, recent studies indicate that a significant fraction of these emissions sometimes may be ion acoustic waves [Wahlund et al., 1997] or electrostatic ion cyclotron waves [Bonnell et al., 1996]. The simple cyclotron heating mechanism can nevertheless be useful when estimating the effects of observed waves on the ions. Assuming that some fraction, which must be less than 100%, of the observed electric field spectral density near the ion gyrofrequency is due to left-hand Alfvén waves is, as a first approximation, the same as assuming that the same fraction of the waves are in resonance with the ions, regardless of wave mode.

The lines in Figure 6a give an estimate of the ion energies expected from the cyclotron resonance mechanism. The expected energies have been estimated from test particle calculations [Chang et al., 1986]. The particles are started 300 km below the satellite, as indicated by the folding of the observed O$^+$ distributions. The initial energy and pitch-angle is taken to be 0.5 eV and 135° respectively, but the final O$^+$ energy is not sensitive to these two parameters. To be able to map the spectral density at the local oxygen gyrofrequency to a lower altitude, the spectral density is approximated by a power law $S_E(f) = S_0 (f_0 / f)^\alpha$ where $S_0$ and $f_0$ are constants, $f$ is the frequency and $\alpha$ is a model spectral slope [Chang et al., 1986]. We always use $\alpha = 1$, but the result is not sensitive to this parameter since the altitude range is small. Following the particles in a dipole geomagnetic field, the three lines in Figure 6a can be obtained. These lines correspond to some fraction (100%, 10% or 1%) of the spectral density being in resonance with the ions.

It is clear that for ion energies above 5-10 eV essentially all data points in Figure 6a are consistent with heating by less than 100% of the observed waves. Thus Figure 6a is consistent with the observed O$^+$ ions being energized by the observed waves with frequencies around $f_{O^+}$.

Ion energies below 3-5 eV in Figure 6a are uncertain due to effects such as spacecraft motion and charging. The fact that several mean energy levels are displayed at exactly one, two or three eV is artificial. There is a threshold at spectral densities of $10^{-3}$ to $10^{-2}$ (mV/m)$^2$/Hz above which ion mean energies below a few eV only rarely occur. This is consistent with the cyclotron resonance mechanism, and also with a study using the cold plasma analyzer on Freja [Knudsen et al., 1997].

So far we have considered only Type 1 and 2 events. All these ion heating events are associated with broadband electric field waves, and these waves include high intensities at 25 Hz. We now instead investigate all Type 3 (both 3(LH) and 3(EMIC)) events. Figure 6b is similar to Figure 6a but shows mean O$^+$ energy as a function of $S_E$ (25Hz) for Type 3 events. It is obvious that many data points can not be explained by the cyclotron reso-nance mechanism. For several other points the cyclotron mechanism might be significant. At least for some points a combination of heating by waves near $f_{O^+}$ and near the lower hybrid frequency may be important [André et al., 1994]. In summary we find that waves near $f_{O^+}$ can not explain all ion energization during Type 3 events.

Above we have concentrated on the heating that can be caused by waves near $f_{O^+}$. Similar studies of EMIC waves near the half the proton gyrofrequency, show that during Type 3(EMIC) events the observed EMIC can cause the observed O$^+$ ion energies. Similarly, investigations of waves near the lower hybrid frequency clearly indicate that during Type 3(LH) events these waves can energize the observed ions. Also, during most Type 3 events, it is not likely that the (often weak) broadband low-frequency waves that are present can cause the ion heating, but the other observed wave emissions are really needed to explain the observed ion energies.

To investigate the relative importance of ion energization by waves at different frequencies, a preliminary statistical study was performed by using Freja data from the November 18, 1993 to March 8, 1994 (orbits 5390 to 6850) [André et al., 1997]. This period gives approximately equal coverage of all magnetic local times. Nearly all events with O$^+$ mean energies above 5-10 eV could be classified as belonging to one of the previously discussed Types of ion energization. In this study, the events of Types 3(EMIC) and 3(LH) could not be separated, and were all categorized as Type 3. Figure 7a shows the

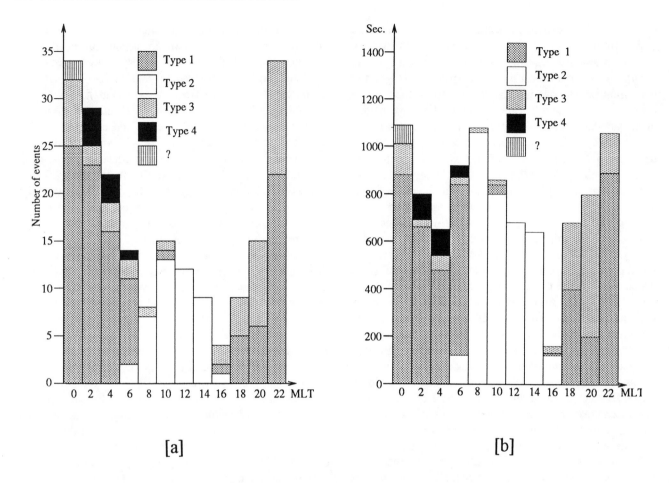

**Figure 7.** Different types of ion heating events obtained by the Freja satellite near 1700 km during approximately 1300 passes over some ground station. (a) Number of events. (b) Number of seconds spent in each type of event. Type 1 and 2 are both associated with broadband low frequency waves. Type 3 is connected with auroral electrons and waves near half of the hydrogen gyrofrequency or waves around the lower hybrid frequency, while type 4 is associated with precipitating $H^+$ or $O^+$ ions and, again, around the lower hybrid frequency.

number of different types of events as a function of Magnetic Local Time (MLT). Figure 7b displays the number of seconds spent by Freja in various types of ion heating events, again as a function of MLT. This latter figure gives an estimate of the overall importance (the linear dimension perpendicular to the geomagnetic field) of the various types of events. This preliminary statistical study clearly indicates that events where broadband low frequency waves are associated with ion heating (Types 1 and 2) are more common than events involving waves near the lower hybrid or EMIC emissions near half the proton gyrofrequency (Type 3 and also Type 4).

The Freja observations are similar to recent sounding rocket observations. Observations by the SCIFER sounding rocket at 1400 km and near 10:00 MLT in the cleft again shows ion energization clearly associated with broadband low-frequency waves (Kintner et al., 1996; Moore et al., 1996; Arnoldy et al., 1996). The energization is associated with clear density depletions with sharp boundaries. These density depletions are tens of kilometers in the perpendicular direction. The depletions are consistent with even larger regions of low plasma density, broadband low-frequency waves and ion heating observed by the Viking (Hultqvist, 1991) and Freja satellites (Lundin et al., 1994c; André et al., 1997) at altitudes up to 13500 km (Viking apogee). These ion heating events occur at various local times, and similar ion energization in a sharply confined spatial region poleward of a nightside auroral arc has recently been observed by the AMICIST sounding rocket at 900 km, near 23:00 MLT (Lynch et al., 1997). The AMICIST rocket also observed heating by lower hybrid emissions confined in narrow density depletions, so-called lower hybrid cavities. However, the ion outflow from the region with broadband low-frequency waves was significantly larger than the outflow associated with lower hybrid waves. Thus both Freja and sounding rocket data indicate that regions with broadband low-frequency waves are most important for ion energization and outflow.

## 6. EPILOGUE

In this write-up, we have endeavored to provide a brief summary of several viable mechanisms for the transverse heating of ionospheric ions in the auroral and cusp/cleft regions. These ideas are brought into focus with data collected by Freja, SCIFER and AMICIST. We have not discussed all the possible heating mechanisms. For example, there are no discussions given for the stochastic heating by coherent waves or random scattering of ions by solitary kinetic Alfvén waves. Our selection of ion heating has been strongly guided by spacecraft observations. Data from various satellites and sounding rockets indicate that resonant heating by waves near the appropriate ion gyrofrequency is the dominant energization mechanism in the auroral region.

New data that are being collected by the recently launched satellites, POLAR and FAST, will certainly provide additional new insights into the fundamental processes of ion heating. Indeed, results from FAST seem to indicate that sporadic localized (linear or solitary) wave modes contribute to the fundamental energization process of the ions as well as electrons. Due to the preliminary nature of the new results, however, we have refrained from discussing these topics. Finally, most of the theories of mesoscale evolution of ion heating assume a given plasma background. Self-consistent calculations that address both the inhomogeneous plasma background and the microscopic interactions are ultimately needed. Some calculations aiming in that direction have been made recently [Ganguli et al., 1992; Horwitz, 1995; Brown et al., 1995; Tam et al., 1995]. In particular, Jasperse [1998] has demonstrated that in downward auroral current regions, it will be necessary to include the self-consistent back-ground electric fields for the detailed calculation of ion distribution functions. This so-called "pressure cooker" effect was originally suggested by Gorney [1985]. It is expected that these field-aligned self-consistent electric fields are intimately related to the recently identified "fast solitary waves" that are observed by FAST, POLAR and other polar-orbiting satellites [Ergun et al., 1998; Franz et al., 1998].

*Acknowledgments.* The authors enjoyed strong interactions with J.M. Retterer, P. Norqvist, J.R. Jasperse, P.M. Kintner, B. Coppi, G.B. Crew, R.L. Arnoldy, A.W. Yau, J.L. Vago, K.A. Lynch, R.L. Lysak, M. Temerin, D.M. Klumpar, W.K. Peterson, L. Eliasson, D.A. Gurnett, M. Mellott, M.K. Hudson, H. Koskinen, I. Roth, J. Maggs, W. Lotko, R. McWilliams, N. Rynn, J. LaBelle, G. Ganguli, S. Ganguli, B. Hultqvist, R. Lundin, B. Basu, J.D. Winningham, K. Papadopoulos, V.D. Shapiro and V.I. Shevchenko, and A. I. Eriksson. We are indebted to P. Norqvist for the detailed statistical analysis of the Freja data. This research is partially supported by grants from the National Science Foundation (ATM-9612349), the National Aeronautics and Space Administration (NAG5-2255), the Air Force Office of Scientific Research (F49620-96-1-0340), and the Phillips Laboratory (F19628-91-K-0043). A portion of this research was completed while the authors were visitors at the International Space Science Institute in Bern, Switzerland. The authors wish to thank J. Geiss and B. Hultqvist for their kind hospitality.

## REFERENCES

André, M., (Ed.), *The Freja Scientific Satellite*, IRP Sci. Rep. 214, 260 pp., Swed Inst. of Space Phys., Kiruna, 1993.

André, M., and T. Chang, Ion heating by low frequency waves, *Phys. of Space Plasmas (1992)*, 35, 12, 1993.

André, M., G.B. Crew, W.K. Peterson, A.M. Persoon, C.J. Pollock, and M.J. Engebretson, Heating of ion conics in the cusp/cleft, *J. Geophys. Res.*, 95, 20809, 1990.

André, M., H. Koskinen, L. Matson, and R. Erlandson, Local transverse ion energization in and near the polar cusp, *Geophys. Res. Lett.*, 15, 107, 1988.

André, M., P. Norqvist, L. Andersson, L. Eliasson, A.I. Eriksson, L. Blomberg, R.E. Erlandson, and J. Waldemark, Ion energization mechanisms at 1700 km in the auroral region, *J. Geophys. Res.*, in press, 1997.

André, M., P. Norqvist, A. Vaivads, L. Eliasson, O. Norberg, A.I. Eriksson, and B. Holback, Transverse ion energization and wave emissions observed by the Freja satellite, *Geophys. Res. Lett.*, 21, 1915, 1994.

André, M., M. Temerin, and D. Gorney, Resonant generation of ion waves on auroral field lines by positive slopes in ion velocity space, *J. Geophys. Res.*, 91, 3145, 1986.

André, M, and A. Yau, Theories and observations of ion energization and outflow in the high latitude magnetosphere, *Space Science Reviews*, in press, 1997.

Arnoldy, R.L., K.A. Lynch, P.M. Kintner, J. Bonnell, T.E. Moore, and C.J. Pollock, SCIFER-structure of the cleft ion fountain at 1400 km altitude, *Geophys. Res. Lett.*, 23, 1869, 1996.

Ball, L., Can ion acceleration by double-cyclotron absorption produce $O^+$ ion conics?, *J. Geophys. Res.*, 94, 15257, 1989a.

Ball, L., Heavy ion acceleration by double-cyclotron absorption: some analytic approximations, *Aus. J. Phys.*, 42, 493, 1989b.

Ball, L., and M. André, Heating of $O^+$ ions in the cusp/cleft: double cyclotron absorption versus cyclotron resonance, *J. Geophys. Res.*, 96, 1429, 1991a.

Ball, L., and M. André, What parts of broadband spectra are responsible for ion conic production?, *Geophys. Res. Lett.*, 18, 1683, 1991b.

Bell, T. F., and H. D. Ngo, Electrostatic lower hybrid waves excited by electromagnetic whistler mode scattering from planar magnetic-field-aligned plasma density irregularities, *J. Geophys. Res.*, 95, 149, 1990.

Boehm, M., G. Paschmann, J. Clemmons, H. Höfner, R. Frenzel, M. Ertl, G. Haerendel, P. Hill, H. Lauche, L. Eliasson, and R. Lundin, The test electron spectrometer and correlator (F7) on Freja, *Space Sci. Rev.*, 70, 509, 1994.

Bonnell, J., P. Kintner, J.-E. Wahlund, K. Lynch, and R. Arnoldy, Interferometric determination of broadband ELF wave phase velocity within a region of transverse auroral ion acceleration, *Geophys. Res. Lett.*, 23, 3297, 1996.

Borovsky, J.E., The production of ion conics by oblique double layers, *J. Geophys. Res.*, 89, 2251, 1984.

Brown, D. G., J.L. Horwitz, and G.R. Wilson, Synergetic effects of hot plasma-driver potentials and wave-driven ion heating on auroral ionospheric plasma transport, *J. Geophys. Res.*, 100, 17499, 1995.

Chang, T., Lower hybrid collapse, caviton turbulence, and charged particle energization in the topside auroral ionosphere and magnetosphere, *Phys. Fluids*, 5, 2646, 1993.

Chang, T., and B. Coppi, Lower hybrid acceleration and ion evolution in the suprauroral region, *Geophys. Res. Lett.*, 8, 1253, 1981.

Chang, T., G.B. Crew, N. Hershkowitz, J.R. Jasperse, J.M. Retterer, and J.D. Winningham, Transverse acceleration of oxygen ions by electromagnetic ion cyclotron resonance with broad-band left-hand polarized waves, *Geophys. Res. Lett.*, 13, 636, 1986.

Chang, T., G.B. Crew, and J.M. Retterer, Electromagnetic tornadoes in space: ion conics along auroral field lines generated by lower hybrid waves and electromagnetic turbulence in the ion cyclotron range of frequencies, *Computer Phys. Comm.*, 49, 61, 1988.

Crew, G.B., and T. Chang, Asymptotic theory of ion conic distributions, *Phys. Fluids*, 28, 2382, 1985.

Crew, G.B., T. Chang, J.M. Retterer, W.K. Peterson, D.A. Gurnett, and R.L. Huff, Ion cyclotron resonance heated conics: theory and observations, *J. Geophys. Res.*, 95, 3959, 1990.

Dovner, P.O., A.I. Eriksson, R. Boström, B. Holback, J. Waldemark, L. Eliasson, and M. Boehm, The occurrence of lower hybrid cavities in the upper ionosphere, *Geophys. Res. Lett.*, 24, 619, 1997.

Eliasson, L., M. André, A. Eriksson, P. Norqvist, O. Norberg, R. Lundin, B. Holback, H. Koskinen, H. Borg, and M. Boehm, Freja observations of heating and precipitation of positive ions, *Geophys. Res. Lett.*, 21, 1911, 1994b.

Eliasson, L., O. Norberg, R. Lundin, K. Lundin, S. Olsen, H. Borg, M. André, H. Koskinen, P. Riihelä, M. Boehm, and B. Whalen, The Freja hot plasma experiment-instrument and first results, *Space Sci. Rev.*, 70, 563, 1994a.

Eriksson, A. I., B. Holback, P.O. Dovner, R. Boström, G. Holmgren, M. André, L. Eliasson, and P. M. Kintner, Freja observations of correlated small-scale density depletions and enhanced lower hybrid waves, *Geophys. Res. Lett.*, 21, 1843, 1994.

Ergun, R.E., C.W. Carlson, J.F. McFadden, F.S. Mozer, G.T. Delory, W. Peria, C.C. Chaston, M. Temerin, I. Roth, L. Muschietti, R. Elphic, R. Strangeway, R. Pfaff, C.A. Cattell, D. Klumpar, E. Shelley, W. Peterson, E. Mobius, and L. Kistler, FAST satellite observations of large-amplitude solitary structures, *Geophys. Res. Lett.*, submitted, 1998.

Erlandson, R. E., L.J. Zanetti, M.H. Acuña, A.I. Eriksson, L. Eliasson, M.H. Boehm, and L.H. Bloomberg, Freja observations of electromagnetic ion cyclotron ELF waves and transverse ion acceleration on auroral field lines, *Geophys. Res. Lett.*, 21, 1855, 1994.

Franz, J.R., P.M. Kintner, and J.S. Picket, POLAR observations of coherent electric field, *Geophys. Res. Lett.*, submitted, 1998.

Ganguli, G., M.J. Keskinen, H. Romero, R. Heelis, T. Moore, and C. Pollock, Coupling of microprocesses and macroprocesses due to velocity shear: an application to the low-altitude ionosphere, *J. Geophys. Res.*, 99, 8873, 1994.

Gorney, D.J., S.R. Church, and P.F. Mizera, On ion harmonic structure in auroral zone waves: the effect of ion conic damping of auroral hiss, *J. Geophys. Res.*, 87, 10479, 1982.

Gorney, D.J., Y.T. Chiu, and D.R. Croley, Trapping of ion conics by downward parallel electric fields, *J. Geophys. Res.*, 90, 4205, 1985

Gurnett, D.A., R.L. Huff, J.D. Menietti, J.L. Burch, J.D. Winningham, and S.D. Shawhan, Correlated low-frequency electric and magnetic noise along the auroral field lines, *J. Geophys. Res.*, 89, 8971, 1984.

Horne, R.B., and R.M. Thorne, Ion cyclotron absorption at second harmonic of the oxygen gyrofrequency, *Geophys. Res. Lett.*, 17, 2225, 1990.

Horwitz, J.L., Multiscale processes in ionospheric plasma outflows, *Phys. of Space Plasmas (1995)*, 14, 227, 1996.

Hultqvist, B., On the acceleration of electrons and positive ions in the same direction along magnetic field lines by parallel electric fields, *J. Geophys. Res.*, 93, 9777, 1988.

Hultqvist, B., Extraction of ionospheric plasma by magnetospheric processes, *J. Atmos. Terr. Phys.*, 53, 3, 1991.

Hultqvist, B., On the acceleration of positive ions by high-altitude electric field fluctuations, *J. Geophys. Res.*, 101, 27111, 1996.

Jasperse, J.R., Ion heating, electron acceleration, and the self-consistent parallel E-field in downward auroral current regions, *Geophys. Res. Lett.*, submitted, 1998.

Johnson, J.R., T. Chang, and G.B. Crew, A study of mode conversion in an oxygen-hydrogen plasma, *Phys. Plasmas*, 2, 1274, 1995.

Johnson, J.R., T. Chang, G.B. Crew, and M. André, Equatorially generated ULF waves as a source for the turbulence associated with ion conics, *Geophys. Res. Lett.*, 16, 1469, 1989.

Kindel, J.M., and C.F. Kennel, Topside current instabilities, *J. Geophys. Res.*, 76, 3055, 1971.

Kintner, P.M., J. Bonnell, R. Arnoldy, K. Lynch, C. Pollock, and T. Moore, SCIFER-transverse ion acceleration and plasma waves, *Geophys. Res. Lett.*, 23, 1873, 1996.

Kintner, P.M., J. LaBelle, W. Scales, A.W. Yau, and B.A. Whalen, Observation of plasma waves within regions of perpendicular ion acceleration, *Geophys. Res. Lett.*, 13, 1113, 1986.

Kintner, P.M., J. Vago, S. Chesney, R.L. Arnoldy, K.A. Lynch, C.J. Pollock, and T. Moore, Localized lower hybrid acceleration of ionospheric plasma, *Phys. Rev. Lett.*, 68, 2448, 1992.

Klumpar, D.M., Transversely accelerated ions: an ionospheric source of hot magnetospheric ions, *J. Geophys. Res.*, 84, 4229, 1979.

Klumpar, D.M., A digest and comprehensive bibliography on transverse auroral ion acceleration, in *Ion Acceleration in the Magnetosphere and Ionosphere*, AGU monograph no. 38, T. Chang et al., eds., (American Geophysical Union, Washington, D. C., 1986), .p. 389.

Knudsen, D. J., J. H. Clemmons, and J.-E. Wahlund, Correlation between core ion energization, suprathermal electron bursts, and broad-band ELF plasma waves, *J. Geophys. Res.*, in press, 1997.

Lundin, R., G. Gustafsson, A.I. Eriksson, and G. Marklund, On the importance of high-altitude low-frequency electric fluctuations for the escape of ionospheric ions, *J. Geophys. Res.*, 95, 5905, 1990.

Lundin, R., G. Haerendel, and S. Grahn, The Freja project, *Geophys. Res. Lett.*, 21, 1823, 1994a.

Lundin, R., G. Haerendel, and S. Grahn, The Freja science mission, *Space Sci. Rev.*, 70, 405, 1994b.

Lundin, R., G. Haerendel, M. Boehm, and B. Holback, Large-scale auroral plasma density cavities observed by Freja, *Geophys. Res. Lett.*, 21, 1903, 1994c.

Lundin, R., and B. Hultqvist, Ionospheric plasma escapes by high-altitude electric fields: magnetic moment pumping, *J. Geophys. Res.*, 94, 6665, 1989.

Lynch, K.A., R.L. Arnoldy, P.M. Kintner, and J. Bonnell, The AMICIST auroral sounding rocket: a comparison of transverse ion acceleration mechanisms, *Geophys. Res. Lett.*, 23, 3293, 1996.

Holback, B., S.-E. Jansson, L. Åhlén, G. Lundgren, L. Lyngdal, S. Powell, and A. Meyer, The Freja wave and plasma density experiment, *Space Sci. Rev.*, 70, 577, 1994.

Lysak., R.L., Ion acceleration by wave-particle interactions, in *Ion Acceleration in the Magnetosphere and Ionosphere*, AGU monograph no. 38, T. Chang et al., eds., (American Geophysical Union, Washington, D. C., 1986), .p. 261.

Lysak, R.L., M.K. Hudson and M. Temerin, Ion heating by strong electrostatic ion cyclotron turbulence, *J. Geophys. Res.*, 85, 678, 1980.

Marklund, G.T., L.G. Blomberg, P.-A. Linqvist, C.-G. Fälthammar, G. Haerendel, F.S. Mozer, A. Pederson, and P. Tanskanen, The double probe electric field experiment on Freja: experiment description and first results, *Space Sci. Rev.*, 70, 483, 1994.

Melatos, A., and P.A. Robinson, Local transit-time damping in a magnetic field, and the arrest of lower-hybrid wave collapse, *Phys. Plasmas*, 3, 1263, 1996.

Miyake, W., T. Mukai and N. Kaya, On the evolution of ion conics along field lines from EXOS D observations, *J. Geophys. Res.*, 7, 11127, 1993.

Moore, T.E., C.J. Pollock, M.L. Adrian, P.M. Kintner, R.L. Arnoldy, K.A. Lynch, and J. Holtet, The cleft ion plasma environment at low solar activity, *Geophys. Res. Lett.*, 23, 1877, 1996.

Norqvist, P., M. André, L. Eliasson, A.I. Eriksson, L. Blomberg, H. Lühr, and J.H. Clemmons, Ion cyclotron heating in the dayside magnetopause, *J. Geophys. Res.*, 101, 13179, 1996.

Palmadesso, P.J., T.P. Coffey, S.L. Ossakow, and K. Papadopoulos, Topside ionosphere ion heating due to electrostatic ion cyclotron turbulence, *Geophys. Res. Lett.*, 1, 105, 1974.

Peterson, W.K., H.L. Collin, M.F. Doherty and C.M. Bjorklund, $O^+$ and $He^+$ restricted and extended (bi-modal) ion conic distributions, *Geophys. Res. Lett.*, 19, 1439, 1992.

Pinçon, J.L., P.M. Kintner, P.W. Schuck, and C.E. Seyler, Observation and analysis of lower hybrid solitary structures as rotating eigenmodes, *J. Geophys. Res.*, in press, 1997.

Potellette, R., R.A. Treumann, and N. Dubouloz, Generation of auroral kilometric radiation in upper hybrid-lower hybrid soliton interaction, *J. Geophys. Res.*, 97, 12029, 1992.

Retterer, J.M., A model for particle acceleration in lower hybrid collapse, *Phys. Plasmas*, 4, 2357, 1997.

Retterer, J.M., T. Chang, and J.R. Jasperse, Ion acceleration by lower hybrid waves in the suprauroral region, *J. Geophys. Res.*, 91, 1609, 1986.

Retterer, J.R., T. Chang, G.B. Crew, J.R. Jasperse, and J.D. Winningham, Monte Carlo modeling of ionospheric oxygen acceleration by cyclotron resonance with broad-band electromagnetic turbulence, *Phys. Rev. Lett.*, 59, 148, 1987.

Retterer, J.M., T. Chang, and J.R. Jasperse, Transversely accelerated ions in the topside ionosphere, *J. Geophys. Res.*, 99, 13189, 1994.

Romero, H., G. Ganguli, and Y.C. Lee, Ion acceleration and coherent structures generated by lower hybrid shear-driven instabilities, *Phys. Rev. Lett.*, 69, 3503, 1992.

Rönnmark, K., and M. André, Convection of ion cyclotron waves to ion-heating regions, *J. Geophys. Res.*, 96, 17573, 1991.

Roth, I., and M. Hudson, Lower hybrid heating of ionospheric ions due to ion ring distributions in the cusp, *J. Geophys. Res.*, 90, 4191, 1985.

Shapiro, V.D., V.I. Shevchenko, G.I. Solov'ev, V.K. Kalinin, R. Bingham, R.Z. Sagdeev, M. Ashour-Abdalla, J. Dawson, and J. J. Su, Wave collapse at lower hybrid resonance, *Phys. Fluids,* 5, 3148, 1993.

Schuck, P.W., C.E. Seyler, J.L. Pinçon, J.W. Bonnell, and P.M. Kintner, Theory, simulation, and observation of discrete eigenmodes associated with lower hybrid solitary structures, *J. Geophys. Res.,* submitted, 1997.

Sharp, R.D., , R.G. Johnson, and E.G. Shelley, Observation of an ionospheric acceleration mechanism producing energetic (keV) ions primarily normal to the geomagnetic field direction, *J. Geophys. Res.,* 82, 3324, 1977.

Seyler, C.E., Lower hybrid wave phenomena associated with density depletions, *J. Geophys. Res.,* 99, 19513, 1994.

Tam, S.W.Y., and T. Chang, The limitation and applicability of Musher-Sturman equation to two-dimensional lower hybrid wave collapse, *Geophys. Res. Lett.,* 22, 1125, 1995.

Tam, S.W.Y., F. Yasseen, T. Chang, and S. Ganguli, Self-consistent kinetic photoelectron effects on the polar wind, *Geophys. Res. Lett.,* 22, 2107, 1995.

Temerin, M., and I. Roth, Ion heating by waves with frequencies below the ion gyrofrequency, *Geophys. Res. Lett.,* 13, 1109, 1986.

Wahlund, J.-A., A.I. Eriksson, B. Holback, M.H. Boehm, J.H. Clemmons, L. Eliasson, D.J. Knudsen, P. Norqvist, and L.J. Zanetti, *J. Geophys. Res.,* in press, 1997.

Winningham, J.D., and J. Burch, Observation of large-scale ion conic generation with DE-1, *Phys. of Space Plasmas (1982-4),* 5, 137, 1984.

Yau, A., and M. André, Sources of ion outflow in the high latitude ionosphere, *Space Science Reviews,* in press, 1997.

Yau, A.W., B.A. Whalen, F. Creutzberg, and P.M. Kintner, Low altitude transverse ion acceleration: auroral morphology and in-situ plasma observations, *Phys. of Space Plasmas (1985-7),* 6, 77, 1987.

Zanetti, L., Magnetic field experiment on the Freja satellite, *Space Sci. Rev.,* 70, 465, 1994.

---

T. Chang, Center for Space Research, Massachusetts Institute of Technology, Room 37-261, 77 Massachusetts Avenue, Cambridge, MA 02139, U.S.A.

Mats André, Swedish Institute of Space Physics, Umeå University, S-901 87 Umeå, Sweden.

# High-Latitude Auroral Boundaries Compared with GEOTAIL Measurements during Two Substorms

W. J. Burke,[1] N. C. Maynard,[2] G. M. Erickson,[3] M. Nakamura,[4] S. Kokubun,[5] B. Jacobsen,[6] and R. W. Smith[7]

GEOTAIL plasma and field data acquired near $X_{GSM}$ = -95 $R_E$ are compared with ground-based optical and magnetic field measurements and observations from low-Earth and geosynchronous satellites to specify relationships between auroral activity and magnetotail dynamics during the expansion phases of two moderate substorms. Multiple-onset phenomena appeared at different latitudes in the auroral oval. Time-delayed plasmoid signatures were detected by GEOTAIL. The observations are consistent with activations of near-Earth $X$ lines (NEXL) after substorm onsets with subsequent ejections of one or several plasmoids down the magnetotail. GEOTAIL data indicate that the plasmoids were limited in the $Y_{GSM}$ direction. Also, reconnection at the NEXL may proceed at variable rates on closed magnetic field lines for significant times before engaging lobe flux. This suggests that the plasma sheet in the near-Earth magnetotail is relatively thick in comparison to its embedded current sheet, and that both NEXLs and distant $X$ lines (DXL) can be active simultaneously. Until reconnection at a NEXL reaches the lobes, the DXL maintains control of the poleward auroral boundary. If a NEXL remains active after reaching the lobes, the boundary expands poleward.

## INTRODUCTION

Variations of the high-latitude ionosphere are more regularly monitored by sensors at ground facilities and on satellites in low-Earth orbit than in magnetospheric source regions. It is useful to develop remote sensing techniques to improve our understanding of magnetospheric processes through ionospheric signatures. This paper synopsizes a detailed analysis by *Maynard et al.* [1997] of auroral boundary dynamics based on simultaneous ground and space measurements taken during two substorms on January 14, 1994. Here greater emphasis is placed on optical measurements and their interpretation. Interplanetary magnetic field (IMF) data from IMP 8 have a gap during the first substorm and were highly variable in all three components during the second. Throughout the day GEOTAIL was at apogee in the magnetotail near $X_{GSM}$ = -95 $R_E$, $Y_{GSM}$ = 9 $R_E$, and a few $R_E$ south of the GSM equatorial plane. Allowing for the 4° aberration of the solar wind, GEOTAIL was slightly to the evening side of the magnetic midnight meridian where it remained in or near the plasma sheet. The CANOPUS and IMAGE systems were operating in the midnight sector during the first and second substorms, respectively. Auxiliary information came from sensors on Defense Meteorological Satellite Program (DMSP) satellites in the ionosphere and by energetic particle detectors

---

[1] Air Force Research Laboratory, Hanscom AFB, Massachusetts
[2] Mission Research Corporation, Nashua, New Hampshire
[3] Center for Space Physics, Boston University, Boston, Massachusetts
[4] Department of Earth and Planetary Physics, University of Tokyo, Tokyo, Japan
[5] Solar Terrestrial Environment Laboratory, Nagoya University, Toyokawa, Japan
[6] Department of Physics, University of Oslo, Oslo, Norway
[7] Geophysical Institute, University of Alaska, Fairbanks, Alaska

Geospace Mass and Energy Flow: Results From the International Solar-Terrestrial Physics Program
Geophysical Monograph 104
Copyright 1998 by the American Geophysical Union

on three Los Alamos National Laboratory (LANL) geosynchronous satellites [*Belian et al.*, 1992].

The poleward boundary of the aurora oval maps near the separatrix between open and closed field lines. With steady convection conditions, the boundary's location should vary smoothly with the potential imposed across the polar cap [*Siscoe*, 1982]. Expansions or contractions of the polar cap indicate that the open flux fraction of the Earth's magnetic field is either increasing or decreasing. These imply that the dayside merging rate is greater or less than the reconnection rate in the magnetotail. Departures from equilibrium excite substorms with onsets marked by brightenings of the equatorwardmost auroral arc, followed by poleward expansions of activity and formation of bulges in the midnight sector [*Akasofu*, 1964]. Optical techniques were developed to monitor the location and motion of the poleward auroral boundary and estimate magnetotail reconnection rates [*Blanchard et al.*, 1996].

A widely accepted model describes the linkage between the magnetosphere and high latitude ionosphere during substorms. After southward turnings of the IMF, merging with the Earth's field is enhanced along the dayside magnetopause. Flux is transferred from the dayside magnetosphere to the nightside. During the growth phase, intensified currents reconfigure field lines of the nightside magnetosphere causing them to stretch from dipolar to taillike. Onset occurs on field lines connecting the inner edge of the plasma sheet to the ionosphere [*Maynard et al.*, 1996]. Field lines rapidly dipolarize, and westward electrojets spanning a few hours of local time develop in the midnight sector [*Singer et al.*, 1983]. Onset is triggered by diversions of tail currents through the ionosphere by field-aligned current filaments [*McPherron et al.*, 1973].

*Hones* [1977] outlined a model of substorm dynamics that considers the consequences of a near-Earth X line (NEXL) forming as activity expands. A magnetic O type neutral line also forms and is expelled down the magnetotail as a plasmoid. Traveling compression regions provide evidence of substorm related plasmoids in the magnetotail [*Slavin et al.*, 1993]. *Moldwin and Hughes* [1993] reported close correspondence between detection of plasmoids in the distant tail and ground onsets. Controversy persists about where NEXLs form and their relation to substorm onsets. Growing evidence indicates that they form tailward of $X_{GSM}$ = -30 $R_E$ after onset [*Cowley*, 1992]. Since the launch of GEOTAIL, many plasmoid detections have been reported [*Kawano et al.*, 1994; *Machida et al.*, 1994; *Nagai et al.*, 1994; *Nishida et al.*, 1994a, b]. Slowly moving or quasi-stagnant plasmoids, first reported by *Nishida et al.* [1986], show that a NEXL and a distant X line (DXL) can be active simultaneously [*Nishida et al.*, 1996; *Hoshino et al.*, 1996; *Kawano et al.*, 1996].

The following section discusses plasma and field signatures detected by satellites on field lines controlled by an active X line and associated emissions in the auroral ionosphere. We then describe ground and space based data taken during two substorms. The final section compares GEOTAIL measurements with predictions of the *Hones* [1977] model, then demonstrates consistency between satellite and auroral observations.

## SIGNATURES OF ACTIVE X LINES

Before examining GEOTAIL and ground measurements we consider plasma and field signatures of active X lines observed both in the distant magnetotail and in the auroral ionosphere. Figure 1a provides the simplest representation of plasma flow near a DXL in its own rest frame. Allowance is made only for the **E** x **B** plasma drift into and away from the reconnection region. Reflecting on the regular detection by satellites of the plasma sheet boundary layer (PSBL), *Schindler and Birn* [1987] showed that the conservation of momentum also requires field aligned components of plasma drifts in the outflow region, represented schematically in Figure 1b. This is a general requirement for magnetic merging, not just for a DXL. Figure 1c represents plasma flows in the vicinity of a NEXL and DXL operating simultaneously [*Nishida et al.*, 1996; *Hoshino et al.*, 1996; *Kawano et al.*, 1996]. Plasma flows away from the two X lines, both toward the Earth and away from it. The convergence of plasma flows between the X lines indicates that this configuration is unstable and cannot continue indefinitely. In the magnetotail the two X lines are widely separated. After the NEXL activation a finite interval is required for information to propagate to the DXL. Depending on reconnection rates at the two X lines, force balance is achieved by a tailward retreat of the DXL and/or slowing of the plasmoid [*Nishida et al.*, 1986]. This continues until the NEXL merges all closed magnetic flux and begins to reconnect open field lines.

Near 95 $R_E$ in the magnetotail GEOTAIL should detect distinctive plasma and field signatures when connected to field lines influenced by processes operating at an active NEXL or DXL. Using standard GSM coordinates, these may be enumerated as follows: (1) Beyond the DXL, plasma flow is mostly antisunward ($V_X$ < 0), $B_Z$ is southward (negative) and $E_Y$ is dawn-to-dusk (positive). (2) Earthward of the DXL, but far from a NEXL, plasma flow is mostly sunward ($V_X$ > 0) with components along and across **B**, $B_Z$ is northward (positive) and $E_Y$ is dawn-to-dusk. (3) In the outer part of a plasmoid, plasma flow is antisunward ($V_X$ < 0), $B_Z$ is positive and $E_Y$ is dusk-to-dawn. (4) Earthward of the plasmoid's O line, flow remains antisunward, but $B_Z$ is southward and $E_Y$ dawn-to-dusk. (5) Earthward of the NEXL, plasma flow is sunward with a component along **B**, $B_Z$ is northward and $E_Y$ dawn-to-dusk. Magnetotail flapping often causes GEOTAIL to

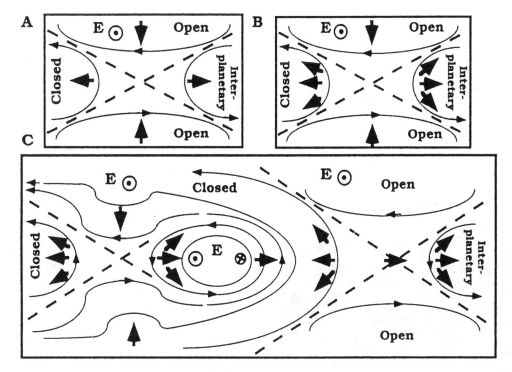

**Figure 1.** Schematic representations of plasma flows near $X$ lines with: (A) $\mathbf{E} \times \mathbf{B}$ drifts, (B) added field-aligned components, and (C) simultaneously active NEXL and DXL.

move quickly from the plasma sheet to the lobes where the plasma density is low. In the PSBL the satellite crosses strongly field aligned flows. If the flow is antisunward then GEOTAIL is tailward of either a DXL or an active NEXL.

Closed field lines magnetically connect the auroral oval to the magnetotail. The separatrix between open and closed field lines maps close to the poleward auroral boundary. Satellite investigations show that velocity dispersed ions from the PSBL intercept the ionosphere in a latitudinally thin strip just poleward of auroral and equatorward of polar rain electron fluxes [*Zelenyi et al.*, 1990; *Burke et al.*, 1994]. Thus, this boundary maps just earthward of whichever $X$ line is reconnecting lobe flux, usually the DXL. In ground optical measurements this transition is also marked by a sharp change in 630.0 nm emissions whose dynamics have been used to estimate the DXL reconnection rate [*Blanchard et al.*, 1996]. The existence of earthward field aligned fluxes from NEXLs suggests that their activations also can be monitored by optical remote sensing techniques. Owing to the large mass difference between electrons and ions, field aligned plasma flows require that they have different pitch angle distributions, field aligned for ions and nearly isotropic for electrons. *Alfvén and Fälthammar* [1963] pointed out that in magnetic mirror configurations above the auroral ionosphere, ions reflect lower than the electrons. To maintain quasi-neutrality, potential drops must develop along field lines, accelerating electrons into the ionosphere where they excite auroral emissions.

## FIRST SUBSTORM: 0600 - 0730 UT

Figure 2 shows magnetic field (top) and optical (bottom) data from CANOPUS [*Rostoker et al.*, 1994] acquired during the first substorm. Measurements of the $X$ magnetic field component along the CANOPUS meridian help identify onset times and electrojet locations. Station invari-ant latitudes are listed on the figure. Emissions at 630.0 nm measured by meridian scanning photometers at Pinawa, Gillam, and Rankin Inlet, presented in a stacked plot format, provide an optical history of the substorm. The start of negative bays at GILL and ISL mark onset at 0629 UT, just south of Gillam. Onset also appears as an intensification of 630.0 nm emissions, commencing at 0629 UT near 65° MLAT then moving poleward. A significant optical enhancement occurred over Ft. Churchill at ~0636 UT along with a sharp decrease in the magnetic $X$ component. Optical activity then progressed poleward to between 73° and 76° near 0641 UT. After the initial activity, amplitudes of the magnetic deflections and the intensity of auroral emissions decreased slightly at higher latitudes. However, both optical and magnetic activity

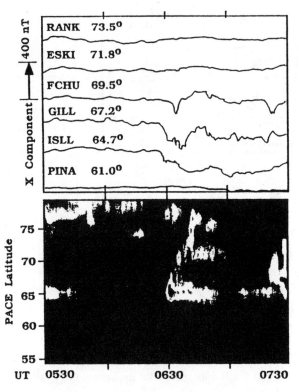

**Figure 2.** Magnetic and 630.0 nm optical data from CANOPUS stations along the Rankin - Pinawa magnetic meridian on January 14, 1994.

were earthward and $B_Z$ was positive. Except for a brief excursion into the PSBL at ~0540 UT, GEOTAIL was in the plasma sheet prior to onset, as indicated by low magnetic fields and elevated particle fluxes (not shown). Thus, GEOTAIL was on closed field lines earthward of a DXL during the growth and early expansion phases.

The ion flow turned tailward at 0637 UT ~1 min after the auroral expansion to 70° MLAT, becoming stronger and more energetic at 0642 UT. A second enhancement was observed at ~0655 UT. The increases in speed and intensity at 0642 and 0655 UT coincided with auroral brightenings near 75° MLAT. During the strong tailward flow, $B_Z$ was mostly northward. $E_Y$ was negative, except when $B_Z$ briefly changed sign, consistent with relatively steady tailward flow lasting > 20 min. Positive $B_Z$ and tailward flow indicate that GEOTAIL was on closed field lines or tailward of a plasmoid's $O$ line.

At 0659 UT flow turned sunward and field-aligned as GEOTAIL entered the PSBL. $E_Y$ and $B_Z$ were both positive, consistent with observed earthward flow. $V_X$ returned to tailward as $B_Z$ briefly reversed sign, then earthward with positive $B_Z$. During the second earthward flow, $V_X$ was field aligned. From 0710 to 0720 UT, $B_Z$ turned negative and plasma velocity became strongly tailward; $E_Y$

continued along the low latitude auroral border near 65°. The aurora intensified near 75° at 0650 UT. After 0710 UT the intensity of optical emissions at 74° dropped to background levels, indicating that the photometer scan crossed the open/closed field line and auroral oval/polar cap boundaries [*Blanchard et al.*, 1996]. We conclude that the auroral brightening at 0641 UT (2340 MLT) marks the polar cap boundary near 79° MLAT, the northernmost extent of optical coverage. The boundary then retreated equatorward. Our determinations of the onset time and open/closed boundary are consistent with data from one LANL and two DMSP satellites [*Maynard et al.*, 1997].

GEOTAIL data come from the magnetic field fluxgate instrument [*Kokubun et al.*, 1994], the electric field detector [*Tsuruda et al.*, 1994], and the low energy particle experiment [*Mukai et al.*, 1994]. Figure 3 displays the magnetic and electric field and ion bulk flow components during the first substorm as functions of UT and GSM locations of GEOTAIL. The top plot gives the magnetic field magnitude (dotted line) and its $X_{GSM}$ component $B_X$. The next two plots show $B_Z$ and $E_Y$. The lower four plots contain the velocity in the $X_{GSM}$ direction $V_X$, the velocity component along **B** (positive toward the Earth), as well as the $X_{GSM}$ and $Y_{GSM}$ velocity components perpendicular to **B**. For most of the hour prior to onset, ion flows at GEOTAIL

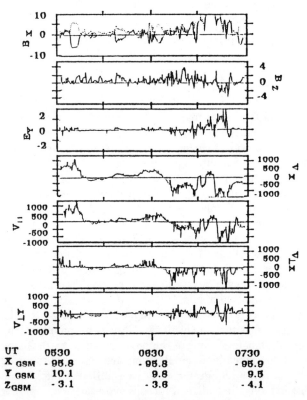

**Figure 3.** Magnetic field, electric field and ion drift velocities measured by GEOTAIL from 0530 to 0730 UT on January 14, 1994 [from *Maynard et al.*, 1997].

remained positive. While the $B_Z$ polarity changed, $V_Y$ was significant and dawn-to-dusk. The most intense fluxes were detected at this time (Plate 2 of *Maynard et al.* [1997]).

## SECOND SUBSTORM: 1900 - 2200 UT

During the second substorm, ground data came from the IMAGE chain [*Lühr et al.*, 1984] which extends from Svalbard into Finland. Magnetic meridian scanning photometer (MSP) and all-sky imager data were taken at Ny Ålesund and Longyearbyn in Svalbard. Since magnetic local time (MLT) is ~3.2 hours ahead of UT, data come from the magnetic midnight sector. The event, which lacked a clean onset, occurred after a southward IMF turning near 1815 UT. Figure 4 gives the variations of the $X$ magnetic field component along the IMAGE meridian chain. Activity began near the SOR and MAS stations in Finland at ~1858 UT then spread poleward, with a second initiation at 1925 UT. A third intensification began near 1950 UT at SOR and MAS and 2002 UT at HOP.

MSP measurements of 630.0 nm emissions from Longyearbyn are shown at the bottom of Figure 4. At 1925 UT, aurora near the southern horizon brightened and gradually moved poleward. After retreating briefly, the region of auroral emissions expanded rapidly after 1955 UT, crossing zenith at ~2000 UT to cover the visible sky until 2005 UT. MSP data from Longyearbyn and Ny Ålesund show a decrease in intensity between 2010 and 2020 UT. At 2025 UT the aurora retreated southward to the Ny Ålesund horizon and south of Longyearbyn's zenith until 2140 UT when a second poleward excursion began.

Complementary 2-D views of auroral activity in the midnight sector are given in Plate 1, which contains eight images taken at 10 min intervals by the all-sky television system at Ny Ålesund. Individual images provide 1 min averages of visible light intensities in a false-color representation. For comparison, the diagonal white line in the upper left image indicates the direction and extent of the Svalbard MSP scans. Prior to 1930 UT no emissions were visible from Ny Ålesund. In qualitative agreement with the MSP data in Figure 4, the aurora first appeared at Svalbard's southern horizon near 1930 UT, progressively stepped poleward until ~2000 UT, then exploded beyond the northern horizon. By 2040 UT, all emissions came from south of the two stations. The critical point is that substorm related variations in the aurora's poleward boundary extended at least three hours in local time near midnight, providing confidence that the ground observations relate to phenomena observed by GEOTAIL in the magnetotail. The DMSP F10 satellite crossed the polar cap boundary into the auroral oval at 2028 UT between Svalbard and Greenland. Its measurements are consistent

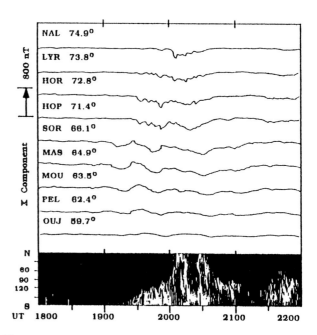

**Figure 4.** Magnetic $X$ component variations from IMAGE stations (top) and 630.0 nm optical measurements from Longyearbyn (bottom) on January 14, 1994.

with our interpretation that the poleward auroral boundary maps the location of open and closed field lines. A LANL satellite near midnight detected numerous injections.

Figure 5 shows GEOTAIL data in the same format as Figure 3. To relate events at GEOTAIL to ground observations of substorm phenomena, we present the data in four segments: (1) 1830 to 1925 UT, (2) 1925 to 2000 UT, (3) 2000 to 2021 UT, and (4) 2021 to 2150 UT.

In segment 1 GEOTAIL entered the plasma sheet from the southern lobe at 1830 UT. Ion flow was weak and variable until 1915 UT, when it increased to ~250 km/s becoming mostly in the $-X_{GSM}$ direction. $B_Z$ was positive, indicating closed field lines, and $E_Y$ became negative (dusk-to-dawn). This first effect at GEOTAIL related to ground activity occurred ~17 min after electrojet activation. The average energies of electrons and ions rose to a few hundred eV and a few keV, consistent with GEOTAIL being in the plasma sheet (Plate 3 of *Maynard et al.* [1997]).

In segment 2 $B_Y$ was the dominant component. From 1932 to 1938 UT $B_X$ was strongly negative, and the flow decreased as GEOTAIL approached the southern PSBL. Primarily flow was toward dusk, with regions of nearly isotropic fluxes. At 1938 UT the polarities of $B_Z$, $E_Y$, and $V_Y$ reversed, $B_X$ returned to near zero, and the plasma flow again became strongly tailward. $B_Y$ was large, negative and dominant in the next three minutes, after which it returned to near zero. $E_Y$ changed from negative to positive unlike the $B_Z$ reversal observed during the previous substorm

134  AURORAL BOUNDARIES DURING SUBSTORMS

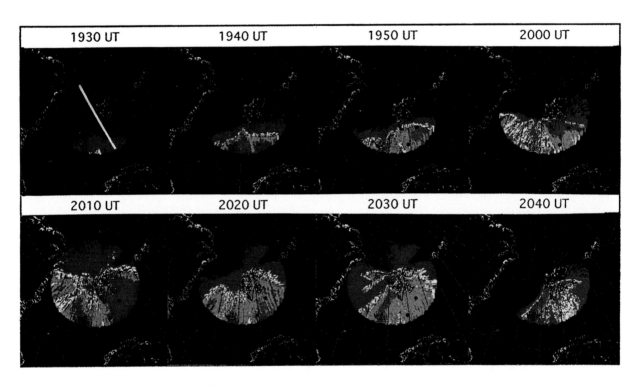

**Plate 1.** All-sky television images at visible wavelengths from Ny Ålesund during the second substorm.

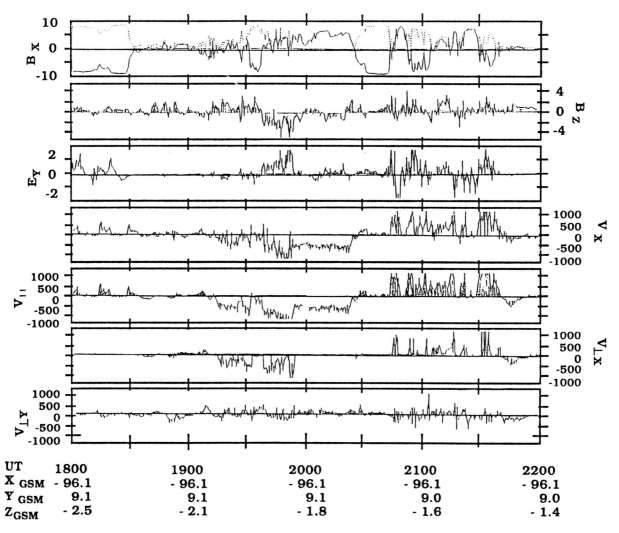

**Figure 5.** Magnetic field, electric field and ion drift velocities measured by GEOTAIL from 1830 to 2230 UT on January 14, 1994 [from *Maynard et al.*, 1997].

which was positive on both sides of the reversal. There was no evidence of significant earthward plasma flow. Between 1950 and 1955 UT, $V_X$ peaked in excess of 1000 km/s; both $E_Y$ and $B_Z$ also maximized. At 1955 UT these quantities quickly decreased and the plasma flow returned to a steady 500 km/s.

In segment 3 $B_X$ became large and positive, indicating that GEOTAIL entered the northern PSBL. Since ion temperatures and densities were high, it did not penetrate the lobe. Attention is directed to the $B_Z$ and $E_Y$ data between 2000 and 2020 UT. For the first 10 min $B_Z$ was positive then reversed polarity, as did $E_Y$, consistent with continued tailward convection. These changes are similar to observations at 1939 UT. The most poleward auroral excursion over Svalbard occurred in this interval.

During segment 4, from 2021 to 2030 UT, $B_Z$ was almost always positive, and after a brief negative excursion $E_Y$ remained positive. From 2028 to 2139 UT the dominant flow was earthward, except for nearly zero velocity episodes as GEOTAIL entered the southern lobe. Earthward flowing plasma and positive $B_Z$ indicate that it was again on closed field lines. Upon reentering the plasma sheet at 2043 UT, GEOTAIL detected variable ion fluxes. A series of earthward flow intensifications appeared at 5 - 8 min intervals. The most severe velocity decreases occurred during excursions into the lobe from the PSBL. While flow appears highly variable at GEOTAIL, it probably did not turn off between detections. Multiple reversals of $B_X$ indicate that the tail was flapping, with the satellite alternately in the northern and southern PSBLs. Peaks in $V_X$ occurred as GEOTAIL crossed the central plasma sheet. These flow enhancements continued until

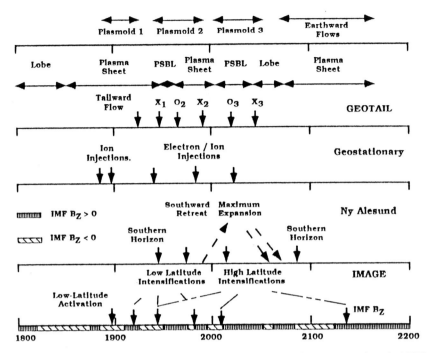

**Figure 6.** Summary of observations from the second substorm [from *Maynard et al.*, 1997].

2140 UT when $V_X$ turned tailward for ~10 min. This time $B_Z$ did not reverse, indicating that the tailward flow occurred on closed field lines caused by processes happening nearer to Earth.

## DISCUSSION

In neither interval did an isolated substorm develop fully from a clear onset through expansion to recovery. Multiple activations occurred to produce substorms, including rapid auroral expansions in the midnight sector within the oval and at its poleward boundary. To specify relationships between auroral activations/expansions and magnetospheric phenomena, we consider GEOTAIL data in the context of substorms and check the consistency of our interpretation against measurements.

Substorm models postulate that growth-phase stretching of field lines results from intensified plasma sheet currents and enhanced transfer of flux from day to night sides of the magnetosphere. *Hones* [1977] proposed that substorm expansion involves NEXL formation in which magnetic merging proceeds on closed field lines until it eats through to the lobes. Open field lines in the lobes of the tail then reconnect at the NEXL. Rapid inclusion of newly reconnected flux in the magnetotail causes the closed field line region, and with it the NEXL, to expand tailward. The NEXL then becomes a DXL until the next substorm cycle.

In Figure 1c we see that for GEOTAIL, located in the distant plasma sheet, the signatures predicted by the NEXL model during strong, isolated substorms are: (1) Shortly after onset, tailward plasma convection should be detected in a region with $B_Z > 0$ and $E_Y < 0$. (2) As the plasmoid's center approaches, the polarities of $B_Z$ and $E_Y$ should reverse. (3) For a brief time $B_Z > 0$ and $E_Y < 0$ should be detected while the $X$ line rapidly propagates tailward past the satellite. (4) As the magnetotail reverts to equilibrium, $B_Z$ and $E_Y$ should be positive and plasma should convect in the sunward direction. We argue that observed signatures are consistent with *Hones* [1977] if we allow significant periods for merging closed field lines before engaging lobe field lines. This allows the DXL to continue adding field lines until merging at the NEXL reaches the lobes.

Comparing GEOTAIL data with auroral observations is difficult since it is impossible to know exactly which regions are magnetically conjugate. We submit the following postulates to constrain our interpretation of events at both locations: (1) Field lines of the nightside auroral oval are closed. (2) Electrons ejected from active $X$ lines with access to the ionosphere produce auroral emissions. (3) Merging at NEXLs proceeds for some time before engaging lobe flux. (4) Until NEXLs engage the lobes, the poleward auroral boundary is controlled by the DXL. (5) NEXL and DXL activities do not necessarily correlate. (6) Onset brightening of the most equatorward arc is not necessarily caused by a NEXL. Although the second substorm is more complex, it is more easily reconciled with the plasmoid model. We consider it before discussing the first data set.

## Second Substorm

Figure 6 gives the second substorm timeline. Beginning at 1858 UT, magnetometers detected an electrojet activation and three distinct intensifications that were followed by electron injections observed by a LANL satellite. A brief delay occurred between dipolarization/ground onset and the arrival of electrons near geostationary altitude [*Maynard et al.*, 1996]. As conditions developed after 1858 UT, a NEXL site formed. Relaxation of the electrojet at ~1909 UT indicates that the NEXL merging rate slowed or stopped. As a result, field lines with $O$ topologies were introduced into the magnetotail and plasmoid #1 migrated tailward. This is consistent with GEOTAIL's detection of tailward flow with $B_Z > 0$ after 1915 UT, followed by a flow reduction at 1932 UT. The second electrojet intensification and particle injection corresponds to reactivation of the NEXL, creating more $O$ type field lines and plasmoid #2. Since no lobe flux reconnected, plasmoid #2 encompassed #1. Plasmoid #2 is marked by an $O$ line passage at 1938 UT and a flow reduction at 1955 UT. Again, merging either slowed or stopped. During the activation at ~1950 UT the NEXL engaged open flux. $O$ and $X$ lines of plasmoid #3 passed GEOTAIL at 2006 and 2021 UT.

Figure 7 sketches events as a sequence of magnetotail states during the four segments of this substorm. Before 1900 UT, convection in the plasma sheet near 95 $R_E$ was earthward under the influence of a DXL. Soon a small plasmoid formed and moved tailward. The first sign of its existence was a reversal of $V_X$ and $E_Y$, with $B_Z$ remaining northward. In segment 2 ion flow remained tailward as plasmoid #2 moved past GEOTAIL. The associated $O$ line crossed the location of GEOTAIL at 1938 UT. In segment

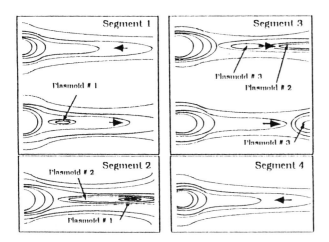

**Figure 7.** Schematic representation of the magnetotail during the second substorm [from *Maynard et al.*, 1997].

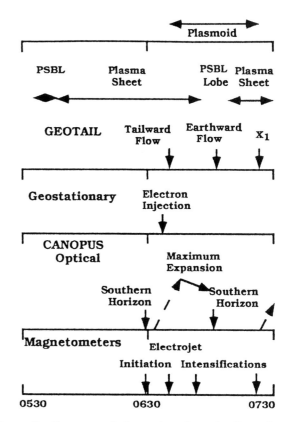

**Figure 8.** Summary of observations from the first substorm [from *Maynard et al.*, 1997].

3 all initially closed field lines were pinched off and open flux reconnected. As plasmoid #3 convected downtail, GEOTAIL moved into the PSBL where it did not fully encounter the $O$ line. It detected the telltale signature of its passage as bipolar responses of $B_Z$ and $E_Y$. The reduced velocity near 2025 UT marked the plasmoid's passage. GEOTAIL then went into the southern lobe of the tail. As GEOTAIL reentered the plasma sheet, $B_Z$, $E_Y$ and $V_X$ were positive, consistent with a DXL tailward of its location.

## First Substorm

Figure 8 is a timeline for the first substorm. Onset at 0629 UT was marked by an electrojet activation and auroral brightening at 65° ILAT. A LANL satellite ~3 hr east of midnight observed energy dispersed electrons at ~0632 UT, indicating dipolarization occurred in the midnight sector. We next investigate when the NEXL formed injecting a plasmoid into the magnetotail and how they relate to the model of *Hones* [1977].

Despite difficulties, GEOTAIL data are consistent with its encountering a plasmoid of finite extent in the $Y_{GSM}$ direction and allow us to extend the *Hones* [1977] model.

138  AURORAL BOUNDARIES DURING SUBSTORMS

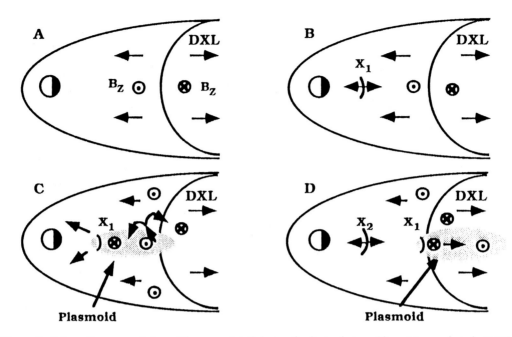

**Figure 9.** Schematic representation of the magnetotail during the first substorm [from *Maynard et al.*, 1997].

S*inger et al.* [1983] showed that after onset dipolarization is confined to a narrow $Y_{GSM}$ sector. Outside this sector the field remains taillike. A scenario that reconciles the NEXL model with GEOTAIL observations is shown in Figure 9. The top left plot shows an equatorial plane projection of the quiet time magnetotail with a DXL. The lower left plot indicates the change when NEXL $X_1$ formed. The upper right plot shows a plasmoid of finite size in the $Y_{GSM}$ direction after it propagated to GEOTAIL and tailward flowing plasma sheet ions were first detected (0638 to 0700 UT). However, instead of observing a $B_z$ polarity reversal, GEOTAIL then moved into a part of the PSBL with earthward flow. These observations indicate that GEOTAIL moved from a plasmoid into a region controlled by a DXL. At 0711 UT, $V_x$ turned strongly tailward and $B_z$ south-ward. After 0718 UT $B_z$ turned northward, but plasma flow remained tailward.

Observations of earthward flow in the PSBL after 0700 UT are critical for any interpretation and can only mean that GEOTAIL temporarily left field lines controlled by a NEXL and moved to those with dynamics influenced by remnants of the DXL. Note that at 95 $R_E$, the magnetotail moves ~1.5 $R_E$ for every 1° change in solar wind direction. Figure 9c shows that GEOTAIL left a plasmoid across its dawnside border. While it is possible that GEOTAIL moved in the Z direction into the PSBL, this is not supported by data. GEOTAIL's magnetometer would have detected a traveling compression region [*Slavin et al.*, 1993]. It did not.

When GEOTAIL returned to the plasma sheet at 0711 UT, both $B_z$ and $V_x$ were negative. Figure 1c allows two interpretations. GEOTAIL either crossed the DXL or returned to the plasmoid earthward of its O line. Plasma data in Plate 2 of *Maynard et al.* [1997] show that the intensity of particle fluxes makes the latter interpretation more plausible. The earthward flowing ions in the PSBL (0700 - 0705 UT) had energies centered at a few-hundred eV. During the $B_z < 0$ interval, the energies of tailward moving ions were >10 keV and mimic intense ion fluxes detected from 1940 to 1952 UT in the $B_z < 0$ portion of plasmoid #2. For the first interpretation to be correct, the DXL must have undergone a sudden activation and moved earthward of $X_{GSM}$ = -95 $R_E$. Figure 2 shows neither enhanced emissions nor new magnetic deflections at the auroral oval's poleward boundary near this time. The poleward auroral boundary continued a steady equatorward retreat. Thus, the polarity reversal of $B_z$ at 0718 UT can only be explained if GEOTAIL returned to the plasmoid earthward of the O line. Its X line passed GEOTAIL near 0720 UT when $B_z$ again turned positive. Persistent tail-ward flows after the X line passage suggest a new NEXL formed near the time of the 0720 UT electrojet activation.

*Auroral Boundary Dynamics*

The magnetotail dynamics observed at GEOTAIL have been interpreted as signs of NEXL generated plasmoids. We believe that during the second substorm, closed flux did not completely pinch off at the NEXL until the third electrojet activation when the auroral boundary expanded poleward (Figure 4 and Plate 1). Subsequently, open flux reconnected at the NEXL and the $X$ line eventually passed GEOTAIL. During the first substorm, GEOTAIL moved outside and then back into a plasmoid that was spatially limited in the $Y_{GSM}$ direction. The poleward auroral boundary moved equatorward throughout this event. We compare these conclusions with the auroral observations.

Assuming that electrons responsible for observed auroral emissions precipitated along closed magnetic field lines, we consider some ionospheric consequences of lobe-flux merging at $X$ lines. In the midnight sector at high latitudes, electric fields cause these field lines to convect with an equatorward drift component of ~1 km/s. In steady state flow, the boundary between open and closed field lines is stationary. Reconnection proceeds at a rate such that the inflow of open field lines to the boundary matches the outflow of newly closed field lines. If a NEXL becomes active on closed magnetic field lines, the ejected particles have different pitch angle distribution functions. Charge separations in the magnetic bottle above the ionosphere lead to the formation of field aligned potential drops [Alfvén and Fälthammer, 1963]. Electrons accelerated crossing the potential drop produce auroral signatures. Until the last closed field line in the magnetotail pinches off, auroral emissions are confined to latitudes equatorward of the DXL mapping. As merging of closed flux at the NEXL proceeds, electrons accelerated in the process gain access to the auroral ionosphere at progressively higher latitudes inside the open/closed field lines. During this time the DXL still adds closed flux to the magnetotail and thus maintains control of the poleward auroral boundary. Poleward movements of bright aurora within the oval are consistent with enhanced NEXL activity. Rapid poleward expansions of the auroral boundary indicate high rates of lobe-flux reconnection, most likely at the NEXL.

Figure 2 shows that auroral emissions brightened at 65° at onset (0629 UT). In the following 12 min the bright aurora expanded to 76°, a distance of ~1200 km, and approached the poleward boundary of the oval. We believe that the auroral enhancement and magnetic deflection beginning at 0636 UT at Churchill (69.5°) marked a NEXL activation. The NEXL formed after onset and well tailward of the initial dipolarization in agreement with Cowley [1992]. Note that there were no significant emissions above Churchill prior to this activation, including onset time. During the expansion phase the poleward auroral boundary steadily migrated equatorward while the active aurora expanded poleward (Figure 1) on the closed field lines. The fact that the poleward auroral boundary moved equatorward throughout the expansion phase indicates that the reconnection of lobe flux at the distant $X$ line continued but at a slower rate than flux transport from the dayside. Based on migration of the active aurora to the poleward boundary, we conclude that reconnection progressed to the lobe magnetic field lines, but did not engage the lobe until after 0720 UT when the poleward boundary began moving poleward. The lack of an auroral enhancement as the pole-ward boundary moved equatorward near 0711 UT is consistent with the DXL not moving earthward to 95 $R_E$. Rather, GEOTAIL first exited then reentered a plasmoid that was spatially restricted in the $Y_{GSM}$ direction.

The second substorm was marked by three electrojet intensifications. The first produced only weak emission at the edge of Longyearbyn's field of view. The second was accompanied by ~1 kR emissions that reached the southern horizon of Ny Ålesund. Only after the intensification of 1950 UT did the aurora expand rapidly poleward across the sky indicating NEXL reconnection of lobe field lines. As the rate of reconnection slowed, the NEXL moved down the tail and the poleward auroral boundary retreated equatorward. That the $X$ line remained active after passing GEOTAIL is evidenced by enhanced earthward flows.

These two events exemplify the magnetotail with two active $X$ lines. Hoshino et al. [1996] reached a similar conclusion in their analysis of an event three days later. The concept of quasi-stagnant plasmoids was originally postulated by Nishida et al. [1986] to explain pre-expansion phenomena. Slow moving structures were often followed by fast moving plasmoids associated with substorm expansions. Our events occurred after the onset of activity. The quasi-stagnant event seen by Kawano et al. [1996] also occurred after a ground onset and was followed by a fast moving plasmoid from a later onset. Our results, complemented by those of Hoshino et al. [1996] and Kawano et al. [1996], show that moderate and variable activity can create slow tailward moving plasmoids within closed field lines being augmented by an active DXL.

Let us return to our original point, the search for remote signatures of magnetospheric processes that can be monitored by ground based sensors of the auroral ionosphere. We noted that measurement techniques exist for identifying the boundary between open and closed field lines and convective flow across it, thus approximating the rate of reconnection [Blanchard et al., 1996]. This paper shows that: (1) Equatorward moving enhancements of activity at the poleward auroral boundary indicate an increased DXL reconnection rate. Here, however, since the polar cap was expanding, reconnection proceeded at a slower rate than flux transfer from the dayside [Siscoe, 1982]. (2) A poleward moving arc within the oval signifies an active NEXL on closed field lines inside the plasma sheet while lobe flux reconnection continues at a DXL. (3) A rapid, poleward expansion of the auroral boundary indicates active reconnection of lobe flux at an *X* line, most likely a NEXL that broke through to the lobe.

*Acknowledgments.* This work was supported in part by the U. S. Air Force Office of Scientific Research task 2311PL013, NSF grant ATM-9401629, and NASA grant NAGW-2627. We thank A. Egeland of the University of Oslo, H. Lühr of the University Braunschweig, T. Mukai, K. Tsuruda and T. Yamamoto of the Institute of Space and Astronautical Science, G. D. Reeves of the Los Alamos National Laboratory, J. C. Samson of the University of Alberta, and D. R. Weimer of Mission Research Corporation for access to and helpful discussions about data used in this paper.

## REFERENCES

Akasofu, S.-I., The development of the auroral substorm, *Planet. Space Sci., 12*, 273, 1964.

Alfvén H., and C.-G. Fälthammer, *Cosmical Electrodynamics*, 2nd ed., Oxford U. Press, Oxford, UK, 161-165, 1963.

Belian, R. D., G. R. Gisler, T. Cayton, and R. Christensen, High-Z energetic particles at geosynchronous orbit during the great solar proton event series of October 1989, *J. Geophys. Res., 97*, 16,897, 1992.

Blanchard, G. T., L. R. Lyons, O. de la Beaujardiere, R. A. Doe, and M. Mendillo, Measurement of the magnetotail reconnection rate, *J. Geophys. Res., 101*, 15,265, 1996.

Burke, W. J., J. S. Machuzak, N. C. Maynard, E. M. Basinska, G. M. Erickson, R. A. Hoffman, J. A. Slavin, and W. B. Hanson, Auroral ionospheric signatures of the plasma sheet boundary layer in the evening sector, *J. Geophys. Res., 99*, 2489, 1994.

Cowley, S. W. H., The role and location of magnetic reconnection in the geomagnetic tail during substorms, *Proceedings of the International Conference on Substorms*, 23-27 March 1992, Kiruna, Sweden, ESA SP-335, 1992.

Hones, E. R., Jr., Substorm processes in the magnetotail: Comments on "On hot tenuous plasma, fireballs, and boundary layers in the Earth's magnetotail" by L. A. Frank, K. L. Ackerson, and R. P. Lepping, *J. Geophys. Res., 82*, 5633, 1977.

Hoshino, M., T. Mukai, A. Nishida, Y. Saito, T. Yamamoto, and S. Kokubun, Evidence of two active reconnection sites in the distant magnetotail, *J. Geomag. Geoelectr., 48*, 515, 1996.

Kawano, H., T. Yamamoto, S. Kokubun, K. Tsuruda, A. T. Y. Lui, D. J. Williams, K. Yumoto, H. Hayakawa, M. Nakamura, T. Okada, A. Matsuoka, K. Shiokawa, and A. Nishida, A flux rope followed by recurring encounters with travelling compression regions: GEOTAIL observations, *Geophys. Res. Lett., 21*, 2891, 1994.

Kawano, H., A. Nishida, M. Fujimoto, T. Mukai, S. Kokubun, T. Yamamoto, T. Terasawa, Y. Saito, S. Machida, K. Yumoto, H. Matsumoto, and T. Murata, A quasi-stagnant plasmoid observed with GEOTAIL on October 15, 1993, *J. Geomag. Geoelect., 48*, 525, 1996.

Kokubun, S., T. Yamamoto, M. H. Acuna, K. Hayashi, K. Shiokawa, and H. Kawano, The GEOTAIL magnetic field experiment, *J. Geomag. Geoelectr., 46*, 7, 1994.

Lühr, H., S. Thürey, and N. Klöcker, The EISCAT magnetometer cross: Observational aspects, first results, *Geophys. Surv., 6*, 305, 1984.

Machida, S., T. Mukai, Y. Saito, T. Obara., T. Yamamoto, A. Nishida, M. Hirahara, T. Terasawa, and S. Kokubun, GEOTAIL low energy particle and magnetic field observations of a plasmoid at $X_{GSM}$ = -142 $R_E$, *Geophys. Res. Lett., 21*, 2995, 1994.

Maynard, N. C., W. J. Burke, E. M. Basinska, G. M. Erickson, W. J. Hughes, H. J. Singer, A. G. Yahnin, D. A. Hardy, and F. S. Mozer, Dynamics of the inner magnetosphere near times of substorm onsets, *J. Geophys. Res., 100*, 7705, 1996.

Maynard, N. C., W. J. Burke, G. M. Erickson, M. Nakamura, T. Mukai, S. Kokubun, T. Yamamoto, B. Jacobsen, A. Egeland, J. C. Samson, D. R. Weimer, G. D. Reeves, and H. Lühr, GEOTAIL measurements compared with the motions of high-latitude auroral boundaries during two substorms, *J. Geophys. Res., 101*, 9553, 1997.

McPherron, R. L., C. T. Russell, and M. P. Aubry, Satellite studies of magnetospheric substorms on August 15, 1968: Phenomenological models for substorms, *J. Geophys. Res., 78*, 3131, 1973.

Moldwin, M. B., and W. J. Hughes, Geomagnetic substorm association of plasmoids, *J. Geophys. Res., 98*, 81, 1993.

Mukai, T., S. Machida, Y. Saito, M. Hirahara, T. Terasawa, N. Kaya, T. Obara, M. Ejiri, and A. Nishida, The low energy particle (LEP) experiment onboard the GEOTAIL satellite, *J. Geomag. Geoelectr., 46*, 669, 1994.

Nagai, T., K. Takahashi, H. Kawano, T. Yamamoto, S. Kokubun, and A. Nishida, Initial GEOTAIL survey of magnetic substorm signatures, *Geophys. Res. Lett., 21*, 2991, 1994.

Nishida, A., M. Scholer, T. Terasawa, S. J. Bame, G. Gloeckler, E. J. Smith, and R. D. Zwickl, Quasi-stagnant plasmoids in the middle tail: A new pre-expansion phase phenomena, *J. Geophys. Res., 91*, 4245, 1986.

Nishida, A., T. Mukai, Y. Saito, T. Yamamoto, H. Hayakawa, K. Maezawa, S. Machida, T. Terasawa, S. Kokubun, and T. Nagai, Transition from slow to fast tailward flow in the distant plasma sheet, *Geophys. Res. Lett., 21*, 2939, 1994a.

Nishida, A., T. Yamamoto, K. Tsuruda, H. Hayakawa, A. Matsuoka, S. Kokubun, M. Nakamura, and H. Kawano, Classification of the tailward drifting magnetic structures in the distant tail, *Geophys. Res. Lett., 21*, 2947, 1994b.

Nishida, A., T. Mukai, T. Yamamoto, Y. Saito, and S. Kokubun, Magnetotail convection in geomagnetically active times: 1. Distance to neutral lines, *J. Geomag. Geoelectr., 48*, 489, 1996.

Rostoker, G., J. C. Samson, F. Creutzberg, T. J. Hughes, D. R. Mc Diarmid, A. G. McNamara, A. V. Jones, D. D. Wallis, and L. L. Cogger, CANOPUS - A ground-based instrument array for remote sensing in the high latitude ionosphere during ISTP/GGS program, *Space Sci. Rev., 71*, 743, 1994.

Schindler, K., and J. Birn, On the generation of field-aligned flow at the boundary of the plasma sheet, *J. Geophys. Res., 92*, 99, 1987.

Singer, H. J., W. J. Hughes, P. F. Fougere, and D. J. Knecht, The localization of Pi 2 pulsations: ground-satellite observations, *J. Geophys. Res., 88*, 7029, 1983.

Siscoe, G. L., Polar cap size and potential: A predicted relationship, *Geophys. Res. Lett, 9*, 672, 1982.

Slavin, J. A., M. F. Smith, E. L. Mazur, D. N. Baker, E. W. Hones, Jr., T. Iyemori, and E. W. Greenstadt, ISEE 3 observations of travelling compression regions in the Earth's magnetotail, *J. Geophys. Res., 98*, 15,245, 1993.

Tsuruda, K., H. Hayakawa, M. Nakamura, T., Okada, A. Matsuka, F. S. Mozer, and R. Schmidt, Electric field measurements on the GEOTAIL satellite, *J. Geomag. Geoelectr. 46*, 693, 1994.

Zelenyi, L. M., R. A. Kovrazkhin, and J. M. Bosqued, Velocity dispersed ion beams in the nightside auroral zone, *J. Geophys. Res., 95*, 12,119, 1990.

---

W. J. Burke, Air Force Research Laboratory, 29 Randolph Road, Hanscom AFB, MA, 01731.

G. M. Erickson, Center for Space Physics, Boston University, 725 Commonwealth Ave., Boston, MA 02215.

B. Jacobsen, Department of Physics, University of Oslo, 1048 Blindern, Oslo, Norway.

S. Kokubun, Solar Terrestrial Environment Laboratory, Nagoya University, Toyokawa 441, Japan.

N. C. Maynard, Mission Research Corporation, One Tara Blvd., Suite 302, Nashua, NH, 03062.

M. Nakamura, Department of Earth and Planetary Physics, University of Tokyo, Tokyo 113, Japan.

R. W. Smith, Geophysical Institute, University of Alaska, Fairbanks, AK 99701

# Energy Characterization of a Dynamic Auroral Event Using GGS UVI Images

G. A. Germany[1], G. K. Parks[2], M. J. Brittnacher[2], J. F. Spann[3], J. Cumnock[4], D. Lummerzheim[5], F. Rich[6], and P. G. Richards[7]

The GGS POLAR satellite, with an apogee distance of 9 Earth radii, provides an excellent platform for extended viewing of the northern auroral zone. Global FUV auroral images from the Ultraviolet Imager onboard the POLAR satellite can be used as quantitative remote diagnostics of the auroral regions, yielding estimates of incident energy characteristics. UVI images have been used previously to study total hemispheric input into auroral regions and local time variations. The same analysis on a higher temporal and spatial scale can be challenging in the presence of dynamic auroral forms which change significantly between images. Here, such an event is analyzed and compared with *in situ* DMSP observations of the same event. The difficulties, and potential remedies, associated with such a study are discussed.

## INTRODUCTION

The principal science objective of the Ultraviolet Imager (UVI) is to provide global information on the flow of energy between the Earth's magnetosphere and its ionosphere [*Torr et. al.*, 1995]. This is accomplished by using far ultraviolet (FUV) auroral images provided by UVI to estimate incident energy flux and average energy of the precipitating auroral particles causing the observed emissions. UVI images have been used previously to study total hemispheric input into auroral regions and local time variations [e.g. *Brittnacher et al.*, 1997; *Doe et al.*, 1997; *Lummerzheim et al.*, 1997] but only limited effort has been expended to analyze UVI image data at high spatial and temporal resolution.

The Ultraviolet Imager is capable of temporal resolutions on the order of 1 second for specialized operating modes. In practice, however, the normal imaging resolution is 36.8 seconds. The spatial resolution varies with orbital altitude from a perigee resolution of about 5 km per pixel to 40 km per pixel at apogee. Since auroral forms can significantly change in brightness on a time scale shorter than the UVI temporal resolution, and since at least two image frames are required for average energy estimation, analysis of dynamic auroral forms can be particularly difficult – thus the relative lack of such studies to date. Here, such an event is analyzed and compared with *in situ* DMSP observations of the same event.

## TECHNIQUE

The technique for using FUV emissions as remote diagnostics depends on energy-dependent loss mechanisms, principally absorption by $O_2$ and has been discussed elsewhere [e.g. *Strickland et al.*, 1983; *Germany et. al.*, 1994a,b; 1990]. The principal emissions within the UVI bandpass (125.0 - 200.0 nm) are atomic oxygen emissions and molecular $N_2$ Lyman-Birge-Hopfield (LBH) emissions. Atomic nitrogen lines also appear. The LBH emissions are of particular interest since they are due solely to electron impact excitation. Thus, in the absence of dayside photoelectrons, LBH emissions are direct diagnostics of the incident auroral flux. The $O_2$ Schumann-Runge absorption continuum peaks within the UVI bandpass, decreasing with

[1]Center for Space Plasma and Aeronomic Research, University of Alabama in Huntsville, Huntsville.
[2]Geophysics Department, University of Washington, Seattle.
[3]NASA Marshall Space Flight Center, Huntsville, AL.
[4]Space Physics Research Laboratory, University of Michigan, Ann Arbor.
[5]Geophysical Institute, University of Alaska, Fairbanks.
[6]Geophysics Dir., USAF Phillips Laboratory, Hanscom AFB, MA.
[7]Computer Science Department, University of Alabama in Huntsville, AL.

Geospace Mass and Energy Flow: Results From the International Solar-Terrestrial Physics Program
Geophysical Monograph 104
Copyright 1998 by the American Geophysical Union

longer wavelength. Auroral emissions viewed from space in this spectral region of strong absorption will thus exhibit some losses, provided the incident auroral energy is high enough to reach lower altitudes where $O_2$ density is greatest. Therefore the short wavelength emissions exhibit an energy-dependent loss mechanism that can be used as a remote diagnostic of the mean incident energy.

The energy characteristics of the precipitating auroral particles are described by stating the flux of incident particles, their mean energy, and how their energy is distributed about that mean. The incident flux is typically specified as an energy flux in $mW/m^2$. The energy distribution about the mean is typically given as either a Gaussian, a Maxwellian, or some other similar distribution. Ideally all three parameters would be derivable from UVI images.

Two of the requisite three parameters, incident energy flux and average energy, are provided from the intensity of the LBHl images and from the ratio of LBHl and LBHs (or 1356) images. The LBH emissions can be divided into two regions: one at shorter wavelengths with significant losses due to $O_2$ absorption (LBHs) and longer wavelength emissions with less loss (LBHl). *Germany et al.* [1994b] studied the possibility that FUV auroral images could be used to specify the incident auroral energy distributions, i.e. the third parameter necessary to completely specify the energy characteristics of the incident auroral particles. They found that the magnitude of changes in column intensities with choice of energy distribution is less than about 30% for a number of different distributions, including Gaussian and Maxwellian. The dependence of modeled FUV auroral emissions on energy distribution is thus much less than their dependence on average energy or energy flux. The determination of these two quantities from UVI images should therefore be relatively unambiguous, since such determinations will not be dependent on an exact knowledge of the incident energy spectrum. Since the estimated errors in the analysis of UVI data (see below) are larger than the variations found by *Germany et al.* [1994b], the sensitivity of this type of analysis is generally not enough to discriminate between different energy distributions based on UVI images. Nevertheless, UVI images provide the two most significant diagnostics - energy flux and average energy.

Modeling of expected emissions can be used to estimate incident energy flux from LBHl intensities. Auroral intensities are modeled with a two-stream energy deposition code [*Richards and Torr, 1990*]. Emission cross sections of $N_2$ LBH are those of *Ajello and Shemansky* [1985] with the downward scaling of *Ajello et al.* [1991]. The average energy can be estimated from the ratio of either OI 1356 or LBHs to LBHl. The ratio is necessary to normalize against changes in incident energy flux. Since the emissions in the LBHs and LBHl filter passbands originate from the same specie, the ratio is nearly independent of compositional changes with season or over a solar cycle [*Germany et al., 1990*].

In the ideal case, analysis of FUV auroral emissions would be performed with only two of the many LBH emission bands - one at the wavelength of peak absorption and the other at a longer wavelength where $O_2$ absorption is negligible. The longer wavelength emission would therefore be essentially independent of average energy and solely dependent on energy flux. This is the case modeled previously by *Germany et. al.* [1994a,b; 1990]. In practice, however, the UVI LBH filters must necessarily include multiple LBH bands and therefore contain a range of loss factors. Thus the LBHl emission is not totally independent of average energy and shows a weak dependence with average energy that is not indicated in the previous work by *Germany et. al.* [1994a,b; 1990]. For example, the LBHl emission intensity produced by a fixed electron energy flux input decreases by about 10% for a change in average energy from 5 to 10 keV. Because this effect is relatively small, however, it is not included in the analysis shown below.

For the period examined here UVI was in an operational mode that employed only the two LBH filters. The discussion below is therefore limited to the analysis of the LBHs/LBHl ratio.

Quantitative photometric analysis of the type attempted here requires a proper knowledge of all error i.e., non-auroral, sources. In particular, auroral images must have instrumental and airglow backgrounds removed before analyzing. Airglow can be estimated by *ab initio* calculations. However, this requires the use of an airglow model that is tailored to meet the specific viewing requirements of the observation, e.g. changes in line of sight enhancements across the image. For the work presented here airglow is approximated by binning the airglow pixels by solar zenith angle, excluding auroral emissions, using a technique similar to that of *Lummerzheim et al.* [1997]. From these binned pixels an airglow surface is approximated and then removed from the original image.

A few caveats must be kept in mind when reviewing energy products derived from UVI images. First of all, the ratioed images used to determine the average energy are not temporally coincident since the imager has only a single optical path allowing only a single filter to be used at a time. Temporal and spatial changes in the auroral morphology between image frames can lead to incorrect energy estimations. This is especially evident in highly dynamic conditions in which the local intensities may change significantly between image times. Also, since the energy estimation utilizes ratios of images it is sensitive to uncertainties in both images. The results presented here have internal quality checks to remove these artifacts from the final products. For the work presented here image processing (binning, smoothing, ratioing, etc.) introduces about 3% uncertainty and the Poisson uncertainty is about 5%. Assuming an accuracy of 25% for the instrumental calibration, 22% for the LBH cross sections [*Ajello and Shemansky*, 1985], and 25% for all other model uncertainties provides a total uncertainty of about 45%.

## DATA

Data analyzed is from May 19, 1996 (96140) with the POLAR spacecraft near an apogee distance of 9 Earth radii. Between 21:41 and 21:46 UT the DMSP F12 satellite passed through the northern auroral region at an altitude of about 800 km. Plate 1 shows a sequence of UVI images taken between 21:42 and 21:44 UT coincident with the DMSP overflight. The DMSP F12 satellite traveled northward through the nightside oval over Greenland. The UVI operations during this time used pairs of images in the following sequence: 2 LBHs, 2 LBHl, 2 LBHs, shutter. The first of each pair of images is an 18 second exposure instead of the normal 36 second exposures. (This is a UVI operational requirement to prevent image smearing during movement of the filter wheel.) The images in Plate 1 are the three 36 second exposures available for this period and are spaced about 1.25 minutes apart. These are the images used in the analysis below.

At 21:15 UT, about 30 minutes before the first LBHs image in Plate 1, the poleward boundary began the intensification that is clearly evident in the image. Activity propogates westward almost 90 degrees in longitude during the 1.5 minutes shown in Plate 1.

The two images in Plate 2 are energy maps derived from LBHl image and the second LBHs image beginning at 21:43:20 UT and 21:44:33 UT, respectively. The images are displayed in MLAT-MLT coordinates where MLAT is computed for an apex magnetic coordinate system and MLT is the corresponding local time. The average energy map (Eavg) is calculated from the ratio of the LBHl and LBHs images and has a time resolution set by the time encompassed by the two images (roughly 1 minute). The energy flux map (Eflux) is calculated from the LBHl image alone and therefore has a higher time resolution than the average energy map. The ground track of the DMSP overflight is shown.

The energy flux map shows concentrated high energy fluxes in a broad arc (in local time) on the poleward boundary of the oval. A secondary arc of lower energy is seen on the equatorward boundary which appears to be the remnant of previous activity beginning at 20:20 UT. The average energy map shows two bands of higher energy electrons colocated with the higher energy fluxes. A trough of lower energy electrons exists between the two arcs on the dusk sector and on the equatorward boundary near midnight. Little information is displayed for the dayside since the signal strength after dayglow removal was deemed too low for analysis.

We wish to compare the *in situ* DMSP observations with those derived from analysis of the UVI images. This is done by plotting DMSP energy parameters along its ground track and then superimposing values derived from UVI images for the same spatial locations. Note that since the two instruments have different time resolutions the comparisons along the ground will be coincident for only a limited time corresponding to the UVI integration time. If we assume that the overall auroral morphology is roughly stable over the 5 minute DMSP overflight we can compare values from the two instruments along the entire overflight using only the LBHs and LBHl images used to compute Plate 2. However, as shown in Plate 1, the morphology is clearly dynmic and changes signifcantly during this period. Nevertheless, for the purpose of discussion, the data from DMSP and the energy derived from UVI are superimposed in Figure 1.

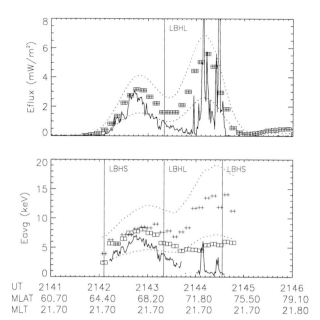

**Figure 1.** Comparisons between energy flux and average energy as determined from DMSP measurements and UVI image analysis. *In situ* observations of incident energy flux and average energy are shown as solid lines. Crosses and boxes represent values from UVI energy maps corresponding to the ground tracks. (Crosses are derived using the first two images in Plate 1; boxes from the second pair of images.) Vertical lines correspond to the beginning of each of the UVI LBHl or LBHs integrations, as appropriate. Each integration is 37 seconds. Dashed lines represent an estimated 45% uncertainty level (shown, for clarity, for only the crosses).

Figure 1 shows comparisons between energy flux and average energy as determined from DMSP measurements and UVI image analysis along the DMSP ground track. *In situ* observations of incident energy flux and average energy are shown as solid lines. Crosses and boxes represent values from UVI energy maps corresponding to the ground tracks. Vertical lines correspond to the beginning of the UVI LBHl or LBHs integrations (36.8 seconds), as appropriate. The data marked by crosses is obtained from analysis of the first two images in Plate 1 while boxed data are derived using the final two images in Plate 1. Dashed lines represent the assumed 45% total uncertainty discussed above.

## DISCUSSION

The derived energy flux from the first pair of UVI images (top of Figure 1) show generally good agreement for the equatorward part of the oval (UT less than 21:44 in the figure). On the poleward boundary the *in situ* DMSP observations show two spikes of energy flux that are significantly higher than is derived from the UVI images. Clearly the UVI data doesn't have the same resolution as teh DMSP observations and tends to smooth all the poleward structure into a single broad arc with its peak near the first (equatorward) brightening seem by DMSP but seemingly not detecting the second (poleward) brightening. Re-

146 ENERGY CHARACTERIZATION OF AN AURORAL EVENT USING UVI IMAGES

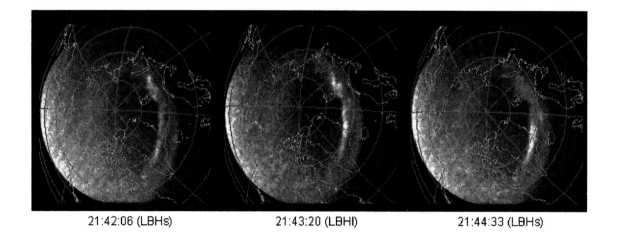

**Plate 1.** Sequence of UVI images taken between 21:42 and 21:44 UT on May 19, 1996 coincident with a DMSP F12 overflight. The DMSP satellite traveled northward through the nightside oval over Greenland during this period.

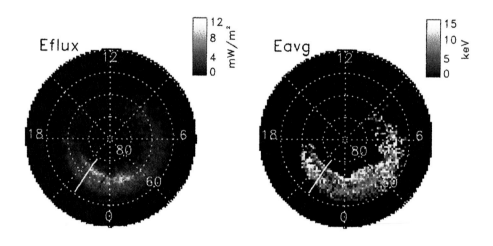

**Plate 2.** Maps of incident energy flux (Eflux) and average energy (Eavg). The ground track of the DMSP satellite is shown near 21:30 MLT.

call, however, that the poleward boundary is brightening and expanding throughout this event. The UVI data is properly coincident only at the time of the LBHl integration which ends by 21:44 UT. Thus, by the time the DSMP satellite reached the northern boundary the incident energy flux had increased over what was seen in the UVI image taken over a minute previously.

This interpretation is further borne out by the average energy comparisons shown in the bottom panel of Figure 1. The average energies derived using the first LBHs images are in general agreement for the equatorward portion of the oval but disagree strongly for the poleward portion. The image ratio used in this calculation is LBHl divided by LBHs. Since the LBHs preceded the LBHl image, a monotonic increase in poleward brightness as seen here results in an overestimate of the average energy. The boxed values derived from the second pair of images shows significantly better agreement, though UVI still overestimates the *in situ* average energy observations, presumably because of dynamic changes between the two image frames used in the analysis. An analysis employing the 18 second integration between the two could help further refine the analysis.

Thus Figure 1 shows both the power and the potential limitations of using FUV auroral images as remote diagnostics of the aurora. The analysis is most limited in highly dynamic situations in which the aurora changes significantly between two UVI images. Plate 2, however, best illustrates the potential of analysis based on Ultraviolet Imager data. The images allow quantitative estimates across extended regions of the auroral zone and are not restricted to a single ground track. Thus imaging data should be viewed as complementary to the various available *in situ* and ground based observations.

In this paper, UVI images have been used to estimate the magnitude of the incident energy flux over the entire auroral zone. The inferred energy fluxes generally agree in magnitude and morphology with selected DMSP overflights. The fluxes inferred from UVI images do not exhibit the same spatial or temporal resolution as the *in situ* measurements but offer a global perspective unattainable from single satellite passes. Average energy maps are also constructed from ratioed images. Care must be taken in interpreting such maps since they are subject to error from temporal and spatial changes as illustrated in Figure 1. Continued comparison with *in situ*, ground based, and other imaging observations is needed to build confidence in this approach. Such studies are currently underway.

*Acknowledgments.* This work was supported, in part, under U. Washington contract 256730 to the University of Alabama in Hunstville and NASA grant NAG5-3170 to the University of Washington. Work at the University of Alaska was supported by NASA grant NAG5-1097. A. Richmond kindly supplied the database upon which the apex coordinates are based. The authors gratefully acknowledge the many helpful comments of the reviewers.

## REFERENCES

Ajello, J. M., and D. E. Shemansky, A reexamination of important $N_2$ cross sections by electron impact with application to the dayglow: The Lyman-Birge-Hopfield band system and NI (119.99nm), *J. Geophys. Res., 90*, 9845, 1985.

Ajello, J. M., D. E. Shemansky, and G. K. James, Cross sections for production of H(2p, 2s, 1s) by electron collisional dissociation of $H_2$, *Astrophys. J., 371*, 422, 1991.

Brittnacher, M., R. Elsen, G. Parks, L. Chen, G. Germany, and J. Spann, A dayside auroral energy deposition case study using the Polar Ultraviolet Imager, *Geophys. Res. Lett., 24*, 991 1997.

Doe, R. A., J. D. Kelly, D. Lummerzheim, G. K. Parks, M. J. Brittnacher, G. Germany, and J. Spann, Initial comparison of POLAR UVI and Sondrestrom IS radar estimates for auroral electron energy flux, *Geophys. Rev. Lett., 24*, 999, 1997.

Germany, G. A., D. G. Torr, P. G. Richards, M. R. Torr, and S. John, Determination of ionospheric conductivities from FUV auroral emissions, *J. Geophys. Res., 99*, 23297, 1994a.

Germany, G. A., M. R. Torr, D. G. Torr, P. G. Richards, Use of FUV auroral emissions as diagnostic indicators, *J. Geophys. Res., 99*, 383, 1994b.

Germany, G. A., M. R. Torr, P. G. Richards, and D. G. Torr, The dependence of modeled OI 1356 and $N_2$ LBH auroral emissions on the neutral atmosphere, *J Geophys. Res., 95*, 7725, 1990.

Lummerzheim, D., M. Brittnacher, D. Evans, G. A. Germany, G. K. Parks, M. H. Rees, and J. F. Spann, High time resolution study of the hemispheric energy flux carried by energetic electrons into the ionosphere during the May 19/20 auroral activity, *Geophys. Rev. Lett., 14*, 987, 1997.

Richards, P. G., and D. G. Torr, Auroral modeling of the 3371 A emission rate: Dependence on characteristic electron energy, *J. Geophys. Res., 95*, 10337, 1990.

Strickland, D. J., J. R. Jasperse, and J. A. Whalen, Dependence of auroral FUV emissions on the incident electron spectrum and neutral atmosphere, *J. Geophys. Res., 88*, 8051, 1983.

Torr, M. R., D. G. Torr, M. Zukic, R. B. Johnson, J. Ajello, P. Banks, K. Clark, K. Cole, C. Keffer, G. Parks, B. Tsurutani, J. Spann, A far ultraviolet imager for the International Solar-Terrestrial physics mission, *Space Sci. Rev., 71*, 329, 1995.

---

M. Brittnacher, G. Parks, Geophysics Program, Box 351650, University of Washington, Seattle WA 98195.

J. Cumnock, Space Physics Research Laboratory, University of Michigan, 2455 Hayward Rm 2517, Ann Arbor, MI 48109.

G. Germany, Center for Space Plasma and Aeronomic Research, OB 348, University of Alabama in Huntsville, AL 35899 (e-mail: germanyg@cspar.uah.edu).

D. Lummerzheim, Geophysical Institute, Fairbanks, AK 99775.

F. Rich, Geophysics Dir., USAF Phillips Laboratory (PL/GPS), Hanscom AFB, MA 01731.

P. Richards, Computer Science Department, OB 348, University of Alabama in Huntsville, AL 35899.

J. Spann, Space Science Laboratory, NASA Mail Code ES83, NASA Marshall Space Flight Center, Huntsville AL 35812.

# Auroral Observations from the POLAR Ultraviolet Imager (UVI)

G. A. Germany[1], J. F. Spann[2], G. K. Parks[3], M. J. Brittnacher[3], R. Elsen[3], L. Chen[3], D. Lummerzheim[4], and M. H. Rees[5]

Because of the importance of the auroral regions as a remote diagnostic of near-Earth plasma processes and magnetospheric structure, spacebased instrumentation for imaging the auroral regions have been designed and operated for the last twenty-five years. The latest generation of imagers, including those flown on the POLAR satellite, extends this quest for multispectral resolution by providing three separate imagers for the visible, ultraviolet, and X ray images of the aurora. The ability to observe extended regions allows imaging missions to significantly extend the observations available from *in situ* or groundbased instrumentation. The complementary nature of imaging and other observations is illustrated below using results from the GGS Ultraviolet Imager (UVI). Details of the requisite energy and intensity analysis are also presented.

## INTRODUCTION

The principal science objective of UVI is to provide global information on the flow of energy between the Earth's magnetosphere and its ionosphere [*Torr et al.*, 1995]. This is accomplished by using auroral images to estimate, on a per-pixel basis, incident energy flux and average energy of precipitating particles. The secondary objective of UVI is to provide a near-continuous monitor of the auroral morphology in the far ultraviolet for both sunlit and dark side conditions, subject only to the constraints of the orbital viewing geometry. Each of these objectives is to be achieved in the full context of the ISTP mission by using observations from the full suite of ISTP missions to provide a detailed understanding of the global nature of the integrated geospace system.

[1] Center for Space Plasma and Aeronomic Research, University of Alabama in Huntsville, Huntsville.
[2] NASA Marshall Space Flight Center, Huntsville, AL.
[3] Geophysics Department, University of Washington, Seattle.
[4] Geophysical Institute, University of Alaska, Fairbanks.
[5] Physics Department, University of Southampton, Southampton, England.

Geospace Mass and Energy Flow: Results From the International Solar-Terrestrial Physics Program
Geophysical Monograph 104
Copyright 1998 by the American Geophysical Union

The UVI science goals directly address the GGS goals of measuring mass, momentum, and energy flow in the coupled solar wind/magnetospheric system; of studying geospace plasma processes; and of assessing their impact on the terrestrial environment [*Acuna et al.*, 1995]. These goals, in turn, drove the design of the Ultraviolet Imager and, in part, the POLAR spacecraft. For example, the requirement to measure energy flow into the auroral regions required ultraviolet spectral resolution to a degree not previously obtained with filtered devices. The requirement that the instrument monitor the global behavior of the entire auroral region argued for a much wider field of view than typically available for such a compact camera. This requirement also called for a high apogee orbit both to allow the POLAR imagers to view extended fields of view and to allow full auroral viewing for extended periods of time. With an apogee distance of 9 earth radii, UVI is capable of viewing the entire auroral region continuously for almost 12 hours out each 18 hour orbit.

The Ultraviolet Imager is a small sophisticated camera with a wide field of view that uses a filter wheel to select one of five available far ultraviolet spectral regions for imaging. One of the major design accomplishments of UVI is the development of the high transmission, high resolution FUV filters used in the camera. Five filters are used to isolate emissions from OI 1304, OI 1356, $N_2$ LBH (short wavelength), $N_2$ LBH (long wavelength), and long wavelength scattered sunlight (Figure 1). These wavelength regimes were chosen to maximize the scientific return from the UVI images. Note that the OI 1304 and 1356 emissions are also resolved by two dedicated filter systems.

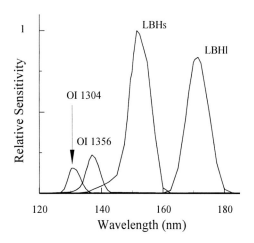

**Figure 1.** UVI spectral sensitivity as a function of wavelength. The curves are shown on a relative scale and the diagnostic solar filter centered longward of 180 nm is not shown.

Extensive laboratory calibrations were conducted to determine absolute and relative photometric and spectral calibrations, as well as the pointing characteristics for each of the two detector systems. These studies have been augmented with data taken from orbit. The absolute photometric calibration allows the instrumental response to be directly related to the brightness of the observed emissions. This is the first requirement for quantitative analysis of image data. In addition, the relative response of the instrument across the image plane must also be determined so the brightness at different parts of the images may be directly compared with the rest of the image. The wide field of view of the Ultraviolet Imager made this part of the calibration especially challenging. Finally the spectral response of the camera for each filter selection was determined. This is especially critical for analysis of the LBH band system as well as for proper determination of out-of-band contamination of the 1356 emissions from the 1304 emission.

## ANALYSIS TECHNIQUE

The technique for using FUV emissions as remote diagnostics has been discussed elsewhere [*Strickland et al.*, 1983; *Germany et. al.*, 1994a,b; 1990; *Lummerzheim et al.*, 1991] and will only be summarized here. The principal emissions within the UVI spectral bandpass (120.0 - 200.0 nm) are atomic oxygen emissions and molecular $N_2$ Lyman-Birge-Hopfield (LBH) emissions. Atomic nitrogen lines also appear. The only significant excitation mechanism is electron impact excitation. Thus, ignoring photoelectrons produced on the dayside, observed LBH intensities are directly proportional to the incident auroral flux provided there is no significant loss of intensity along the line of sight. Some of the LBH emissions will be absorbed by $O_2$ along the imaged line of sight. The amount of absorption is determined by the $O_2$ Schumann-Runge absorption continuum which peaks within the UVI bandpass. Since $O_2$ absorption is significant only for altitudes below about 150 km, loss will only be observed for the more energetic electrons that cause emission from these altitudes. It is this altitude-dependent production and loss with mean auroral energy that allows the emissions to be used as remote diagnostics of mean energy.

The $N_2$ LBH emissions are thus divided into two regions: one at shorter wavelengths with significant losses due to $O_2$ absorption (denoted LBHs and extending roughly from 140 to 160 nm) and longer wavelength emissions with less loss (LBHl, 160 to 180 nm), as shown in Figure 1. Modeling of expected emissions can be used to estimate incident energy flux from LBHl intensities and the average energy from the ratio of either OI 1356 or LBHs to LBHl. (The OI 1356 emission exhibits the same absorption losses as LBHs and can therefore also be used to estimate average energy.) The 1356 emission, however, is dependent on changes in the atomic oxygen density and can therefore exhibit significant variability. On the other hand, since the emissions in the LBHs and LBHl filter passbands originate from the same specie, their ratio is nearly independent of compositional changes with season or over a solar cycle [*Germany et al.*, 1990]. For these reasons, the LBH ratio is used almost exclusively for average energy analysis for UVI images. The following discussion is therefore limited to the analysis of the LBHs/LBHl ratio.

Since absorption by $O_2$ is the principal physical mechanism used to estimate average auroral energy, it can be expected that there will be variation with changes in $O_2$ relative to $N_2$, similar to the variability seen in OI 1356 with changes in O. [*Germany et al.*, 1990] examined this possibility and found that the LBH ratio is also stable with changes in composition from a variety of sources. Molecular oxygen dependence is generally not considered significant since the $O_2/N_2$ ratio changes little at the lower altitudes of significance here. This dependence can be included, however, as a special analysis though it is not part of normal UVI analysis.

In the ideal case, UVI average energy analysis would be performed with two LBH bands - one at the wavelength of peak absorption and the other at a longer wavelength where $O_2$ absorption is negligible. The longer wavelength emission would be essentially independent of average energy and solely dependent on energy flux. This is the case modeled previously by *Germany et. al.* [1994a,b; 1990]. In practice, the UVI LBH filters must necessarily include multiple LBH bands and therefore contain a range of loss factors. Thus the LBHl emission shows a weak dependence with average energy that is not indicated in the previous work by *Germany et. al.* [1994a,b; 1990].

Figure 2 shows modeled vertical column brightnesses for LBH emissions convolved with the UVI instrumental bandpass. Thus each curve now contains contributions for all LBH bands and not just the idealized case of only two bands modeled previously. As noted above, the LBHl emission intensity produced by a fixed electron energy flux input is not entirely independent of mean energy and varies by about 10% between 1 and 10 keV. Thus the dependence on energy flux is, in turn, dependent on a knowledge of mean energy. This is illustrated in Figure 3 where vertical brightnesses as a function of incident energy flux is presented for a range of mean energies. In addition to this dependence on mean energy there will also be dependences on diurnal and seasonal changes as well as changes in magnetic activity, principally by changes in the neutral atmospheric composition. These variations were examined previously by Germany et. al. [1990] for the idealized cases of only two LBH bands. Table 1 shows sample variations for the case of the entire LBH band system convolved with the UVI instrumental response. The largest variation is the seasonal variation of about 6% which is slightly larger than the 3-4% reported by Germany et. al. [1990]. Since each of these effects is relatively small they are not included in the energy analysis shown below. Instead a fixed value for the brightness to energy flux conversion is used.

Two of the requisite three parameters, incident energy flux and average energy, are provided from the intensity of the LBHl images and from the ratio of LBHl and LBHs (or 1356) images. Germany et al. [1994b] studied the possibility that FUV auroral images could be used to specify the incident auroral energy distributions, i.e. the third parameter necessary to completely specify the energy of the incident auroral flux. They found that the sensitivity of this type of analysis is not enough to discriminate between different energy distributions. Nevertheless, the two derivable diagnostics - energy flux and average energy - are excellent probes of the observed auroral processes.

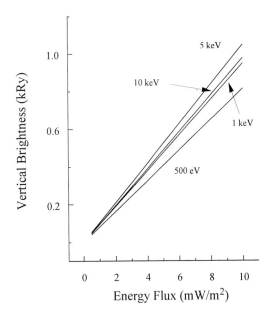

**Figure 3.** Modeled vertical column brightnesses using the UVI LBHl filter for Gaussian energy distributions with mean energies of 0.5, 1, 5, and 10 keV.

Since UVI images include auroral and airglow emissions, the images must have both instrumental and airglow backgrounds removed before analyzing. Airglow is approximated from the image by assuming pixels with similar values of solar zenith angle will have similar airglow emissions. The underlying airglow surface is estimated by binning image pixels by solar zenith angle, excluding auroral emissions.

Additional corrections are also made for brightness variations with changing spacecraft view angle (the angle from the local zenith to the spacecraft). These are illustrated in Figure 4 which shows the increased slant path brightness (relative to the vertical brightness) for a range of

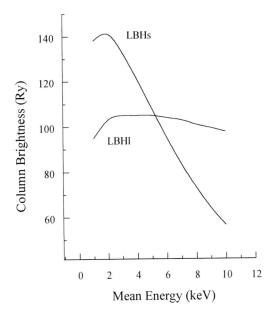

**Figure 2.** Modeled vertical column brightnesses for LBH emissions convolved with the UVI instrumental bandpass. The convolution is over all bands of the LBH band system. Emissions are modeled for a Gaussian energy distribution with an incident energy flux of 1 mW/m².

**Table 1.** Sample variations of the LBH band system convolved with the UVI instrumental response. Each of the test cases represents a perturbation from the reference case.

| Case | Brightness | Comments |
|---|---|---|
| Reference | 105 Ry | January 18, 1997, 0 UT |
| | | 70 degrees geographic latitude, |
| | | 0 degrees geographic longitude, |
| | | F107=75.0, F107a=75.0, Ap=4 |
| | | 1mW/m² energy flux |
| | | 5keV Gaussian |
| Noon | 106 Ry | 12:00 UT |
| Lat50 | 106 Ry | 50 degrees geographic latitude |
| Active | 104 Ry | F107=200, F107a=200, Ap=100 |
| Summer | 111 Ry | July 17, 1997 |

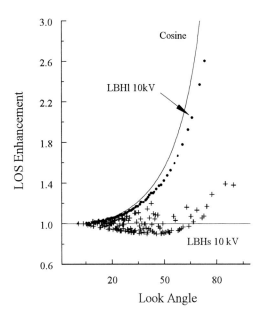

**Figure 4.** Slant path brightness enhancements for a range of spacecraft angles, where the spacecraft look angle is the angle between the local zenith at point of emission and the direction to the spacecraft. (Look angle equals zero degrees when viewing directly down toward the center of the Earth and equals 90 degrees when viewing the limb.) Brightness is shown relative the nadir direction (look angle equal zero degrees). Both the LBHl and LBHs emissions are modeled for auroral emissions with a mean energy of 10 keV. Each modeled point represents an observation corresponding to a single pixel of the UVI field of view. For clarity, only a fraction of all points are shown on the plot. The line labeled 'Cosine' is the enhancement expected from purely geometric considerations.

spacecraft angles. For the LBHl images the enhancement is nearly the cosine enhancement that would be expected from purely geometric concerns. The departure from pure cosine behavior is due to $O_2$ absorption from some of the shorter wavelength LBH bands which are observed by the LBHl filter. These shorter bands are partially absorbed by $O_2$, introducing an additional, nongeometric, factor into the line of sight analysis. The enhanced absorption along the slant path is competing with the increased emission resulting in a brightness less than expected from the line of sight geometry. This is a small effect for the LBHl filter since the absorbed bands are only a small fraction of the total observed bands. For LBHs, however, this effect has significant consequences as seen in the figure. Not only is the slant path enhancement countered by the increased absorption along the view path, but the absorption is a function of the neutral composition of $O_2$ which can vary significantly across the UVI field of view due to local time effects. The result is a surface dependent on both solar zenith angle and spacecraft look angle. Note that this behavior is significant only for higher energy LBHs auroral emissions. LBHs observed dayglow, originating from higher in the thermosphere, will not be as strongly affected by $O_2$ absorption and will have behavior more similar to the LBHl data shown in Figure 4.

For average energy calculations, the image data must be registered onto a regular magnetic latitude-MLT grid before image ratios can be calculated. This ensures the pixels used in the image ratios correspond to the same location even though the two images are not temporally coincident. The fact that the two ratioed images are not coincident can lead to analysis problems in cases where the auroral morphology is significantly changing between the two observations. In a companion paper, *Germany et al.* [1997] analyze a dynamic auroral event showing how energy characteristics can be extracted under such conditions.

For the work presented here typical uncertainties can be estimated. Image processing (binning, smoothing, ratioing, etc.) introduces about 3% uncertainty and the Poisson uncertainty is about 5%, dependent on the brightness of the observed features. Assuming an accuracy of 25% for the instrumental calibration, 22% for the LBH cross sections [*Ajello and Shemansky*, 1985], and 25% for all other model uncertainties provides a total uncertainty of about 45%.

There is also a well-known wobble in the POLAR despun platform that affects the imagers on that platform. The wobble is seen in UVI images as a vertical smearing of about 0.5 degree, or about 10 pixels or so. (There is no distortion in the transverse direction.) The effect is clearly seen for fixed bright sources such as stars but is not always present in auroral displays. We interpret this as evidence that the observed auroral emissions are changing over the course of the image integration and don't result in a smeared signature.

## OBSERVATIONS AND RESULTS

In the discussion below UVI images are used to illustrate how global imaging can be used to augment *in situ* and ground based observations, including the capability of observing sunlit aurora. This is then followed with two case studies based on UVI images. The first study is an examination of energy influx during the onset of the magnetic storm of January 10, 1997. The second study is a coordinated ISTP study involving multiple spacecraft and groundbased observations.

### Global Versus In Situ Observations

One of the principal uses of spacebased imagery is to provide a larger context for measurements that are often limited in temporal and spatial extent. For example, in Figure 5, taken from *Doe et al.* [1997], a one hour time history of incident auroral flux is shown corresponding to the location of the Sondrestrom radar. Also shown are energy flux values derived from four radar measurements during the same period. (The mean and maximum curves are used to investigate the relative influence of the binned 0.5 x 1 degree bins.) The principal feature to note from the figure is that the energy flux data derived from the image data shows much more temporal detail than is available from the ground based observations.

Another example of this is given in Figure 6 from *Lummerzheim et al.* [1997] which compares hemispheric power input derived from UVI images with the same parameters derived from energetic electron and ion precipitation measured during passes of the NOAA/TIROS satellites over the aurora [*Evans*, 1987; *Fuller-Rowell and Evans*, 1987]. The power derived from the satellites is obtained by fitting the *in situ* observations to a statistical database derived from the entire auroral oval. Possible shortcomings of such an approach are seen in the figure at 19:20 and 22:45 UT where there are significant disagreements between UVI and NOAA/TIROS power estimates. Examination of UVI data for these periods reveals that the satellite pass at these times sampled locally brighter areas of the oval, thus leading to an overestimate of the total power.

Both of these examples examine total hemispheric power input and necessarily ignore smaller scale structures in the auroral zone. As discussed above, however, UVI images can be used to estimate the magnitude of the incident energy flux over the entire auroral zone on a per pixel basis. Plate 1 shows a map of incident energy flux derived from a UVI LBHl image from May 19, 1996. The UVI image data have been rebinned and registered in a mlt-magnetic latitude coordinate system using apex coordinates. Superimposed on the energy map is the ground track of a DMSP satellite as it passed through the region. The inferred energy fluxes generally agree in magnitude and morphology with selected DMSP overflights (Figure 7). This is a more challenging comparison due to differences in spatial and temporal resolution between UVI and DMSP as well as the fact that this is a dynamic event in which the poleward boundary was changing rapidly on a time scale of less than a minute. The fluxes inferred from UVI images do not exhibit the same spatial or temporal resolution as *in situ* measurements but offer a global perspective unavailable from single satellite passes. This event is studied in more detail by *Germany et al.*, [1997].

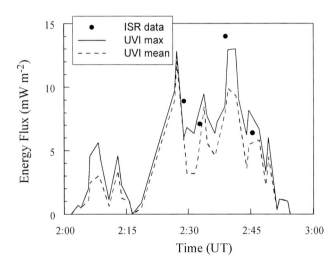

**Figure 5.** Time history of energy flux derived from UVI and Sondrestrom observations. The line labeled 'maximum' corresponds to the brightest pixel within a bin size of 0.5 (lat) x 1 (lon) degree; the line labeled 'mean' is the mean of all pixels within the binned area. From *Doe et al.* [1997].

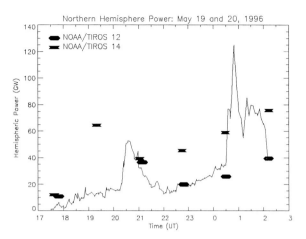

**Figure 6.** Total hemispheric energy flux derived from UVI images on May 19/20, 1997. Thick horizontal bars show the hemispheric power derived from NOAA/TIROS satellites. From *Lummerzheim et al.* [1997].

In summary, when spacebased imagery is available from a high altitude the information content provided is significantly greater than can be supplied by either *in situ* or groundbased observations alone. However, as seen in Figure 7, *in situ* observations are often available with a resolution that exceeds that of the imager. (UVI spatial resolution varies from 4 km at perigee to 40 km at apogee.) hence the two observation techniques must be viewed as complementary.

*Dayside Aurora*

A significant consequence of the ultraviolet wavelength discrimination capabilities of UVI is its ability to monitor dayside aurora. When coupled with the extended field of view of the Ultraviolet Imager this allows simultaneous observations of auroral activations throughout the auroral zone on both dark and sunlit conditions. Plate 2 illustrates this with two views of northern auroral activity observed on April 9, 1996 at 14:35 and 14:45 UT.

The dayside is to the right in both images. The auroral oval straddles the FUV terminator with dayside intensifications seen in both images. The circular field of view of UVI is seen outlined on the map. The period from 13:45 to 14:00 UT, prior to the images in Plate 2, represents the recovery phase of a minor substorm on the predawn nightside over Russia and Alaska. By 14:00 UT the nightside intensifications had faded away. However, throughout this period the dayside auroral zone showed continuous activity covering local time from prenoon to almost dusk. From 14:00 to 14:30 UT the auroral activity is confined totally to the dayside. The situation at 14:34 UT is shown in the first image of Plate 2. At 14:39 UT a substorm began on the nightside, roughly premidnight. Over the next 40 minutes this activity will expand to encompass the full nightside oval. Observations over the 1.5 hours monitored here clearly show that on this day the dayside auroral activity is much

154 AURORAL OBSERVATIONS FROM THE POLAR ULTRAVIOLET IMAGER (UVI)

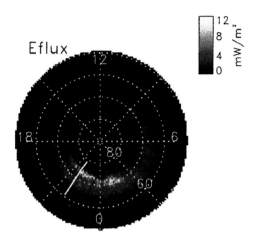

**Plate 1.** Map of incident energy flux derived from a UVI LBHl image. The UVI image data has been rebinned and registered in a mlt-magnetic latitude coordinate system using apex coordinates. The image is from 21:44 UT May 19, 1996 and is coincident with a DMSP F12 overflight as shown by the colored track near 21:00 MLT.

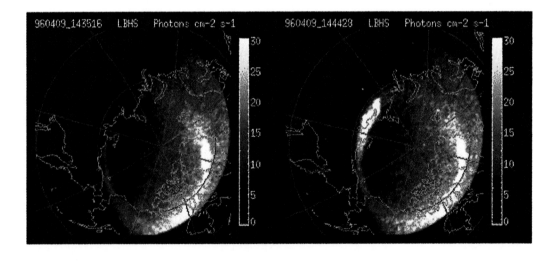

**Plate 2.** Observations of auroral activations throughout the auroral zone on both dark and sunlit conditions on April 9, 1996 at 14:34 and 14:44 UT. The dayside is to the right in both images. The dayside and nightside activities are essentially decoupled with continuous dayside activity coincident with distinct quiet and substorm periods on the nightside. The scale is in units of photons $cm^{-2}$ $s^{-1}$.

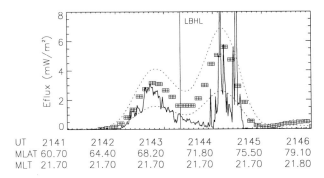

**Figure 7.** Comparisons between incident energy flux determined from coincident DMSP measurements and UVI image analysis. *In situ* observations of incident energy flux are shown as solid lines. Boxed crosses represent values from the UVI energy map corresponding to the DMSP ground track. The vertical line corresponds to the beginning of the UVI LBHl integration of 37 seconds. Dashed lines represent an estimated 45% uncertainty level.

less variable than is the nightside activity, a fact also noted by *Brittnacher et al.*, [1997]. In that study the total power deposition was estimated in the dayside and nightside halves of the auroral region (divided along the 0600 to 1800 MLT boundary) for two quiet days in spring (April 7, 1996) and summer (July 23, 1996). The results are reproduced here in Figure 8 showing that the power deposited on the dayside aurora varied significantly less than did that on the nightside.

In the image shown in Plate 2 the dayside and nightside activities are essentially decoupled with continuous dayside activity coincident with distinct quiet and substorm periods on the nightside. In other cases, however, the two regions appear to be coupled with simultaneous night and dayside activations or with one type of activity leading to the other. *Brittnacher et al.*, [1996], for example, observed dayside precipitation that subsequently propagated eastward toward local midnight. This particular event is especially interesting in that it accompanied an unusually strong dipolarization as observed by ISTP spacecraft yet did not proceed to a classical substorm development.

*Onset of Magnetic Storm On January 10, 1997*

One principal advantage of having a continuously operating camera is the ability to observe unexpected or rapidly changing auroral morphologies. An example of this capability occurred during the preparation of this manuscript with the magnetospheric response of the solar coronal mass ejection event observed at Earth on January 10, 1997. A shock front was observed around 01:00 UT with a magnetic cloud following around 03:40 UT. UVI images from the period show extremely distorted ovals with multiple, distorted transpolar arcs evident (Plate 3). The principal storm is typically identified as beginning near 03:40 UT. However, analysis of power deposition derived from UVI images (Figure 9) for the 3 hours preceding this time clearly shows significant activity beginning before 01:10 UT. The surface in Figure 9 is shifted so that local midnight occurs in the center of the axis with dawn to the right and dusk to the left. Near 01:10 UT there is significant enhancement in both the dawn and dusk sectors. The dawn enhancement persists in some form until the main storm onset. At 1:30 UT there is a very localized, intense brightening at midnight. This brightening then expands into the poleward arcs shown in Plate 3. Near 02:50 UT there is a brightening throughout most of the dusk sector that expands into the dawn sector beginning near 03:15 UT. Storm onset is indicated at 03:35 with an initially localized intense brightening at midnight that quickly broadens into the full storm expansion by 04:00 UT.

Figure 10 shows the instantaneous power input and the total energy deposition during this period. It shows the initial energy onset near 01:00 UT, the precursor buildup from 03:00 to 03:30 UT followed by the main storm onset between 03:30 and 04:00 UT.

*Coordinated Observations*

The hallmark of the ISTP program is the ability to simultaneously observe disparate parts of geospace and analyze the data from this unique perspective. One of the first opportunities for this type of coordinated analysis, using UVI images, occurred early in the operational lifetime of the POLAR spacecraft on March 27, 1996. This corresponded to a perigee pass of the WIND spacecraft which placed both WIND and GEOTAIL in the earth's magnetotail. The two spacecraft were in the dusk flank near the

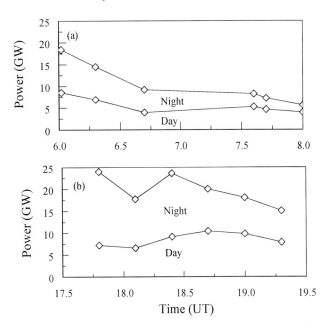

**Figure 8.** Dayside and nightside incident power for (a) April 7, 1996 and (b) July 23, 1996. The diamonds show the times at which the images were acquired. Of note is that the nightside precipitation shows much more variation than does the dayside precipitation. From *Brittnacher et al.* [1997].

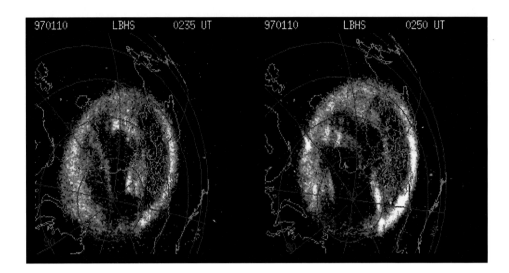

**Plate 3.** Multiple distorted transpolar arcs seen in response to the solar coronal mass ejection event observed at Earth on January 10, 1997. The image color table has been modified to emphasize the structure of these faint arcs. Peak intensity is about 10 photons cm$^{-2}$ s$^{-1}$ for both images.

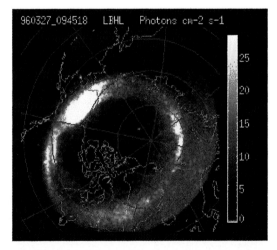

**Plate 4.** An auroral substorm beginning at 09:45 UT as seen by UVI using the LBHl filter. The substorm is located over northern Alaska with the auroral arc extending over CANOPUS magnetometer stations which indicated substorm activity. Airglow is seen in the images in the bottom right hand portion of the image over Greenland and northern Europe.

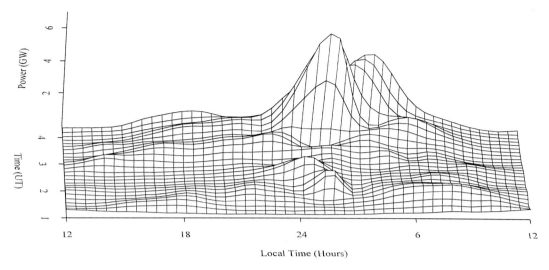

**Figure 9.** Incident power as a function of local time at the onset of the magnetic storm on January 10, 1997. The local time scale has been shifted so midnight is at the center of the axis.

Z=0 plane. Both WIND and GEOTAIL were within 700 km of each other in the Z direction. At the same time IMP8 and INTERBALL-TAIL were upstream of the bow shock sampling the solar wind and the interplanetary magnetic field. POLAR was operational with auroral imaging available. Added to these platforms were NOAA, DMSP, and GOES spacecraft as well as ground observations. Taken all together this has served as an excellent opportunity to characterize substorm events from multiple perspectives [e.g. *Angelopoulos et al*, 1997; 1996; *Germany et al.*, 1996].

Ground magnetometer data from the CANOPUS Magnetometer and Riometer Array (MARIA) [*Rostoker et. al.*, 1995] indicate auroral activity near 10:00 UT and near 13:00 UT over central and western Canada. Smaller events near 06:00 and 12:00 UT are also indicated. UVI image data shown in Plate 4 illustrates an auroral substorm beginning at 09:45 UT over northern Alaska with the auroral arc extending over the active magnetometer stations which had indicated the substorm activity. Airglow is seen in the images in the bottom right hand portion of the image over Greenland and northern Europe. Some dayside auroral activity is also seen near the dusk terminator. The temporal development of the 09:45 UT substorm is shown in Plate 5 as a mosaic of six images from 09:44 to 10:34 UT. (The image times for each image, from top left to bottom right, are 09:44, 09:52, 10:06, 10:18, 10:24, and 10:34 UT.) The substorm begins with an intensification near 22:00 magnetic local time. This is then followed by a classic expansion eastward and poleward until near 10:34 UT when the auroral is in recovery leaving a boundary arc on the poleward boundary of the expansion. The substorm advances westward in a series of discrete intensifications with each new brightening taking place westward of the previous one. The edge of the UVI field of view is seen near 06:00 LT

IMF and magnetometer data are shown in Figure 11 on a common plot with auroral activity shown qualitatively by horizontal bars. The bars indicate the beginning and duration of a specific auroral intensification. Where necessary, the horizontal bars are displaced vertically to avoid overlap. Some auroral activity is clearly correlated with the magnetometer data. However, at 11:00 UT UVI images reveal auroral intensifications that do not clearly correspond to enhanced magnetic activity in the magnetometer data underneath the aurora. Conversely, at 12:00 UT the magnetometer data indicate magnetic activity that does not correlate with auroral activity in the images, at least not with the intensity of the other activations. Correlating auroral activity with the IMF is more difficult because of the processes in the intervening magnetosphere and the associated time delay. The IMF data indicate general activity through the period but it is difficult to correlate individual features to specific auroral events in the images without detailed modeling.

This type of multi-instrument coordinated analysis has been one of the principal modes of study since the POLAR launch with UVI images being used to provide a reference for multiple magnetospheric and groundbased measurements [e.g. *Brittnacher et al.*, 1996; *Elsen et al.*, 1996; *Horwitz et al.*, 1996; *Nagai et. al.*, 1996].

## FUTURE DIRECTIONS

Each of the examples discussed above were basically event driven. That is, a narrow period of time was selected and then the image data was processed with significant interactive support. In particular, the location of the oval is critical for much of the analysis and must either be estimated based on statistical principals or be specified in an interactive fashion. This significantly slows down the image analysis effort. Furthermore, it restricts analysis to event driven studies since there is no way to automatically search available image data for specific events.

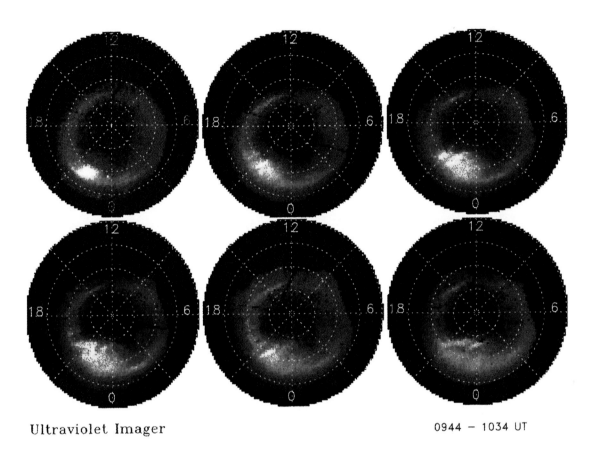

**Plate 5.** Temporal development of the 09:45 UT substorm as a mosaic of six images from 09:44 to 10:34 UT. The image times for each image, from top left to bottom right, are 09:44, 09:52, 10:06, 10:18, 10:24, and 10:34 UT. Each image is independently scaled in brightness to emphasize the relative morphological changes with time.

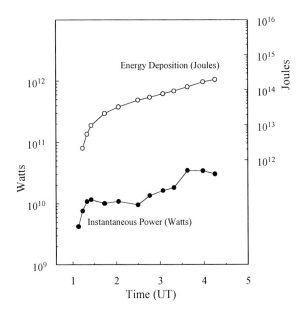

**Figure 10.** Total deposition for the onset of the magnetic storm on January 10, 1997. The bottom plot shows the total power for a given image frame in Watts. The top plot shows the time integration of this quantity in Joules.

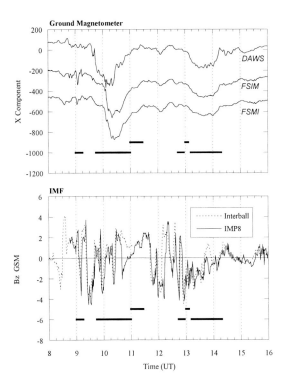

**Figure 11.** Interplanetary magnetic field and ground magnetometer data with auroral activity shown qualitatively by horizontal bars. The bars indicate the beginning and duration of a specific auroral intensification. Where necessary, the horizontal bars are displaced vertically to avoid overlap.

To address this problem, UVI researchers are developing techniques to automatically identify the oval location in image data. An existing neural network algorithm with proven utility in medical and defense image analysis applications is being adapted for use with image data from the Ultraviolet Imager. The goal is to use the algorithm not only to locate the auroral oval but also to enable automated recognition of auroral forms in UVI image data. Once developed for use with images from the Ultraviolet Imager the algorithm can be used with image data from multiple auroral imagers. The purpose of this study is to develop methods to process large volumes of auroral image data from past, current, and future image programs.

*Acknowledgments.* We are grateful to G. Rostoker for providing CANOPUS magnetometer data and for several helpful comments. The CANOPUS instrument array was constructed and is maintained and operated by the Canadian Space Agency for the Canadian scientific community. A. Skalsky kindly supplied the INTERBALL magnetometer data from the MIF instrument (S. Romanov PI). We are also grateful to R. Lepping for sharing IMP8 magnetometer data. This work was supported, in part, under U. Washington contract 256730 to the University of Alabama in Hunstville. The authors gratefully acknowledge the helpful comments of the reviewers.

## REFERENCES

Acuna, M. H., K. W. Ogilivie, D. N. Baker, S. A. Curtis, D. H. Fairfield, and W. H. Mish, The Global Geospace Science program and its investigations, *Space Sci. Rev. 71*, 5, 1995.

Angelopoulos, B., T. D. Phan, D. E. Larson, F. S. Mozer, R. P. Lin, K. Tsuruda, H. Hayakawa, T. Mukai, S. Kokubun, T. Yamamoto, d. J. Williams, R. W. McEntire, R. P. Lepping, G. K. Parks, M. Brittnacher, G. Germany, J. Spann, H. J. Singer, and K. Yumoto, Magnetotail flow bursts: association to global magnetospheric circulation, relationship to ionospheric activity and direct evidence for localization, *Geophys. Res. Lett., 24*, 2271, 1997.

Angelopoulos, V., T. Phan, D. Larson, F. S. Mozer, R. P. Lin, M. Moyer, T. Mukai, K. Tsuruda, R. W. McEntire, D. J. Williams, G. Parks, R. Lepping, K. Yumoto, and E. Friis-Christensen, Correlative IMP8-WIND-GEOTAIL-POLAR measurements of magnetotail substorms, EOS Trans. AGU, 77(46), Fall Meet. Suppl., F609, 1996.

Brittnacher, M. J., G. K. Parks, L. J. Chen, R. Elsen, J. Spann, and G. Germany, Dipolarization of the plasma sheet observed without a classical substorm injection, EOS Trans. AGU, 77(46), Fall Meet. Suppl., F633, 1996.

Brittnacher, M., R. Elsen, G. Parks, L. Chen, G. Germany, and J. Spann, A dayside auroral energy deposition case study using the Polar Ultraviolet Imager, *Geophys. Res. Lett., 24*, 991, 1997.

Doe, R. A., J. D. Kelly, D. Lummerzheim, G. K. Parks, M. J. Brittnacher, G. Germany, and J. Spann, Initial comparison of POLAR UVI and Sondrestrom IS radar estimates for auroral electron energy flux, *Geophys. Rev. Lett., 24*, 999, 1997.

Elsen, R. K., G. K. Parks, M. J. Brittnacher, L. J. Chen, S. M. Petrinec, R. M. Skoug, R. M. Winglee, G. A. Germany, J. F. Spann, Mapping UVI images into the magnetosphere with empirical and global MHD magnetospheric models, EOS Trans. AGU, 77(46), Fall Meet. Suppl., F633, 1996.

Germany, G. A, G. K. Parks, M. J. Brittnacher, J. F. Spann, J. Cumnock, D. Lummerzheim, F. Rich, and P. G. Richards,

Global Auroral Remote Sensing Using GGS UVI Images, AGU Monograph "Encounter Between Global Observations and Models in the ISTP Era", Jim Horwitz, Dennis Gallagher, and Bill Peterson, ed., 1997.

Germany, G. A., G. K. Parks, M. Brittnacher, L. Chen, R. Elsen, J. F. Spann, J. Cumnock, P. G. Richards, and F. Rich, Characterization of an auroral intensification using multiple spacecraft observations, EOS Trans. AGU, 77(46), Fall Meet. Suppl., F623, 1996.

Germany, G. A., M. R. Torr, D. G. Torr, and P. G. Richards, Use of FUV auroral emissions as diagnostic indicators, *J. Geophys. Res.*, 99, 383, 1994a.

Germany, G. A., D. G. Torr, P. G. Richards, M. R. Torr, and S. John, Determination of ionospheric conductivities from FUV auroral emissions, *J. Geophys. Res.*, 99, 23297, 1994b.

Germany, G. A., M. R. Torr, P. G. Richards, and D. G. Torr, The dependence of modeled OI 1356 and $N_2$ LBH auroral emissions on the neutral atmosphere, *J. Geophys. Res.*, 95, 7725, 1990.

Hirahara, M., J. L. Horwitz, T. E. Moore, J. F. Spann, G. Germany, W. Peterson, E. Shelley, M. Chandler, G. Giles, P. Craven, C. Pollock, J. Scudder, D. Gurnett, J. Pickett, A. Persoon, N. Maynard, F. Mozer, M. Brittnacher, and T. Nagai, Correlation of ionospheric ion outflows with auroras, waves, and convection observed with the POLAR satellite: Case studies during active and quiet intervals, in press, *J. Geophys. Res.*, 1997.

Hirahara, M., J. L. Horwitz, G. Germany, T. E. Moore, J. F. Spann, M. O. Chandler, and B. L. Giles, Properties of upflowing ionospheric ion conics and magnetosheath proton precipitation at 5000 km altitude over cusp/cleft auroral forms: Initial observations from the TIDE and UVI instruments on POLAR, Huntsville 96 Workshop 'Encounter Between Global Observations and Models in the ISTP Era' (Guntersville, AL) September 15-20, 1996.

Horwitz, J. L., G. Kunin, M. Hirahara, G. Germany, D. G. Brown, J. F. Spann, T. Nagai, and R. Lepping, Auroral and magnetospheric activity during intervals of steady southward interplanetary magnetic field from ISTP POLAR and WIND observations, EOS Trans. AGU, 77(46), Fall Meet. Suppl., F613, November 1996.

Lummerzheim, D., M. Brittnacher, D. Evans, G. A. Germany, G. K. Parks, M. H. Rees, and J. F. Spann, High time resolution study of the hemispheric energy flux carried by energetic electrons into the ionosphere during the May 19/20 auroral activity, *Geophys. Rev. Lett.*, 24, 987, 1997.

Lummerzheim, D., M. H. Rees, J. D. Craven, and L. A. Frank, Ionospheric conductances derived from DE-1 auroral images, *J. Atmos. Terr. Phys.*, 53, 281, 1991.

Moore, T. E., C. R. Chappell, M. O. Chandler, S. A. Fields, C. J. Pollock, D. L. Reasoner, D. T. Young, J. L. Burch, N. Eaker, J. H. Waite, Jr., D. J. McComas, J. E. Nordholdt, M. F. Thomsen, J. J. Berthelier, and R. Robson, The Thermal Ion Dynamics Experiment and Plasma Source Instrument, *Space Sci. Rev., 71*, 409, 1995.

Nagai, T., Y. Saito, T. Yamamoto, T. Mukai, A. Nishida, S. Kokubun, M. Hirahara, G. Germany, J. L. Horwitz, J. Spann, and M. Brittnacher, GEOTAIL-POLAR simultaneous observations of substorm onsets, EOS Trans. AGU, 77(46), Fall Meet. Suppl., F631, 1996.

Richards, P. G., and D. G. Torr, Auroral modeling of the 3371 A emission rate: Dependence on characteristic electron energy, *J. Geophys. Res.*, 95, 10337, 1990.

Rostoker, G., H.C. Samson, F. Creutzberg, T.J. Hughes, D.R. McDiarmid, A.G. McNamara, A. Vallance Jones, D.D. Wallis and L.L. Cogger, CANOPUS - a ground-based instrument array for remote sensing the high latitude ionosphere during the ISTP/GGS prgram, *Space. Sci. Rev., 71*, 743, 1995.

Strickland, D. J., J. R. Jasperse, and J. A. Whalen, Dependence of auroral FUV emissions on the incident electron spectrum and neutral atmosphere, *J. Geophys. Res.*, 88, 8051, 1983.

Torr, M. R., D. G. Torr, M. Zukic, R. B. Johnson, J. Ajello, P. Banks, K. Clark, K. Cole, C. Keffer, G. Parks, B. Tsurutani, and J. Spann, A far ultraviolet imager for the International Solar-Terrestrial physics mission, *Space Sci. Rev., 71*, 329, 1995.

---

M. Brittnacher, L. Chen, R. Elsen, G. K. Parks Geophysics Program, Box 351650, University of Washington, Seattle WA 98195.

G. Germany, Center for Space Plasma and Aeronomic Research, OB 348, University of Alabama in Huntsville, AL 35899 (e-mail: germanyg@cspar.uah.edu).

D. Lummerzheim, Geophysical Institute, Fairbanks, AK 99775.

F. Rees, Physics Department, University of Southampton, Southampton, SO17 1BJ, England.

J. Spann, Space Science Laboratory, NASA Mail Code ES83, NASA Marshall Space Flight Center, Huntsville AL 35812.

# Field Line Resonances, Auroral Arcs, and Substorm Intensifications

J.C. Samson, R. Rankin, and I. Voronkov

*Department of Physics, University of Alberta, Edmonton, Alberta, Canada*

Ultra low frequency (ULF, 1-5 mHz) narrow band, shear Alfvén, field line resonances (FLRs) are commonly found on the nightside, on field lines threading the equatorward region of the auroral oval. The nightside FLRs have been seen in HF-radar data, magnetometer data, and optical data. These FLRs have ionospheric, latitudinal scale sizes comparable to the scale sizes of the inverted V electron spectra associated with larger scale arc structures (10s of kilometers). During part of the FLR cycle, upward field aligned currents can be of the order of several microamperes and have latitudinal scale sizes of several kilometers, matching observations for auroral arcs. Optical and radar observations also show that the ionospheric flow and auroral luminosity produced by FLRs can evolve from a stable arc configuration to large scale vortices (100s of kilometers), particularly before and near local midnight. The paper outlines the MHD theory for the formation of the discrete frequency FLRs, mode conversion, and aspects of the nonlinear evolution of the FLRs. Ponderomotive forces and spatial harmonic generation might lead to rapid fine scale, radial (latitudinal) structuring, the onset of electron inertia effects through mode conversion and the formation of field aligned potential drops. Nonlinear, magnetohydrodynamic instabilities in the equatorial region of the FLR might lead to the large scale vortex structures seen in the auroral ionosphere. We also show observations of auroral emissions produced by FLRs.

## 1. INTRODUCTION

Field line resonances (FLRs) are the name given to standing shear Alfvén waves which exist on magnetic shells in the dipole-like inner magnetosphere. For a highly conducting ionosphere these FLRs have approximate electric field nodes at the ionosphere, with the fundamental having an antinode in the equatorial plane. The radial scale size of these FLRs in the equatorial plane is a fraction of an $R_E$, perhaps as small as a tenth of an $R_E$, mapping to 10s of km in the auroral ionosphere. The FLRs are formed by the coupling of monochromatic compressional energy to shear Alfvén waves on magnetic shells where $\omega^2/v_A^2 = k_\parallel^2$, $\omega$ is the frequency of the compressional driver, $v_A$ is the local Alfvén speed, and $k_\parallel$ is the wave number in the direction of the magnetic field [*Southwood, 1974; Chen and Hasegawa, 1974*]. In the auroral ionosphere, these FLRs have frequencies in the range 1 to 4 mHz and are evident in HF-radar observations of velocity fields in the F-region [*Ruohoniemi et al., 1991; Samson et al., 1991; Walker et al., 1992*], in ground-based magnetometer observations, and in optical measurements [*Samson et al., 1991; Xu et al., 1993; Samson et al., 1996a*].

The monochromatic energy or monochromatic compressional waves which drive low azimuthal wavenumber ($m<10$, where the azimuthal dependence is of the form $e^{im\phi}$) FLRs associated with some auroral arcs can come from two sources. One source is a surface wave associated with sharp changes in plasma parameters, for example, the Alfvén velocity. *Hasegawa* [1976] has conjectured that the inner edge of the plasma sheet might be one location for these surface waves. A second source of the monochromatic compression waves is through formation of MHD cavity or waveguide modes. *Samson and Rankin* [1994] give a relatively complete review of these modes. In this paper we shall not discuss these monochromatic sources, but shall emphasize the details about the resonance, and possible influences these resonances might have on auroral precipitation.

The growth of FLRs is limited by a number of factors. In the linear regime, ionospheric dissipation or mode conversion to electron inertia or kinetic Alfvén waves can cause saturation in the growth. In the nonlinear regime, ponderomotive forces and harmonic generation stabilize the growth through frequency "detuning" and Alfvén velocity profile modification [*Rankin et al.*, 1995]. Tearing modes near the auroral ionosphere, where j-parallel is very large in the FLR, can lead to small scale (kilometer) vortex structuring and dissipation. In the equatorial plane of the fundamental mode of the FLRs, fluid instabilities, including nonlinear Kelvin-Helmholtz instabilities and nonlinear shear flow-ballooning [*Voronkov et al.*, 1997] instabilities are possible. The shear flow-ballooning scenario might occur near the Earthward edge of the plasma sheet, where strong pressure gradients are prevalent, particularly during the substorm growth phase.

In this paper, we discuss a number of these features of FLR with a particular emphasis on the prediction of auroral structures, including field aligned currents. We also show a number of observations of auroral precipitation modulated by FLRs.

## 2. CONFIGURATION OF THE FIELD LINE RESONANCE

To describe FLRs we use the complete set of MHD equations, plus a generalized Ohm's law:

$$\mathbf{E} + \mathbf{v} \times \mathbf{B} = \frac{m_e}{ne^2} \frac{\partial \mathbf{J}}{\partial t} + \eta \mathbf{J} \quad (1)$$

where **v** is the flow velocity, **J** is the current density, **B** is the magnetic field, **E** is the electric field, $m_e$ is the electron mass, $e$ is the electron charge, and $\eta$ is the resistivity.

We use a Cartesian geometry and linearize the MHD equations, neglecting for now the electron inertia term in the Ohm's law. Assuming gradients in the x-direction and solving for the radial displacement vector, $\xi_x$, gives the equation

$$\frac{d^2 \xi_x}{dx^2} + \frac{F'}{F} \frac{d\xi_x}{dx} + G\xi_x = 0 \quad (2)$$

where

$$F = \frac{\rho_0(\omega^2 - k_z^2 V_A^2)}{G},$$

$$G(x) = \frac{\omega^2[\omega^2 - (k_y^2 + k_z^2)V^2] + k_z^2(k_y^2 + k_z^2)V_S^2 V_A^2}{\omega^2 V^2 - k_z^2 V_S^2 V_A^2},$$

$$V = (V_A^2 + V_S^2)^{1/2},$$

where $k_y$ is the azimuthal wave number, $k_z$ is the B-field aligned wave number and $\omega$ is the wave frequency, and $V_A$ and $V_S$ are, respectively, the Alfvén and sound speed.

Near the position of the shear Alfvén resonance, $F=0$ and $\omega^2/V_A^2(x_r) = k_z^2$, the equation has the approximate form

$$\frac{d^2 \xi_x}{dx^2} + \frac{1}{x - x_r + i\gamma} \frac{d\xi_x}{dx} - k_y^2 \xi_x = 0 \quad (3)$$

where $\gamma = \text{Im}\varepsilon(x_r)/(d\,\text{Re}\varepsilon/dx)_{x=x_r}$,

$$\text{Re}\,\varepsilon = \frac{\omega^2}{V_A^2} - k_z^2$$

and the imaginary component of $\varepsilon$ takes into account losses. The solution is

$$\xi_x = \xi_0 \ln(k_y(x - x_r + i\gamma)). \quad (4)$$

The solution for the radial component of the electric field shows the true singularity at the resonance (when $\gamma = 0$) and is given by

$$E_x(x) = \frac{ik_y E_o}{(\text{Re}\,\varepsilon - k_y^2)(x - x_r + i\gamma)}. \quad (5)$$

The profiles of $|E_x|$ and the phase of $E_x$ for an ionosphere with a reflectivity of 0.7 are given in Figure 1. Note the

180° phase shift across the resonance. The reflectivity and width of the resonance were chosen to be compatible with the observations to be shown later.

Figure 2 shows a "snapshot" of the radial electric field in the equatorial plane, where y is the azimuthal direction. Azimuthal phase propagation is in the +y direction. At one longitude, y, the maximum in the E-field starts at the smallest radial distance, and moves outward with time, followed by the beginning of a new maximum closer to the Earth. Mapped to the ionosphere, this gives the poleward moving bands of maximum E-field seen in coherent radar data.

More important, for the formation of auroral emissions, are the large field aligned currents in the FLR just above the auroral ionosphere (see *Samson et al.*, 1996a, Figure 14). In Figure 3 we have plotted upward j-parallel assuming a dipole topology for the resonance profile shown in Figure 1. We have plotted these data in a latitude-time format in order to be compatible with the meridian scanning photometer data to be shown later. Maximum upward field aligned currents in typical FLRs are much greater than $10^{-6} A/m^2$, indicating that these FLRs should lead to the acceleration of auroral electrons. Note the poleward moving bands of upward j-parallel, which we will later correlate with auroral luminosities.

To clearly show that FLRs lead to some types of auroral arcs, we have included one example from CANOPUS [*Rostoker et al.*, 1994] magnetometer (Figure 4) and meridian scanning photometer data (Figure 5). FLRs are prevalent in much of the magnetometer data, including the 00 UT to ~0600 UT interval prior to a number of substorm intensifications, and in the interval 1000-1400 following the substorms. The FLRs in the interval before the substorm (see the GILL magnetometer) are embedded in a region of Hβ prior to 0600 UT. The equatorward

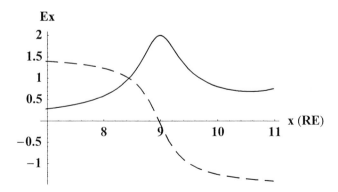

**Figure 1.** The amplitude (solid line) and phase (dotted line) of the radial component of the electric field of a FLR in the equatorial plane.

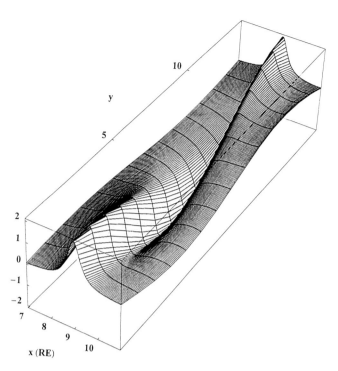

**Figure 2.** A snapshot of a the radial electric field of a FLR, showing the radial and azimuthal (one cycle) dependence in the equatorial plane.

motion of these Hβ emissions is indicative of a substorm growth phase, with field line stretching occurring in the near Earth region [*Samson et al.*, 1992]. Although difficult to see in these MPA plots, the FLR produces modulation of the 5577 Å emissions. Auroral arcs associated with FLRs embedded in this evening region of Hβ have already been studied in some detail [*Samson et al.*, 1996a], and consequently, we shall turn our attention to the emissions associated with the morning sector FLRs.

The luminosity fluctuations associated with these morning sector FLRs are found over a broad latitudinal region (~68° to 75°) and at a variety of frequencies. We shall look at a particularly clear example in the 6300 Å emissions, occurring between ~10:50 and 11:50 UT (Figures 6 and 7). A comparison of Figures 3 and 6 indicate that the modulated auroral emissions have many of the features of a modulated auroral arc which might be produced by a FLR. The periodic poleward moving bands of luminosity, which would be associated with poleward moving arcs in the region of the FLR, are quite clear. The time series in Figure 7, and the power spectrum in Figure 8, illustrate the almost monochromatic nature of the luminosity fluctuations, further supporting a FLR source for the auroral features.

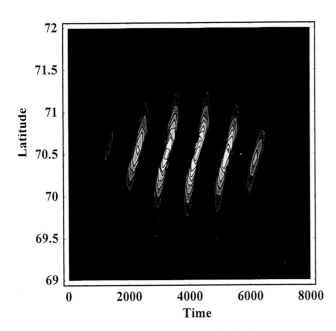

**Figure 3.** The upward ionospheric field aligned currents associated with a 5000 second wave packet of a FLR. These data are plotted in a format to be compatible with meridian scanning photometer data.

## 3. FINITE ELECTRON ION INERTIA EFFECTS

When a FLR narrows to a width which is several electron inertia lengths, $\lambda_e = (m_e/\mu_0 n e^2)^{1/2}$, then mode conversion to electron inertia waves is possible [*Goertz*, 1984; *Wei et al.*; 1994; *Streltsov and Lotko*, 1996]. This process is most likely to occur at altitudes of 2-3 $R_E$ above the auroral ionosphere. Although "phase mixing" times, the time it takes for a FLR to narrow to several $\lambda e$, can be rather long in the magnetosphere [*Wright and Allan*, 1996], these times can be as short as several cycles in the low density regions found above the auroral ionosphere [*Persoon et al.*, 1988]. In fact, some of these low density regions might be produced by ponderomotive forces in FLRs, as we shall show later.

*Goertz* [1984] has shown that near the resonance, when finite electron inertia is included, the equation for the potentials associated with the fields is given by

$$\nabla_\perp(\varepsilon \nabla_\perp \phi_\perp) + O_P \phi_\perp = O \quad (7)$$

where $E_\perp = E_x = -\nabla \phi_\perp$.

The first term on the right hand side of (7) is equivalent to Equation (3) and has a singularity at the resonance when there is no damping. The second term is a fourth order derivative in x and includes electron inertia effects leading to mode conversion and a propagating mode (see *Goertz*, 1984, for details). Following Goertz, we chose a particularly simple example and assume an Alfvén velocity profile of the form

$$\frac{1}{V_A^2(x)} = \frac{1}{V_0^2}[1+\kappa x]. \quad (8)$$

Then Equation (7) has a particularly simple form given by

$$\lambda_e^2 \frac{d^2 E_x}{dx^2} - \kappa x E_x = 0. \quad (9)$$

A numerical solution of (9) is shown in Figure 9 and illustrates the characteristic Airy function form of the wave fields. The characteristic perpendicular wavelength

$$\lambda_\perp \approx (3\pi)^{2/3} (\lambda_e^2 / \kappa)^{1/3} \quad (10)$$

is of the order of the order of several kilometers in the auroral accelerator region.

Perpendicular E-fields with patterns very similar to Figure 8 have been observed by the S3-3 satellite (see *Mozer et al.*, 1981, Figure 14). Mode conversion also leads to parallel electric fields and potential drops with

$$\delta \phi_\parallel = \frac{k_\perp^2 \lambda_e^2}{1 + k_\perp^2 \lambda_e^2} \delta \phi_\perp . \quad (11)$$

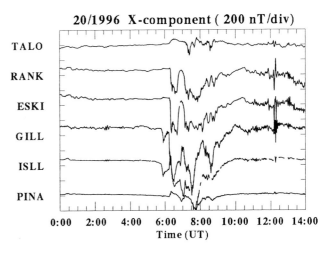

**Figure 4.** Magnetometer data from the CANOPUS array. ULF pulsations associated with FLRs are very clear at GILL (0200-0400 UT) and GILL, ESKI, and RANK (~1030-1400 UT). Local magnetic time is approximately UT 7 hours.

## 4. PONDEROMOTIVE FORCES IN FLRs

Ponderomotive forces are nonlinear effects arising from gradients in the fields associated with the FLR. Averaged over a number of wave periods, there is a net ponderomotive force accelerating plasma toward the equatorial plane of the FLR. This ponderomotive force follows from a time average of the second order terms in the equation of motion [*Allan*, 1993], given by

$$F_{pond} = \left\langle \rho \frac{\partial \mathbf{v}}{\partial t} + \rho_{01}(\mathbf{v} \cdot \nabla)\mathbf{v} + \nabla\left(\frac{\mathbf{b} \cdot \mathbf{b}}{2\mu_0}\right) - \frac{(\mathbf{b} \cdot \nabla)\mathbf{b}}{2\mu_0} \right\rangle \quad (12)$$

where $\langle \rangle$ denotes time average. The last two terms on the right hand side are due to nonlinear $\mathbf{j} \times \mathbf{B}$ forces. The plasma accelerated toward the equatorial plane leads to a local reduction of the Alfvén velocity and movement in the position of the FLR [*Rankin et al.*, 1995]. A nonlinear phase mismatch of the compressional and shear waves can lead to nonlinear saturation of the FLR, nonlinear narrowing, and radial structuring. This nonlinear structuring might contribute to an early onset of electron inertia effects.

Figure 10 illustrates the change in plasma pressure along a FLR in a dipolar magnetosphere. Note the decrease in pressure near the ionospheres (L=0, 25) and the increase in the equatorial plane. The largest decreases in density and pressure occur at altitudes of 2-3 $R_E$. The latitudinal scale sizes of the density cavity should be on the order of 100 km (based on measured widths of FLRs) suggesting that some of the density cavities seen by *Persoon et al.* [1988] might have been produced by FLRs.

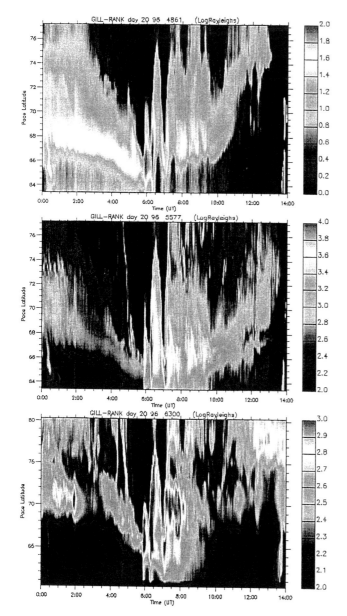

**Figure 5.** Merged meridian scanning photometer data from Gillam and Rankin Inlet in the CANOPUS array. Top: 4861 Å. Center: 5577 Å. Bottom: 6300 Å. Data are plotted in PACE coordinates [*Baker and Wing, 1989*].

*Samson et al.* [1996b] showed, using this relation, that FLR can have parallel potential drops of a number of keV. These numbers suggest that mode conversion associated with FLRs might lead to the smaller, kilometer scale, discrete arcs seen in auroral forms. In fact, *Trondsen et al.* [1997] have shown a number of optical measurements of discrete arcs which show this characteristic Airy function pattern.

## 5. FLRs, SHEAR FLOW BALLOONING INSTABILITIES, AND SUBSTORM INTENSIFICATIONS

*Samson et al.* [1996a] have shown all-sky images of auroral arcs produced by FLRs in the morning sector. At times these arcs developed large scale vortex structures (100s of km) similar to the substorm surge vortex, leading Samson et al. to suggest that nonlinear Kelvin-Helmholtz instabilities (KHI) in the large shear flows in the equatorial plane of the FLR [*Rankin et al.*, 1993] might lead to the vortex formation. In fact, these FLRs were embedded in a region of strong Hβ emissions indicating that the FLRs mapped to the Earthward edge of the plasmasheet, where strong Earthward pressure gradients

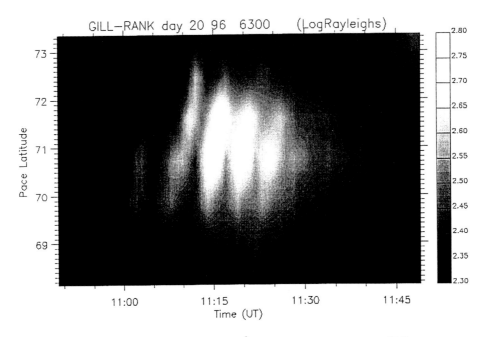

**Figure 6.** Photometer data for 6300 Å emissions associated with a FLR.

exist. These pressure gradients indicate that ballooning modes could play a role in the extraction of potential energy stored in the near Earth magnetotail. Assuming a configuration that is near ballooning stable, *Samson et al.* [1996b] suggested that the FLR might first develop a nonlinear KHI which then couples to and drives a hybrid shear flow ballooning mode. This leads to a rapid growth (10s of seconds) in the kinetic energy of plasma flows. *Samson et al.* [1996b] also conjectured that this hybrid instability might play a major role in triggering substorm intensifications in the near Earth magnetotail.

In a later publication, *Voronkov et al.* [1997] have shown, through nonlinear MHD computer models, that this nonlinear hybrid instability should evolve in the near Earth magnetotail. The process starts as a KHI-like mode which then drives the ballooning through the radial flows of the KH-vortices. Some results of the computer models are shown in Figure 11.

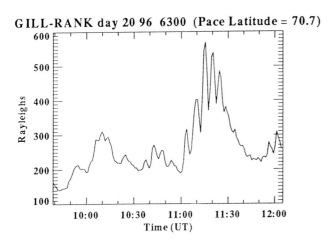

**Figure 7.** Time series of the 6300 Å photometer data in Figure 5, taken at a latitude of 70.7°.

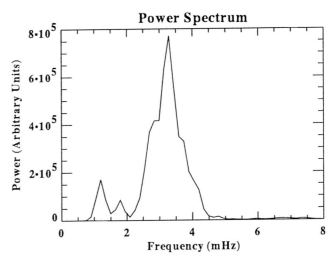

**Figure 8.** Power spectrum of the data in Figure 7.

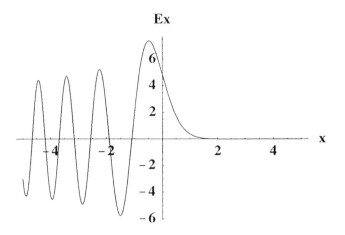

**Figure 9.** Numerical solution of $E_x$ given by Equation (9).

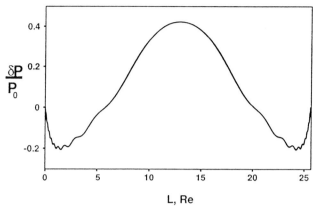

**Figure 10.** Non linear MHD computer models of the change in pressure due to ponderomotive forces in a FLR. $\beta = 2.65 \cdot 10^{-2}$ in the equatorial plane. L is the distance along the field line in $R_E$. The ionosphere is at L=0 and L=26.

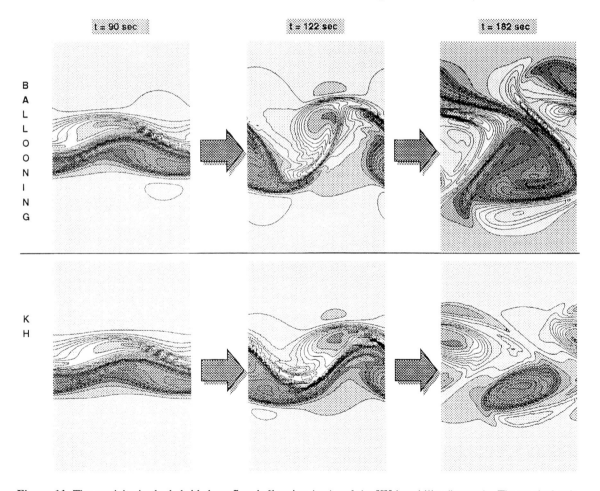

**Figure 11.** The vorticity in the hybrid shear flow ballooning (top) and the KH instability (bottom). The vertical axis is toward the Earth, and the horizontal is in the azimuthal direction. The simulation box is 0.5 x 1.5 $R_E$. The pressure varies from 1 to 5 nPa in the Earthward direction, with a scale size of 1 $R_E$. The flow channel is 0.17 $R_E$ wide with a Gaussian profile and a maximum speed of 100 km/s.

## 6. CONCLUSIONS

Observations now show that some larger scale (10 km thickness) auroral arcs and auroral structures are produced by FLRs. Ponderomotive forces in FLRs might also play a role in producing the density cavities seen above the auroral ionosphere. Mode conversion in the FLRs might possibly lead to the formation of smaller scale size (kilometer scale) discrete auroral arcs and there exists some preliminary optical evidence for this process. Further studies comparing higher resolution (less than a kilometer) optical measurements with larger scale (10s of km) optical measurements and HF-radar data would be very useful. The large scale optical and radar data could be used to identify the FLR, while the high resolution optical studies would give evidence for mode conversion.

The hybrid, shear flow ballooning instability in FLRs looks like a promising mechanism for triggering some substorm intensifications in the near Earth magnetotail. This instability takes into account the important observational constraints of auroral arc brightening to begin the intensification, and the fact that these arcs thread the regions of high pressure gradients at the inner edge of the plasmasheet. Nevertheless, further computational models which incorporate a more realistic magnetic field topology are required. We also need further experiments and observational studies which can compare optical features with radar measurements of ionospheric plasma flows in the FLRs, auroral arcs, and substorm surge vortex.

*Acknowledgements.* This research has been supported by the Natural Science and Engineering Council of Canada and by the Canadian Space Agency.

## REFERENCES

Allan, W., the ponderomotive force of standing Alfven waves in a dipolar magnetosphere, *J. Geophys. Res., 98,* 1409, 1993.

Baker, K. B. and S. Wing, A new magnetic coordinate system for conjugate studies at high latitudes, *J. Geophys. Res., 94,* 9139, 1989.

Chen, L., and A. Hasegawa, A theory of long period magnetic pulsations, 1, Steady excitation of field line resonances, *J. Geophys. Res., 79,* 1024, 1974.

Goertz, C. K., Kinetic Alfven waves on auroral field lines, *Planet. Space Sci., 32,* 1387, 1984.

Hasegawa, A., Particle acceleration by MHD surface wave and the formation of aurora, *J. Geophys. Res., 81,* 5083, 1976.

Mozer, F. S., The low altitude electric field structure of discrete auroral arcs, in *Physics of Auroral Arc Formation,* edited by S.-I. Akasofu and J. R. Kan, *AGU Geophysical Monograph 25,* 136, 1981.

Persoon, A. M., D. A. Gurnett, W. K. Peterson, J. H. Waite, Jr., J. L. Burch, and J. L. Green, Electron density depletions in the nightside auroral zone, *J. Geophys. Res., 93,* 1871, 1988.

Rankin, R., B. G. Harrold, J. C. Samson, and P. Frycz, The nonlinear evolution of field line resonances in the Earth's magnetosphere, *J. Geophys. Res., 98,* 5839, 1993.

Rankin, R., P. Frycz, V. T. Tikhonchuk, and J. C. Samson, Ponderomotive saturation of magnetospheric field line resonances, *Geophys. Res. Lett., 22,* 1741, 1995.

Rostoker, G., J. C. Samson, F. Creutzberg, T. J. Hughes, D. R. McDiarmid, A. G. McNamara, A. Vallance Jones, D. D. Wallis, and L. L. Cogger, CANOPUS - A ground based instrument array for remote sensing in the high latitude ionosphere during ISTP/GGS program, *Space Sci. Rev., 71,* 743, 1994.

Ruohoniemi, J. M., R. A. Greenwald, K. B. Baker, and J. C. Samson, HF radar observations of Pc5 field line resonances in the midnight/early morning MLT sector, *J. Geophys. Res., 96,* 15697, 1991.

Samson, J.C. and R. Rankin, The coupling of solar wind energy to MHD cavity modes, waveguide modes, and field line resonances in the Earth's magnetosphere, *AGU Geophysical Monograph, 81,* 253, 1994.

Samson, J. C., T. J. Hughes, F. Creutzberg, D. D. Wallis, R.A. Greenwald, and J.M. Ruohoniemi, Observations of a detached discrete arc in association with field line resonances, *J. Geophys. Res., 96,* 15683, 1991.

Samson, J. C., D. D. Wallis, T. J. Hughes, F. Creutzberg, J. M. Ruohoniemi, and R. A. Greenwald, Substorm intensifications and field line resonances in the nightside magnetosphere, *J. Geophys. Res., 97,* 8495, 1992.

Samson, J. C., L. L. Cogger, and Q. Pao, Observations of field line resonances, auroral arcs, and auroral vortex structures, *J. Geophys. Res., 101,* 17,373, 1996a.

Samson, J. C., A. K. MacAulay, R. Rankin, P. Frycz, and I. Voronkov, Substorm intensifications and resistive shear flow-ballooning instabilities in the near-Earth magnetotail, in Proc. Third International Conference on Substorms (ICS-3), *ESA SP-389,* 399, 1996b.

Southwood, D. J., Some features of field line resonances in the magnetosphere, *Planet. Space Sci., 22,* 483, 1974.

Streltsov, A., and W. Lotko, The fine structure of dispersive, non radiative field line resonance layers, *J. Geophys. Res., 101,* 5343, 1996.

Trondsen, T. S., L. L. Cogger, and J. C. Samson, Observations of a symmetric multiple auroral arc, *Geophys. Res. Lett., in press,* 1997.

Voronkov, I., R. Rankin, P. Frycz, V. T. Tikhonchuk, and J. C. Samson, Coupling of shear flow and pressure gradient instabilities, *J. Geophys. Res., 102,* 9639, 1997.

Walker, A. D. M., J. M. Ruohoniemi, K. B. Baker, R. A. Greenwald, and J. C. Samson, Spatial and temporal behavior of ULF pulsations observed by the Goose Bay HF radar, *J. Geophys., Res., 97,* 12187, 1992.

Wei, C. Q., J. C. Samson, R. Rankin, and P. Frycz, Electron inertial effects on geomagnetic field line resonances, *J. Geophys. Res., 99,* 11,265, 1994.

Wright, A. N., and W. Allan, Are two fluid effects relevant to ULF pulsations, *J. Geophys. Res., 101,* 24,991, 1996.

Xu, B.-L., J. C. Samson, W. W. Liu, F. Creutzberg, and T. J. Hughes, Observations of optical aurora modulated by resonant Alfvén waves, *J. Geophys. Res., 98,* 11,531, 1993.

---

R. Rankin, J. C. Samson, and I. Voronkov, Physics Department, University of Alberta, Edmonton, Alberta, Canada, T6G 2J1.

# A New Approach to the Future of Space Physics

S.-I. Akasofu and W. Sun

*Geophysical Institute, University of Alaska Fairbanks*

Ground-based space physics has reached the stage on which it is possible to infer the distribution of electric currents, field-aligned currents, equipotential contours (the **ExB** flow lines) in the ionosphere with a time resolution of ~5 minutes. This progress has become possible because a powerful inversion algorithm KRM for ground-based magnetometer data from an extensive network has been developed and tested. Further, a method is being developed and tested to project some of the ionospheric quantities thus obtained onto the equatorial plane. Thus, although there is much to improve in the method, a closer integration of satellite-based and ground-based observations than in the past has become possible. Since any disagreements between them will reveal incompleteness of our understanding magnetosphere processes and modeling, the study will bring satellite-based and ground-based researchers, theorists and modelers together in understanding magnetospheric substorm processes.

## 1. INTRODUCTION

The KRM algorithm and its applications have been described in a number of papers [*Kamide et al., 1981; Kamide et al., 1996*]. Further, the method to project the ionospheric quantities onto the equatorial plane has been developed and extensively tested [*cf. Akasofu et al, 1981; Akasofu et al. 1990; Akasofu, 1992, 1993; Sun et al., 1996*]; for a comprehensive review of these studies, see [*Akasofu and Kamide 1997*]. In the present paper, we apply the method thus developed to one substorm event on March 19, 1978. This event was chosen because magnetometer data from 71 stations in the arctic region are available, allowing an ideal use of the KRM method.

The AE index during this particular event is shown in Figure 1. A distinct increase of the AE index occurred at about 10:35 UT on March 19, 1978, which may be interpreted as the beginning of the growth phase. The onset of the expansion phase began at about 11:35 UT, as indicated by a sharp increase of the AE index. The AE index reached its peak value at about 12:10 UT and decreased until about 12:40 UT. High activity continued afterward. Figure 2a and 2b show the development and decay of the ionospheric currents obtained by the KRM method. One can recognize easily the development of the two-cell pattern between 10:35 and 11:30 UT. As is well known, the 'axis of symmetry' of the pattern is rotated by ~ 45° clockwise with respect to the noon-midnight line. At 11:35 UT, an intense westward current began to develop along the auroral oval. The enhancement became large in the late evening sector at 11:40 UT and extended toward the midnight and early morning sectors until about 12:00 UT. Then, a new enhancement began at about 12:05 UT in the evening sector, presumably associated with the formation of a westward traveling surge. Again, this activity extended toward the midnight and the morning sectors.

## 2. CHANGES OF THE EQUATORIAL CURRENTS

We have developed a method to project the ionospheric current vectors thus obtained onto the equatorial plane of the magnetosphere. The basic principle in this projection method was given by [*Boström, 1964*]. Practically, we divide ionospheric current vectors into the radial and azimuthal components and project them onto the equatorial plane, using a magnetospheric model. Here, we employ Tsyganenko's

---

Geospace Mass and Energy Flow: Results From the International Solar-Terrestrial Physics Program
Geophysical Monograph 104
Copyright 1998 by the American Geophysical Union

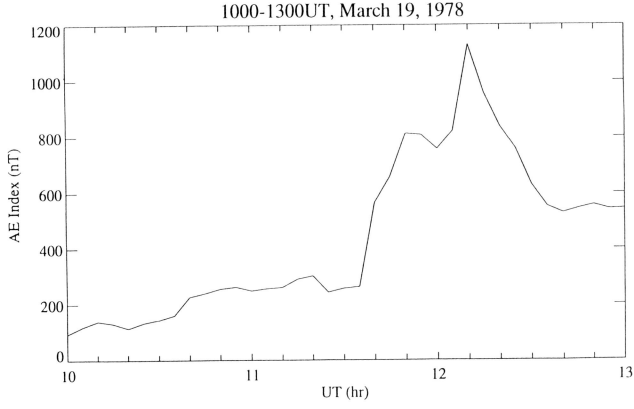

**Figure 1.** The AE index during 1000-1300 UT March 19, 1978.

model (T89M) in tracing the magnetic field lines for four different magnetic activities (Kp = 0, 1, 2) for the quiet time, the growth phase, the expansion and recovery phases; for the peak of substorm, the Kp = 4 model is chosen; for details of the projection, see [*Sun et al., 1996*].

Figure 3a and 3b shows the equatorial currents corresponding to Figure 2a and 2b. The equatorial current until about 11:30 UT was dominated by the cross-tail current, so that it is not easy to recognize the growth phase effect. However, one prominent change is an enhancement of the westward currents in the day sector. A few other important changes were a strong concentration of the current in the evening sector and also the development of the outward component in the late evening sector. A definitive change of the current distribution began at 11:40 UT. A westward current at a distance of 5~8 $R_E$ appeared in the late evening sector and lasted until 11:55 UT, so that the current direction was reversed from the dominant eastward direction of the cross-tail current during this period. This feature may be identified as the "disruption of the cross-tail current" at substorm onset. However, this phenomenon ceased by 12:00~12:05 UT. Then, at about 12:05 UT, a large outward current developed in the evening sector. It is quite likely that this is associated with the westward traveling surge. It appears that the location of this outward component shifted gradually toward the midnight sector after 12:15 UT.

### 3. DIPOLARIZATION

The dipolarization of the magnetic field at about a geosynchronous distance is one of the prominent features of substorm onset. Thus, we examine whether or not the changes of the currents in the ionosphere and the equatorial plane, together with the field-aligned currents, are associated with the dipolarization. Figure 4 shows the configuration of the magnetic field lines in the midnight sector at several chosen epochs. One can easily see that the dipolarization does occur in association with the changes of the currents elucidated in this paper. This feature suggests that the development of the substorm current systems, in particular the loop currents which connect ionospheric currents and the equatorial currents, is closely associated with the disruption of the cross-tail current and dipolarization. A study of the cause-effect relationship among these processes is expected to reveal substorm onset processes. It should be noted that the projection described in section 2 and the changes of the field line configuration are consistent because both results are iterated several times to obtain the converging results.

## IONOSPHERIC CURRENT VECTORS
### 1000-1300UT March 19, 1978

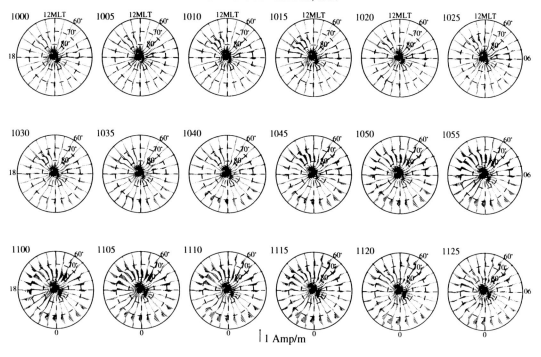

## IONOSPHERIC CURRENT VECTORS
### 1000-1300UT March 19, 1978

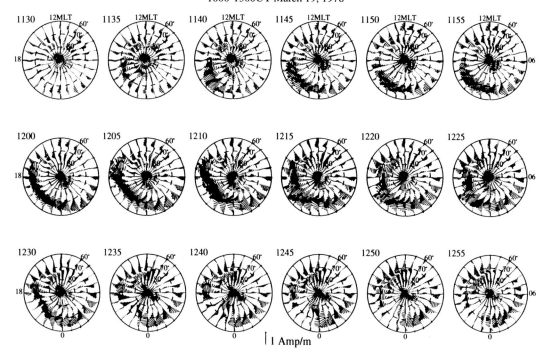

**Figure 2a and 2b.** The ionospheric current vectors during 1000-1300 UT March 19, 1978 which obtained by the KRM inversion algorithm on the basis of six IMS meridian chains magnetometer records.

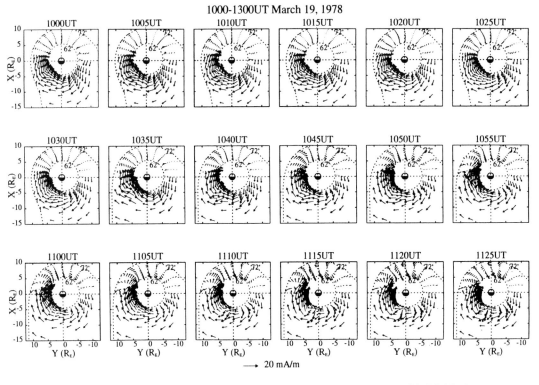

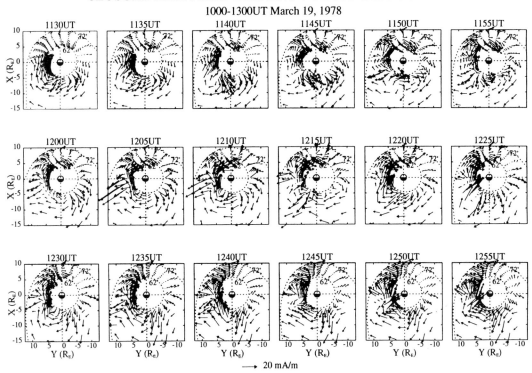

**Figure 3a and 3b.** The current vectors on the equatorial plane during 1000-1300 UT March 19, 1978 which are obtained from the sum of the cross-tail current and projected current from the ionospheric current vectors in Figure 2a and 2b.

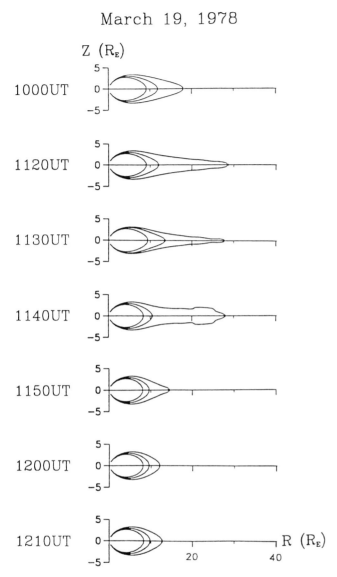

**Figure 4.** The configuration of the magnetic field lines in the we 23 MLT sector between 1000 and 1210 UT on March 19, 1978. The field lines start from the Earth at 67°, 68° and 69° of latitudes.

## 4. DISCUSSION

This is only the first attempt to apply the projection method we developed to an individual substorm, so that a number of substorm must be analyzed in order to determine common features among substorms. It is hoped that a similar analysis can also be conducted when data from satellites on the equatorial plane are available. Since satellites do not measure the electric currents directly, the comparison between the satellite results and the projected results is of great interest, although the comparison is satisfactory for the statistical results [*Akasofu, 1992, 1993*].

It is possible to project the ionospheric equipotential contour lines and electric fields onto the equatorial plane in a similar way. Since satellites can measure directly the flows and electric fields, this comparison will provide an opportunity to calibrate the projection.

These comparisons will, in turn, provide us with an opportunity to understand substorm processes, in particular the cause and extent of the electric field associated with the dipolarization. This is because any discrepancy must be due to incompleteness of our knowledge on substorm processes and of the modeling. Thus, the study described in this paper will be able to bring closer than before satellite-based and ground-based researchers, theorists and modelers together in advancing substorm research.

*Acknowledgments.* We would like to thank Professor Y. Kamide and his discussion of the results. The work reported here was supported in part by a National Science Foundation Grant 93-11474.

## REFERENCES

Akasofu, S.-I., Y. Kamide, J. R. Kan, L. C. Lee, and B.-H. Ahn, Power transmission from the solar wind-magnetosphere dynamo to the magnetosphere and to the ionosphere: analysis of the IMS Alaska meridian chain data, *Planet. Space Sci.*, 69, 721, 1981.

Akasofu, S.-I., D. Weimer, T. Iijima, B.-H. Ahn, and Y. Kamide., Agreements between ground-based and satellite-based observations, *Planet. Space Sci.*, 38, 1,533, 1990.

Akasofu, S.-I., A confirmation of the validity of the electric current distribution determined by a ground-based magnetometer network, *Geophys. J. Int.*, 109, 191, 1992.

Akasofu, S.-I., Linking ground- and space-based observations in space physics, *EOS*, 74, 259, 1993.

Akasofu, S.-I. and Y. Kamide, Toward a closer integration of magnetospheric research: magnetospheric currents and convection pattern inferred from ground-based magnetic data, *J. Geophys. Res.*, Invited, 1997.

Boström, R., A model of the auroral electrojets, *J. Geophys. Res.*, 69, 4983, 1964.

Kamide, Y., A. D. Richmond, and S. Matsushita, Estimation of ionospheric electric field ionospheric currents, and field-aligned currents from ground magnetic records, *J. Geophys. Res.*, 86, 801, 1981.

Kamide, Y, W. Sun, and S.-I. Akasofu, The average ionospheric electrodynamics for the different substorm phases, *J. Geophys. Res.*, 101, 99, 1996.

Sun W., Y. Kamide, and S.-I. Akasofu, Substorm current in the equatorial plane of the magnetosphere deduced from ground-based magnetometer chain records, *J. Geophys. Res.*, 101, 24655, 1996.

---

S.-I. Akasofu and W. Sun, Geophysical Institute, University of Alaska Fairbanks, 903 Koyukuk Drive, P.O. Box 757320, Fairbanks, AK 99775-7320. (e-mail: sakasofu@gi.alaska.edu)

# Reconnections at Near-Earth Magnetotail and Substorms Studied by a 3-D EM Particle Code

K.-I. Nishikawa

*Department of Space Physics and Astronomy, Louisiana State University*
*Baton Rouge, Louisiana*

After a quasi-steady state is established with an unmagnetized solar wind we switch on a southward IMF, which causes the magnetosphere to stretch and allows particles to enter the cusps and nightside magnetosphere. Analysis of magnetic fields near the Earth confirms a signature of magnetic reconnection at the dayside magnetopause, and the plasma sheet in the near-Earth magnetotail clearly thins. Later magnetic reconnection also takes place in the near-Earth magnetotail. Arrival of southward IMF near the front of the magnetosphere causes a sunward velocity in the dayside magnetosphere, as required to feed flux tubes into the dayside reconnection process. Sunward flow near the equatorial plane of the magnetosphere implies a dawn-to-dusk electric field. Initially, the velocity in the distant tail is not much affected by the southward turning. Therefore, the dawn-dusk electric field increases in the sunward direction, which causes $B_z$ to decrease with time in the near-Earth magnetotail. The cross-field current also thins and intensifies, which excites a kinetic (drift kink) instability along the dawn-dusk direction. Due to this instability the electron compressibility effect appears to be reduced and allows the collisionless tearing to grow rapidly with the reduced $B_z$ component. Later the nightside magnetic fields are dipolarized and a plasmoid is formed tailward. We find that due to the reconnection, particles are injected toward the Earth from the neutral line (X-line). In our simulations kinetic effects self-consistently determine the dissipation rate in the magnetopause associated with reconnection.

## 1. INTRODUCTION

In MHD codes [*e.g., Fedder et al.*, 1995] the microscopic processes can be represented by statistical (macroscopic) constants such as diffusion coefficients, anomalous resistivity, viscosity, temperature, equation of state, and the adiabatic constant. The near-Earth magnetotail is one of the regions where kinetic effects are critical and particle simulations become very important [*e.g., Birn et al*, 1996; *Pritchett et al*, 1996].

Recently, 3-D particle simulations indicate that the collisionless tearing instability grows rapidly due to the reduction of electron compressibility caused by the persistent drift kink mode in the presence of a small $B_z$ component [*Pritchett and Coroniti*, 1996; *Pritchett et al.*, 1996]. These simulations study only the near-Earth region, therefore, the resolutions in space and time are better than those in the global simulations [*Buneman et al.*, 1992, 1995]. However, effects from the dayside magnetopause, self-consistent stretched magnetic field by the solar wind with an IMF in the magnetotail, and particle entry from low-latitude boundary layer (flank) in the presence of $B_y$ component of IMFs [e.g., *Rostoker*, 1996] are not included, although they are important for the magnetotail dynamics (see, for example, [*Baker et al.*, 1996b; *Sergeev et al.*, 1996]).

## 2. A 3-D EM PARTICLE SIMULATION MODEL

The increase of available core memory and speed on supercomputers such as the CRAY C90 now enables us to perform three-dimensional particle simulations with reasonably realistic parameters. We use a 3-D electromagnetic particle code [*Buneman*, 1993]. Several new features have been implemented in order to increase the speed and versatility of the code.

For the simulation of solar wind-magnetosphere interaction, the same boundary and initial conditions are used for the particles and fields (for details see [*Buneman* 1993; *Buneman et al.* 1992, 1995; *Nishikawa et al.* 1995; *Nishikawa*, 1997]).

In order to bring naturally disparate time- and spacescales closer together in this simulation of phenomena dominated by ion inertia and magnetic field interaction, the natural electron mass was raised to 1/16 of the ion mass and the velocity of light was lowered to twice the incoming solar wind velocity. This means that charge separation and anomalous resistivity phenomena are accounted for qualitatively but perhaps not with quantitative certainty. Likewise, radiation related phenomena (e.g. whistler modes) are covered qualitatively only.

## 3. SIMULATION RESULTS

A first test exploring the solar wind-magnetosphere interaction was run on the CRAY-YMP at NCAR using a modest 105 by 55 by 55 grid and only 200,000 electron-ion pairs [*Buneman et al.*, 1992]. We also have reported on our second test run on the CRAY-2 at NCSA using a larger 215 by 95 by 95 grid and about 1,500,000 electron-ion pairs [*Buneman et al.*, 1995]. Initially, these fill the entire box uniformly and drift with a velocity $v_{sol} = 0.5c$ in the $+x$ direction, representing the solar wind without an IMF. The electron and ion thermal velocities are $v_{et} = (T_e/m_e)^{1/2} = 0.2c$, and $v_{it} = (T_i/m_i)^{1/2} = 0.05c (= v_s = (T_e/m_i)^{1/2})$, respectively, while the magnetic field is initially zero. A circular current generating the dipole magnetic field is increased smoothly from 0 to a maximum value reached at time step 65 and kept constant at that value for the rest of the simulation. The center of the current loop is located at $(70.5\Delta, 47.5\Delta, 48\Delta)$ with the current in the $xy$-plane and the axis in the $z$ direction. The initial expansion of the magnetic field cavity is found to expel a large fraction of the initial plasma. The injected solar wind density is about 0.8 electron-ion pairs per cell, the mass ratio is $m_i/m_e = 16$, and $\omega_{pe}\Delta t = 0.84$.

The results of an unmagnetized solar wind plasma streaming past a dipole magnetic field show the formation of a magnetopause and a magnetotail, the penetration of energetic particles into cusps and radiation belt and dawn-dusk asymmetries [*Buneman et al.* 1992, 1995; *Nishikawa et al.*, 1995].

After a quasi-steady state is established with an unmagnetized solar wind we switch on a southward IMF, which causes the magnetosphere to stretch and allows particles to enter the cusps and nightside magnetosphere. Analysis of magnetic fields near the Earth confirms a signature of magnetic reconnection at the dayside magnetopause, and the plasma sheet in the near-Earth magnetotail clearly thins. Later magnetic reconnection also takes place in the near-Earth magnetotail.

At step 768 [*Buneman et al.*, 1995; *Nishikawa et al.*, 1995] a southward IMF ($B_z^{IMF} = -0.4$) is switched on gradually (for comparison, average $B$ at the dayside magnetopause ($\approx 10 R_E$) is nearly 2.8), and the southward-IMF front reaches about $x = 120\Delta$ at step 1280. The Alfvén velocity with this IMF is $v_A/c = 0.1(\bar{n}_i)^{-1/2} = 0.1$ for the average ion density $\bar{n}_i = 1$.

To display magnetic reconnection at the dayside magnetopause and in the magnetotail, Figure 1 shows the magnetic field lines in the noon-midnight meridian plane for four different times [*Nishikawa*, 1997]. (Simulation coordinates are transformed to Geocentric solar magnetospheric (GSM) coordinates in only this figure). For example, $x = -15$ corresponds to $x = 85\Delta$) At time step 1024, the solar wind with the southward IMF starts to interact with the dipole magnetic field at the dayside magnetopause (Fig. 1a). Figure 1b shows the X-line at the magnetopause (time step 1088). The southward IMF is bent by the magnetosphere as shown in Fig. 1c. Figure 1c, which corresponds to time step 1216, displays an interesting magnetic structure near the subsolar magnetopause. Three-dimensional analysis shows that the reconnection occurs three-dimensionally in the dayside magnetopause along the equator [see, also, *Walker and Ogino*, 1996]. At the same time the magnetic fields are stretched in the magnetotail, which leads to growth of a tearing instability there. Figure 1d shows magnetic reconnection occurring at time step 1280, with the X-line located near $x = 85\Delta$ (about $-15R_E$) [*Pritchett and Coroniti*, 1996; *Pritchett et al.*, 1996].

We have checked the electron density in the $xz$-plane containing the dipole center at time steps 1280 and 1344

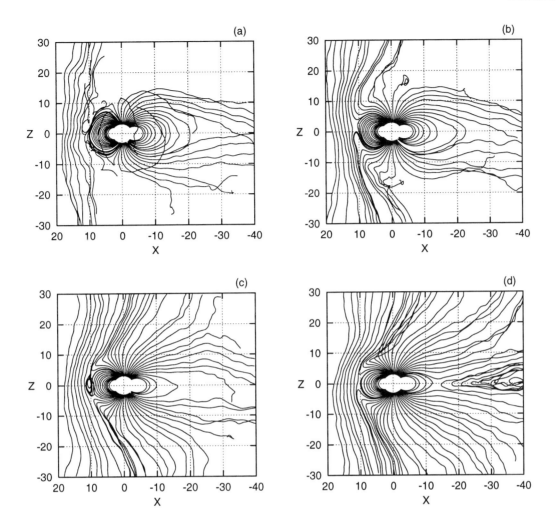

**Figure 1.** Magnetic field lines in the $xz$-plane containing the dipole center at step 1024 (a), 1088 (b), 1216 (c), and 1280 (d). The magnetic field lines are traced from near the Earth ($r = 3\Delta (\approx 3R_E)$) and subsolar line in the dayside and the magnetotail. Some magnetic field lines are moved toward dawn or duskward. The tracing was terminated due to the preset number of points or the minimum strength of total magnetic field.

with the southward IMF. (The time step used in the simulations corresponds to about 4 seconds in nature.) In the process, the dipole field is compressed on the side facing the plasma wind and is extended to a long tail on the down-wind side, similar to the Earth's magnetic field in the solar wind [Buneman et al., 1992, 1995; Nishikawa et al., 1995; Nishikawa, 1997]. More particles penetrate into the magnetotail near the Earth in the case of southward IMF than in the case without the IMF. At time step 1216 the near-Earth magnetotail clearly thins and a tearing instability starts to grow near $x = 85\Delta (\approx -15R_E)$, and the density is bunched in the magnetotail owing to this instability (substorm breakup). At time step 1344, electrons are evacuated from the X-line region due to reconnection (expansion phase) as found in previous simulations [Pritchett and Coroniti, 1996; Pritchett et al., 1996]. Some electrons are trapped in the magnetic island around $x = 108\Delta$.

The enhanced near-Earth cross-field current density [Pritchett and Coroniti, 1995] also occurs at time step 1280, as Lui [1991, 1996] has suggested, which may lead to the current disruption. We found that the cross-field current in the near-Earth tail ($85 < x/\Delta < 90$) is about four times more intense at time step 1280, under the full influence of southward IMF, than at time step 768, which corresponds to zero IMF. We infer that

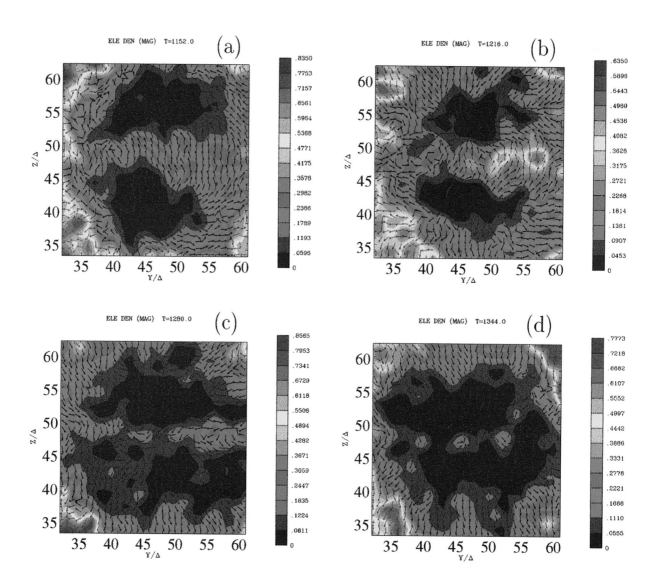

**Plate 1.** Electron density in the $yz$-plane at $x = 85\Delta$ with the magnetic fields by arrows at step 1152 (a), 1216 (b), 1280 (c), and 1344 (d). The kinetic (drift kink) instability grows and lead to the tearing instability.

the weak $B_z$ field and the kinetic (drift kink) instability [*Lui*, 1991, 1996] may excite the tearing instability [*Pritchett and Coroniti*, 1996; *Pritchett et al.*, 1996].

The magnetic reconnection takes place around $x = 85\Delta$ (see Fig. 1d). The tearing instability is excited by the combined effects of the reduced $B_z$ and the kinetic (drift kink) instability. In order to make sure that the kinetic instability is excited along the $y$ (dawn-dusk) direction, we check the time evolution of the plasma sheet in the dawn-dust cross section at $x = 85\Delta$. plate 1 shows the electron density on the $yz$-plane with the magnetic field $(B_y, B_z)$ by arrows at time steps 1152 (a), 1216 (b), 1280 (c), and 1344 (d). The magnitude of the field has been scaled in order to make the field direction clearer for weak fields, so that the length of the vectors is not a true representation of the field magnitude. As shown in Plate 1a and 1b, the plasma sheet is kinked as shown in the previous localized 3-D simulations [*Zhu and Winglee*, 1996; *Pritchett and Coroniti*, 1996; *Pritchett et al.*, 1996]. This structure is similar to Fig. 3 in *Lyons* [1996]. The electron compressibility effect appears to be decreased due to the transport of plasma across flux tubes caused by the drift kink instability [*Pritchett and Coroniti*, 1996; *Pritchett et al.*, 1996]. Consequently, the tearing instability grows rapidly. The northward closed magnetic field (see Plate 1a) is disturbed by the drift kink instability as shown in Plate 1b. Finally, the X-line is created at $x = 82\Delta$ and the plasmoid is formed tailward (see, for example, [*Birn et al.*, 1996; *Walker and Ogino*, 1996]). Therefore, the magnetic field at $x = 85\Delta$ is reversed to southward as shown in Plate 1c and 1d. At the same time, due to the reconnection the electrons are pushed away from the X-line region. We will examine the precise mechanisms responsible for this substorm breakup in the near future.

## 4. DISCUSSION

The results presented here show that even with the modest grid-size of 215 by 95 by 95 cells, our 3-D fully kinetic model is able to generate a complete magnetosphere with the basic characteristics associated with the IMF. For southward IMF, the simulation results show reconnection taking place at the dayside magnetopause, and increased particle entry into the magnetosphere. Southward IMF also causes magnetic-field stretching in the near-Earth plasma sheet. The cross-field current also thins and intensifies, which excites a kinetic (drift Kink) instability. The reduced $B_z$ with a kinetic (drift kink) instability in the central plasma sheet apparently allows the collisionless tearing to grow [*Pritchett and Coroniti*, 1996]. The plasma transport across tubes caused by the kinetic (drift kink) instability appears to reduce the electron compressibility effect and to allow the collisionless tearing instability to grow rapidly. Due to this collisionless tearing instability, magnetic reconnection (X-line) is formed in the near-Earth magnetotail. At the same time, the nightside magnetic fields are dipolarized and a plasmoid is formed tailward. A thin, intense current sheet is disrupted, which is observed during substorm breakup and expansion [*Lui*, 1991, 1996].

In order to examine the evolution of substorms and associated particle entry, we need to investigate the dynamics of magnetosphere including $B_y$ component of IMF. Recent simulation results show that with a dawnward IMF, a reconnection groove occurs from the cusps through the flanks which facilitates plasma penetration into the inner magnetosphere and the near-Earth magnetotail [*Baker et al.*, 1996a; *Fujimoto et al.*, 1996, 1997]. This new results will be reported elsewhere.

The results reported here suggest that full 3-D electromagnetic particle simulations will become an important tool for the theoretical understanding of Earth's magnetosphere in the not-so-distant future. It is clearly necessary to use more realistic values of $m_i/m_e$ with a larger system and more particles in a cell, which relies on the development of future more powerful and faster supercomputers. This would help to establish more precisely the nature of the magnetic reconnection and associated phenomena and to clarify their relation to the observations. Multi-processor machines such as CRAY T3E and ORIGIN2000 in the future will allow even greater improvement, since the code allows simultaneous updating of many cells, which is in progress.

*Acknowledgments.* We thank T. W. Hill, R. A. Wolf, C. Ding, D. Cai, H. Okuda, A. T. Y. Lui, F. R. Toffoletto, and J. Büchner for useful discussions. Support for this work was provided by NSF grants ATM-9106639, ATM-9121116, ATM-9122310, and ATM-9730230. The development of the simulation code was performed at the National Center for Supercomputing Applications, University of Illinois at Urbana-Champaign and the production runs were performed at Pittsburgh Supercomputing Center. Both centers are supported by the National Science Foundation.

# REFERENCES

Baker, D. N., T. I. Pulkkinen, P. Toivanen, M. Hesse, and R. L. McPherron, A possible interpretation of cold ion beams in the Earth's tail lobe, *J. Geomag. Geoelectr, 48*, 699, 1996a.

Baker, D. N., T. I. Pulkkinen, V. Angelopoulos, W. Baumjohann, and R. L. McPherron, Neutral line model of substorms: Past results and present views, *J. Geophys. Res., 101*, 12,975, 1996b.

Birn, J., M. Hesse, and K. Schindler, MHD simulation of magnetotail dynamics, *J. Geophys. Res., 101*, 12,939, 1996.

Buneman, O., TRISTAN: The 3-D, E-M Particle Code, in *Computer Space Plasma Physics, Simulation Techniques and Software*, edited by H. Matsumoto and Y. Omura, Terra Scientific, Tokyo, p. 67, 1993.

Buneman, O., T. Neubert, and K.-I. Nishikawa, Solar wind-magnetosphere interaction as simulated by a 3D, EM particle code, *IEEE Trans. Plasma Sci., 20*, 810, 1992.

Buneman, O., K.-I. Nishikawa, and T. Neubert, Solar wind-magnetosphere interaction as simulated by a 3D EM particle code, *Space Plasmas: Coupling Between Small and Medium Scale Processes*, ed. M. Ashour-Abdalla, T. Chang, and P. Dusenbery, AGU Geophys. Monograph, *86*, p. 347, 1995.

Fedder, J. A., S. P. Slinker, J. G. Lyon, and R. D. Elphinstone, Global numerical simulation of the growth phase and the expansion onset for a substorm observed by Viking, *J. Geophys. Res., 100*, 19.083, 1995.

Fujimoto, M., A. Nishida, T. Mukai, Y. Saito, T. Yamamoto, and S. Kokubun, Plasma entry from the flanks of the near-Earth magnetotail: GEOTAIL observations in the dawnside LLBL and the plasma sheet, *J. Geomag. Geoelectr, 48*, 711, 1996.

Fujimoto, M., T. Mukai, A. Matsuoka, A. Nishida, T. Terasawa, K. Seki, H. Hayakawa, T. Yamamoto, S. Kokubun, and R. P. Lepping, Dayside reconnected field lines in the south-dusk near-tail flank during an IMF $B_y > 0$ dominated period, *Geophys. Res. Lett., 24*, 931, 1997.

Lui, A. T. Y., A synthesis of magnetospheric substorm models, *J. Geophys. Res., 96*, 1849, 1991.

Lui, A. T. Y., Current disruption in the Earth's magnetosphere: Observations and models, *J. Geophys. Res., 101*, 13,067, 1996.

Nishikawa, K.-I., Particle entry into the magnetosphere with a southward IMF simulated by a 3-D EM particle code, *J. Geophys. Res., 102*, 17,631, 1997.

Nishikawa, K.-I., O. Buneman, and T. Neubert, Solar Wind-Magnetosphere Interaction as Simulated by a 3-D EM Particle Code, *Plasma Astrophysics and Cosmology*, edited by A. T. Peratt, Kluwer Academic Pub., p. 265, 1995.

Pritchett, P. L. and F. V. Coroniti, The role of the drift kink modes in destabilizing thin current sheets, *J. Geomag. Geoelectr., 48*, 833, 1996.

Pritchett, P. L. and F. V. Coroniti, Formation of thin current sheets during plasma sheet convection, *J. Geophys. Res., 100*, 23,551, 1995.

Pritchett, P. L., F. V. Coroniti and V. K. Decyk, Three-dimensional stability of thin quasi-neutral current sheets, *J. Geophys. Res.*10127,4131996

Rostoker, G., Phenomenology and physics of magnetospheric substorms, *J. Geophys. Res.*10112,9551996

Sergeev, V. A., T. I. Pulkkinen, and R. J. Pellinen, Coupled-mode scenario for the magnetospheric dynamics, *J. Geophys. Res., 101*, 13,047, 1996.

Walker, R. J. and T. Ogino, A global magnetohydrodynamic simulation of the origin and evolution of magnetic flux ropes in the Magnetotail, *J. Geomag. Geoelectr., 48*, 765, 1996.

Zhu, Z., R. M. Winglee, Tearing instability, flux ropes, and the kinetic current sheet kink instability in the Earth's magnetotail: A three-dimensional perspective from particle simulations, *J. Geophys. Res.*1014885 1996

---

Ken-Ichi Nishikawa, Department of Physics and Astronomy, Louisiana State University, Baton Rouge, LA 70803

# Magnetospheric-Ionospheric Activity During an Isolated Substorm: A Comparison between Wind/Geotail/IMP 8/CANOPUS Observations and Modeling

R. M. Winglee[1], A. T. Y. Lui[2], R. P. Lin[3], R. P. Lepping,[4]

S. Kokubun[5], G. Rostoker[6], and J. C. Samson[6]

The dynamics of an isolated substorm observed on Feb. 9, 1996 using is examined using (1) insitu observations from Wind, Geotail and IMP 8, (2) ground-based observations from Canadian Auroral Network for the OPEN Program Unified Study (CANOPUS) and (3) global simulations. The CANOPUS data show that the substorm has the classical signatures of the expansion of the auroral oval during the growth phase, and the intensification and poleward motion of the high latitude portion of the auroral oval. The enhanced convection of dayside magnetic field into the nightside that drives the growth phase also produces the growth of the lobes in the north-south direction, forcing the high-latitude magnetopause and bow shock to move outwards. This motion causes the bow shock crossing of IMP 8 at the beginning of the growth phase. The tail dynamics as observed by Geotail are very much more complicated. Small (-1 nT) $B_z$ values are observed in the tail during the growth phase, when Geotail at about 35 $R_E$ down the tail makes several crossings of the current sheet. These results as well as the global simulations suggest that for this substorm a near-Earth neutral point can form during the growth phase but not trigger substorm onset. Additional processes appear to be necessary to either locally disrupt the current sheet earthward of Geotail or to move the existing neutral point earthward in order to produce the large (-5 nT) negative $B_z$ values observed at substorm onset.

---

[1] Geophysics Program, University of Washington, Seattle.
[2] Applied Physics Laboratory, John Hopkins University, Laurel, Maryland.
[3] Space Science Laboratory, University of California, Berkeley, California
[4] NASA Goddard Space Flight Center, Greenbelt, Maryland.
[5] Solar-Terrestrial Environment Laboratory, Nagoya University, Japan.
[6] Department of Physics, University of Alberta, Edmonton, Canada

Geospace Mass and Energy Flow: Results From the International Solar-Terrestrial Physics Program
Geophysical Monograph 104
Copyright 1998 by the American Geophysical Union

## 1. INTRODUCTION

Magnetic reconnection can play a vital role in the dynamics of the magnetosphere [*Hones*, 1979]. Over the last decade, global magnetohydrodynamic (MHD) models have provided important insight into the generation of plasmoids and the reconfiguration of the magnetosphere for idealized configurations [e.g., *Ogino*, 1986; *Fedder and Lyon*, 1987; *Watanabe and Sato*, 1990; *Hesse and Birn*, 1991; *Kageyama et al.*, 1992; *Usadi et al.*, 1993; *Ogino et al.*, 1992, 1994; *Raeder et al.*, 1995]. Recently, there have been direct comparisons made between global simulations and satellite observations, including Geotail crossings into the magnetosheath [*Frank et al.*, 1995], Viking auroral imaging [*Fedder et al.*, 1995; *Winglee* and Menietti, 1997], and the influence of varying interplanetary magnetic field (IMF) conditions on motion of the magnetopause/magnetotail as observed by Wind/Geotail/IMP 8 [Winglee et al., 1997a].

However, to date quantitative comparisons between modeling and observed reconnection signatures in association with substorm activity have not been made. Such studies are only now becoming possible due to the development of the International Solar Terrestrial Physics (ISTP) program which is providing simultaneous solar wind data, comprehensive insitu magnetospheric observations and extensive ground-base coverage. In this paper, we use this comprehensive data set to investigate magnetic reconnection signatures observed in the tail during an isolated substorm that was observed on Feb. 9, 1995. These observations are combined with global fluid simulations to develop a fully 3-D picture for the substorm dynamics including a relatively long (> 1 hr) growth phase observed by ground-based stations, and reconnection signatures in the magnetotail observed by Geotail.

As discussed in section 2, the simulation model uses a higher-order Ohm's law that includes finite ion inertial effects (i.e. processes of the order of the ion skin depth) which can cause the diversion of current from the tail-current sheet into the ionosphere. Due to the resistivity of the ionosphere, these processes produce an effective resistivity that can control the reconnection rate. The model also uses all three magnetic field components of the IMF to drive the simulations. The presence of a significant $B_x$ field component produces substantial timing delays for the arrival of IMF turnings between the dawn and dusk flanks (section 3). These delays are confirmed by a comparison of Wind and IMP 8 observations, and produces significant flapping of the magnetotail.

The Canadian Auroral Network for the OPEN Program Unified Study (CANOPUS) data shows the classical growth-phase signature of the equatorward motion of the oval (section 4). The growth phase last about 1.5 hrs before onset which is mark by the rapid intensification and poleward motion of the poleward edge of the auroral oval. This intensification is driven by the convection of reconnected dayside field lines at the magnetopause into the nightside. These processes produce two observable effects. The first is that it causes the high latitude magnetopause and bow shock to move outward, as seen in the global simulations. As a result, IMP 8 crosses the high latitude bow shock shortly after the southward turning of the IMF that marks the beginning of the growth phase.

The second is that it causes the thinning of the tail current sheet which causes $B_z$ observed by Geotail to essentially vanishing and even turn slightly (-1 nT) negative (section 5). In the vicinity of substorm onset, Geotail observed a flux rope signature with a large negative (-5 nT) $B_z$ and a strong core ($B_y = 4$ nT) component. The thinning of the current sheet is seen in the global models as the formation of a near-Earth neutral point at around 20 - 30 $R_E$. The small negative perturbations appear at localized enhancements in the reconnection rate driven by pressure variations in the solar wind and which cause the neutral point to move around in the above range in tail distances. Both the observations and modeling suggested that the formation of the near-Earth neutral point while possibly necessary for substorm onset is not sufficient.

Additional processes appear to be needed to explain both the large negative $B_z$ signature observed by Geotail at substorm onset as well as the smaller negative $B_z$ signatures seen during the growth phase. It is possible that kinetic processes in the inner magnetosphere as proposed by Lui [1996] or some field-line resonances lead to the additional disruption of the thin, growth-phase current sheet that leads to the creation or motion of additional neutral point earthward of Geotail at onset. Other studies between Geotail and CANOPUS data [Maynard et al., 1997] also show that reconnection particularly on closed field-lines can occur without substorm onset, the later being associated with reconnection of lobe field lines.

## 2. SIMULATION MODEL

The fluid dynamics of a plasma are described by

$$\frac{\partial \rho}{\partial t} + \nabla \cdot (\rho \mathbf{V}) = 0 \qquad (1)$$

$$\rho \frac{d\mathbf{V}}{dt} = -\nabla P + \frac{\mathbf{J} \times \mathbf{B}}{c} + \rho_q \mathbf{E} \qquad (2)$$

$$\frac{\partial}{\partial t}\varepsilon + \nabla \cdot (\mathbf{V}(\varepsilon + P)) = \mathbf{J} \cdot \mathbf{E} \qquad (3)$$

$$\frac{\partial \mathbf{B}}{\partial t} = -c \nabla \times \mathbf{E} \qquad (4)$$

where $\varepsilon = \rho \mathbf{V}^2/2 + P/(\gamma - 1)$ The space-charge term $\rho_q \mathbf{E}$ where $\nabla \cdot \mathbf{E} = 4\pi \rho_q$ is in general small but is retained for self-consistency since $\rho_q$ is required if the cross-polar cap potential is to be calculated [cf. Winglee et al., 1997b].

These equations can be closed by Ampere's law that relates current to the magnetic field which is usually approximated by $\mathbf{J} = \nabla \times \mathbf{B}$ and the generalized Ohm's law [e.g., Krall and Trivelpiece, 1986]. The generalized Ohm's law contains all the electron and ion dynamics including finite electron inertial effects and finite gyroradius effects. However, an exact solution of these equations requires a very small time step in order to resolve electron processes. Thus, approximations for the generalized Ohm's law have to be made to attain a tractable solution.

Ideal MHD retains the two dominant terms of the generalized Ohm's law, i.e.,

$$E + \frac{V \times B}{c} = 0 \ . \quad (5)$$

In this approximation the only electric field generated in the plasma is the convective electric field which is always orthogonal to $V$ and $B$. For this electric field, the electrons and ions move as a single fluid under $E \times B$ drift motion. The configuration of the tail is such that the electric field lies essentially in the direction of the tail current, which is primarily in the $y$ direction.

However, within the tail and particularly at the plasma sheet boundary layer there can be additional electric fields in the $x$ and $z$ plane [e.g., *Cattell and Mozer*, 1984]. Such fields can, under the right conditions pull additional currents in the $x$ and $z$ plane. *Winglee* [1994] has shown that these currents can lead to a current diversion from the tail current sheet into the ionosphere. With the latter's resistivity, these additional processes can be thought of as producing an effective plasma resistivity on the magnetotail and an unphysical anomalous or numerical resistivity is not required to drive the magnetic reconnection.

Incorporation of these processes is most easily attained by noting that the time step in fluid simulations is of the order of a few seconds. On this time scale, the electrons can travel significant distances along a magnetic field line so that they can be assumed to be approximately in steady-state, i.e. $DV_e/Dt = 0$, or

$$E + \frac{V_e \times B}{c} + \frac{1}{en_e}\nabla P_e = 0 \ . \quad (6)$$

The $\nabla P_e$ term in the presence of a non-uniform density, such as at boundary layers, can lead to the above current diversion from the magnetotail and magnetopause.

The electron momentum equation (6) can be cast into the more usual terms of an Ohm's law by noting that $J = en(V_i - V_e)$ and $V \simeq V_i$. Making these substitutions into (6) the next higher order form of the generalized Ohm's law is obtained with

$$E + \frac{V \times B}{c} \simeq (\frac{J \times B}{ecn_e} - \frac{1}{en_e}\nabla P_e) \ . \quad (7)$$

A perturbative solution was used in *Winglee* [1994] to show that the higher order Ohm's law can lead to additional field-aligned currents in the noon-midnight meridian and which map into the auroral region. This perturbative expansion is equally valid even in 3-D simulations. However, to avoid any ambiguities that arise from perturbative solutions, the following work is based on an exact solution to (7) with the assumption that the electron pressure $P_e$ is equal to one half of the total plasma pressure. This equipartition in temperatures is implicit in ideal MHD and is therefore an appropriate starting point to look at differences arising from non-ideal MHD processes.

In the real magnetosphere there can be regions where the electron temperature can be much less than the ion temperature. The effect on the system in this case can be estimated by assuming that the ion fluid is in approximate equilibrium, i.e. $J \times B \simeq \nabla P$. In this case, the right hand side of (9) reduces to $(1/2)J \times B$ for equal electron and ion temperatures and to $J \times B$ when the electron temperature is much less than the ion temperature. This approximation is not made in the simulations but goes to show that the effects from the corrections to the Ohm's law are still present even when the electron temperature is relatively low.

In principal the higher order corrections in (7) can give rise to high-frequency/short-wavelength whistlers. These waves are not resolved in the present model (nor are they incorporated in MHD) and restrictions that they might impose on the time step and spatial scales do not appear to be relevant. Comparison with *in situ* measurements with Geotail and IMP 8 show that the model is able to produce reasonable agreement with observed magnetic fields in a variety of regions of the magnetosphere [sections 3 and 5; see also *Winglee et al.*, 1997a].

The grid spacing for this case study is variable with a minimum $0.2\ R_E$ in the inner magnetosphere and $3\ R_E$ in the distant tail. The simulations extent $77\ R_E$ on the flanks and $200\ R_E$ into the tail. The above corrections to the ideal Ohm's law in dimensionless units are proportion to the ratio of the grid spacing ($\Delta$) to the ion skin depth ($c/\omega_{pi}$). In the present case, it is assumed to be 8 which would be representative of a plasma that has both $H^+$ and $O^+$ components.

The inner radius of the simulations is set at $4.0\ R_E$, which is typical of most global simulations [e.g. *Ogino et al.*, 1994; *Fedder et al.*, 1995]. Around this inner boundary, the model incorporates a resistive layer that acts like the ionosphere. In this region the Ohm's law (7) is modified such that

$$E + \frac{V \times B}{c} \simeq (\frac{J \times B}{ecn_e} - \frac{1}{en_e}\nabla P_e) + \eta_{iono}(r)J \ . \quad (8)$$

Note that while only a scalar resistivity is incorporated in (8), the actual conductivity is a tensor when the retention of the Hall term is taken into account. By assuming that the magnitude of $\eta_{iono}$ decreases rapidly with radial distance (in practice it is only significant within 3 cells of the inner radius), the Ohm's law allows a smooth transition from the collisional plasma of the ionosphere to the collisionless plasma of the magnetosphere. Placing the resistive layer inside the simulation system also has the advantage that the resistivity along the path that the current closes is not fixed, but rather can float since the currents can close at different altitudes depending on the forcing from the solar wind conditions.

The other feature of the ionospheric resistivity is that it is allowed to vary with both latitude and longitude due to the presence of (1) a daylight ionosphere and (2) enhanced ionization from particle precipitation. For the daylight component the ion-neutral collision frequency at the inner radius is assumed to equal the ion-cyclotron frequency at the equator and then to decrease to 1/10 of the value at the day-night terminator. The nightside collision frequency at the inner boundary is assumed to be constant at 1/10 of the ion-cyclotron frequency. Effects from particle precipitation are included by the inclusion of an additional component that is centered at 65° magnetic latitude with a half width of 5°. The peak ion-collision frequency arising from this component is set at the ion-cyclotron frequency. Experiments with different forms of resistivity show that the magnetospheric configuration is not very sensitive to the actual form of the resistivity imposed on the system except possibly for the shape of the auroral potential at mid-latitudes. Much of the results presented in the following would be attained if just a weak uniform resistive layer were assumed.

The last feature of the simulations is that a dense ($200$ cm$^{-3}$) plasma is assumed around the inner radius. The density of this component then falls off as $1/r^5$ in the simulations system. This plasma in many ways acts like the ionosphere by providing a sink for energy flowing in from the magnetosphere. At the same time it provides a source of plasma that can be used to populate the tail, as is it shrinks and grows with activity. This plasma component also limits the Alfvén speed so that the time step of the simulations is allowed to be of the order of a second.

## 3. SOLAR WIND CONDITIONS

The solar wind conditions as observed by Wind on Feb. 8-9, 1995 are shown in Figure 1. These

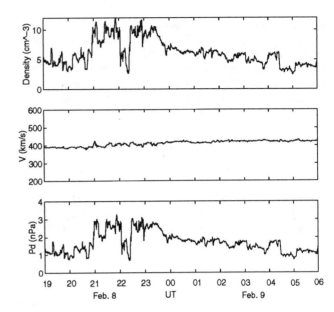

Figure 1. Time histories of the solar wind density and speed observed by Wind. The solar wind density and dynamic pressure reach moderate values at the end of Feb. 8 and then slowly decline to nominal solar wind conditions of 4 cm$^{-3}$ and 420 km/s.

observations were taken at (193, -40, -26) $R_E$ which is near the libration point. As shown in Figure 1, the density reaches moderately high values of 10 cm$^{-3}$ in the latter part of Feb. 8, but then on average slowly declines from about 7 cm$^{-3}$ to 5 cm$^{-3}$ on Feb. 9. The observed substorm occurs around 4:37 UT on Feb. 9.

The corresponding IMF conditions are shown in Figure 2. This event was chosen because of the prolonged northward IMF (of the order of 5 nT) on Feb. 8 prior to the substorm. During this northward IMF, $B_y$ IMF was only slightly positive so that there is only minimal torquing of the magnetosphere, and the auroral activity (section 4) was low. The southward turning of the IMF at Wind is seen at about 2 UT, and lasts for about 2 hrs although at the beginning of the turning, $B_z$ IMF fluctuates between -2.5 and 0 nT. This southward turning drives the growth phase of the substorm. It is also accompanied by an increase in $B_y$ IMF.

A large $B_x$ component is also seen throughout the period of interest. This component is incorporated in the simulations by adding the initial value (in this case 3.5 nT) of $B_x$ IMF to the initial magnetic field. In other words, the initial magnetic field configuration consists of the terrestrial dipole plus a uniform component from the heliospheric current sheet. Temporal changes in $B_x$ IMF are then incorporated by allowing the $B_y$ and $B_z$ components to have variations in $y$ and $z$ along the solar wind boundary. These variations are estimated by assuming that the solar wind field lines are straight with a direction vector $(B_x, B_y, B_z)$ so that the distance to the solar wind boundary differs along the field line from the observing position. The corresponding time delay is then estimated using the observed solar wind speed.

In order to show that the IMF is being properly incorporated into the system, Figure 3 shows the magnetic field observed by IMP 8 which was located at (-12.5, 17, 32) $R_E$ and that predicted by the model. The simulations were started at 20 UT from a ground state with the IMF conditions corresponding to those observed at 20 UT. Some of the discrepancy at early times is due to the magnetosphere relaxing to a true equilibrium with the observed solar wind conditions. It is seen that IMP 8 for most of the time is in the solar wind, except near the beginning of the period where pressure fluctuations in the solar wind allow IMP 8 to cross the bow shock (as evidenced by the periods of large increases in the magnitude of the magnetic field). Near the end of the period there are additional crossings are due to the reconfiguration of the magnetosphere induced by the southward turning of the IMF, as shown in the following section. The bow shock crossings in the simulations are not as sharp as the observations due to the fact that the grid resolution at IMP was 1.6 $R_E$. Nevertheless, it is seen that the overall increase

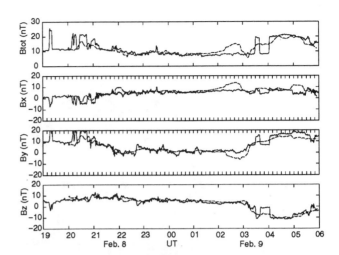

Figure 3. Time histories of the observed (solid lines) and predicted (dashed lines) for the magnetic field at IMP 8. The bow shock crossings predicted by the model are not quite as sharp as the observed crossing due to the limited resolution at IMP 8. The bow shock crossings that appear around 3:30 UT occur shortly after the arrival of the southward turning of the IMF.

of $B_x$ in the solar wind from 2 nT to 7 nT is well incorporated into the model.

The gradients in $y$ and $z$ imposed on the solar wind boundary in order to incorporate a variable $B_x$ also has the effect that it causes the appearance of the southward turning to be different between the dawn and dusk flanks. The is delay for the present case can be of the order of 15 to 20 min. For example, using only the difference in $x$ and the observed solar wind speed, disturbances should take 54 minutes to move from Wind to IMP 8. It is seen from Figures 2 and 3 that the northward spike just prior to the southward turning that is seen at Wind at 1:50 UT is not observed by IMP 8 until 3:05 UT. The peak southward value of the IMF does not experience as much delay because the $B_y$ component becomes the dominant component and the field lines are directed almost parallel to the $y$–$z$ plane.

## 4. GROWTH PHASE PROCESSES

As already noted, IMP 8 makes bow shock crossings (around 3:30 UT and 4:00 UT) shortly after the southward turning is observed by IMP 8. The bow shock crossings at these times are not just a coincidence but are in fact a signature of the growth phase of a substorm due to the reconfiguration of the magnetosphere. This reconfiguration is illustrated in Figure 4 which shows cross-sections of the plasma

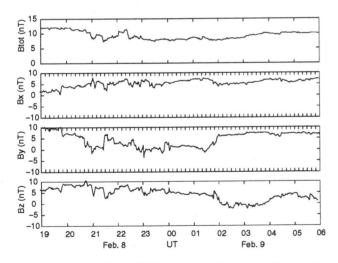

Figure 2. The IMF corresponding to Figure 1. There is a long period of northward IMF, followed by a 2 hr southward turning starting around 02 UT Feb. 9. There is about an 1 hr delay between solar wind features observed by the Wind and their arrival at the magnetosphere.

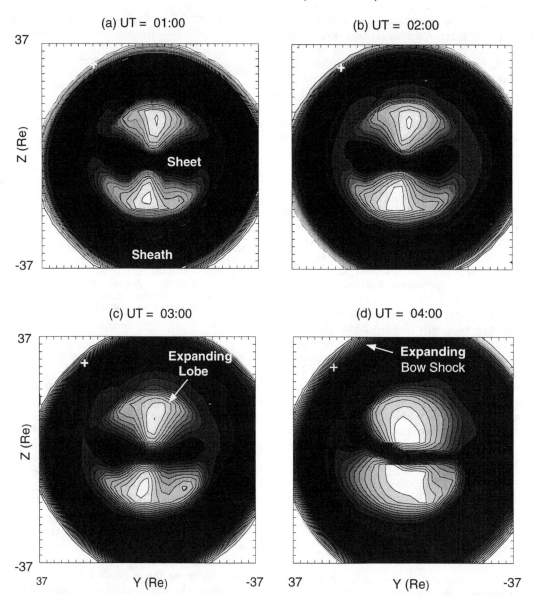

Figure 4. Cross-section of the plasma pressure through IMP 8 at $x = -11\ R_E$. The bow shock crossing of IMP 8 coincides with the thinning of the current sheet and the increased flaring of the lobes and magnetopause produced by the southward turning of the IMF.

pressure at the position of IMP 8. The high pressure regions indicate the magnetosheath and the plasma sheet while the low pressure regions indicate the lobes. At the earliest time, the IMF is northward and the magnetosheath has a circular cross-section and the current sheet is relatively thick with full width of about 4 $R_E$ near the center and about 8 $R_E$ on the flanks.

As the IMF becomes less northward, the lobes are seen to expand and the current sheet thins (Figures 4b and 4c). The expansion of the lobes causes the high latitude magnetopause and magnetosheath to move outwards by between 2 to 4 $R_E$, while the equatorial magnetopause and bow shock hardly move. With the actual southward turning the lobes increase to more than twice their initial cross-sectional area forcing the magnetopause and bow shock to move outward beyond the IMP 8 position (Figure 4d). At the same time the current sheet thins to less than 2 $R_E$.

The expansion of the high-latitude magnetopause and bow shock in conjunction with the thinning of the tail current sheet is entirely consistent with the classical model of substorms. In particular, as the IMF becomes more southward, there is more draping or convection of dayside field lines into the nightside. This enhanced convection leads to the thinning of the current sheet, as well to an increase in the lobe-field strength. If the solar wind has nearly steady dynamic pressure, then pressure balance requires that the high-latitude magnetopause to move out. As it does so it also pushes the bow shock outwards, causing the IMP 8 bow shock crossings. The reason that there are multiple crossings in this case is that the southward turning is modulated so that there are small periods of northward IMF mixed in with the southward turning. The simulation model is sufficiently sensitive even with 1.6 $R_E$ resolution to see the multiple crossings actually made by IMP 8.

In addition to the thinning of the current sheet, the other classical signature of the growth phase is the expansion of the auroral oval. This expansion was observed by CANOPUS as illustrated in Figure 5 which shows data from the meridian-scanning photometer. The top panel shows the emissions from the 5577 Å line which is primarily due to ion precipitation while the bottom panel shows the 6300 Å line which is primarily due energetic electron precipitation. The latitudinal dependence is determined from observations at Gillam and Rankine Inlet. The intensity for both lines have the same overall features except that the electron emissions tend to be about 2° in latitude higher that those from the ions.

Both lines at 03:00 UT show that the peak of the particle precipitation is at relatively high latitudes being at about 70° for the 5577 and 72° for 6300. The position of this peak emission is then seen to move to lower latitudes reaching a minimum of about 66.5° and 69°, respectively. At 04:35 UT, there is an intense brightening of the auroral emissions and the poleward expansion of the oval to latitudes of 73-75°. This time will be used to indicate substorm onset, although it should be noted that because the data are restricted in MLT, substorm onset in different regions could appears offset by a few minutes. The equatorward edge after onset moves poleward very much more slowly, reaching only about 66° at 08:00 UT.

Some idea of the global properties of the auroral oval can be attained by superimposing the position of the above CANOPUS stations on the map of the auroral field-aligned currents as determined from the simulations as in Figure 6. In an effort to provide some calibration of the simulation predictions we

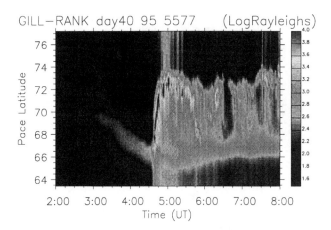

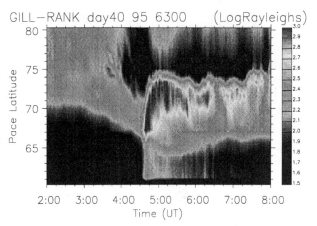

Figure 5. CANOPUS meridian scanning photometer data from Gillam and Rankin Inlet showing the expansion of the oval during the growth phase and substorm onset and subsequent polar expansion at around 04:37 UT. At this time the MLT of the stations is 21:25.

have previously undertaken comparative case studies [*Winglee et al.*, 1997b] with the assimilative mapping of ionospheric electrodynamics (AMIE) method and other semi-empirical models for other events, and shown that the quantitative agreement could be attained.

One key feature that is seen in the simulations that appears to be consistent with the CANOPUS data is that position of the region 1 currents that intersects the CANOPUS meridian move to low latitudes during the growth phase. As a result, the electron precipitation lies consistently on the poleward edge of the region 1 currents while the ion precipitation lies on the equatorward edge. The positioning of the electron precipitation on the poleward edge of the region 1 currents is also consistent our comparisons with AMIE in that the nightside particle boundary for the separatrix between open and closed field lines appears to lie on the poleward edge of the region 1 currents.

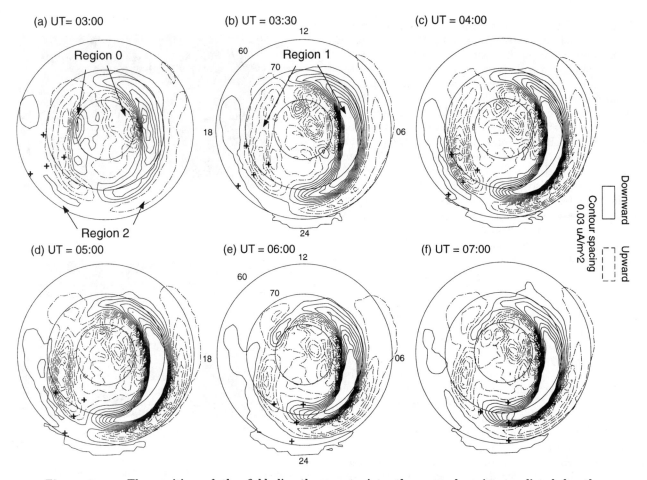

Figure 6. The position of the field-aligned currents into the auroral region predicted by the model. The CANOPUS chain of Rankin Inlet, Gillam and Pinawa are shown by the three '+' moving from high latitude to low latitude, respectively. The fourth '+' indicates the position of Fort Smith for reference. The 6300 CANOPUS emissions lie consistently on the poleward edge of the region 1 currents while the 5577 lie on the equatorward edge. The current intensifies through out the growth phase and they reach their peak around the time of substorm onset.

The overall magnitude of the model currents shown in Figure 6 approximately follows the optical intensifications seen by CANOPUS. In particular he field-aligned currents in Figure 6a support a total of about 0.8 MA prior to substorm on set. The model currents then peak between 04:20 and 05:00 UT (Figure 6d) at about 2.2 MA, after which they slowly decay reaching about 1.8 MA by 07:00 UT (Figure 6f) and 1.3 MA by 08:00 UT (not shown). It is hard to differentiate space and time variations in the model currents near the CANOPUS chain at around substorm onset because of its proximity to the Harang discontinuity.

## 5. RECONNECTION AT SUBSTORM ONSET

Figure 7 shows the tail magnetic field as observed by Geotail (solid lines) and predicted by the simulations (dashed lines). Geotail was moving slowly earthward being at $(-35, 11, 0)$ $R_E$ at 19 UT on Feb. 8 and at $(-32, 6, -1.5)$ $R_E$ by 04 UT on Feb. 9. It is seen that even though Geotail is moving fairly slowly it nevertheless makes several crossings of the center of the current sheet as evidenced by the changes in the sign of $B_x$. The simulations capture several of these crossing although the first crossing from the southern hemisphere into the northern hemisphere occurs about an 1 hr earlier than observed and the final crossing occurs about an 1 hr later than observed. It turns out that these timing errors correspond to a difference in the position of the current sheet by only about 0.4 - 0.8 $R_E$. Given that the tail is flapping such an error should be considered reasonable.

The overall trend for $B_y$ predicted by the simulations closely matches the Geotail observations except

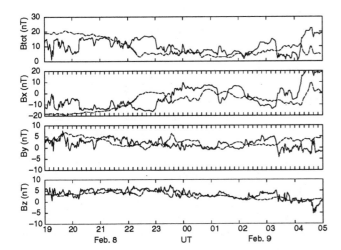

Figure 7. Time histories of the observed (solid lines) and predicted (dashed lines) for the magnetic field at Geotail. The tail is subject to significant flapping due to inhomogeneities in the solar wind which causes Geotail to make several crossings of the current sheet, as seen by the changes in sign in $B_x$. $B_z$ slowly declines as the IMF becomes more southward and small negative (-1 nT) $B_z$ is seen during the growth phase between 03:20 and 04:30 UT. Substorm onset is seen by a large (-5 nT) $B_z$ which is not captured by the modeling.

after about 03:25 UT when Geotail observes some negative $B_y$ while the simulations indicate that it remains positive. The predicted $B_z$ is in excellent agreement with the observed $B_z$. During the period of strong northward IMF observed by Wind, the average value of $B_z$ in the tail is of the order of 5 nT. As the northward IMF decreases so does the normal magnetic field component. The model also captures the small (1 nT) negative $B_z$ components observed by Geotail that first appear around 03:30 to 03:50 UT and continue to reappear quasi-periodically until substorm onset around 04:35 UT. This period corresponds to the growth phase as identified by CANOPUS.

The presence of the small negative $B_z$ components well before substorm onset is very important. The development of the tearing mode in the tail has often been thought to be suppressed by the presence of a positive $B_z$ in the tail. Here the external conditions are such that the tail has undergone very strong thinning and yet explosive reconnection does not occur until very late.

Some indication of the tail dynamics during the growth phase is illustrated in Figure 8 which shows the tracing of magnetic field lines. At the first time shown there are many closed field lines in view. At 03:25 UT many of the high latitude field-lines that were originally closed appear as open field-lines. Geotail at this stage is still on a closed field line.

At 03:35 UT (Figure 8c) Geotail appears on an open flux rope structure. The flux rope is due to the presence of a positive $B_y$ field in the tail which appear to arise from the penetration of the $B_y$ component of the IMF. At 03:50 UT a near-Earth neutral line appears to form at about 20 $R_E$ and this coincides with one of the -1 nT $B_z$ events seen by Geotail. Shortly afterwards the neutral point moves out to about 30 $R_E$ near Geotail. The neutral point repeats this motion between 20 $R_E$ and 30 $R_E$ in the model until 04:45 UT when it is finally ejected down the tail.

At this point several possible scenarios are possible. One can always suggest that the near-Earth neutral did not form as predicted by the global model, despite the very close agreement between the predicted and observed $B_z$. In this case some other arguments have to made as to why negative $B_z$ were observed at 30 $R_E$ in the tail and why reconnection has to be excluded. Geotail makes several crossing through the center of the current sheet at this time (as evidenced by the reversals in $B_x$) so that simple necking of the tail current sheet without reconnection cannot be easily invoked. The alternative is to suggest that maybe a near-Earth neutral line formed as suggested by the model. Then the conclusion would be that the formation alone is not sufficient to produce substorm onset, since the negative $B_z$ signatures are seen 1 hr prior to substorm onset.

The generation of the large -5 nT $B_z$ observed by Geotail at substorm onset could be due a variety of processes. One is that processes in the inner magnetosphere cause the neutral point to simply move much closer in to the Earth. It is seen from the magnetic field mapping in Figure 8e for example that much larger negative $B_z$ appear behind the Geotail position so that should the structure move more earthward, Geotail would see the large negative signature. The other alternative is that a new, neutral point forms earthward of Geotail. Such a neutral point could be produced by kinetic processes not included in the global model [cf. *Lui*, 1996] or possibly by coupling to ionosphere via changes in the ionospheric conductivity, by the outflow of heavy ionospheric ions, or by field-line resonances. Irrespective of the processes involved is that the global model predicts dipolarization (increase of $B_z$ to 1-2 nT positive) and the expulsion of the near-Earth neutral point by 5 UT, similar to the observations.

## 6. SUMMARY

In this paper, processes associated with the development of an isolated substorm has been investigated

190 MAGNETOSPHERIC-IONOSPHERIC ACTIVITY

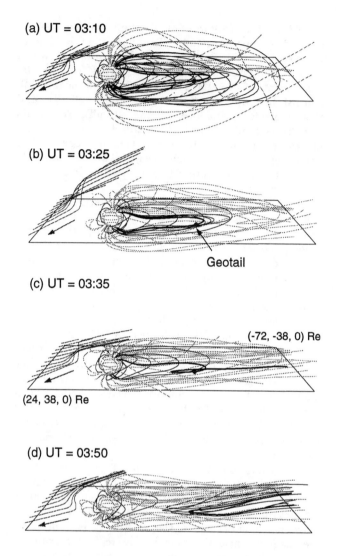

Figure 8. Magnetic field line tracing during the growth phase of the substorm. The thin dark lines represent either closed magnetospheric field lines or reconnected field lines. The thick dark lines shows the instantaneous Geotail field line. The dotted lines indicated open terrestrial field lines while the dashed line and arrow indicate solar wind field lines. The magnetosphere moves from a closed configuration to an increasingly open configuration and Geotail eventually appears on a reconnected field line associated with a shallow near-Earth neutral line slightly earthward of the spacecraft position.

through (1) *in situ* observations from Wind, Geotail, and IMP 8, (2) ground-based observations from CANOPUS, and (3) global simulations. The isolated substorm which occur around 04:35 UT on Feb. 9, 1996 followed a long period of northward IMF followed by a period of weak (-2 nT) southward IMF that lasted for about 2 hrs. The substorm evolved along classical lines with the equatorward expansion of the auroral oval during the growth phase, and the rapid brightening and polarward motion of the poleward edge of the oval at substorm onset. The global model shows a slow increase and expansion in the auroral field-aligned currents during the growth phase, with the 6300 emissions observed by CANOPUS being consistently on the poleward edge of the region 1 currents and the 5577 emissions on the equatorward edge.

The global model also shows that the enhanced convection of dayside magnetic field into the night that drives the growth phase can be seen as an expansion of the lobes in primarily the north-south direction. As a result, the high-latitude magnetopause and bow shock move outward. This outward motion can account for the apparent high latitude bow shock crossing made by IMP 8 at the beginning of the growth phase.

The corresponding dynamics of the tail as observed by Geotail is very complicated. First, the tail is experiencing significant warping or flapping produced by inhomogeneities in the solar wind. The global model is able to account for several of the crossing of the current sheet made by Geotail in the 10 hrs treated in the event produced by this flapping. The average value of $B_z$ in tail is initially fairly strong at 5 nT positive during the period of northward IMF. However, it slowly declines as the IMF becomes more southward.

The most surprising result is that the Geotail sees several small negative $B_z$ events during current sheet crossings up to about an hour before substorm onset. The global model is able to account for the observed changes in $B_z$. The small negative $B_z$ events appear to be associated with the formation of a shallow near-Earth neutral point between 20 and 30 $R_E$. This result is suggestive that the formation of a near-Earth neutral line is not sufficient for substorm onset. The very large (-5 nT) negative $B_z$ event observed by Geotail in association with substorm onset is not produced in the model. The hypothesis is that processes in the inner magnetosphere either enhances the reconnection rate to make the existing neutral point deeper and possibly closer to the Earth or a new, compact or local neutral point forms earthward of Geotail. Kinetic processes [cf. *Lui*, 1996] or ionospheric coupling through changes in ionospheric conductivity, heavy ion outflow, or field-line resonances are not incorporated in the global model but any of them could play important roles in enhancing the reconnection rate.

The global influences are such that while the global model does not capture the explosion reconnection at substorm onset, recovery of the tail magnetic field after substorm onset is incorporated in a manner consistent with the observations. The implication is that there is clearly a coupling between local or internal processes with global or external conditions. In another study, we will show that in an active period, the global model can capture about half the large negative $B_z$ events observed by Geotail at about the same position down the tail, and all of these events are associated with small scale structures.

*Acknowledgments.* Support for the 3D Plasma and Energetic Particle instrument was provided by NASA grant NAS5-2815 at the Univ. of California, Berkeley and by NASA grant NAG5-2813 to the Univ. of Washington. The simulations were supported by NSF grant ATM-9321665 and NASA grant NAGW-5047 to the Univ. of Washington and by the Cray C-90 at the San Diego Supercomputing Center which is supported by NSF. The authors also wish to thank Y. Saito and T. Yamamoto for assistance in processing the Geotail magnetic field and plasma data, and G. K. Parks for many valuable discussions. CANOPUS which was constructed by and is maintained and operated by the Canadian Space Agency.

## REFERENCES

Cattell, C. A., F. S. Mozer, Substorm electric fields in the earth's magnetotail, in *Magnetic Reconnection in Space and Laboratory Plasmas*, Geophys. Monogr., vol. 30, edited by E. W. Hones, Jr., AGU, Washington, D.C., 1984.

Fedder, J. A., and J. G. Lyon, The solar wind-magnetosphere-ionosphere current voltage relationship, *Geophys. Res. Lett.*, 14, 880, 1987.

Fedder, J. A., S. P. Slinker, J. G. Lyon, and R. D. Elphinstone, Global numerical simulation of the growth phase and the expansion onset for a substorm observed by Viking, *J. Geophys. Res.*, 100, 19,083, 1995.

Frank, L. A., et al., Observations of plasmas and magnetic fields in Earth's distant magnetotail: Comparison with a global MHD model, *J. Geophys. Res.*, 100, 19,177, 1995.

Hesse, M., and J. Birn, Magnetospheric-ionospheric coupling during plasmoid evolution: First results, *J. Geophys. Res.*, 96, 11513, 1991.

Hones, E. W., Jr., Plasma flow in the magnetotail and its implication for substorm theories, in *Dynamics of the Magnetosphere* edited by S. I. Akasofu, p. 545, D. Reidel Publ. Co., 1979.

Kageyama, A., K. Watanabe, and T. Sato, A global simulation of the magnetosphere with a long tail: No interplanetary magnetic field, *J. Geophys. Res.*, 97, 3929, 1992.

Krall, N. A., and A. W. Trivelpiece, *Principles of Plasma Physics*, San Francisco Press, Inc., San Francisco, 1986.

Lui, A. T. Y., Current disruption in the Earth's magnetosphere: Observations and models, *J. Geophys. Res.*, 101, 13,067, 1996.

Maynard, N. C., et al., Geotail measurements compared with the motions of high-latitude auroral boundaries during two substorms, *J. Geophys. Res.*, 102, 9553, 1997.

Ogino, T., A three-dimensional MHD simulations of the interaction of the solar wind with the Earth's magnetosphere: The generation of field-aligned currents, *J. Geophys. Res.*, 91, 6791, 1986.

Ogino, T., R. J. Walker, and M. Ashour-Abdalla, A global magnetospheric simulation of magnetosheath and magnetosphere when the interplanetary magnetic field is northward, *IEEE, Trans. Plasma Sci*, 20, 817, 1992.

Ogino, T., R. J. Walker, and M. Ashour-Abdalla, A global magnetospheric simulation of the response of magnetosphere to a northward turning of the interplanetary magnetic field, *J. Geophys. Res.*, 99, 11,027, 1994.

Raeder, J., R. J. Walker, and M. Ashour-Abdalla, The structure of the distant geomagnetic tail during long periods of northward IMF, *Geophys. Res. Lett.*, 22, 349, 1995.

Usadi, A., A. Kageyama, K. Watanabe and T. Sato, A global simulation of the magnetosphere with a long tail: Southward and northward interplanetary magnetic field, *J. Geophys. Res.*, 98, 7503, 1993.

Watanabe, K., and T. Sato, Global simulation of the solar wind-magnetosphere interaction : The importance of its numerical validity *J. Geophys. Res.*, 95, 75, 1990.

Winglee, R. M., 1994, Non-MHD Influences on the Magnetospheric Current System, *J. Geophys. Res.*, 99, 13,437, 1994.

Winglee, R. M., and J. D. Menietti, Auroral Activity Associated with Pressure Pulses and Substorms : A Comparison Between Global Fluid Modeling and Viking UV Imaging, *J. Geophys. Res.*, submitted, 1997.

Winglee, R. M., et al., Modeling of upstream energetic particle events observed by Wind, *Geophys. Res. Lett.*, 23, 1227, 1996.

Winglee, R. M., R. M. Skoug, R. K. Elsen, M. Wilber, R. P. Lin, R. L. Lepping, T. Mukai, S. Kokubun, H. Reme, and T. Sanderson IMF induced changes to the nightside magnetotail: A comparison between Wind/Geotail/IMP 8 observations and modeling, *Geophys. Res. Lett.*, 24, 947, 1997a.

Winglee, R. M., V. O. Papitashvili, and D. R. Weimer, Comparison of the Auroral Currents and Potential During the January 1992 GEM Campaign, *J. Geophys. Res.*, in press, 1997b.

---

R. P. Lepping, Code 695, NASA Goddard Space Flight Center, Greenbelt, MD 20771.

R. P. Lin, Space Science Laboratory, University of California - Berkeley, Berkeley, CA 94720.

A. T. Y. Lui, Applied Physics Laboratory, John Hopkins University, Laurel, MD 20723-6099.

S. Kokubun, Solar-Terrestrial Environment Laboratory, Nagoya University, 3-13 Honohara, Toyokawa Aichi 442, Japan.

G. Rostoker, and J. C. Samson, Department. of Physics, University of Alberta, Edmonton AB T6G 2J1, Canada.

R. M. Winglee, Geophysics Program, Box 351650, University of Washington, Seattle, WA 98195-1650.

# Sporadic Localized Reconnections and Multiscale Intermittent Turbulence in the Magnetotail

Tom Chang

*Center for Space Research, Massachusetts Institute of Technology, Cambridge, MA, and International Space Science Institute, Bern, Switzerland*

A multiscale, intermittent turbulence theory is suggested for the description of the Earth's magnetotail. The full dynamical behavior is characterized by the sporadic, localized merging and diffusion of coherent structures (flux ropes/plasmoids.) The onset of such a stochastic state may be viewed as a nonclassical, nonlinear instability which naturally results in an inhomogeneous, multiscale, multi-fractal fluctuation spectrum. Adoption of this approach circumvents the conventional assumption of unrealistically high value(s) of "anomalous resistivity" to some restricted region(s) of the magnetotail plasma domain in order to achieve MHD reconnection(s).

## 1. INTRODUCTION

There is sufficient experimental evidence indicating that the reconnection signatures associated with the magnetotail are localized and occur sporadically throughout the near-Earth and distant tail regions. For example, bursty bulk flows and associated plasmoids/flux ropes have been detected [Angelopoulos et al., 1996; and references contained therein]. In addition to being localized in nature, the energy and time scales of each of these detected events are generally much smaller than those that would have been expected for the overall dynamics of the entirety of the plasma sheet and magnetotail. Most of the observed reconnection signatures seems to be associated with thin current sheets [Sergeev et al., 1990; Pulkkinen et al., 1992; and references contained therein] and three-dimensional magnetic field geometries, indicating that much of the observed localized merging processes are three-dimensional as well as kinetic (or, at least, partially kinetic) in nature.

Thus, a reasonable scenario for the energy conversion and particle acceleration processes in the magnetotail should probably be expressed in terms of a stochastic description. The observed merging (or reconnection) of organized magnetic structures and simultaneous appearance of flux ropes/plasmoids indicate that coherent plasma structures of varied sizes are constantly generated, mingled, combined and dispersed, suggesting that the underlying turbulent behavior in the magnetotail must be multi-scale as well as intermittent in character.

In this paper, we shall discuss the framework for an intermittent turbulence model involving the generation, dispersing and merging of multiscale coherent plasma structures to describe the dynamic behavior of the magnetotail. The model necessarily incorporates a multitude of cross-scale interactions (kinetic merging and MHD mixing) within the entire dynamic region. We shall demonstrate below such a dynamic process will generally require the existence of an underlying three-dimensional magnetic field geometry; and that coherent flux structures and localized merging can set in wherever the parallel component of the local macroscopic magnetic fluctuation vanishes. (A more detailed theoretical reasoning supporting this statement is given in the Appendix.) The onset of such a stochastic merging process is the consequence of a nonclassical nonlinear instability; while individually, each localized merging process is associated with some classical linear or nonlinear plasma instability process or processes (e.g., Huba et al., 1977; Lui, 1996; Galeev et al., 1986; Ugai, 1996; and references contained therein.) An appealing attribute of adopting such a view point is the non-necessity of assigning unrealistically large values of "anomalous resistivities" to the plasma medium in the magnetotail.

## 2. STOCHASTIC MIXING OF COHERENT STRUCTURES AT MULTISCALES

Measurements of reconnection signatures in the "neutral sheet" region of the magnetotail indicate that most of the individual localized merging processes occur at intermediate or microscales. On the other hand, the dimensions of the full domain that is responsible for the transferring of energy and momentum from the solar wind to the Earth's magnetotail generally involve time and spatial scales much larger than those characterized by the microscopic plasma parameters such as the ion gyroradius, skin depth and Debye length. The transport processes at the two ends of the dynamic spectrum can involve characteristic parameters differ by an order of magnitude or higher, suggesting that the underlying dynamics of the magnetotail is intrinsically multiscale. To describe the overall energy exchange, it will be necessary to consider the gross mixing of fluctuations at macroscopic (or roughly speaking, MHD) scales, while the merging of coherent structures near the "neutral sheet" are affected generally by kinetic (or at least, partially kinetic) effects. It is the interplay of the kinetic, intermediate and MHD scale fluctuations and the relevant methodology required to address such questions that form the central theme of this paper.

As discussed in the Appendix, propagation of magnetic fluctuations (particularly those related to the macroscopic scales such as shear Alfvén waves) requires the propagation vector to contain a field-aligned component, i.e., $\mathbf{B} \cdot \nabla \rightarrow i\mathbf{k} \cdot \mathbf{B} \neq 0$. At sites where the parallel component of the propagation vector vanishes, $k_\parallel = \mathbf{k} \cdot \mathbf{B} = 0$, energies are localized and the field lines may be distorted effortlessly. Just away from these singular (resonance) sites, the forces that arose from the fluctuations, $\delta\mathbf{B} \cdot \nabla$, will tend to restore the field lines towards the resonance sites, thereby forming island-like magnetic structures. For sheared magnetic field geometry found in the "neutral sheet" region of the magnetotail, these island structures would be in the form of twisted tubes aligned primarily in

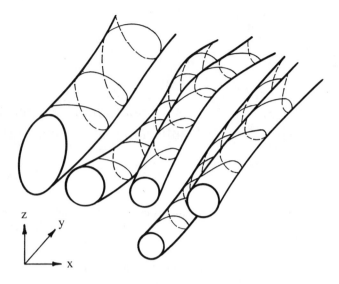

**Figure 1.** Illustrative configuration of twisted flux tubes near the "neutral sheet".

the direction of the cross-tail current with field lines wound tightly around them, Figure 1. In general, there will be a constellation of such islands floating around in the magnetotail.

As these coherent structures migrate toward each other, they will either annihilate or merge together and form larger islands depending on the polarity of the current densities characterizing these structures. When they merge (for parallel polarity), localized magnetic reconnection sets in. As the structures merge, new coherent structures are formed and at the same time new fluctuations are generated. The new fluctuations will then spontaneously set up new resonance sites. Thus, an interesting scenario of intermittent turbulent mixing and merging sets in. This type of intermittent turbulence is anisotropic, inhomogeneous, and multiscale in the magnetotail.

In regions close to the "neutral sheet," the coherent structures could be of the size of the local ion gyroradius. These structures generally exhibit the shape of twisted flux tubes primarily aligned with the polarity of the cross-tail current and they are more densely packed. Thus, it is expected that localized merging will occur frequently in these regions and the reconnection signatures will generally be kinetic (or partially kinetic.) Away from the "neutral sheet," the flux tubes (or more complicated island structures) can be oriented in arbitrary directions with equal probability of varied polarities and they are more loosely packed. The sizes of these structures generally will be much larger than the ion gyroradius and the

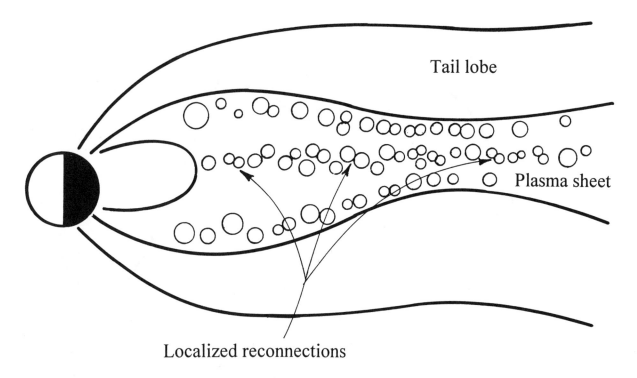

**Figure 2.** Multiscale intermittent turbulence in the magnetotail.

merging process (when it occurs) can be simply the stochastic mixing of magnetic field lines. There is, of course, a multitude of regions where the phenomena of mixing and merging are at intermediate scales. [See Figure 2.] We conclude this section by noting that there are many examples of resonance overlap in terms of MHD (See, e.g., Waddell et al., 1978 ; Tetreault, 1992) and kinetic descriptions (See, e.g., Karney, 1978; Galeev et al., 1986; Horton and Ichikawa, 1996; and references contained therein).

## 3. NONCLASSICAL, NONLINEAR INSTABILITY AND MULTI-FRACTALS

For the random onset of multiple magnetic islands (flux tubes/plasmoids) discussed above, it has been suggested that the propagation vector **k** of the macroscopic magnetic fluctuations must exhibit multiple resonance sites at which $k_\parallel = \mathbf{k} \cdot \mathbf{B} = 0$. Therefore, the background magnetic field geometry must be generally three-dimensional in nature and there must be pre-existing finite magnetic fluctuations. Under favorable conditions (e.g., with the availability of a free energy source such as the cross-tail current), such a state of random finite magnetic fluctuations may grow by producing new and larger coherent structures and fluctuations as well as new resonance sites. This type of instability, by definition, is genuinely "nonlinear." [See Figure 3.]

For the onset of a classical nonlinear instability, there generally exists a prescribed minimum finite amplitude disturbance (measured, for example, by the root-mean-square of fluctuations) beyond which the fluctuations can grow provided that there is an available abundance of efficient convertible free energy. Much attention has been paid recently to the onset of substorms associated with the phenomenon of "magnetic reconnection". During the onset of a substorm, it has been recognized, both from in-situ observations and dimensional estimates based on nonlinear dynamics calculations (Baker et al., 1996; Klimas et al., 1992; Sharma et al., 1992; and references contained therein), that the non-equilibrium plasma state in the magnetotail is near criticality (similar to the critical point for equilibrium liquid/gas phase transitions.) At such a dynamic state (commonly referred to as a state of self-organized criticality), the effect of the fluctuations themselves becomes an important factor in determining the critical threshold of onset. In this case, the nonlinear instability is no longer described by its classical threshold, Figure 4. And it is expected that the resulting fluctuation spectrum will generally exhibit fractal structures. [See, e.g., Chang, 1992; Chang et al., 1992.]

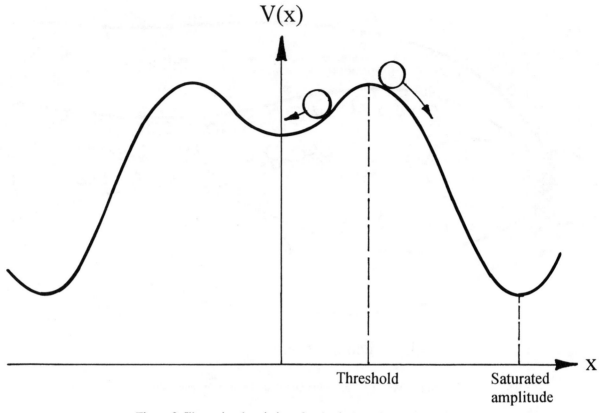

**Figure 3.** Illustrative description of a classical nonlinear instability.

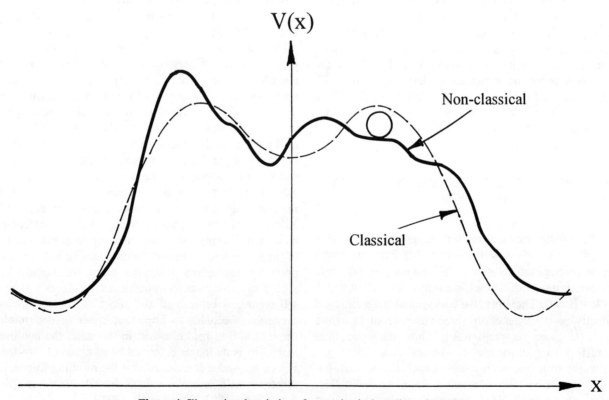

**Figure 4.** Illustrative description of a nonclassical nonlinear instability.

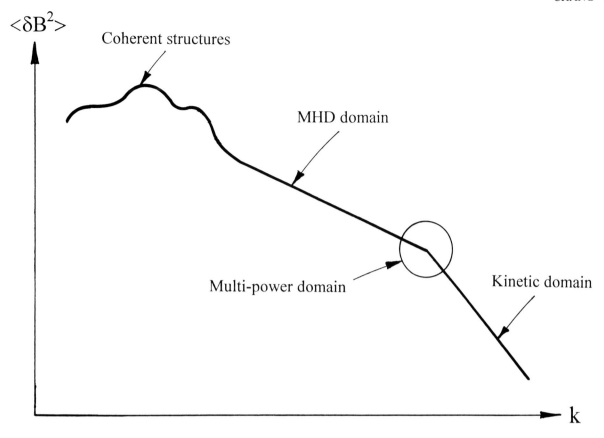

**Figure 5.** Fluctuation spectrum near the "neutral sheet".

For the magnetotail, in those regions where the fluctuations and merging dimensions are much larger than that of the local ion gyroradius, the spectrum is expected to exhibit two distinguishable parts: a domain characterized by the larger scale coherent structures and a fractal domain characterized by the predominantly MHD fluctuations (and generally with nonclassical spectrum slopes). On the other hand, in regions close to the "neutral sheet," there generally will be at least three distinguishable parts for the fluctuation spectrum: the coherent domain, an MHD fractal domain and a kinetic fractal regime whose fractal dimension (and therefore its spectrum slope) generally depends on the type of microscopic fluctuations and microinstabilities that are relevant for the merging and diffusion processes. Such a fluctuation spectrum has been recently observed [Hoshino et al., 1994; Milovanov et al., 1996; and references contained therein]. There are also intermediate regimes where the fractal properties are much more complicated. The scaling laws for these intermediate regions are expected to exhibit multiple-power characteristics. [See Figure 5.]

## 4. SUMMARY

An anisotropic, inhomogeneous, multiscale, intermittent turbulence theory is suggested for the description of the dynamics of the magnetotail. A salient feature of this theory is the non-necessity of assigning unrealistically high values of "anomalous resistivity" to the magnetotail plasma medium and of requiring the relevant magnetotail dynamics to occur only after some magical "magnetic reconnection" had set in. Instead, the full dynamics is characterized by the sporadic, localized merging and diffusion of coherent structures (flux ropes/plasmoids.)

The onset of such a stochastic state is characterized by a nonclassical, nonlinear instability which naturally results in an inhomogeneous, cross-scale, multi-fractal magnetic fluctuation spectrum.

We conclude this paper with a conceptual flow chart [Figure 6] depicting how the plasma inflow from the lobe is channeled through the plasma sheet of the magnetotail.

*Acknowledgment.* The author wishes to acknowledge useful discussions with A.T.Y. Lui, C.C. Wu, C.F. Kennel, H.E.

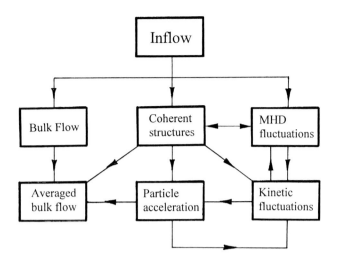

**Figure 6.** Flow chart depicting how the plasma inflow from the lobe is channeled through the plasma sheet of the magnetotail.

Petschek, D. Tetreault, M. Kivelson, L. Kepko, L. Zelenyi, V. Angelopoulos, M. Hoshino, and M. Ugai. This research is partially supported by NASA, NSF, AFOSR and the Phillips Laboratory. A portion of this research was completed at the International Space Science Institute, Bern, Switzerland.

## APPENDIX - MACROSCOPIC LOCAL RESONANCES

The magnetotail domain, in general, is vast and should be describable in terms of the bulk flow velocity and other macroscopic field variables. Near the neutral sheet, on the other hand, kinetic effects such as finite gyroradius and wave-particle interaction should be considered. Thus, it is appropriate to adopt a middle of the road approach in viewing the dynamics of the plasma medium in the magnetotail. We shall assume that the dynamics of the plasma medium is primarily characterized by the basic MHD variables with an anisotropic pressure tensor. To bring in some of the possible kinetic effects, we shall relate the pressure tensor to the particle distribution functions in terms of the appropriate moments.

Assuming that the collisional effects are negligible, the basic equations are, therefore, (a more restrictive formulation is given in Cheng, 1991),

$$\frac{d\rho}{dt} + \rho \nabla \bullet \mathbf{V} = 0, \quad (1)$$

$$\rho \frac{d\mathbf{V}}{dt} = -\nabla \bullet \mathbf{P} + \mathbf{j} \times \mathbf{B}, \quad (2)$$

$$\nabla \bullet \mathbf{B} = 0, \quad (3)$$

$$\frac{\partial \mathbf{B}}{\partial t} + \nabla \times \mathbf{E} = 0, \quad (4)$$

$$\mathbf{E} + \mathbf{V} \times \mathbf{B} = 0, \quad (5)$$

$$\mathbf{j} = \nabla \times \mathbf{B}, \quad (6)$$

where all notations are standard and the pressure tensor is given by

$$\mathbf{P} = \sum_i m_i \int (\mathbf{v} - \mathbf{V})(\mathbf{v} - \mathbf{V}) f_i(\mathbf{v}) d\mathbf{v}, \quad (7)$$

and the particle distribution functions $f_i(\mathbf{x}, \mathbf{v}, t)$ satisfy the Vlasov equations.

Combining Eqs. (3)-(6), we find that

$$\frac{d\mathbf{B}}{dt} = (\mathbf{B} \bullet \nabla)\mathbf{V} + \cdots, \text{ and} \quad (8)$$

$$\rho \frac{d\mathbf{V}}{dt} = (\mathbf{B} \bullet \nabla)\mathbf{B} + \cdots, \quad (9)$$

where the ellipsis represent compressibility and anisotropic pressure effects, etc. It is clear from above that one of the propagating modes allowed by these equations is the shear Alfvén wave. At localities where $k_\parallel = \mathbf{k} \bullet \mathbf{B} = 0$, however, such type of macroscopic modes cannot propagate and energies are localized, leading to the concept of shear Alfvén resonance discussed in the main text of this paper. Because of the other terms represented by the ellipsis in Eqs. (8-9), there exists also the possibility of other macroscopic as well as kinetic resonances. Nevertheless, the shear Alfvén resonances would dominate the formulation of the macroscopic islands that are responsible for the formation of the larger scale magnetic islands. To consider the merging of such magnetic islands, particularly near the "neutral sheet" region, it will be necessary to include the effects of the pressure tensor and the associated particle kinetics. These discussions are beyond the scope of this scenario paper.

## REFERENCES

Angelopoulos, V., F.V. Coroniti, C.F. Kennel, M.G. Kivelson, R.J. Walker, C.T. Russell, R.L. McPherron, E. Sanchez, C.I. Meng, W. Baumjohann, G.D. Reeves, R.D. Belian, N. Sato,

E. Fris-Christensen, P.R. Sutcliffe, K. Yumoto and T. Harris, Multi-point analysis of a BBF event on April 11, 1985, *J. Geophys. Res.*, 101, 4967, 1996.

Baker, D., A.J. Klimas, D. Vassiliadis and T.I. Pulkkinen, The magnetospheric dynamical cycle: role of microscale and mesoscale processes in the global substorm sequence, *Physics of Space Plasmas (1995)*, 14, 41, 1996.

Chang, T., Low-dimensional behavior and symmetry breaking of stochastic systems near criticality - can these effects be observed in space and in the laboratory?, *IEEE Trans. on Plasma Science*, 20, 691, 1992.

Chang, T., D.D. Vvedensky and J.F. Nicoll, Differential renormalization-group generators for static and dynamic critical phenomena, *Physics Reports*, 217, 279, 1992.

Cheng, C.Z., A kinetic-magnetohydrodynamic model for low-frequency phenomena, *J. Geophys. Res.*, 96, 21159, 1991.

Galeev, A.A., M.M. Kuznetsova and L.M. Zeleny, Magnetopause stability threshold for patchy reconnection, *Space Science Reviews*, 44, 1, 1986.

Horton, W. and Y.H. Ichikawa, "Chaos and structures in nonlinear plasmas," *World Scientific*, Singapore, 1996.

Hoshino, M., A. Nishida, T. Yamamoto, and S. Kokubun, Turbulent magnetic field in the distant magnetotail: Bottom-up process of plasmoid formation?, *Geophys. Res. Lett.*, 21, 2935, 1994.

Huba, J., N.T. Gladd and K. Papadopoulos, The lower-hybrid drift instability as a source of anomalous resistivity for magnetic field reconnection, *Geophys. Res. Lett.*, 4, 125, 1977.

Karney, C.F.F., Stochastic ion heating by a lower hybrid wave, *Phys. Fluids*, 21, 1584, 1978.

Klimas, A.J., D.N. Baker, D.A. Roberts, D.H. Fairfield and J. Büchner, A nonlinear dynamical analogue model of geomagnetic activity, *J. Geophys. Res.*, 97, 12253, 1992.

Lui, A.T.Y., Current disruptions in the Earth's magnetosphere: observations and models, *J. Geophys. Res.*, 101, 4899, 1996.

Milovanov, A., L. Zelenyi and G. Zimbardo, Fractal structures and power law spectra in the distant Earth's magnetotail, *J. Geophys. Res.*, 101, 19903, 1996.

Pulkkinen, T.I., D.N. Baker, R.J. Pellinen, J. Büchner, H.E.J. Koskinen, R.E. Lope, R.L. Dyson and L.A. Frank, Particle scattering and current sheet stability in the geomagnetic tail during substorm growth phase, *J. Geophys. Res.*, 97, 19283, 1992.

Sergeev, V., P. Tanskanen, K. Mursula, A. Korth and R.C. Elphic, Current sheet thickness in the near-Earth plasma sheet during substorm growth phase, *J. Geophys. Res.*, 95, 3819, 1990.

Sharma, A.S., D. Vassiliadis and K. Papadopoulos, Reconstruction of low-dimensional magnetospheric dynamics by singular spectrum analysis, *Geophys. Res. Lett.*, 20, 335, 1993.

Tetreault, D., Turbulent relaxation of magnetic fields: 1. coarse-grained dissipation and reconnection, *J. Geophys. Res.*, 97, 8531, 1992.

Ugai, M., Computer studies on dynamics of a large-scale magnetic loop by the spontaneous fast reconnection model, *Phys. Plasmas*, 3, 4172, 1996.

Waddel, B.V., B. Carreras, H.R. Hicks and J.A. Holmes, Nonlinear interaction of tearing modes in highly resistive tokamaks, *Phys. Fluids*, 22, 896, 1979.

---

T. Chang, Center for Space Research, Massachusetts Institute of Technology, Room 37-261, 77 Massachusetts Avenue, Cambridge, MA 02139

# Geotail Observations of Current Systems in the Plasma Sheet

W. R. Paterson and L. A. Frank

*Department of Physics and Astronomy, The University of Iowa, Iowa City, Iowa*

S. Kokubun

*Solar-Terrestrial Environment Laboratory, Nagoya University, Japan*

T. Yamamoto

*Institute of Space and Astronautical Science, Sagamihara, Japan*

Plasma instrumentation on the Geotail spacecraft acquires measurements of the velocity distributions of electrons and ions with sensitivity sufficient to sense currents as small as approximately 10 nA/m$^2$ on the basis of differences in the bulk flow velocities of the negative and positive charge carriers. In addition, measured pressure tensors provide information regarding local stresses that are associated with currents. We present two cases that illustrate the instrumental capabilities for the analysis of current systems. The first case is an observation of Birkeland currents in the plasma sheet on 10 February 1995 at a radial distance from Earth ~20 $R_E$. The measured current densities are associated with field-aligned electron beams and the simultaneous magnetic field measurements indicate the presence of variable currents. The second example is acquired at a radial distance 132 $R_E$ as Geotail encounters the plasma mantle, the plasma sheet boundary layer, and the plasma sheet. The ion velocity distributions in the plasma sheet are found to be complex and the measured pressure tensors suggest several mechanisms that are likely to contribute to the balance of stresses associated with the reversal of the magnetic field in the distant current sheet.

## INTRODUCTION

Global schematics of Earth's magnetosphere are sometimes drawn starting with the major systems of currents that must exist to support the observed configuration of magnetic fields. Although the magnetosphere is not a collection of wires, pictures of this type provide a degree of conceptual simplification that is necessary for an overview of this large and complex system. The fact that electric currents are continuous is a useful property that provides a connection between regions that otherwise often seem to be disparate. For example, the Region 1 and Region 2 field-aligned Birkeland current systems must close at low altitude within the ionosphere and at high altitude on the night side within the plasma sheet or its boundary layer.

For a current to exist there must be a difference in the bulk motion of the positive and negative charge carriers in the plasma. Magnetospheric plasmas have been surveyed repeatedly since the 1970s, and there is considerable information available regarding average values and variability of some basic plasma parameters such as number densities, temperatures, and bulk flow velocities. However, the locations and strengths of magnetospheric current systems are deduced almost entirely on the basis of magnetic field observations. Direct measurement of currents with spacecraft-borne plasma instrumentation is a difficult task and reports in the literature are quite limited in number [Frank et al., 1984; 1997]. There is a reason for this. Although the total current in a given magnetospheric region may be enormous, the currents are typically distributed over large volumes and the local current densities that are to be sensed with plasma instrumentation are generally quite small. Thus, the differences in the flow velocities of electrons and ions are generally small in comparison with the thermal velocities of the particle velocity distributions, and therefore detection of a current requires a sensitive measurement.

To accomplish this task the instrumentation must determine the phase-space densities of both electrons and ions with high resolution of energies and angles in three dimensions, and cross calibration of sensors must be good over the entire range of energies to within a few percent. In addition, photoemission and secondary electron emission from spacecraft surfaces causes charging. The effects of the photosheath and the potential must be understood and corrected for in the analysis. At energies-per-charge below the spacecraft potential the electron population observed by a plasma analyzer is a combination of photoelectrons and secondary electrons with total density hundreds per $cm^3$. This density is to be compared to the typically much smaller densities of the ambient magnetospheric plasma populations that can be 0.1 /$cm^3$ or less. Thus it is essential to separate the spacecraft electrons from the ambient population. Though somewhat less critical, correction of the observed particle energies for the potential is also important when computing plasma parameters for both the electron and ion components of the plasmas. If the corrections are successful and the calibrations are good then detection of currents is limited by statistical accuracy of the measured bulk flows, and in the low density plasma sheet, for example, current densities as small as ~10 $nA/m^2$ can be detected.

In general, currents in the magnetosphere flow in response to mechanical stresses, and various possible contributing terms are identified through analysis of the momentum equation [Rich et al., 1972, Vasyliunas 1984]. Detection of a difference in the bulk flow of electrons and ions, when possible, may not be illuminating unless the physical mechanisms that produce the current can also be determined. Evaluation of stresses can be accomplished through analysis of plasma pressures including the dynamic pressure and the static pressure tensor. Pressure tensors can be measured at the location of the spacecraft with 3 dimensional plasma analyzers, but analysis of stresses generally requires knowledge of gradients. With measurements from a single spacecraft, determination of gradients is ambiguous at best, and theory or modeling is required to provide the missing information on spatial dependencies. Nevertheless, intercomparison of locally measured pressure terms can provide insight. For example, if dynamic pressures in the plasma sheet are always found to be small, then it is unlikely that gradients of this term will play an important role.

In the remaining sections of this paper we discuss observations of plasmas and magnetic fields in the geomagnetic tail observed with instrumentation on board the Geotail spacecraft. These observations are chosen to illustrate present capabilities for detection of currents and analysis of associated stresses.

First, we present one example of field-aligned currents in the plasma sheet at a distance from Earth approximately 20 $R_E$. These currents may be related to closure of Birkeland currents from the ionosphere in the central plasma sheet and hence are intrinsically quite interesting. In addition, these observations provide a practical test of the instrumental capabilities for direct detection of currents. The measured electron velocity distributions exhibit field-aligned beam-like features at low energies that are consistent with the measured current densities. This identification of current carriers is important because it strengthens the case for a current that is measured as a difference in flow of the electrons and the ions. The simultaneous magnetic field measurements also suggest the presence of currents. Together, the field-aligned beams and the magnetic perturbations affirm the detection of currents and substantiate the analysis of plasma flows. This is significant because it allows us to extend our analysis of bulk flows and currents to other cases where the current is not easily associated with a particular feature in the distributions, as may be the case for cross-tail currents in the current sheet.

The second example considered in this paper is an encounter with the plasma sheet at a radial distance from Earth equal to 132 $R_E$. In this case we present an analysis of ion pressure terms that include the field-aligned pressure anisotropy, the dynamic pressure, and off-diagonal elements of the pressure tensor. Each of these may contribute to the balance of stresses associated with the reversal of the Bx component of the magnetic field across the current sheet [Rich et al., 1972, Ashour-Abdalla et al., 1993]. These terms are found to be significant, and in addition the ion velocity distributions suggest plausible origins for the source of these terms.

## INSTRUMENTATION AND MEASUREMENTS

Plasma measurements discussed here come from the Hot Plasma (HP) analyzer that is one element of the Comprehensive Plasma Instrumentation on the Geotail spacecraft [Frank et al., 1994]. The HP analyzer is designed to provide three-dimensional observations of electrons and ions with resolution of energies and angles sufficient to characterize velocity distributions typical of the plasma sheet and its boundary layer. Charged particles with velocity vectors directed into a fan-shaped region parallel to the spin axis are intercepted by one of the electron or ion sensors that divide the field of view into nine approximately equal polar sections. A programmed sequence of 24 logarithmically spaced voltage steps applied to the electrostatic deflection plates selects particles with energies-per-charge in the range approximately 20 V to 50 kV. Azimuth angles are determined by sectioning the spin into either 8 or 16 equal sectors. A total 1728 (8 sectors) or 3456 (16 sectors) samples of the velocity distribution are acquired in six spins or approximately 18 s. Every seventh spin is devoted to a special mode that provides samples at energies-per-charge as low as 1 V. The low-energy electron measurements are used to compute spacecraft potentials by means of an algorithm that compares the densities of return photoelectrons and secondary electrons on the daylit and shadowed sides of the spacecraft. This estimate of the potential is used to correct the particle energies for the effects of the satellite potential sheath. Sensor calibrations are verified in-flight by comparison of measured densities with densities inferred from the plasma frequency observed with the Geotail Plasma Wave Instrument [Matsumoto et al., 1994]. Cross calibration of sensors is verified by comparison of measured spectra in plasmas of the central plasma sheet near Earth. Moments are computed as weighted sums of the measured phase space densities. For this computation ions are assumed to be protons. Moments given in this study are computed using the sum of three consecutive determinations of the velocity distributions and temporal resolution therefore is about one minute. Magnetic field data are 3 s average vectors from the Geotail Magnetic Field Experiment (MGF) [Kokubun et al., 1994]. When magnetic fields and plasma measurements are combined to compute parameters such as plasma beta or to determine components of current in parallel and transverse directions one-minute average vectors are computed to match the timing of the plasma measurements.

### Birkeland Currents on 10 February 1995

The Birkeland currents discussed here were observed on 10 February 1995 between 0000 and 0140 UT in the central plasma sheet on the dawn side of the magnetotail. During this time the Geotail spacecraft was in the plasma sheet between midnight and dawn, and in motion towards the dawn flank. In Earth-centered solar magnetospheric coordinates the Z coordinate was nearly constant and equal to $-4$ $R_E$. The X coordinate ranged from $-20.5$ to $-18.9$ $R_E$ and the Y coordinate from $-5.2$ to $-6.3$ $R_E$. Thus Geotail was at a radial distance that is near the apogee distances of the ISEE 1, ISEE 2, and AMPTE/IRM spacecraft which have provided extensive contributions to our understanding of the plasma sheet near Earth [Frank, 1985; Baumjohann et al., 1993, Huang and Frank, 1994]. A record of the plasma parameters for both electrons and ions is given in Figure 1. Figure 1 also displays the measured components of the magnetic field and the plasma

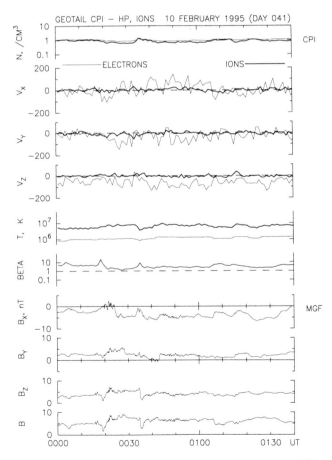

**Figure 1.** From top to bottom respectively, electron and ion densities, components of the bulk velocities, temperatures, plasma beta, and components of the magnetic field in the near-Earth plasma sheet between 0000 and 0140 UT on 10 February 1995.

beta. Number densities are nearly constant and approximately equal to 1 /cm$^3$. Temperatures for ions and electrons are approximately $5\times10^6$ and $1\times10^6$ K, respectively. These parameters indicate that the spacecraft is in the plasma sheet [Frank, 1985]. Also, the small variability of the ion bulk flow which does not exceed 50 km/s during this interval and the plasma beta which is in the range 1 to 10 indicate that this is the central part of the plasma sheet and not its boundary layer. In comparison with the ion bulk flow, the electron flow is found to be generally larger. The Vx, Vy, and Vz components of electron flow are variable with magnitudes of each component ranging from 0 to ~ 100 km/s. It is important to note that the components of electron flow do not fluctuate randomly about zero as might be expected if the apparent velocities were due to statistical uncertainties. Also, the offsets from zero are not constant as they might be if the measurements were subject to some systematic bias. Instead, there appear to be regions of nonzero flow with variable direction but the direction of the observed flow is generally constant for several consecutive minutes. Although we do not here provide detailed analysis of the magnetic fields, the fluctuations are sufficient so as not to preclude the possibility of perturbation by currents. In fact, several of the larger temporal variations in the field do correlate with changes in the direction or the magnitude of the electron velocity, for example near 0023, 0034, and 0042 UT.

Eight representative samples of the electron velocity distributions are shown in Plate 1. Each is an approximately one-minute average comprised of three consecutive determinations of the distribution. These two-dimensional cuts are given for the plane that includes the Earth-sun direction along Vx, and the one-minute average magnetic field for which the direction and magnitude are given in each panel. The velocity distributions indicate a variable degree of field alignment with core electrons generally counterstreaming along the direction of the field with approximately equal intensities along $+\vec{B}$ and $-\vec{B}$. However, several distributions, notably those at 0027:58, 0041:36, 0052:06, and 0103:39 UT indicate an imbalance with a discrete or nearly-discrete beam-like feature at a speed approximately 6000 km/s directed along $-\vec{B}$. These beams have a characteristic dimension in velocity space approximately 2000 km/s and a phase-space density $\sim 5\times10^{-28}$ s$^3$/cm$^6$. If the beams were totally unbalanced by particles flowing in the opposite direction they would contribute current densities along the direction of the field of approximately +40 nA/m$^2$.

Measured current densities are given in Figure 2. These are computed as

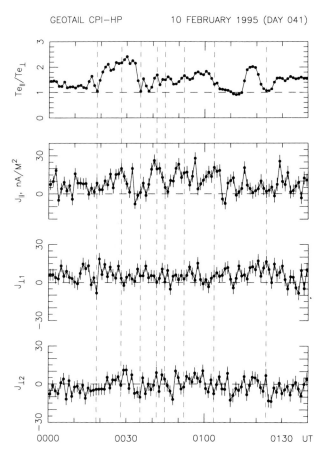

**Figure 2.** Ratio of the parallel to the perpendicular electron temperature, and the three components of the current density in a field-aligned coordinate system for 10 February 1995. Vertical lines are drawn at times that correspond to the electron velocity distributions given in Plate 1.

$$\vec{j} = nq(\vec{V}_i - \vec{V}_e)$$

where n is the number density, q is the elementary charge, and $\vec{V}_i$ and $\vec{V}_e$ are the velocities of ions and electrons respectively as determined from the plasma moments. Because we expect charge neutrality to hold we utilize a common density, in this case the density measured for electrons. The components shown in Figure 2 have been transformed to a coordinate system with one axis along the measured direction of the magnetic field. This system is not constant, but varies with the direction of the field. The error bars for the measured components of the current density are the one-sigma level computed from the uncertainties in the ion and the electron measurements based on counting statistics. For reference, the top panel of Figure 2 shows the ratio of T$_{e\parallel}$ to T$_{e\perp}$ For much of this sequence

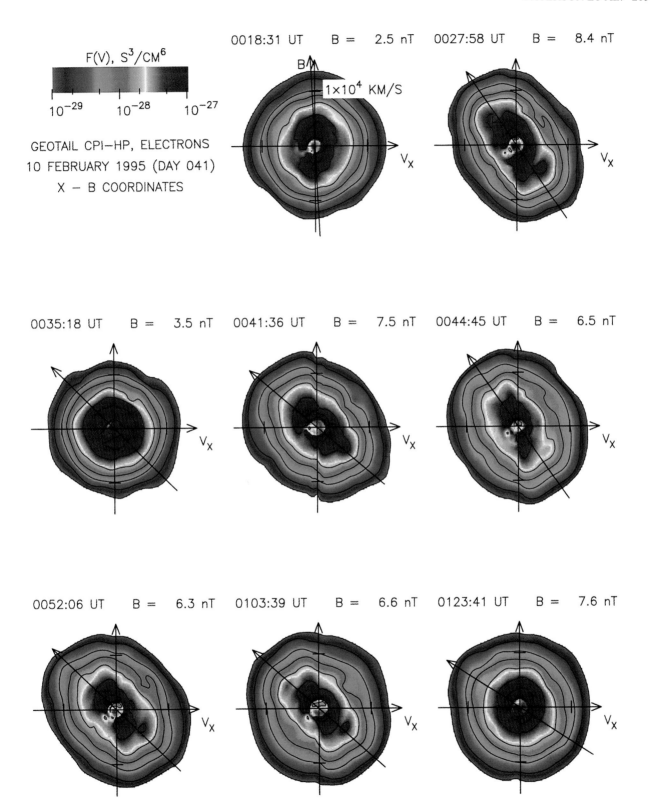

**Plate 1.** Electron velocity distributions observed in the plasma sheet on 10 February 1995. Each distribution is given in a plane that includes the direction of the magnetic field. The Vx axis is along the Earth-sun line.

the electron velocity distributions are field aligned according to this measurement, and this can be verified through comparison with the electron velocity distributions given in Plate 1. Times that correspond to the velocity distributions of Plate 1 are marked with vertical lines in Figure 2. As noted above, the velocity distributions at 0027:58, 0041:36, 0052:06, and 0103:39 UT exhibit evidence of a beam along $-\vec{B}$ that might be expected to carry a current with current density as large as 40 nA/m$^2$. In fact, the measured current densities at these times have positive field-aligned components between 10 and 25 nA/m$^2$ and these measurements meet the test of statistical significance by several sigma. Perpendicular components of the current density at these times are considerably smaller and significance for these is at or below about one sigma. This measurement of a net field-aligned current in combination with an identifiable field-aligned beam in the electron velocity distributions provides a degree of assurance that the moments analysis is in fact correct. In-depth analysis of the magnetic field measurements during this time period is beyond the scope of this paper. However, a comparison of the timing and variability of the current densities given in Figure 2 with the magnetic fields shown in Figure 1 seems to indicate that a strong case for currents can also be made on the basis of the magnetic perturbations.

Although the currents noted above are field aligned, during the interval 0117 to 0131 UT there is persistent evidence of a perpendicular current in the panel of Figure 2 that displays $J_{\perp 2}$. The direction of this component is perpendicular to the local direction of the magnetic field and also perpendicular to the solar-magnetospheric X axis. The nature of this current is unknown. We do note that during this interval there is a marked change in the ratio of $T_{e\parallel}$ to $T_{e\perp}$ and the velocity distributions apparently change from field aligned to isotropic and back to field aligned. However, there is no dramatic change in either the densities or in $T_i$ or $T_e$ and there appears to be no reason why our confidence in the measured current densities should not extend to this period. Thus it seems likely that these perpendicular currents should also be considered significant.

The measurements on 10 February 1995 provide a practical test of the capabilities of the Geotail plasma instrumentation for the direct determination of currents. If we take 10 nA/m$^2$ as a benchmark value for our lower limit then we can make some simple predictions regarding possible detection of cross-tail currents in the current sheet. To reverse a lobe field of 15 nT the current in the current sheet of scale height h must be $j_y \times h = 20$ mA/m. If h = 10,000 km (1.5 R$_E$), then $j_y = 1.5$ nA/m$^2$ and the current will be undetectable with this instrumentation. However, if the sheet is thin, ~1000 km, then $j_y$ will be a factor of 10 larger and within the range of detection. Thus, a search for thin current sheets in the tail on the basis of measured current densities may be a practical possibility.

*Plasma Sheet Observations on 30 March 1993*

As a second demonstration of the utility of the plasma measurements for the diagnosis of current systems we examine measurements acquired on 30 March 1993 during an encounter with the plasma sheet at a radial distance from Earth equal to 132 R$_E$. The plasma momentum equation provides a useful point of departure for studies of the current sheet. Rich et al., [1972] analyzed this equation for the steady state and concluded that the magnetic stresses associated with the current sheet are most likely balanced by one of three mechanisms.
1) A pressure gradient along the axis of the tail.
2) A gradient of the flow kinetic energy (dynamic pressure) along the tail axis.
3) A pressure anisotropy, $P_\parallel - P_\perp > 0$.

In addition, acceleration of plasmas at the current sheet may produce non-gyrotropic ion velocity distributions. Gradients of the off-diagonal elements of the pressure tensor associated with these distributions also can contribute to the balance of stress and should be considered in this analysis [Ashour-Abdalla et al., 1993]. Thus we also consider:
4) Gradients of $P_{lm}$, $l \neq m$ where the $P_{lm}$ are elements of the pressure tensor.

Pressure gradients are likely to be important in the dipolar field region near Earth. On average, there is a significant increase in particle pressure with decreasing distance in the near-Earth plasma sheet [Huang and Frank, 1994; Paterson and Frank, 1994]. At radial distances less than about 30 R$_E$ the cross-tail current often may be dominated by gradient-drift currents and the current sheet may often be thick. In the more distant tail the lobe field becomes essentially constant with X and gradients of the static scalar pressure are less likely to play a significant role. Thus, for these observations in the distant tail we examine measured plasma parameters that are relevant to 2), 3), and 4) above.

Measured plasma parameters and components of the magnetic field are given in Figure 3 for the period 0040 to 0200 UT on 30 March 1993. The Bx component of the magnetic field provides useful information regarding the spacecraft location during this interval. Times when the Bx component is relatively constant and is at its maximum value, approximately 15 nT, correspond to times when the spacecraft is in either the plasma mantle or the plasma sheet boundary layer. If only magnetic field measurements were available these times might be grouped as encounters with the lobe, for example during the interval between 0033 and 0055 UT. The distinction between mantle and plasma sheet boundary layer is made on the basis of measured ion velocity distributions and some relevant examples are to be given below. Ion temperatures in the plasma sheet boundary layer exceed temperatures in the mantle, and the temperatures given in the fifth panel of Figure 3 also provide a method to distinguish these two different regions. Times when the Bx component departs significantly from 15 nT correspond to encounters with the plasma sheet. This present analysis concentrates on the most extended of these encounters that occurs during the interval between approximately 0108 and 0140 UT.

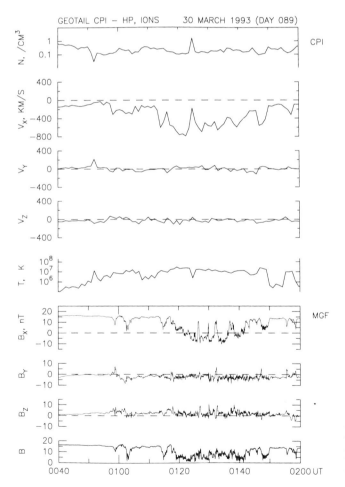

**Figure 3.** Ion number density, components of the ion bulk flow, ion temperature, and magnetic fields observed on 30 March 1993 between 0040 and 0200 UT.

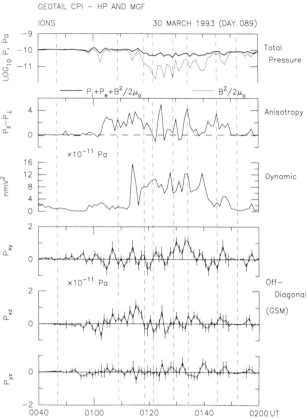

**Figure 4.** Particle and magnetic pressures measured on 30 March 1993 as the spacecraft traversed the plasma mantle, the plasma sheet boundary layer, and the plasma sheet. Vertical lines are drawn at times that correspond to the ion velocity distributions given in Plate 2.

At this time the spacecraft was located at X,Y,Z = (−132, 11.3, 6.0 $R_E$) in geocentric solar magnetospheric coordinates.

Pressure terms relevant to this analysis are given in Figure 4. The top panel displays the total scalar pressure (magnetic plus plasma) as well as the magnetic pressure only. From top to bottom respectively, subsequent panels show the pressure anisotropy, the dynamic pressure, and the three independent off-diagonal elements of the pressure tensor. Evidence of encounters with the plasma sheet are clearly indicated in the first panel by the decreases in the magnetic pressure, $B^2/2\mu_o$. The pressure anisotropy, $P_\parallel - P_\perp$, is generally found to be appreciable within the plasma sheet boundary layer, for example between 0104 and 0117 UT, and is sometimes significant within the plasma sheet as well. The dynamic pressure measured within the plasma sheet is large, between $6\times10^{-11}$ and $12\times10^{-11}$ Pa, and is comparable to the magnetic pressure in the lobe, $B^2/2\mu_o = 10^{-10}$ Pa. The largest off-diagonal pressures are quite small in comparison, $\sim 1\times 10^{-11}$ Pa. However, these terms are statistically significant as indicated by the error bars which show the one-sigma level based on counting statistics. It is important to note that it is the gradients of the off-diagonal pressure elements that affect the balance of stresses, and although the magnitudes are small these terms will be important if the gradients are sufficiently steep.

Some representative ion velocity distributions are shown in Plate 2. These are acquired at the times indicated by vertical lines in the previous figure, Figure 4. The plane of these two-dimensional cuts includes the Earth-sun line which is along the Vx axis, and the magnetic field which is indicated by a vector. The cold tailward-flowing beam observed in the distribution at 0046:46 UT at a velocity Vx = −200 km/s is characteristic of the plasma mantle. The measured density and temperature of this plasma are approximately 0.5 /cm$^3$ and $3\times 10^6$ K, respectively. Distributions similar to this are observed between 0040 and 0050 UT and again between 0150 and 0153 UT. At 0108:56 UT a similar low-speed beam is observed, but in addition there is a hot high-speed beam comprised of ions with a Vx component between approximately −1400 and −900 km/s. This distribution is typical of the plasma sheet boundary layer at this distance from Earth, and the relatively large parallel pressure that produces the pressure anisotropy observed within this region is a consequence of the separation along the field of the cold and the hot components. The mantle and plasma sheet boundary layer distributions prior to the encounter with the plasma sheet given in the first two panels of Plate 2 are similar to distributions observed when the spacecraft exits the plasma sheet. Examples of these distributions at 0144:49 and 0152:12 UT are shown in the last two panels of Plate 2.

Within the plasma sheet the total pressure is dominated by the particles and the magnetic fields are variable and fluctuating. Four examples of ion velocity distributions within the plasma sheet between 0118:26 and 0134:16 UT are also given in Plate 2. Although these four distributions are not identical they exhibit several features in common. Particle velocities are directed tailward, and for the higher speeds, above approximately 500 km/s, the distributions appear to be essentially symmetric about the Vx axis. This is consistent with the bulk velocity shown in Figure 3 which is substantial and directed along −X independent of the direction of the magnetic field. Each of these plasma sheet distributions also has components at lower speeds, between 0 and $\sim$ −500 km/s. These particles have rather complex distributions, and two-dimensional cuts do not reproduce all relevant features. It can be noted that within the plasma sheet the low-speed and high-speed components are not well separated as they are in the plasma sheet boundary layer.

The low speed particles within the plasma sheet appear to be consistent with a mantle distribution that has entered the current sheet but not yet gained substantial energy through acceleration across the tail. This may suggest recent entry. However, the history of these particles is really not clear and it is possible that this is a low speed population that has been trapped within the current sheet for a substantial period of time [Ashour-Abdalla et al., 1993]. The hot beams within the plasma sheet are similar to the kind of distribution that would be observed as a consequence of meandering motion and acceleration by a cross-tail electric field in a process like that described by Speiser [1965] and modeled by Lyons and Speiser [1982]. The distributions with two well-separated components that are observed in the plasma sheet boundary layer appear to be consistent with ejection of the fast plasma sheet component into the plasma mantle. The fact that the lower-speed populations and the higher-speed particles are not clearly separated, but appear to form a continuum, is probably an indication that this is an active site of particle acceleration and a region where cold source plasmas are heated and accelerated.

The positive pressure anisotropies observed within the plasma sheet, i.e. $P_{\parallel} - P_{\perp} > 0$, occur when the cold and the hot components happen to be aligned more or less along the magnetic field as they are at 0118:26 and 0134:16 UT. As noted previously, this anisotropy can account for part of the stress associated with reversal of the magnetic field at the current sheet. A positive anisotropy is observed most commonly in the plasma sheet boundary layer, and curvature drift currents associated with the distributions there may be important to the development of the current sheet. Of the four mechanisms for stress balance noted above, the pressure anisotropy is the only one that does not depend explicitly on gradients. If the other terms are unimportant then one expects

$$P_{\parallel} - P_{\perp} = B_l^2/\mu_o$$

where $B_l$ is the field in the lobe [Rich et al., 1972]. For a 15 nT lobe field the anisotropy would be $18\times 10^{-11}$ Pa which exceeds the measured maximum value by a factor approximately 4. It is possible that there are systematic errors associated with measurement of the colder features in the distributions that could lead to measured anisotropies that are somewhat lower than the true values. Although the mantle plasmas are often detected with the

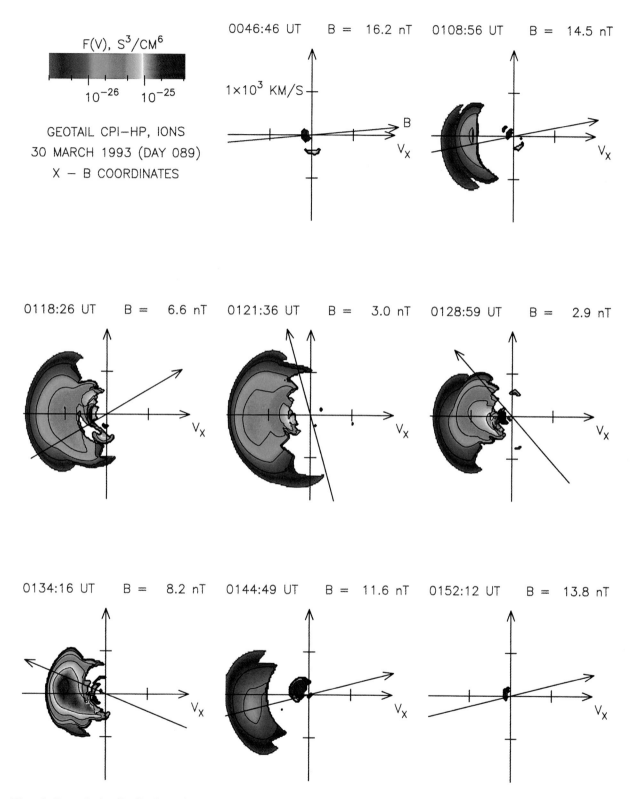

**Plate 2.** Ion velocity distributions observed in the plasma mantle, the plasma sheet boundary layer, and the plasma sheet on 30 March 1993. Each distribution is given in a plane that includes the direction of the magnetic field. The Vx axis is along the Earth-sun line.

CPI-HP analyzer, this instrumentation is not really designed to provide high resolution measurements of these cold low-speed beams. However, in view of the fact that the dynamic pressure and also the off-diagonal pressure elements are significant and apparently rather variable, either or both of these terms may well contribute to the balance of stresses as well. The dynamic pressures are clearly due to the large bulk flow along $-X$. The features in the distributions that lead to significant off-diagonal pressures are not easily discerned from the two-dimensional cuts presented here, though it is clear that the combined cold and hot components exhibit a considerable degree of complexity. Because the stresses associated with the off-diagonal pressures and the dynamic pressure do depend critically on gradients, it is difficult to assess their importance on the basis of these measurements. Nevertheless, this sequence of observations provides strong evidence that these terms generally cannot be ignored as they are plausibly quite important to the development of the current sheet in the distant tail.

## SUMMARY AND DISCUSSION

In the past, analysis of magnetospheric current systems has been based primarily on observations of magnetic fields which indicate global topologies or local perturbations that imply the presence of currents. Although plasmas in the near-Earth magnetotail have been surveyed repeatedly, direct observation of currents is a challenging task and only a few reported attempts are to be found in the literature [Frank et al., 1984; 1997]. The observations of Birkeland currents in the plasma sheet on 10 February 1995 demonstrate the ability of the CPI-HP analyzer to detect current densities as small as about 10 nA/m$^2$. The beam-like features in the distributions that are identified as the current carriers, and the general correlation with variability of the magnetic fields serve to substantiate these inherently difficult measurements. Although thin current sheets may sometimes develop in the plasma sheet near Earth, it is likely that the cross-tail currents there are often carried by gradient drifts in a thick sheet. In that case the current densities often may be too small to be detected with plasma instrumentation. At larger radial distances the likelihood that the current sheet is thin probably increases and chances for observation of currents will increase. Thus the measurements acquired with Geotail may be suitable for an investigation of thin current sheets that may sometimes develop in the near-Earth plasma sheet and may occur more commonly deep in the tail.

As compared to the near tail, the plasma sheet beyond about 20 R$_E$ has been observed less extensively, and Geotail is the first mission to provide a systematic survey of both the electron and ion components of the plasmas there. Although the ability to measure current densities directly on the basis of ion and electron bulk velocities is important, other parameters measured with the plasma instrumentation can also provide insight into the development of current systems. The observations of pressure tensors and dynamic pressures in the plasma sheet on 30 March 1993 indicate that the balance of stress in the distant plasma sheet may be complex and dependent on several different physical mechanisms. The ion velocity distributions, however, do provide some degree of simplification. The velocity distributions suggest that the dynamic pressure is due to ions accelerated within the current sheet. The pressure anisotropy and off-diagonal pressure elements may arise from the mixing of cold source plasmas with hot plasmas that have gained substantial energy in the current sheet. The precise way in which the local distributions develop to support the global field topology is a challenging problem. Although the distributions appear to retain considerable information regarding the histories of the particles, it is likely that the specifics of a given set of observations will be understood only with the input of global modeling to determine the topology and configuration of the tail and determine the histories of the particles.

*Acknowledgements.* This work was supported in part at The University of Iowa under NASA grant NAG5-2371.

## REFERENCES

Ashour-Abdalla, M., J. P. Berchem, J. Büchner and L. M. Zelenyi, Shaping of the magnetotail from the mantle: Global and local structuring, *J. Geophys. Res.*, *98*, 5651-5676, 1993.

Baumjohann, W., The near-Earth plasma sheet: An AMPTE/IRM perspective, *Space Sci. Rev.*, *64*, 141-163, 1993.

Frank, L. A., C. Y. Huang and T. E. Eastman, Currents in Earth's magnetotail, in *Magnetospheric Currents*, edited by T. A. Potemra, pp. 147-157, American Geophysical Union, Washington D.C., 1984.

Frank, L. A., Plasmas in Earth's Magnetotail, *Space Sci Rev.*, *42*, 211-240, 1985.

Frank, L. A., K. L. Ackerson, W. R. Paterson, J. A. Lee, M. R. English and G. L. Pickett, The Comprehensive Plasma Instrumentation (CPI) for the Geotail mission, *J. Geomag. Geoelctr.*, *46*, 23-37, 1994.

Frank, L. A., W. R. Paterson, S. Kokubun, T. Yamamoto, R. P. Lepping and K. W. Ogilvie, Observations of a current pulse in the near-Earth plasma sheet associated with a substorm onset, *Geophys. Res. Lett.*, 24, 967-970, 1997.

Huang, C. Y. and L. A. Frank, A statistical survey of the central plasma sheet, *J. Geophys. Res.*, *99*, 83-95, 1994.

Kokubun, S., T. Yamamoto, M. H. Acuña, K. Hayashi, K. Shiokawa and H. Kawano, The Geotail magnetic field experiment, *J. Geomag. Geoelctr.*, *46*, 7-21, 1994.

Lyons, L. R., and T. W. Speiser, Evidence for current-sheet acceleration in the geomagnetic tail, *J. Geophys. Res.*, *87*, 2276-2286, 1982.

Matsumoto, H., I. Nagano, R. R. Anderson, H. Kojima, K. Hashimoto, M. Tsutsui, T. Okada, I. Kimura, Y. Omura, and M. Okada, Plasma wave observations with Geotail spacecraft, *J. Geomag. Geoelctr.*, *46*, 59-95, 1994.

Paterson, W. R. and L. A. Frank, Survey of plasma parameters in Earth's distant magnetotail with the Geotail spacecraft, *Geophys. Res. Lett.*, *21*, 2971-2974, 1994.

Rich, F. J., V. M. Vasyliunas, and R. A. Wolf, On the balance of stresses in the plasma sheet, *J. Geophys. Res.*, *77*, 4670-4676, 1972.

Speiser, T. W., Particle trajectories in model current sheets, *J. Geophys. Res.*, *70*, 4219-4226, 1965.

Vasyliunas, V. M., Fundamentals of current description, in *Magnetospheric Currents*, edited by T. A. Potemra, pp. 63-66, American Geophysical Union, Washington D.C., 1984.

---

S. Kokubun, Solar-Terrestrial Environment Laboratory, Nagoya University, 3-13 Honohara, Toyokawa, Aichi 442, Japan. (e-mail: kokubun@stelab.nagoya-u.ac.jp)

W. R. Paterson and L. A. Frank, Department of Physics and Astronomy, The University of Iowa, Van Allen Hall, Iowa City, IA 52242-1479. (e-mail: paterson@iowasp.physics.uiowa.edu; frank@iowasp.physics.uiowa.edu)

T. Yamamoto, Institute for Space and Astronautical Science, Sagamihara, Kanagawa 229, Japan. (e-mail: yamamoto@gtl.isas.ac.jp)

# Near-Earth Plasma Sheet Behavior During Substorms

T. Nagai

*Department of Earth and Planetary Sciences, Tokyo Institute of Technology, Tokyo, Japan*

R. Nakamura and S. Kokubun

*Solar-Terrestrial Environment Laboratory, Nagoya University, Toyokawa, Japan*

Y. Saito, T. Yamamoto, T. Mukai, and A. Nishida

*Institute of Space and Astronautical Science, Sagamihara, Japan*

We have studied plasma sheet behavior in the near-Earth magnetotail during the substorm expansion and recovery phases using Geotail plasma and magnetic field observations. In association with onset of the expansion phase, tailward plasma flows form with southward magnetic fields in the plasma sheet beyond 20 $R_E$. These tailward flows do not necessarily continue during the whole expansion phase. After the tailward flows, plasma sheet plasmas become almost stationary. These observations imply that magnetic reconnection ceases in or near the onset site. In the recovery phase, fast earthward flows appear intermittently in the plasma sheet/tail lobe boundary. The source of these earthward flows appears to be disconnected from the near-Earth magnetic reconnection process. Stationary plasmas, which are often cold ($< 2$ keV) and dense ($> 0.4$ cm$^{-3}$), are observed near the equatorial current sheet of the plasma sheet in the expansion and recovery phases. These plasmas are likely transported from the dawnside and/or the duskside of the plasma sheet just after the end of magnetic reconnection.

## 1. INTRODUCTION

Plasma sheet dynamics during substorms have been studied at radial distances of 15-30 $R_E$ using various spacecraft observations [e.g., *Hones*, 1977, 1980, 1984; *Hones et al.*, 1976, 1986; *Paschmann et al.*, 1985]. In association with an onset of the expansion phase, fast tailward plasma flows with southward magnetic fields are frequently observed in the near-Earth plasma sheet. These tailward flows usually terminate when the spacecraft exits the plasma sheet, implying that the plasma sheet becomes very thin in the expansion phase. When the plasma sheet reappears in the recovery phase of substorms, the spacecraft observes fast earthward flows with northward magnetic fields. These features are usually interpreted as an expansion of the plasma sheet in the recovery phase. *Baumjohann et al.* [1991] studied plasma characteristics during substorms in the plasma sheet at 10-19 $R_E$. In this region, plasma flows are earthward. The flow speed increases and the magnetic field becomes dipolar in the expansion phase. Furthermore, the number density of plasmas decreases and ion temperature increases in the expansion and recovery phases. On the other hand, it is known that the plasma sheet becomes cold and dense during geomagnetically quiet periods [e.g., *Fairfield et al.*, 1981]. Typically, the number density exceeds 0.4 cm$^{-3}$ and ion temperature becomes less than 2 keV near the neutral sheet during the prolonged northward IMF $B_z$ periods, which correspond to quiet periods [*Terasawa et al.*, 1997].

Therefore, substorms produce flow activity and change plasma characteristics in the plasma sheet.

The spacecraft Geotail made an extensive survey of the plasma sheet at radial distances of 20-30 $R_E$ in the period from November 1995 to March 1996. This was the first survey in the near-Earth magnetotail after the apogee of Geotail was changed to 30 $R_E$. The novelty provided by Geotail was that the spacecraft could stay inside the plasma sheet almost continuously during substorms. Therefore, Geotail observations have provided us with important measurements of plasma sheet plasmas in the whole course of a substorm. In this paper, we report on the salient features of plasma sheet behavior in the expansion and recovery phases of substorms, using Geotail observations.

## 2. DATA

The magnetic field data from Geotail were obtained with the magnetic field experiment MGF [*Kokubun et al.*, 1994]. The basic time resolution of measurements is 1/16 s; however, we use 12-s average $B_x$, $B_y$, and $B_z$ values. We calculate the total magnetic field $B_t$. We use the geocentric solar magnetospheric (GSM) coordinate system in this paper. The 0-40 keV ion and electron data from Geotail were obtained with the low energy particle experiment LEP [*Mukai et al.*, 1994]. The time resolution of the plasma data is 12 s. Plasma moment data, i.e., three velocity components ($V_x$, $V_y$, and $V_z$), number density, and ion temperature, are calculated on the spacecraft, assuming that all ions are protons. These plasma moment data and two-dimensional velocity distribution function data (in the equatorial plane) are available for all intervals. Three-dimensional velocity distribution function data are available only for time intervals when the data are received at the Japanese station. The three-dimensional data are presented in the BCE coordinate system. In this system, the B-C plane contains the magnetic field. B is the direction of the local magnetic field, C is the direction of the plasma velocity perpendicular to the magnetic field, and E completes the right-hand coordinate system.

Ground magnetic field data were used for determining substorm phases. We examined magnetic field data from midlatitude stations (Kakioka, Hermanus, San Juan, Fredericksburg, Boulder, and Honolulu), and from auroral zone stations (Kiruna, Leirvogur, Poste-de-la-Balein, Fort Churchill, Yellowknife, College, and Tixie Bay). We used 1-s values of the magnetic field at Kakioka to monitor Pi 2 activity. The geomagnetic latitude of Kakioka is 26.9° and magnetic midnight is 1500 UT. We also examined the data from the 210° magnetic meridian chain and from CANOPUS.

## 3. OBSERVATIONS

### 3.1. The February 12, 1996, Event

Figure 1 shows selected ground magnetic field data for the period 1200-1500 UT. A westward electrojet

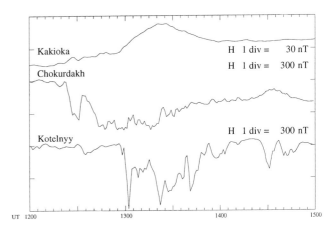

**Figure 1.** Magnetic field data on the ground for the period 1200-1500 UT on February 12, 1996.

started near 1220 UT at Chokurdakh (64.7° geomagnetic latitude, 2105 MLT). A sharp negative H bay started near 1300 UT at Kotelnyy (69.9°, 2110 MLT), and the amplitude of the westward electrojet reached 600 nT. Pi 2 pulsations are seen after 1220 and 1252 UT. This event is defined as a double-onset substorm, with the onsets at 1220 and 1252 UT. The midlatitude positive bay at Kakioka reached its peak near 1325 UT and ended near 1400 UT, and its behavior was well correlated with the activity of the westward electrojet. Therefore, the expansion phase continued until 1325 UT, and the recovery phase ended near 1400 UT. A new activity started near 1420 UT at Kotelnyy.

Figure 2 shows magnetic field and plasma data from Geotail for the period 1200-1500 UT. We calculated $V_\perp$, which is the plasma flow velocity perpendicular to the magnetic field, namely, the convection velocity of the magnetic field line. The plasma flow velocity is presented by ($V_x$, $V_y$, and $V_z$) with thin curves, and the convection velocity is presented by ($V_{\perp,x}$, $V_{\perp,y}$, and $V_{\perp,z}$) with thick curves. Here, $V_{\perp,x}$ is the $x$ component of the convection velocity $V_\perp$. Plasma pressure is calculated from number density and ion temperature. Plasma pressure is underestimated when ions with energies higher than 40 keV are abundant. Total pressure is the sum of plasma pressure and magnetic pressure. When plasma pressure is significantly less than total pressure, Geotail is considered to be in the tail lobe. Plate 1 presents selected ion distribution functions in the B-C plane. Three-dimensional distribution function data are available until 1415 UT for this event. Geotail was located at (-26.3, +13.9, -0.9 $R_E$) at 1300 UT.

Although the substorm activity had started on the ground, no fast tailward flows with southward $B_z$ were observed before 1248 UT. Fast tailward flows with southward $B_z$ continued in the time interval 1248-1300 UT. In particular, a strong convection flow was observed near 1258 UT, when $B_z$ became -10 nT, implying that the southward magnetic fields were transported tail-

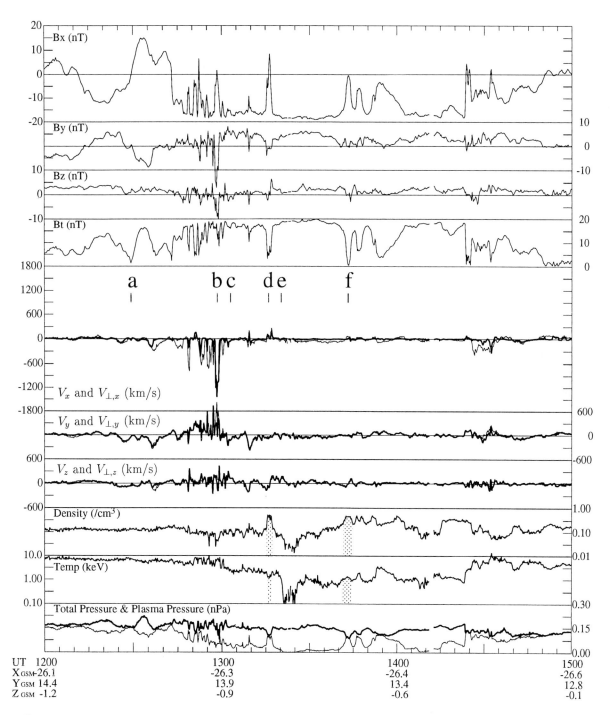

**Figure 2.** Geotail magnetic field and plasma data for the period 1200-1500 UT on February 12, 1996. Ion distribution functions are examined at times a - f. Cold and dense plasmas near the neutral sheet are observed in hatched time intervals.

ward. Ions at 1258:28 UT show a single-component distribution with a bulk motion, as seen in Plate 1b. Since the +C direction is approximately tailward at 1258:28 UT, the bulk motion is attributable to a tailward convection motion. For this strong convection flow, slight acceleration of electrons was observed; see the electron energy-time diagram in Plate 1. After these tailward flows, Geotail did not observe any strong flows. Plasmas

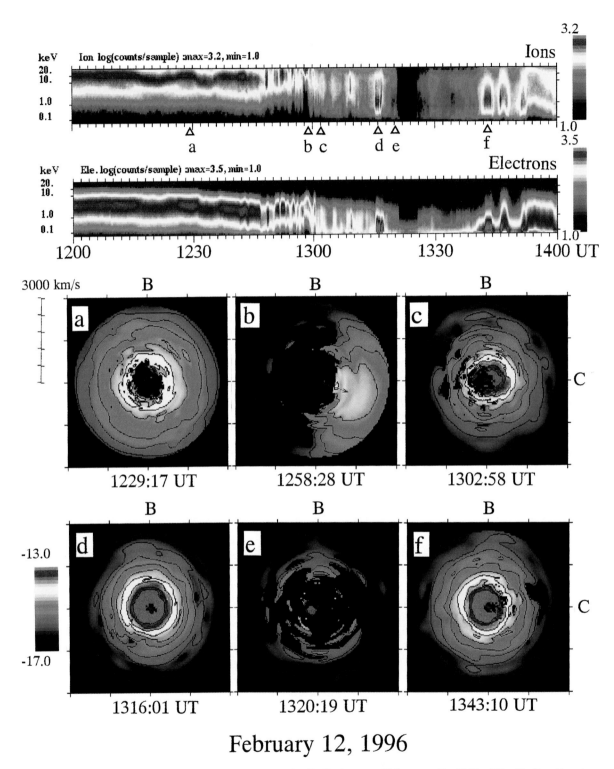

**Plate 1.** Ion velocity distribution functions in the B-C plane on February 12, 1996. The B direction is the local magnetic field, and the C direction is the convection direction. The upper two panels are the ion energy-time diagram (counts/sample) and the electron energy-time diagram (counts/sample).

were almost isotopic and basically stationary. Number density was near 0.1 cm$^{-3}$ and ion temperature was near 2 keV. Therefore, although number density did not change significantly, ion temperature decreased in comparison with ion temperature prior to the onset. A dawnward convection flow was present, and this convection direction is the same as the direction prior to the fast tailward flows. Ions at 1302:58 UT, just after the fast tailward flows, are basically isotropic and show a slight dawnward convection, see Plate 1c. Geotail entered the tail lobe in the period 1320-1325 UT. Even when Geotail exited and reentered the plasma sheet, there was neither strong flow nor a strong counterstreaming feature near the plasma sheet/tail lobe boundary. The ion distribution function at 1320:19 UT is presented in Plate 1e. Ions do not show any intense flow characteristics.

Geotail made some equatorial current sheet crossings in this event. Here, we define the equatorial current sheet of the plasma sheet as the position at which $B_x$ becomes zero. In this period, $B_y$ became near zero in most crossings. At the crossing near 1211 UT before the onset, plasmas were almost stationary, with a number density of 0.14 cm$^{-3}$ and an ion temperature of 7.0 keV. At the crossing near 1229 UT, even after the first onset, plasmas were almost stationary, with a number density of 0.13 cm$^{-3}$ and an ion temperature of 7.0 keV; see the ion distribution function at 1229:17 UT in Plate 1a. At the crossing near 1243 UT, plasmas were still almost stationary, with a number density of 0.17 cm$^{-3}$ and an ion temperature of 6.3 keV. Therefore, the plasma characteristics were not changed until the second onset. At the crossing near 1316 UT, which took place after the fast tailward flows, plasmas showed earthward flows at a speed of 100-250 km/s. Earthward flows with a speed of less than 250 km/s are common in the near-Earth tail, irrespective of substorm activity. The earthward flows near 1316 UT might be associated with an intensification of the westward electrojet at Kotelnyy (Figure 1). It is important to note that the earthward flow speed is low near the equatorial current sheet and that the number density was near 0.44 cm$^{-3}$ and the ion temperature near 1.3 keV there. The ion distribution function at 1316:01 UT is presented in Plate 1d. The ions are almost isotropic and the earthward speed is less than 100 km/s. Geotail approached the current sheet near 1343 UT and observed almost isotropic plasmas; see the ion distribution function at 1343:10 UT in Plate 1f. The number density was near 0.45 cm$^{-3}$ and the ion temperature was near 1.2 keV. Therefore, after the fast tailward flows in association with the second onset, the cold and dense plasmas existed near the current sheet in the expansion and recovery phases of this substorm. No strong earthward flow was observed in the recovery phase and tailward flows started in association with the intensification of the westward electrojet near 1425 UT.

In summary, fast tailward flows with southward $B_z$ appeared only briefly in association with the second onset. In other time intervals, plasmas were almost stationary. Plasmas near the current sheet were cold (< 1.2 keV) and dense (> 0.4 cm$^{-3}$) in the expansion and recovery phases. There were no intense earthward flows even near the plasma sheet/tail lobe boundary in the recovery phase of this substorm. It appears that hot plasmas are replaced by cold plasmas in association with substorm onset.

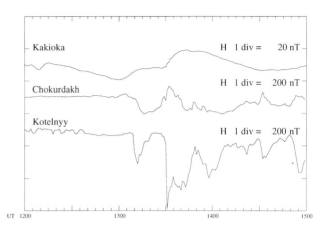

**Figure 3.** Magnetic field data on the ground for the period 1200-1500 UT on February 18, 1996.

### 3.2. The February 18, 1996, Event

Figure 3 shows selected ground magnetic field data for the period 1200-1500 UT. Although a midlatitude positive bay at Kakioka showed a double-onset signature, there were several onsets in electrojet activity. The first onset, at 1308 UT, was characterized by a Pi 2 burst and a development of the westward electrojet at Kotelnyy. The second onset, at 1329 UT, was associated with a large Pi 2 burst and a westward electrojet that had an amplitude of 400 nT at Kotelnyy. At 1346 UT, a Pi 2 pulsation burst was seen, and a westward electrojet developed at Kotelnyy. There was a small onset near 1432 UT in Pi 2 pulsation activity and westward electrojet activity. For this substorm, the recovery phase is thought to be the time interval 1400-1430 UT.

Figure 4 shows magnetic field and plasma data from Geotail for the period 1200-1500 UT. Three-dimensional distribution function data are available until 1345 UT for this event. Geotail was located at (-25.6, +5.8, -1.0) $R_E$ at 1330 UT. Geotail stayed almost continuously in the plasma sheet during the course of this substorm. Just prior to the first onset at 1308 UT, Geotail was located near the equatorial current sheet. Plasmas show a duskward convection at a speed of 80 km/s and have a number density of 0.4 cm$^{-3}$ and an ion temperature of 2.3 keV. The plasma characteristics are similar to those at the current sheet crossing at 1202 UT, well before the substorm onset. For the first onset at 1308 UT, Geotail observed tailward flows with southward $B_z$ for the

218 PLASMA SHEET BEHAVIOR DURING SUBSTORMS

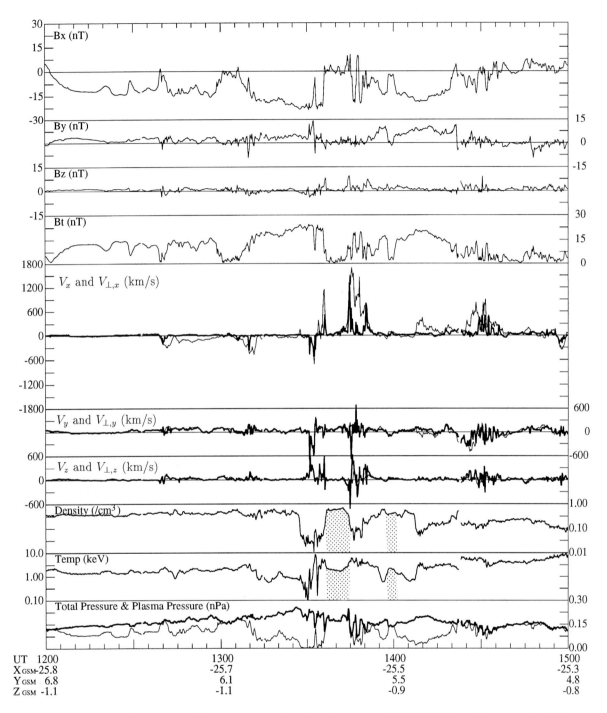

**Figure 4.** Geotail magnetic field and plasma data for the period 1200-1500 UT on February 18, 1996.

period 1305-1313 UT, although these flows were mostly field-aligned. After these flows, Geotail stayed near the plasma sheet/tail lobe boundary and was engulfed with cold stationary plasmas.

Just prior to the second onset at 1329 UT, Geotail entered the tail lobe and observed southward $B_z$ after 1325 UT. For this onset, Geotail observed tailward flows after 1331 UT and entered the tail lobe briefly. Then, Geotail observed earthward flows, although these were mostly field-aligned. $B_z$ was mostly southward for the tailward flow period and northward for the earthward flow period. Acceleration of electrons was observed for this flow activity [Nagai et al., 1998], implying that magnetic reconnection was in progress near Geotail. For this flow reversal, the magnetic field showed rapid fluctuations in the 1/16-s data (not shown here), and it is difficult to interpret the 12-s plasma distribution function data correctly, so we do not discuss these data here. It is important to note that the flow activity continued for only 6 min. After this activity, Geotail stayed near the current sheet and observed almost stationary plasmas. These plasmas have a number density of 0.6 cm$^{-3}$ and an ion temperature of 1.7 keV.

For the third onset, at 1332 UT, Geotail observed strong earthward flows. These flows had a sharp enhancement in $B_z$ at their front, so that they were accompanied by large flux transport. The flow activity continued for only 9 min, after which the plasmas became almost stationary.

For the recovery phase of this substorm, Geotail approached the current sheet and then stayed near the plasma sheet/tail lobe boundary. Near the current sheet at 1400 UT, the number density was 0.4 cm$^{-3}$ and the ion temperature was 2.1 keV. These values are comparable to those observed before the substorm. After 1407 UT, Geotail observed field-aligned earthward flows near the plasma sheet/tail lobe boundary.

In summary, fast flows formed in association with each onset in the multiple-onset sequence, and plasmas near the current sheet became almost stationary between two successive onsets. Earthward flows in the recovery phase were disconnected with earthward flows in the expansion phase.

### 3.3. The February 18, 1995, Event

To complement the study of the above two events, we examine the substorm that took place at 0940 UT on February 18, 1995. This substorm seems to be rather simple. Figure 5 shows ground magnetic field data from the pre-midnight station Ewa Beach (EWA) in Hawaii (21.6°, 2306 MLT at 1000 UT). Figure 6 shows magnetic field and plasma data from Geotail. Three-dimensional distribution function data are not available for this event. A midlatitude positive bay started near 0940 UT and reached its peak near 1015 UT. In the auroral zone, the westward electrojet started near 0940 UT, and an amplitude of the westward electrojet reached 300 nT at Yellowknife near 1020 UT (the digital data are not available, and are not shown here). The electrojet activity ended near 1200 UT. These auroral zone signatures are consistent with a midlatitude positive bay. Geotail was located at (-29.8, +6.8, -1.6 $R_E$) in the pre-midnight sector at 1000 UT. Before the onset, Geotail stayed near the current sheet and observed stationary plasmas with a number density of 0.5 cm$^{-3}$ and an ion temperature of 1.5 keV. Geotail observed fast tailward flows with southward $B_z$ during the time interval 0940-1000 UT. However, plasmas became almost stationary in the late expansion phase and re-

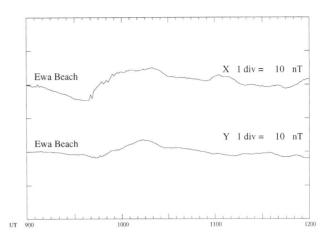

**Figure 5.** Magnetic field data on the ground for the period 0900-1200 UT on February 18, 1995.

covery phase, irrespective of the regions of the plasma sheet.

Just after the fast tailward flows, Geotail made a current sheet crossing near 1006 UT. Geotail observed almost stationary plasmas, having a number density of 0.9 cm$^{-3}$ and an ion temperature of 0.7 keV. Geotail approached the current sheet near 1035 UT in the recovery phase and observed plasmas having a number density of 0.7 cm$^{-3}$ and an ion temperature of 0.7 keV. Therefore, the plasmas were cold and dense near the current sheet in the late expansion phase and recovery phase. Near 1010 UT, Geotail appeared to approach the plasma sheet/tail lobe boundary; however, no intense flows were detected. After 1055 UT, earthward flows appeared and plasma sheet plasmas became hot and tenuous. However, this flow activity might be associated with a new activity, since there was a small positive H change at Ewa Beach (Figure 5).

In summary, this event seems to be a single-onset substorm. Fast tailward flows continued in only a part of the expansion phase. Cold and dense plasmas existed near the current sheet in the late expansion phase and recovery phase.

### 3.4. The 1502 UT, November 22, 1995, event

Figure 7 shows plasma and magnetic field data from Geotail for the period 1400-1800 UT. For this substorm event, Pi 2 activity started near 1502 UT, and there were several Pi 2 pulsations until 1600 UT. An amplitude of the westward electrojet reached 360 nT at Tixie Bay (61.3°, 0020 MLT at 1600 UT). The recovery phase of the substorm started near 1600 UT. Geotail was located at (-25.3, -1.3, -3.0 $R_E$) at 1600 UT. Geotail observed tailward flows with southward $B_z$ in association with the onset and entered the tail lobe. When the plasma sheet reappeared near 1610 UT in the recovery phase, earthward flows were observed. These earthward flows were mostly field-aligned and fairly variable.

220 PLASMA SHEET BEHAVIOR DURING SUBSTORMS

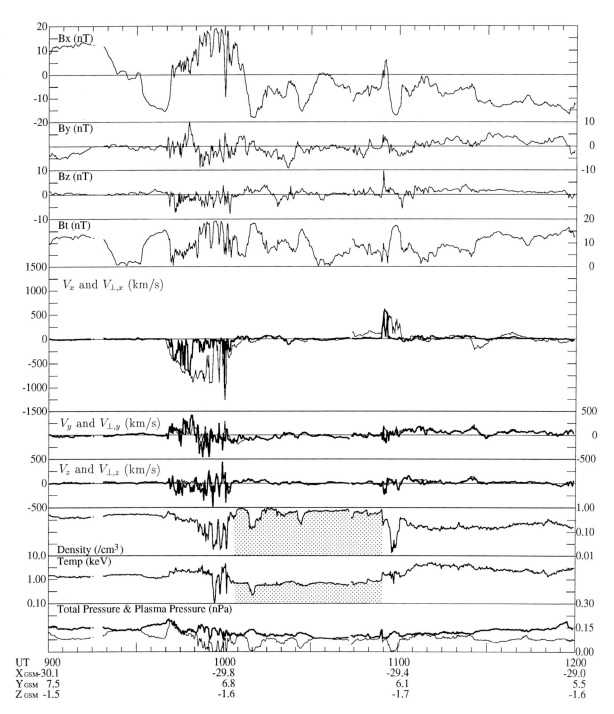

**Figure 6.** Geotail magnetic field and plasma data for the period 0900-1200 UT on February 18, 1995.

Geotail entered the tail lobe in the time intervals 1635-1639 UT, 1646-1650 UT, and 1656 UT. In these cases, earthward flows were stronger during the exit from the plasma sheet than during entry into the plasma sheet.

We examine plasmas near the current sheet. We note that $B_x$ became almost zero in the time interval 1624-1626 UT. Plate 2 shows selected ion distribution function data in the B-C plane on November 22, 1995. We examine the ion distribution function at 1623:59 UT (Plate 2c), when the total magnetic field was 0.4 nT. The B direction was approximately earthward and the C direction was approximately duskward. Ions show a

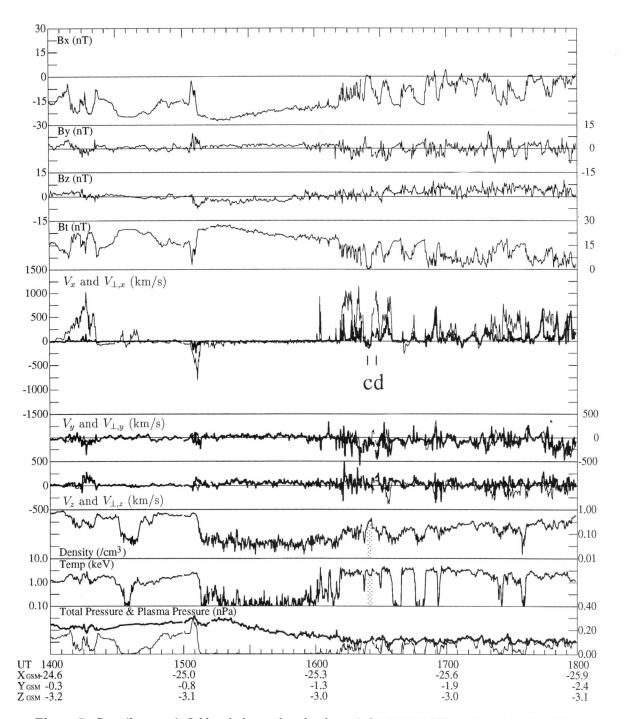

**Figure 7.** Geotail magnetic field and plasma data for the period 1400-1800 UT on November 22, 1995. Ion distribution functions are examined at times c and d.

single-component distribution with the dawnward convection. The ions were almost isotropic, not showing any counterstreaming signatures. There was a significant quantity of cold ions. At 1627:56 UT (Plate 2d), the total magnetic field became 12 nT, so that Geotail was off the current sheet. The B direction was approximately tailward and the C direction was approximately dawnward. Ions have two components: (1) warm ions moving earthward along the magnetic field at a speed of 1200 km/s, and (2) cold ions showing dawnward convection.

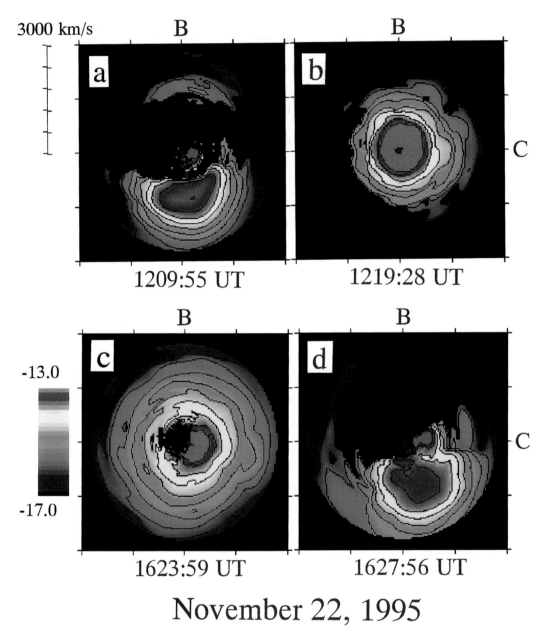

**Plate 2.** Ion velocity distribution functions in the B-C plane on November 22, 1995. The B direction is the local magnetic field, and the C direction is the convection direction.

In summary, the "classical" signature of plasma sheet behavior was seen during this substorm. Tailward flows formed in association with onset, and then the spacecraft exited the plasma sheet in the expansion phase. The plasma sheet reappeared with intense earthward flows in the recovery phase. These earthward flows were highly variable. Stationary plasmas existed near the equatorial current sheet, and these plasmas showed dawnward convection, not any strong earthward convection.

### 3.5. The 1108 UT, November 22, 1995, substorm

Figure 8 shows magnetic field and plasma data from Geotail for the period 1000-1300 UT on November 22, 1995. A substorm started at 1108 UT with a Pi 2 burst. The recovery phase of this substorm started near 1200 UT and ended near 1245 UT. Geotail was located at (-23.8, 0.8, -3.6 $R_E$) at 1200 UT. There was no evident substorm activity in the period 0200-1100 UT, and the IMF $B_z$ became strongly northward in the period 0500-0930 UT on November 22, 1995. Geotail was near the equatorial current sheet near 1100 UT, just prior to the onset, and it observed almost stationary plasmas with a number density of 1.2 cm$^{-3}$ and an ion temperature of 1.4 keV. Geotail observed tailward flows after 1111 UT and exited the plasma sheet at 1120 UT.

Geotail entered the plasma sheet and observed field-aligned earthward flows and counterstreaming features in the time interval 1203-1212 UT in the recovery phase. The ion distribution function at 1209:55 UT is presented in Plate 2a. Ions show counterstreaming features and a cold component showing the convection. In Plate 2a, the +B direction is tailward and the -B direction is earthward. Geotail entered the tail lobe again.

At 1218 UT, Geotail entered the plasma sheet and observed counterstreaming features briefly in its boundary region. Geotail was then engulfed with stationary plasmas. Because $B_x$ was -10 nT, Geotail was not exactly in the current sheet. Ions had an isotropic distribution with a number density of 1.8 cm$^{-3}$ and an ion temperature of 0.5 keV. The ion distribution function at 1219:28 UT is presented in Plate 2b. Cold ions dominate, and no intense flow exists in the tailward-earthward direction. Plasmas had a high density (> 1.0 cm$^{-3}$) and a low temperature (< 1.0 keV) in the time interval 1218-1256 UT, which corresponded to the recovery phase of the substorm.

### 3.6. The February 9, 1995, substorm

To complement the study of the 1108 UT, November 22, 1995, event, we examine the substorm starting at 0432 UT on February 9, 1995. The IMF $B_z$ was northward after 1700 UT on February 8, 1995, and it turned southward near 0200 UT on February 9, 1995. No substorm activity was evident in the northward IMF $B_z$ period. A well-defined onset was seen at 0432 UT, and the recovery phase of this substorm started near 0500 UT. Figure 9 shows magnetic field and plasma data from Geotail. Geotail was located at (-32.1, 5.5, -1.6 $R_E$) at 0500 UT. Three-dimensional distribution function data are not available for this event. Although Geotail observed tailward flows in association with the onset, it was mostly in the tail lobe in the expansion phase. Geotail entered the plasma sheet near 0522 UT. Although there were weak earthward flows near the plasma sheet/tail lobe boundary, the plasmas became stationary. Since $B_x$ was 5 nT, Geotail was not exactly in the current sheet. Plasmas had a number density of 1.3 cm$^{-3}$ and an ion temperature of 0.5 keV. Geotail made a current sheet crossing near 0408 UT, before the onset, and observed plasmas having a number density of 1.3 cm$^{-3}$ and an ion temperature of 0.6 keV. Therefore, the characteristics of the plasmas in the recovery phase were very similar to those near 0400 UT, well before the substorm. In this substorm, fast earthward flows appeared after Geotail observed the cold and dense plasmas inside the plasma sheet.

To summarize the 1108 UT, November 22, 1995, event and the February 9, 1995, event, cold and dense plasmas existed inside the plasma sheet in the recovery phase. These plasmas were almost stationary. It is noteworthy that the plasmas were observed in the substorms in which the plasma sheet showed a significant thinning.

## 4. DISCUSSION

*Nagai et al.* [1998] have examined changes in the plasma sheet in association with each Pi 2 onset in the near-Earth plasma sheet. Tailward flows are identified as a signature of the onset in the plasma sheet at radial distances larger than 20 $R_E$. The start time of tailward flows precedes the ground onset in the pre-midnight sector at 20-30 $R_E$, implying that magnetic reconnection is the cause of the substorm. Plasmas do not show any changes in the dawnside and duskside of the plasma sheet for most onsets, so that the dawn-dusk extent of magnetic reconnection is limited in the tail. Furthermore, the occurrence distribution of tailward flows suggests that the dawn-dusk extent of magnetic reconnection covers only the pre-midnight sector in most cases [see also *Nagai et al.*, 1997].

In this paper, we examine the termination of tailward flows in individual events. In the plasma sheet, tailward flows cease, and plasmas that are almost stationary appear even in the expansion phase. In the 1329 UT, February 18, 1996, event, reversal from tailward flows to earthward flows was observed; however, the duration of the earthward flows was only 6 min. Therefore, magnetic reconnection ceased in or near the site where it started [see also *Nagai et al.*, 1998].

In the recovery phase, earthward flows are observed. Earthward flows are highly variable and intermittent, and they are mostly confined in the plasma sheet/tail lobe boundary. The confinement of the flows in the boundary region has been known since the time of early spacecraft observations [e.g., *DeCoster and Frank,*

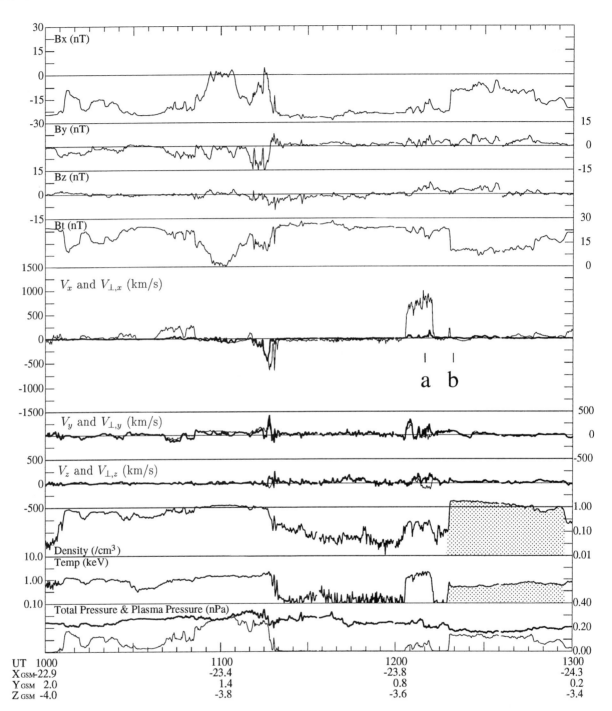

**Figure 8.** Geotail magnetic field and plasma data for the period 1000-1300 UT on November 22, 1995. Ion distribution functions are examined at times a and b.

1979]. Near the equatorial current sheet, plasmas are almost stationary, and they can be cold and dense. It is likely that these stationary plasmas are the same as those observed in the expansion phase.

The stationary plasmas usually have a number density of $> 0.4$ cm$^{-3}$ and an ion temperature of $< 2$ keV. These plasmas are very similar to those observed in prolonged northward IMF $B_z$ periods. Furthermore, it is known that plasmas are colder and denser in the dawn and dusk flanks of the plasma sheet than in the central region [e.g., *Terasawa et al.*, 1997]. When magnetic reconnection takes place in the limited dawn-dusk extent

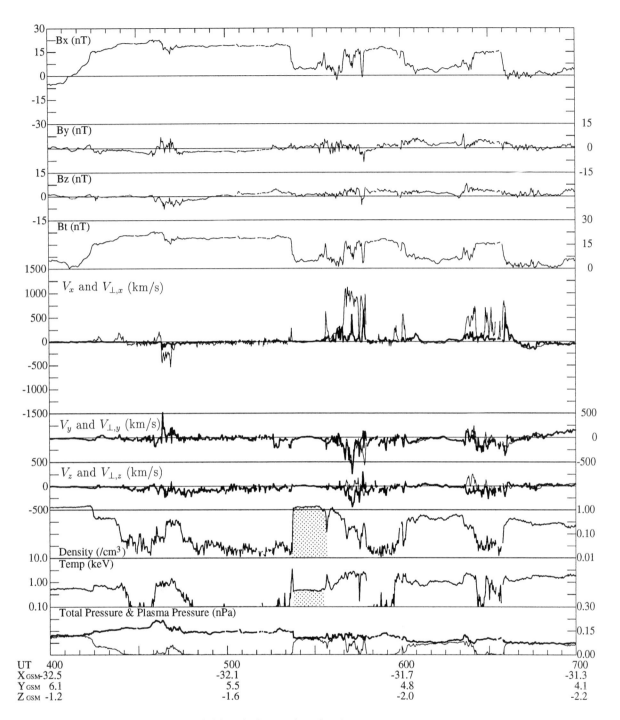

**Figure 9.** Geotail magnetic field and plasma data for the period 0400-0700 UT on February 9, 1995.

in the near-Earth plasma sheet, a significant number of plasmas, which are often cold and dense, remain in the dawnside and duskside of the plasma sheet. When magnetic reconnection ceases in the site where it starts, plasmas can be transported into the past reconnection site because of low total pressure in this site. Duskward or dawnward convection is observed in some cases. It is likely that the transport of plasmas from the dawnside and/or duskside is a source mechanism for stationary plasmas in the equatorial current sheet of the plasma sheet.

The retreat of an X-type neutral line has been proposed as an essential constituent of the recovery phase in the near-Earth neutral line model of substorms [e.g., *Hones*, 1977]. In this model, one would observe that fast tailward flows with southward $B_z$ change to fast earthward flows with northward $B_z$. The present observations show that fast tailward flows in the expansion phase are not connected to fast earthward flows in the recovery phase. The earthward flows in the recovery phase are probably produced by magnetic reconnection that newly takes place in the distant tail beyond 100 $R_E$ [*Nagai et al.*, 1997], and earthward flows are mostly field-aligned in the plasma sheet/tail lobe boundary in the near-Earth plasma sheet. The disconnection between the near-Earth and distant tail magnetic reconnections is supported by *Nishida et al.* [1996].

## 5. CONCLUSIONS

Geotail has provided important information on plasma sheet plasmas in the course of substorms. Fast tailward flows with southward $B_z$ do not necessarily continue in the whole expansion phase, and these flows are usually disconnected from the flows appearing in the recovery phase. Stationary plasmas, which are often cold and dense, exist near the equatorial current sheet in the expansion and recovery phases. It is likely that these plasmas are transported from the duskside and/or dawnside of the plasma sheet after magnetic reconnection ceases.

*Acknowledgments.* The digital magnetic field data from Kakioka were provided by Kakioka Magnetic Observatory. We thank K. Yumoto and S. I. Solovyev for supplying magnetic field data from the 210° magnetic meridian chain stations. Other magnetic field data were supplied by WDC-C2, Kyoto University.

## REFERENCES

Baumjohann, W., G. Paschmann, T. Nagai, and H. Lühr, Superposed epoch analysis of the substorm plasma sheet, J. Geophys. Res., *96*, 11605-11608, 1991.

DeCoster, R. J., and L. A. Frank, Observations pertaining to the dynamics of the plasma sheet, J. Geophys. Res., *84*, 5099-5121, 1979.

Fairfield, D. H., R. P. Lepping, E. W. Hones Jr., S. J. Bame, and J. R. Asbridge, Simultaneous measurements of magnetotail dynamics by IMP spacecraft, J. Geophys. Res., *86*, 1396-1414, 1981.

Hones, E. W., Jr., Substorm processes in the magnetotail: Comments on 'On hot tenuous plasmas, fireballs, and boundary layers in the Earth's magnetotail' by L. A. Frank, K. L. Ackerson, and R. P. Lepping, J. Geophys. Res., *82*, 5633-5640, 1977.

Hones, E. W., Jr., Plasma flow in the magnetotail and its implications for substorm theories, in *Dynamics of the Magnetosphere*, edited by S.-I. Akasofu, pp. 545-562, D. Reidel, Norwell, Mass., 1980.

Hones, E. W., Jr., Plasma sheet behavior during substorms, in *Magnetic Reconnection in Space and Laboratory Plasmas, Geophys. Monogr. Ser.*, vol. 30, edited by E. W. Hones Jr., pp. 178-184, AGU, Washington, D. C., 1984.

Hones, E. W., Jr., S. J. Bame, and J. R. Asbridge, Proton flow measurements in the magnetotail plasma sheet made with Imp 6, J. Geophys. Res., *81*, 227-234, 1976.

Hones, E. W., Jr., T. A. Fritz, J. Birn, J. Cooney, and S. J. Bame, Detailed observations of the plasma sheet during a substorm on April 24, 1979, J. Geophys. Res., *91*, 6845-6859, 1986.

Kokubun, S., T. Yamamoto, M. H. Acuña, K. Hayashi, K. Shiokawa, and H. Kawano, The GEOTAIL magnetic field experiment, J. Geomagn. Geoelectr., *46*, 7-21, 1994.

Mukai, T., S. Machida, Y. Saito, M. Hirahara, T. Terasawa, N. Kaya, T. Obara, M. Ejiri, and A. Nishida, The low energy particle (LEP) experiment onboard the GEOTAIL satellite, J. Geomagn. Geoelectr., *46*, 669-692, 1994.

Nagai, T., R. Nakamura, T. Mukai, T. Yamamoto, A. Nishida, and S. Kokubun, Substorms, tail flows, and plasmoids, Adv. Space Res., *20*, 961-971, 1997.

Nagai, T., M. Fujimoto, Y. Saito, S. Machida, T. Terasawa, R. Nakamura, T. Yamamoto, T. Mukai, A. Nishida, and S. Kokubun, Structure and dynamics of magnetic reconnection for substorm onsets with Geotail observations, J. Geophys. Res., *103*, 4419-4440, 1998.

Nishida, A., T. Mukai, T. Yamamoto, Y. Saito, and S. Kokubun, Magnetotail convection in geomagnetically active times. 1. Distance to the neutral lines, J. Geomagn. Geoelectr., *48*, 489-501, 1996.

Paschmann, G., N. Sckopke, and E. W. Hones Jr., Magnetotail plasma observations during the 1054 UT substorm on March 22, 1979 (CDAW 6), J. Geophys. Res., *90*, 1217-1229, 1985.

T. Terasawa, M. Fujimoto, T. Mukai, I. Shinohara, Y. Saito, T. Yamamoto, S. Machida, S. Kokubun, A. J. Lazarus, J. T. Steinberg, and R. P. Lepping, Solar wind control of density and temperature in the near-Earth plasma sheet: WIND/GEOTAIL collaboration, Geophys. Res. Lett., *24*, 935-938, 1997.

---

T. Nagai, Department of Earth and Planetary Sciences, Tokyo Institute of Technology, Tokyo 152-8551, Japan. (e-mail: nagai@geo.titech.ac.jp)

R. Nakamura and S. Kokubun, Solar-Terrestrial Environment Laboratory, Nagoya University, Toyokawa 442-8507, Japan. (e-mail: rumi@stelab.nagoya-u.ac.jp, kokubun@stelab.nagoya-u.ac.jp)

Y. Saito, T. Yamamoto, T. Mukai, and A. Nishida, Institute of Space and Astronautical Science, Sagamihara 229-8510, Japan. (e-mail: saito@gtl.isas.ac.jp, yamamoto@gtl.isas.ac.jp, mukai@gtl.isas.ac.jp, nishida@gtl.isas.ac.jp)

# Separation of Directly-Driven and Unloading Components in the Ionospheric Equivalent Currents During Substorms by the Method of Natural Orthogonal Components

W. Sun

*Geophysical Institute, University of Alaska Fairbanks*

W.-Y. Xu

*Institute of Geophysics, Chinese Academy of Sciences, Beijing, China*

S.-I. Akasofu

*Geophysical Institute, University of Alaska Fairbanks*

It is important to separate objectively the directly driven and unloading components in substorm processes, particularly in an attempt to identify processes associated with the two components. This paper describes an attempt in this particular effort by applying the method of Natural Orthogonal Components (MNOC). A time series of the ionospheric equivalent current function with time resolution of five minutes during March 17-19, 1978 is calculated on the basis of IMS six meridian chains magnetometer data in order to obtain the fundamental orthogonal basis set. The first and second modes of the set thus obtained dominate over the rest of the modes. The first mode is found to have a two-cell pattern which is well known to be associated with global plasma convection in the magnetosphere. It is enhanced during the growth phase and expansion phase of substorms and decays during the recovery phase of substorms. Further, it varies with the e parameter and thus may be identified as the directly driven component. The second mode reveals itself as an impulsive enhancement of the westward electrojet around midnight between 65° and 70° of latitudes and is much less correlated with the e parameter than the first one. Thus, as a first approximation, we identify it as the unloading component. A cross-correlation was also performed between the two mode components and e function, confirming the above identifications. It is shown that the directly driven component tends to dominate over the unloading component except for a brief period soon after substorm onset and that the process associated with the unloading component is short-lived compared with the lifetime of substorms. This study establishes the first clear determination of the time-profile of the unloading component.

## 1. INTRODUCTION

It had long been believed during the earliest days of substorm research in the 1960s that the magnetospheric substorm was nothing more that a spontaneous, sudden conversion of magnetic energy which had been continuously accumulated and stored in the magnetotail [cf. *Axford*, 1967; *Siscoe and Cummings*, 1969]. However, many researchers found in the 1970s that the north-south component (Bz) of the interplanetary magnetic field plays a vital role in causing substorms [cf. *Foster et al.*, 1971; *Arnoldy*, 1971; *Burton et al.*, 1975; *Perreault and Akasofu*, 1978] on the basis of a high correlation between individual substorms and individual southward turnings of the Bz component. Thus, *Akasofu* [1979] proposed that the magnetospheric substorm consists of two processes, the directly driven component and the unloading component. This concept of two-component process was recognized by *Rostoker et al.* [1987]. The two components may be described as follows:

(a) *Driven Component*: This is the component in which time variations in the rate of energy derived from the solar wind are approximately the same as the sum of time variations of energies deposited directly in the ionosphere, ring current and elsewhere with appreciable time delay. The equivalent current pattern for the driven system features two vortices and has been referred in the past by the terms DS, $S_q^p$ and likely incorporates the high latitude portion of DP2.

(b) *Unloading Component*: The unloading component refers to deposition of the energy stored in the magnetotail into the auroral ionosphere and into the ring current. The equivalent current pattern associated with the unloading has a single vortex involving a longitudinally confined westward ionospheric electrojet located in the midnight sector. This equivalent system has been referenced to as DP1.

The equivalent current provides a mathematical representation method of ground magnetic disturbances and assumes to flow only in a spherical shell, the ionosphere [*Chapman*, 1935]. The DP1 and DP2 current system [*Akasofu et al.* 1965; *Nishida*, 1968] are the equivalent current representation of two modes of magnetic disturbances in the polar region. *Clauer and Kamide* [1985] tried to separate the DP1 and DP2 current systems by means of a differential technique. The technique is accomplished by choosing a time interval T = T2 - T1. The current at T1 is assumed to remain constant during the interval T while an additional current system develops during same interval and is estimated by subtracting the effects of the currents at time T1 from the total currents. Thus, the separation of the DP1 and DP2 in this method had to assume that the current at T1 is kept constant during the interval. They found that both DP1 and DP2 currents develop during the course of the substorm activity. Recently, *Kamide* and *Kokubun* [1996] suggested that the two components change as a result of the relative changes of electric fields and conductivities in the auroral electrojet. However, this method of separation requires the knowledge of both electric field and conductivity over the entire polar region as a function of time.

Many other authors examined also both the directly driven and the unloading components. *Baker et al.* [1981] found that the correlation of the AE index to the ε function and interplanetary electric field (VBs) has a peak value of 0.54 and 0.6 with a time lag of 40 minutes. *Bargatze et al.* [1985] discussed the magnetospheric impulse responses measured by the AL index to the solar wind parameter VBs and found that two response pulses has 20 min and 60 min time lag, respectively. They considered that the 20-min pulse represents directly driven component in solar wind-magnetosphere coupling process and the 60-min pulse represents unloading component.

In the present paper, we introduce the method of Natural Orthogonal Components in analyzing the equivalent current function during a substorm and obtain the fundamental orthogonal basis set. We show that the first mode and the second mode are dominant and found to have well known characteristics of the directly driven component and the unloading component, respectively. As a data set, we use magnetic records which were extensively collected during the IMS, the total number of stations being 71 [*Kamide et al.* 1982a]. The equivalent current functions during March 17-19, 1978 with time resolution of 5 minutes are calculated on the basis of the IMS data [*Kamide, et. al.*, 1982b]. Figure 1 shows an example of the equivalent current function at 0230 UT on March 18, 1978. The contour interval is 20 kA. The total current is the integrated current over all contour intervals. The total of 864 patterns of the equivalent current function during three days is used to be input data for analysis of the Natural Orthogonal Components.

## 2. DESCRIPTION OF THE METHOD OF NATURAL ORTHOGONAL COMPONENTS

The method of Natural Orthogonal Components (MNOC) [*Kendall and Stuart*, 1976] is often used to separate the structure of geomagnetic field into longer and shorter wavelength parts [*Frynberg*, 1975; *Pushkov et al.*, 1976, *Rotanova et. al.*, 1982]. MNOC was also used to separate geomagnetic field variation into quiet and disturbed components [*Golovkov et. al.*, 1978]. The separated Sq variation was applied to improve the calculation of the K indices at a local station [*Papitashvili et. al.*, 1992].

In MNOC the fundamental orthogonal basis set is obtained during the calculation procedure, so that it is different from the spherical harmonic and Fourier analysis in which the fundamental orthogonal basis set is set previously. We assume the values of the equivalent current function $E_{ij}$ at time $t_i$, i=1.2....,m and locations $r_j$, j=1,2,...n, which is expressed by a matrix $\mathbf{E}_{m \times n}$. In the present calculation, the period of time is 3 days with a resolution of 5 min, hence m= 864. In the distribution of the equivalent current function shown in Figure 1, the latitudinal range is from 50° to 90° with 2° interval and the

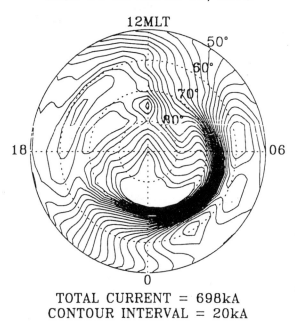

**Figure 1.** The ionospheric equivalent current function at 0230 UT on March 18, 1978. The total current is the integrated current over all contour intervals.

longitudinal range from 0° to 360° with 15° interval. Thus the space grid points n = 20×24 = 480. If $F_{m \times n}^k$ indicates $k^{th}$ independent factor contributing to $E_{m \times n}$ ($k=1,2,...,h \le n_t$) where h is number of modes, each factor can be expanded in terms of $T_{m \times 1}^k (X_{n \times 1}^k)^T$, where $X_{n \times 1}^k$ corresponds to the configuration of spacial distribution for $F_{m \times n}^k$, and $T_{m \times 1}^k$ indicates the magnitude of $X_{n \times 1}^k$ in time sequence.

$$E_{m \times n} = \sum_{k=1}^{h} F_{m \times n}^k = \sum_{k=1}^{h} T_{m \times 1}^k (X_{n \times 1}^k)^T \quad (1)$$

A covariance matrix $A_{n \times n}$ can be constructed by $E_{m \times n}$

$$A_{n \times n} = (E_{m \times n})^T E_{m \times n} = \sum_{l=1}^{h} \sum_{p=1}^{h} X_{n \times 1}^l (T_{m \times 1}^l)^T T_{m \times 1}^p (X_{n \times 1}^p)^T \quad (2)$$

If assuming that $X_{n \times 1}^k$ is orthogonal and normalized

$$(X_{n \times 1}^l)^T X_{n \times 1}^p = 0 \quad \text{for } l \ne p$$
$$= 1 \quad \text{for } l = p \quad (3)$$

and $T_{m \times 1}^k$ is orthogonal

$$(T_{m \times 1}^l)^T T_{m \times 1}^p = 0 \quad \text{for } l \ne p$$
$$= \lambda_p \quad \text{for } l = p \quad (4)$$

Then the eigen equation of the covariance matrix $A_{n \times n}$ becomes

$$A_{n \times n} X_{n \times 1}^k = \lambda_k X_{n \times 1}^k \quad (5)$$

This equation can be solved by standard method. Once $X_{n \times 1}^k$ is found $T_{m \times 1}^k$ can be calculated on the basis of equation (1)

$$T_{m \times 1}^k = E_{m \times n} X_{n \times 1}^k \quad (6)$$

## 3. RESULTS

(1) Figure 2 shows the patterns of first six modes together with the values of $\lambda_k$ which indicate the importance of the corresponding mode. Figure 3 shows the total current and the contributions from the first six modes $(T_{m \times 1}^k (X_{n \times 1}^k)^T, k=1,2,...6)$. We found that the first two modes dominate over the rest. The sum of the contributions of the first mode and the second mode $(T_{m \times 1}^1 (X_{n \times 1}^1)^T + T_{m \times 1}^2 (X_{n \times 1}^2)^T)$ is, on average, 83% of the total current $(E_{m \times n} = \sum_{k=1}^{h} T_{m \times 1}^k (X_{n \times 1}^k)^T)$ for the three days. Therefore, we believe that the first two modes represent the most important and common properties of substorms. Certainly, any individual substorm has its temporal and local variation which can be discussed in details by combining all modes. Hereafter we consider only the first two modes.

As shown in Figure 2, the first mode has a two-cell convection pattern, while the second mode indicates a westward electrojet around the midnight sector between 65° and 70° of latitudes. Thus, the patterns of the first two modes resemble the equivalent current patterns of DP2 and DP1, respectively.

(2) An individual substorm event during 0900-1300 UT, March 19, 1978 was selected to examine the contributions of the first mode and the second mode in details. The top panel in Figure 4 shows the total current and the AE index. The magnetic condition was a quiet period before 1035 UT. The growth phase started at 1035 UT and continued until 1135 UT. The expansion onset of the substorm was at 1135 UT and the peak period of the substorm was at 1210 UT. The bottom panel in Figure 4 shows the contributions of the first and second mode. It is interesting to note that first mode, as shown by the dashed curve, is enhanced during the growth phase from 1035 UT to 1135 UT. Then, it is further intensified after the onset of the expansion phase and reached its maximum value about ten minutes after the peak time of the substorm. The contribution from second mode (solid curve) remained to be near zero, then increased sharply at the expansion onset and reached its maximum value at the peak time of the substorm as shown in the total current. Note that the level of negative values can be considered the level of uncertainty in this analysis.

## NORMALIZED MODES

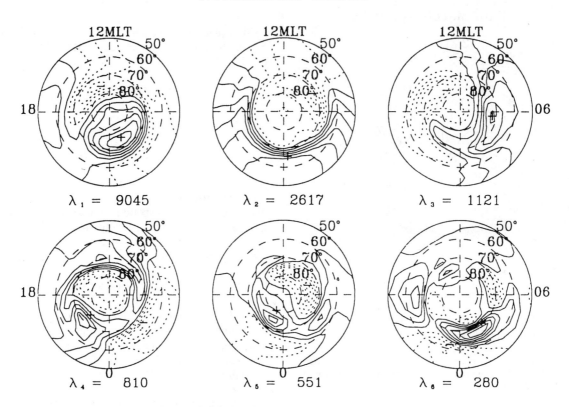

**Figure 2.** The patterns of the first six normalized modes.

The first row in Figure 5 shows the patterns of current function at some particular epochs, 1000 UT, 1030 UT, 1100 UT, 1130 UT, 1200 UT, 1230 UT and 1300 UT on March 19, 1978. The second row shows the patterns of the contribution of the first mode ($\mathbf{T}_{m\times 1}^1 (\mathbf{X}_{n\times 1}^1)^T$). The third row shows the second mode ($\mathbf{T}_{m\times 1}^2 (\mathbf{X}_{n\times 1}^2)^T$). One can see that the first mode shows clearly the two-cell convection pattern during the growth phase. Further, it becomes enhanced during the expansion phase. On the other hand, the second mode shows little indication of its presence during the growth phase, but is greatly enhanced during the expansion phase.

(3) The upper panel of Figure 6 shows the solar wind-magnetosphere coupling function ε with one hour resolution during March 18 - 19, 1978, which is calculated on the basis of the solar wind and IMF data observed by the satellite IMP-8. The second and bottom panels show the contributions from the first and second modes, respectively, during same period. One can see that there is fair correlation between ε function and the first mode. On the other hand, the second mode tends to be impulsive and large mainly at about substorm onset of three substorms at 09 UT on March 18, 22 UT and 11 UT on March 19. Mode 2 shows some negative changes. When we add all the mode values, the result is same as the total current. Therefore, so long as we deal with the first two modes, the magnitude of the negative values are considered to indicate the level of uncertainty in this analysis is ~ ±200 kA.

In order to examine the correlation in more detail, we plot the correlation between coupling function ε and the first mode (and second mode) in Figure 7 by using hourly-mean values. An one hour time delay was considered when taking the correlation between the second mode and the ε function. The correlation coefficient between ε function and the first mode is 0.63, and the correlation coefficient between ε function and the second mode is 0.24. This result may be considered to be another identification that the first mode represents the directly driven component, in addition to having the two-cell

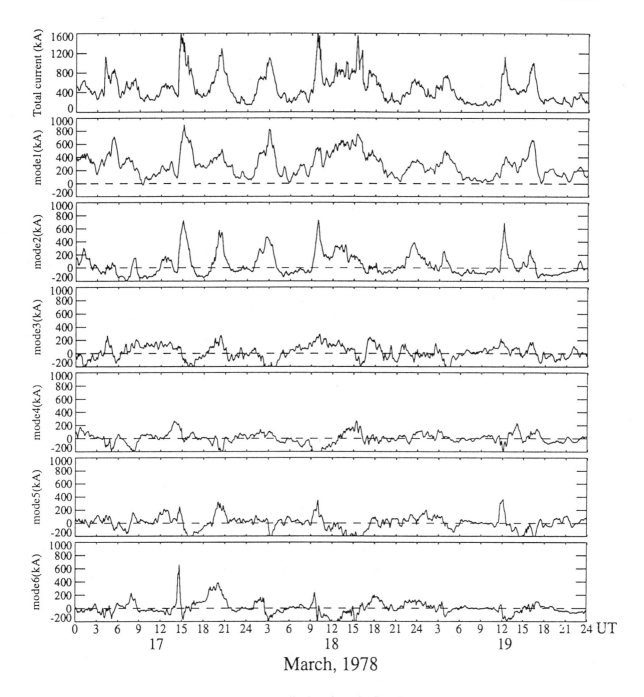

**Figure 3.** The total current and the contributions from the first six modes ( $\mathbf{x}_{n\times 1}^k$ $\mathbf{T}_{m\times 1}^k$, k=1,2,...6).

pattern. The second mode has a single cell pattern centered around the midnight sector and presents impulsively during the expansion phase of substorms. Thus, it may be concluded that it is related to the unloading process.

## 4. SUMMARY AND DISCUSSION

In the present paper, we introduced a mathematical method called the Natural Orthogonal Components in

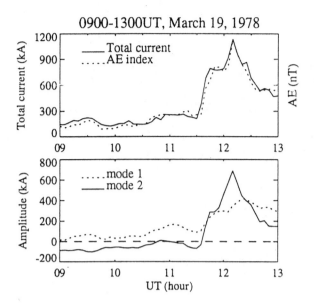

**Figure 4.** The upper panel shows the total current and the AE index for the substorm during 09 - 13 UT on March 19, 1978. The bottom panel shows the contributions from the first and second modes, respectively, during the substorm event.

analyzing the equivalent current function during substorms and obtain the fundamental orthogonal basis set. We find that the first two modes dominate over the rest of them. The first mode has a two-cell pattern and is enhanced during the growth phase and the expansion phase, and decays during the recovery phase of substorms; it has fair correlation with the ε coupling function with a correlation coefficient of 0.63. Therefore, it may be concluded that the first mode represents, as first approximation, the directly driven component during substorms. The pattern of the second mode displays a westward electrojet around the midnight sector between 65° and 70° of latitudes and is much less correlated with the ε coupling function. Thus, it may be concluded that the second mode represents, as first approximation, the unloading component.

It is of great interest that the particular mathematical method we employed in this paper successfully identified the two major modes, of which characteristics agree with the known directly driven component and unloading component, respectively. It is most interesting that the identification of the two components by *Baumjohann* [1983] is in good agreement with our results. Strictly speaking, it may not be possible to separate the two components. For example, an enhanced conductivity

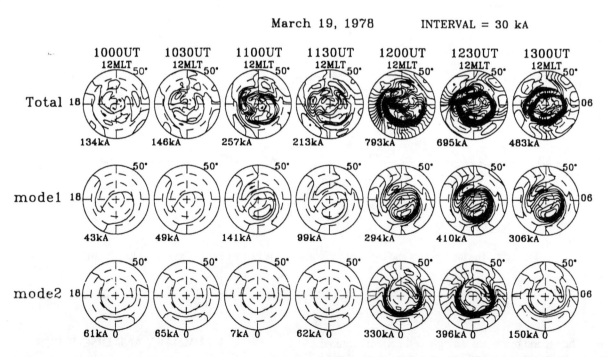

**Figure 5.** The top row shows the patterns of the equivalent current functions at different time during the substorm from 1000 UT to 1300 UT on March 19, 1978. The second and bottom rows show the patterns of the contributions from the first and second modes, respectively.

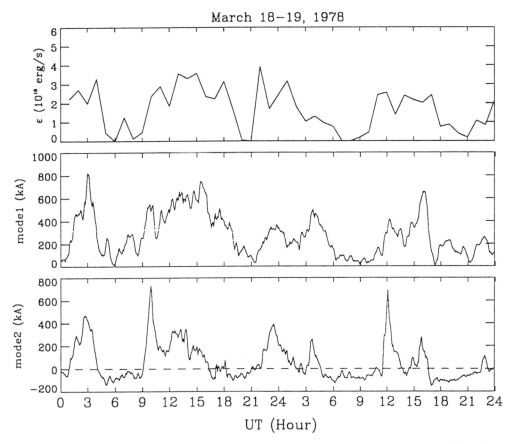

**Figure 6.** The top panel shows the ε coupling function as a function of time during March 18-19, 1978 with one hour time resolution. The second and bottom panels show the contributions of the first and the second modes during same period.

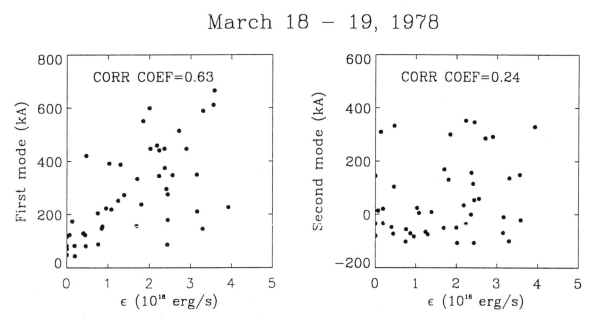

**Figure 7.** The left panel shows the cross-correlation between the ε coupling function and the first mode during March 18 - 19, 1978. The right panel shows the cross-correlation between the ε coupling function and the second mode.

caused by the unloading process could enhance also the directly driven component. The fact that we can roughly separate the two components may be due to the fact that the lifetime of electrons in the E region of the ionosphere (~ 20 min) is relatively short compared with the average life time of substorms, so that effects of the impulsive unloading are short lived and separable from the directly driven component.

We believe that we have succeeded in separating roughly the two components. Our results suggest that the direct effect of the unloading component manifests itself impulsively (~ 1 hour) at substorm onset. Thus, knowing the characteristics and its time profile, we can now discuss physics of the unloading component and its cause. This was not possible until now. The impulsive nature of the unloading component suggests the stress caused during the growth phase is limited to a relatively short distance in the X-coordinate, perhaps where the thin current sheet forms. Thus, the stress can be released in a relatively short period. It remains to be seen that the period of the unloading coincides approximately with the period during which the front of the expanding bulge reaches the highest latitude and the thin current region disappears.

The original definition of recovery onset is the time when the expanded bulge begins to contract [*Akasofu*, 1964]. This does not begin immediately after the front of the expanding bulge reaches the highest latitude. In fact, it takes often a few hours before the contraction begins [*Snyder and Akasofu*, 1972]. The directly driven component must be dominant during the period when the expanded bulge remains at its highest latitude location. The recovery phase begins when the directly driven component begins to subside. Note that the identification of recovery onset based on the AE index is very crude; in fact, the AE index starts to decrease as the center line of the westward electrojet shifts poleward with the expanding bulge.

*Acknowledgments.* We would like to thank Dr. Y. Kamide for his discussion. This work was supported in part by a NSF grant ATM-93-11474.

## REFERENCES

Akasofu, S.-I., The development of the auroral substorm, *Planet. Space Sci.*, 12, 273, 1964.

Akasofu, S.-I., S. Chapman, and C.-I. Meng, The polar electrojet, *J. Atmos. Terr. Phys.*, 27, 1275, 1965.

Akasofu, S.-I., What is a magnetospheric substorm?, p447, *Dynamics of the magnetosphere*, D. Reidel Pub. Co., Dordrecht, Holland, 1979

Arnoldy, R. L., Signature in the interplanetary medium for substorms, *J. Geophys. Res.*, 76, 5189, 1971.

Axford, W. I., Magnetic storm effects associated with the tail of the magnetosphere, *Space Sci. Rev.*, 2, 149, 1967.

Baker, D. N., E. W. Hones, Jr., J. B. Payne, and W. C. Feldman, A high time resolution study of interplanetary parameter correlation with AE, *Geophys. Res. Lett.*, 8, 179, 1981.

Bargatze, L. F., D. N. Baker, R. L. McPherron, and E. W. Hones Jr., Magnetospheric impulse response for many levels of geomagnetic activities, *J. Geophys. Res.*, 90, 6387, 1985.

Baumjohann, W., Ionospheric and field-aligned current system in the auroral zone: A concise review, *Adv. Space Res.*, 2, 55, 1983.

Burton, R. K., R. L. McPherron, and C. T. Russell, An empirical relationship between interplanetary conditions and Dst, *J. Geophys. Res.*, 80, 4204, 1975.

Chapman, S., The electric current systems of magnetic storms, *Terr. Magn. Atm. Elect.*, 40, 347, 1935.

Clauer, C. R. and Y. Kamide, Dp1 and Dp2 current systems for the March 22, 1979 substorms, *J. Geophys. Res.*, 90, 1343, 1985.

Foster, J. C., Fairfield, K. W. Ogilvie and T. J. Rosenburg, Relationship of interplanetary parameters and occurrence of magnetospheric substorms, *J. Geophys. Res.*, 76, 6971, 1971.

Frynberg, E. B., Separation of the geomagnetic field into a normal and anomalous part, *Geomagn. Aeron. Engl. Trans.*, 15, 117, 1975.

Golovkov, V. P., N. E. Papitashvili, Yu. S. Tyupkin, and E. P. Kharin, Separation of geomagnetic field variations on the quiet and disturbed components by the MNOC, *Geomag. Aeron.*, 18, 511, 1978.

Kamide, Y., H. W. Kroehl, A. D. Richmond, B.-H. Ahn, S.-I. Akasofu, W. Baumjohann, E. Friis-Christensen, S. Matsushita, H. Maurer, G. Rostoker, R. W. Spiro, J. K. Walker, and A. N. Zaitzev, Changes in the global electric fields and currents for March 17 - 19, 1978 from six IMS meridian chains of magnetometers, *Rep. UAG-87, World Data Center A*, Boulder, Colo., 1982a

Kamide, Y., H. W. Kroehl, A. D. Richmond, B.-H. Ahn, S.-I. Akasofu, W. Baumjohann, E. Friis-Christensen, S. Matsushita, H. Maurer, G. Rostoker, R. W. Spiro, J. K. Walker, and A. N. Zaitzev, Global distribution of ionospheric and field-aligned currents during substorms as determined from six IMS meridian chains of magnetometers: Initial results, *J. Geophys. Res.*, 87, 8228,, 1982b..

Kamide, Y., and S. Kokubun, Two-component auroral electrojet: Importance for substorm studies, *J. Geophys. Res.*, 101, 13,027, 1996.

Kendall, M. G. and A. Stuart, The advanced theory of statistics, Vol.3, Chap. 35., *Charles Griffin, High Wycombe*, 1976.

Nishida, A., Geomagnetic DP 2 fluctuations and associated magnetospheric phenomena, *J. Geophys. Res.*, 73, 1,795, 1968.

Papitashvili, N. E., V. O. Papitashvili, B. A. Belov, and L. Hakkinen, and C. Sucksdorff, Magnetospheric contribution to K-indices, *Geophys. J. Int.*, 111, 348, 1992.

Perreault, P. and S.-I. Akasofu, A study of geomagnetic storms, *Geophys. J. Roy. Astron. Soc.*, 54, 547, 1978.

Pushkov, A. N., E. B. Frynberg, E. B. Chernova, T. A., and M. V. Fiskina, Analysis of the space-time structure of the main geomagnetic field by expansion into natural orthogonal components, *Geomagn. Aeron.* Engl. Trans., 16, 196, 1976.

Rotanova, N. M., N. E. Papitashvili, and A. N. Pushkov, Use of Natural Othogonal Components to distinguish and analyze the 60-yr geomagnetic field variations, *Geomag. Aeron.*, 22, 821, 1982.

Rostoker, G., S.-I. Akasofu, W. Baumjohann, Y. Kamide, and R. L. McPherron, The roles of directly input of energy from the solar wind and unloading of stored magnetotail energy in driving magnetospheric substorms, *Space Sci. Rev.*, 46, 93, 1987.

Siscoe, G. L. and W. D. Cummings, On the cause of geomagnetic bays, *Planet. Space Sci.*, 17, 1795, 1969.

Snyder, A. L. and S.-I. Akasofu, Observations of the auroral oval by the Alaskan meridian chain of stations, *J. Geophys. Res.*, 77, 3419, 1972.

---

W. Sun, Geophysical Institute, University of Alaska Fairbanks, Fairbanks, Alaska 99775-7320, USA

W.-Y. Xu, Institute of Geophysics, Chinese Academy of Sciences, P. O. Box 9701, Beijing 100101, China

S.-Y. Akasofu, Geophysical Institute, University of Alaska Fairbanks, Fairbanks, Alaska 99775-7320, USA

# Simulation of the March 9, 1995 Substorm and Initial Comparison to Data

R. E. Lopez, C. C. Goodrich, M. Wiltberger, K. Papadopoulos

*Department of Astronomy, University of Maryland, College Park, Maryland*

J. G. Lyon

*Department of Physics, Dartmouth College, Hannover, New Hampshire*

Abstract In this study we examine a period of substorm activity that occurred on March 9, 1995. Using solar wind data as input, the event has been simulated using a 3-D MHD code. To determine how well the simulation fared, we have compared the simulation results to data. This comparison has taken two forms. The first is a comparison to individual spacecraft in the magnetotail. The second is a comparison of global energy storage and release, where we have compared the auroral heating to data-based estimates of auroral energy dissipation, as well as evaluating the variation of the open flux in the simulation. There is a generally good level of agreement between the simulation results and data. Thus our results show that MHD simulations can be used to model at least one magnetospheric substorm.

## INTRODUCTION

Magnetospheric substorms represent the episodic dissipation of energy stored in the geomagnetic tail that was previously extracted from the solar wind. This energy release produces activity throughout the entire magnetosphere-ionosphere system, and it results in a wide variety of phenomena such as auroral intensifications [e.g., Akasofu, 1964], the generation of new current systems [e.g., McPherron et al., 1973], and particle acceleration up to MeV levels [e.g., Baker 1984; Lopez and Baker, 1994]. Gaining a fuller understanding substorms is an important element in characterizing the space environment, and thus is critical to the National Space Weather Initiative.

Substorms are believed to be a global magnetospheric response to the solar wind, so it seems reasonable to investigate substorms using a global simulation of the solar wind-magnetosphere interaction. Therefore we have simulated one well-observed event that occurred on March 9, 1995. The simulation code used in this study is an improved version of previously developed codes [e.g., Fedder and Lyon, 1987]. It solves the full 3-D time-dependent MHD equations over the whole magnetosphere. The simulations are not a model in the usual sense, since there are no *a priori* assumptions made about the structure of the magnetosphere. The only free parameters in the simulations are the solar wind input and the ionospheric conductivity.

To understand whether the simulation results have any relationship to reality it is important to compare those results to data. This comparison can range from a detailed comparison with individual spacecraft in the magnetosphere to more global comparisons of factors related to solar wind-magnetosphere coupling, such as estimates for energy storage and release. The ability of a simulation to reproduce individual, spatially-localized variations is a key factor to its potential use as a useful predictive tool. At the same time, even if individual observations are well-modeled, unless the global interaction is handled properly, one can have little confidence in local correspondences. In this paper we will make both kinds of comparisons for a particular well-defined isolated substorm.

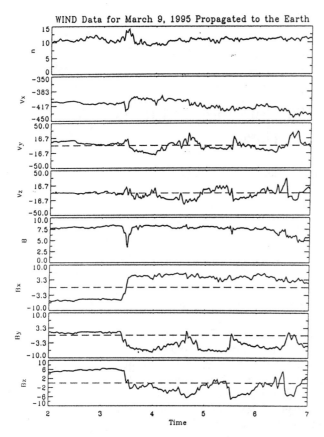

**Figure 1.** Wind data from March 9, 1995 shifted by 54 minutes to reflect the propagation time to Earth.

The paradigm for understanding the storage and release of energy in the magnetospheric system is centered on magnetic reconnection. When the interplanetary magnetic field is directed southward, closed flux on the dayside merges with solar wind magnetic field. This flux is transported to the tail as open polar cap flux, the polar cap grows and the flaring angle increases as the tail energy content grows [Holzer et al., 1986]. At the same time the directly-driven system dissipates some of the energy directly in the magnetosphere-ionosphere system; the input energy not dissipated is stored in the increasingly stressed configuration of the magnetotail. A global simulation that correctly includes the basic physics of solar-wind magnetosphere coupling should be able to reproduce this energy storage quite adequately.

During the substorm expansion phase, energy previously stored in the tail is rapidly unloaded. This unloading of energy must be invoked to account for the energy dissipation seen during substorms, since the output power levels can exceed the input power levels and the time history of the input and output power levels are not related by a simple time delay as one would expect in a purely driven system [Baker et al., 1986]. It is generally thought that nightside reconnection plays a key role in the dissipation of magnetotail tail flux during the substorm [e.g., Hones, 1984]. The observed variation of the open flux polar region requires reconnection [Holzer et al., 1986; Lopez 1994], though at what point in the evolution of the substorm the reconnection of lobe magnetic flux occurs is controversial [Lui, 1991; Lopez, 1994; Lopez et al., 1994; Baker 1996]. In the auroral zone the released energy is dissipated primarily through joule heating from the substorm electrojets [Akasofu, 1981]. In addition, some energy may be deposited in the ring current, while some manifests itself as plasmasheet heating.

## OVERVIEW AND SIMULATION COMPARISON TO SPACECRAFT DATA

The event in question occurred on March 9, 1995, and it was identified for study by Alan Rogers of the British Antarctic Survey. Solar wind data from Wind propagated to the Earth (that is shifted by 54 minutes) are shown in Figure 1; these data were used to drive the simulation. During the last hours of March 8 and the early hours of March 9 the IMF was northward. At about 0230 UT, Wind recorded a rapid rotation of the solar wind magnetic field, including a southward turning of the IMF, that was associated with a crossing of the heliospheric plasma sheet. The southward IMF arrived at the Earth about an hour later, at 0330 UT, as seen in Figure 1. This produced a growth phase in the magnetosphere that was observed by a number of ground stations.

We have used H component data from the ground stations of the CANOPUS array using stations located near the auroral zone (63.3° to 67.3° eccentric dipole geomagnetic latitude) to construct a CL and CU, presented in Figure 2. The local field values at the start of the day as the quiet time values. Prior to substorm onset, the data do not indicate the presence of a growth phase westward electrojet. However, there was a growing (though weak)

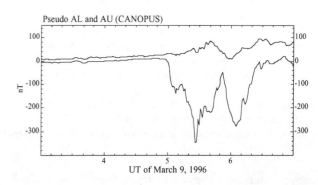

**Figure 2.** CANOPUS CL and CU indexes using those stations whose eccentric dipole geomagnetic latitudes range from 63.3° to 67.3°.

eastward electrojet, which is seen in the CL index as well as CU. Other ground stations (not shown) recorded a stronger growth phase signature. The data show a very clear onset at 0500 UT, an intensification at 0514 UT, some recovery, then another onset at 0552 UT, with a subsequent recovery. Through a detailed examination of CANOPUS data (along with near-by stations) we have determined that the activity was centered on the CANOPUS region. Thus these times for substorm onset and intensification are global times, even though the stations cover a limited local time sector.

Two spacecraft, Geotail and IMP 8, provide information about what happened in the magnetotail during the event. Geotail was at (-12.7, -5.2, -1.7) and IMP 8 was at (-30.3, 2.2, -9.4) in GSM coordinates. Thus Geotail was in the near-Earth dawn region, and IMP 8 was in the cislunar midnight magnetotail. Figures 3 and 4 show magnetic field and plasma data, respectively, from Geotail (dotted lines). Figure 5 shows the IMP-8 magnetic field (dotted lines). Included in all three figures are the simulations results (solid lines), which will be described in more detail below.

Figures 3 and 4 show that Geotail went through a rapid transition of the current sheet at 0458 UT, at which point

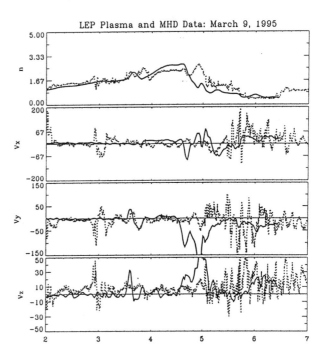

**Figure 4.** Geotail plasma data (dotted line) compared to simulation results (solid line) interpolated to the Geotail position.

the plasma density peaked. Just after the current sheet crossing, Geotail recorded the onset of a perturbation in the Y component that may be due to the field-aligned current in the current wedge that formed at the time with the onset of the substorm at 0500 UT. However, there was no significant dipolarization of the field, or any fast (>100 km/s) flow. This suggests that Geotail was just outside of the active sector at substorm onset. Moreover, since Geotail was on the dawn side of the active region, and in the southern hemisphere when it recorded the negative Y perturbation, the field-aligned currents producing that perturbation must have been located equatorward of the satellite - flowing on field lines that cross the neutral sheet earthward of the field lines that threaded Geotail. And since Geotail was still in the plasmasheet, the source of the substorm field-aligned currents had to be on closed field lines in the near-Earth region. Around 0525 UT, Geotail finally recorded the beginning of the local dipolarization of the field, accompanied by a number of plasma flow bursts.

Figure 5 presents magnetic field data from IMP-8, which shows a classic in the X component during the growth phase. It is interesting to note that the increase in X component began some 12 minutes after the arrival of the southward IMF on the dayside. This twelve minute delay is essentially the amount of time it would take newly merged flux to be convected to the IMP-8 location by the solar wind. At 0500 UT, a significant southward field was recorded, which is often interpreted as being due to the formation of a reconnection region earthward of the

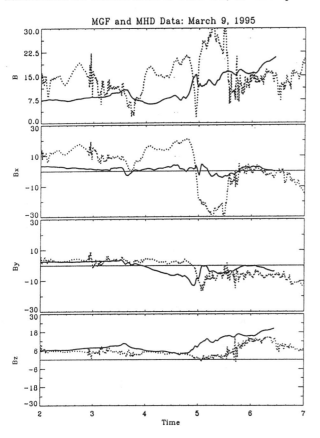

**Figure 3.** Geotail magnetic field data (dotted line) compared to simulation results (solid line) interpolated to the Geotail position.

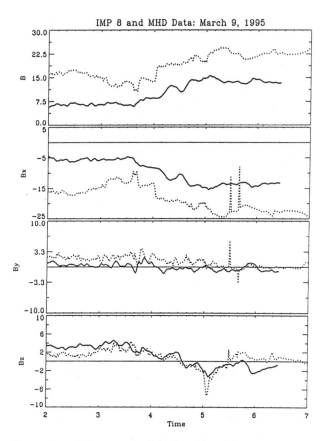

**Figure 5.** IMP 8 magnetic field data (dotted line) compared to simulation results (solid line) interpolated to the IMP 8 position.

satellite. However, given the proximity of IMP-8 to the lobe (as determined by the large X component), this southward field could also be due to the flaring or flapping of the tail. Interestingly enough, the filed did not weaken at this time, even though a decrease in the lobe field is often taken to signify the onset of the unloading phase of a substorm. The field did begin to weaken at 0520 UT, just after the intensification on the ground began at 0514 UT, but it began to grow again at about 0545 UT. Thus while the initial onset, or more precisely the intensification, appears to be correlated with variations at IMP-8, the 0552 UT onset does not appear to have had any significant corollary in the IMP-8 field data.

The period during which this above described substorm activity took place is well suited to simulation. There was a long period of northward IMF, which allows the simulation to relax to a "ground state", and the substorm itself was isolated and well-observed. In order to better approximate reality, such factors as dipole tilt (a constant value corresponding to 0500 UT) and the aberration due to the orbital motion of the Earth have been included. An example of the simulation result is Figure 6, which shows the density in the X-Z plane at midnight at 0455 UT.

Animated sequences of the results have been produced and are available at http://www.spp.astro.umd.edu under "Global simulations". To compare the simulation with the spacecraft data, we have interpolated the simulation results to the spacecraft positions as a function of time. The simulation data are plotted as solid lines in Figures 3, 4, and 5.

The X component in Figure 3 shows apparently little agreement between data and simulation until after 0530 UT, however this initial impression is somewhat misleading. If we inspect the current sheet crossing just before 0500 UT, we see that the rapid nature of the crossing from essentially north lobe to south lobe suggest that Geotail encountered a thin current sheet a fraction of an Earth radius in north-south extent. That kind of a current sheet cannot be modeled by the simulation, whose spatial grid in the relevant region is coarser that the actual current sheet thickness. Therefore it is no surprise that the simulated magnetic field does not produce the extreme variations seen in the data, since in the simulation Geotail is embedded in a much broader current sheet.. However, we note that roughly at the time of the real current sheet crossing, the simulated current sheet center also moved past the Geotail position, so that in the simulation Geotail moved from the northern hemisphere to the southern hemisphere at about the right time. In the light of this consideration, the discrepancies seen the upper panel (field magnitude) become more reasonable.

The simulated Y and Z components show a much better agreement with the data, though there is an important exception. The local onset of the substorm (as marked by dipolarization) is a bit early in the simulation when compared to the data. The Y component shows a deflection that we associate with the field-aligned current in the substorm current wedge. That deflection starts a bit earlier in the simulation than in the data, though the magnitude and shape of the deflection are very similar. The rise in the Z component we associate with the dipolarization due to the substorm current wedge. Again this begins earlier in the simulation that in reality. In fact, Geotail did not see a the onset of the dipolarization until about 0525 UT, which we associate with the substorm intensification at 0515 UT.

Both the simulation results and the data are consistent with the formation of a current wedge just to the west of Geotail (in accord with the ground data) that spread eastward as the substorm intensified. It is just that the simulated current wedge formed a bit early and arrived at Geotail a bit early. Some of this timing discrepancy could be due to a spatial resolution effect, since in reality the current wedge was more spatially confined than in the simulation. Nonetheless, after the substorm has subsided by 0600 UT, there is remarkably good agreement between the simulation and all three components of the magnetic field. It is at this time that one would expect that the current sheets were no longer thin, supporting our interpretation that the

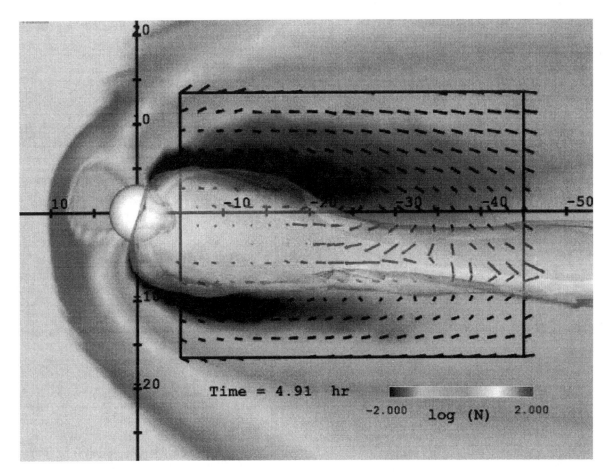

**Figure 6.** A sample of the simulation results. The figure shows the plasma density in the X-Z plane in the midnight meridian at 0455 UT. The arrows within the box show the flow field. The white circle denotes the inner edge of the simulation at about 3 Earth radii. The open-closed field line boundary is shown as a transluscent surface.

agreement is really rather good in a qualitative sense, and that the source of quantitative discrepancy between the Geotail field observations and the simulation has to do with issues of spatial resolution and the inability of the current code to reproduce the thin current sheets observed in nature.

The Geotail plasma data, along with the simulation results, are presented in Figure 4, and again the simulation is the solid line. The overall agreement is good, especially for the density. The high-speed, short duration flow events are not well-represented by the simulation, but again given spatial resolution issues one would not expect to see them. The biggest discrepancy concerns the onset of the activity at and before 0500 UT. The simulation predicts large Y and Z velocities that were not seen, although velocities of similar magnitude, if not duration, were observed a bit later. This discrepancy is probably related to the current wedge issues discussed above, since the curl of the velocity field drives field-aligned currents.

Figure 5 shows the IMP 8 magnetic field and the simulation results (solid line). The agreement with the Z component is remarkable, and the Y component is also very good. The simulated X component, however, is always of lesser magnitude than the data, which in turn drives the total field magnitude to be lower than observed. Part of this discrepancy may be due to the same issue of spatial resolution discussed above that produces a thicker current sheet, and hence a smaller X component at the IMP 8 position. However, this is not the entire story. The maximum lobe field strength in the simulation at the IMP 8 distance down the tail just before the onset is only about 18 nT, which is still lower that the observed lobe field of 25 nT at the same time. This suggests that the flaring of the magnetosphere in the simulation was less than in reality, which in turn suggests that the amount of flux transported into the lobes was less in the simulation than actually occurred. In fact, we have a possible resolution to this discrepancy (discussed in the next section), namely that extra energy was dissipated in the ionosphere and so this energy did not appear as tail lobe flux.

Despite the discrepancies, we feel that the simulation has done a very credible job in reproducing point measurements in the magnetosphere. We also feel that we understand the major contributions to these discrepancies, and that improvements in the code will lessen their impact. We now turn to a comparison between the global evolution of the code and what can be determined from data, using the evolution of the total energy budget as the organizing principle for that comparison.

## ENERGY INPUT AND DISSIPATION: OBSERVATIONS

We may estimate the energy input into the magnetosphere during the event by using the epsilon parameter [Akasofu, 1981], shown in Figure 7. The solar wind data and resulting energy input have been lagged by 54 minutes to reflect the propagation time to Earth, and a merging line length of 7 Re was assumed. By 0630 UT, we estimate that 760 gigawatt-hours had been provided to the magnetosphere by the solar wind.

Estimates of the power dissipated in the auroral zone by electrojet joule heating have used AE as the indicator of electrojet strength [Akasofu, 1981; Baumjohann and Kamide, 1984]. Because our observations are from a limited local time sector, and thus do not reflect the entire auroral zone, our CANOPUS-derived AE index is very likely an underestimate of electrojet activity, since the maximum in the eastward or westward electrojet may be outside of the CANOPUS array. However, we may use the CANOPUS data to estimate the heating at a given time, provided that we can safely presume that the CANOPUS station do in fact lie in the region of maximum eastward or westward electrojet. During the substorm onset and expansion we are very confidant that this was very likely the case with regard to the westward electrojet, and we will use CANOPUS CL to calculate the dissipated energy. During the growth phase this does not seem to be the case with regard to either the eastward or westward electrojets. However, the response of the stations was more due to the eastward, rather than to the westward, electrojet, and so for the growth phase we will use CANOPUS CU. After 0630 UT, the CE we can derive from CANOPUS seems adequate.

Baumjohann and Kamide [1984] found that the joule heating (in gigawatts) produced by the eastward and westward electrojets is 0.42 AU and 0.25 AL, respectively, while the overall joule heating is 0.32 AE. Using these results we can estimate that the joule heating from both eastward and westward electrojets is approximately 0.54 AL. Thus from 0500 UT until 0630 UT, we estimate that 127 gigawatt-hours were dissipated as joule heating in the northern hemisphere. In the growth phase the CANOPUS

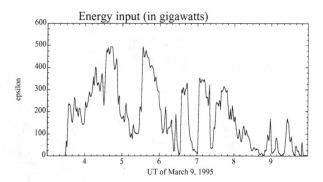

**Figure 7.** Solar wind energy input (epsilon) for March 9, 1995 lagged by 54 minutes to reflect the propagation time to Earth.

chain was responding primarily to the eastward electrojet. We now use Baumjohann and Kamide [1984] to estimate that the total heating from both eastward and westward electrojets is approximately 0.77 AU. This yields an energy dissipation in the northern auroral zone from 0330 UT to 0500 UT of 28 gigawatt-hours (prior to 0330 UT there was essentially no dissipation). Thus the total joule dissipation in the northern hemisphere from 0300 UT to 0630 UT can be estimated at 155 gigawatt-hours, which we in fact regard as a lower limit.

To get an estimate for the total dissipation, north and south, we could simply double the northern estimate, yielding a total dissipation of 310 gigawatt-hours. However, during the event there was a significant dipole tilt, making the southern hemisphere the sunlit hemisphere. Because of the enhanced conductivity, the simulation produces southern auroral zone currents that are bigger that the northern ones, and the energy dissipation in the southern ionosphere is almost twice as large as that in the northern ionosphere. We have no southern hemisphere data which we can directly compare to the CANOPUS data, however, if this current asymmetry was indeed the case, then the total auroral zone dissipation from 0330 UT to 0630 UT was roughly 465 gigawatt-hours.

Another major sink for energy is ring current injection. Using the pressure-corrected Dst values [Akasofu, 1981; Zwickl et al., 1987], the power dissipated may be estimated by

$$K\left(\frac{Dst^{pc}}{\tau} + \frac{\partial Dst^{pc}}{\partial t}\right)$$

where $K=4\times10^{13}$ joules/nT, and $\tau$ is the ring current decay time, which we take to be six hours. We have obtained the provisional hourly Dst from the WDC-C2 for Geomagnetism, Kyoto University, web site (from 0000 UT to 0800 UT, hourly Dst values were 3, 4, 8, 8, 5, 1, -3, -4). Using the above equation we find that the energy dissipated into the ring current from 0330 UT to 0630 UT was roughly 130 gigawatt-hours. Thus in our event the auroral dissipation was the largest sink for magnetospheric energy.

Combining the energy input to the ring current with the dissipation of energy in the auroral zone energy suggests that from 0330 UT to 0630 UT, roughly between 440 and 595 gigawatt-hours were dissipated in the magnetosphere, depending on whether you take the lower or higher estimate for the auroral dissipation. Our estimate of the energy input during the same time period is 760 gigawatt-hours. Thus between 58% and 78% of the input energy appears to have been dissipated. It might be tempting to assume that the remaining energy was expelled as a plasmoid. However, subsequent activity in the magnetosphere ran a significant energy deficit. From about 0715 UT to 0915 UT there was a period of much stronger activity that in the northern ionosphere alone dissipated 280 gigawatt-hours (using our proxy AL and the same assumptions as the above calculation). In the same period the estimate for energy input using epsilon in 230 gigawatt-hours. Thus it is unlikely that the initial activity used up all of the input energy, since some storage until the subsequent period of activity is needed to account for the energy budget from 0715 UT to 0915 UT. Given these considerations, there does not seem to be to much room for a plasmoid to contribute significantly to the energetics of this event.

## ENERGY CONSIDERATIONS: COMPARISON WITH THE SIMULATION

Figure 8 presents the simulated polar cap flux (where the polar cap is defined as the open-closed field line boundary and the flux is integrated over both hemispheres), and Figure 9 shows the polar cap boundaries along the noon and midnight meridians. At 0335 UT the polar cap began to grow. The growth was initially on the dayside, a direct response to dayside merging. The midnight boundary did not move substantially equatorward until about 25 minutes later. If we consider that newly merged field lines are anchored at one end in the solar wind (which was flowing at about 400 km/s), a 25 minute time delay suggests that the field line have been convected about 100 Re downstream (roughly the nominal tail length) before the

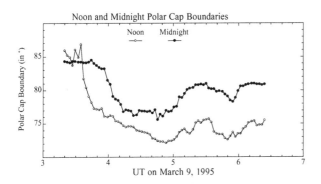

**Figure 8.** Simulated polar cap flux (total, both hemispheres).

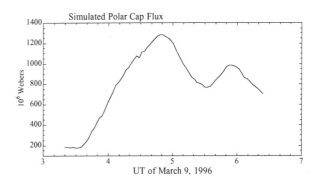

**Figure 9.** The noon and midnight latitudes of the northern open-closed field line boundary.

growth of the polar cap hits midnight. The polar cap continued to add flux until about 0454 UT, when the polar cap began to shrink, although reconnection of lobe field lines (and thus onset) began a couple of minutes earlier. The rate of flux decrease increased dramatically at 0503 UT and continued until about 0530 UT, when the polar cap began to grow again. At 0556 UT the polar cap began to shrink once more.

Each of these episodes of polar cap shrinkage correspond to the unloading of stored energy as represented by magnetic flux. Comparing to the data in Figure 2 we see that in fact the initial onset (0500 UT in the data, 0452 UT in the simulation) the intensification (0514 UT in the data, 0503 UT in the simulation), and the second onset (0553 UT in the data, 0556 UT in the simulation) were all captured by the simulation, though the temporal correspondence is not perfect. Another issue is that the slight decrease in polar cap flux at 0440 UT (which we call a pseudobreakup) did not correspond to any real activity. On the other hand, it is clear that at 0630 UT, not all of the stored energy has been released, as was surmised from the data-based estimate above. This is consistent with the suggestion that stored energy must be carried over to the later period of activity in order to account for the inferred energy budget. It is also consistent with the fact that the simulation did not develop much of a plasmoid. Thus in terms of both qualitative and quantitative behavior of the polar cap flux, our simulation is substantially in accord with the observations.

We have also carried out calculations of the joule heating for both auroral zones, and these are presented in Figure 10. There are areas of agreement, and of discrepancy, when we compare to the dissipation estimates made from the data. One general feature is that both the simulation and the data point to two episodes of significant heating corresponding to the two major onsets. Both show that the first period of activity was of longer duration that the first, and the average level of power dissipation was slightly greater during the second period (the data averages are 90 gigawatts versus 93 gigawatts). However, substorm

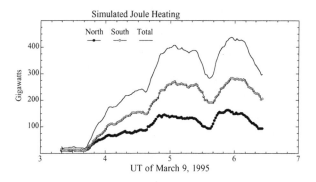

**Figure 10.** Simulated joule heating for northern and southern hemispheres, along with the total dissipation.

onset in the simulation began several minutes before substorm onset in the data. As pointed out above, in the simulation lobe reconnection began before the onset at 0500, and plasmasheet reconnection began even earlier. Reconnection in the simulation is driven numerically, and therefore is to some degree dependent upon the scale-size of the cells used in the computation. As the spatial resolution is increased, the development of reconnection regions can be slowed. Thus we expect that the exact time at which lobe reconnection would occur in a higher-resolution runs (which we plan to do) would likely be delayed.

Another issue regards the exact power levels. The simulation indicates that from 0330 UT to 0630 UT, 839 gigawatt-hours were dissipated, about twice as much as the data suggest. However, a close inspection of the simulation suggests that there is an effect (which we also believe to be related to spatial resolution issues) that produces an eastward electrojet that is much too large. Prior to the onset the dissipation power of around 225 gigawatts was driven primarily by this large eastward electrojet. An interesting point to consider is that if the simulation is artificially dissipating too much energy in the ionosphere then there must be correspondingly less energy stored in the lobes. This might be related to the discrepancy noted above between the observed IMP 8 magnetic field magnitude and the simulated lobe field at the IMP 8 position. If extra energy is dissipated in the ionosphere, there will be less flux in the lobes, resulting in a smaller flaring angle, and thus changing the pressure balance conditions that determine the lobe field magnitude, producing a weaker lobe field.

As an ad hoc correction we subtract 175 gigawatts from the level of ionospheric dissipation after the onset, and hold the dissipation level during the growth phase to be 50 gigawatts (similar to that observed) to estimate the actual power dissipated. The energy dissipated from 0330 UT to 0630 UT now becomes 382 gigawatt-hours, which is very close to the number derived from the data. With this correction we see that the agreement with the CANOPUS data is quite reasonable. For example, the energy dissipated in the northern hemisphere according to CL from 0500 UT to 0552 UT was 78 gigawatt-hours, yielding a total dissipation in both hemisphere of between 156 and 234 gigawatt-hours, whereas the corrected simulated northern dissipated energy was 71 gigawatt-hours and the energy dissipated in both hemispheres was 196 gigawatt-hours. Similar agreement is found for other periods, and by 0630 UT the observed northern auroral power dissipation (using CANOPUS CL) was about 35 gigawatts, which compares well with a corrected northern auroral simulated power dissipation at 0630 UT of 33 gigawatts. However, at no point does the simulation produce a peak corrected northern hemisphere power dissipation of 184 gigawatts, as was observed at 0526 UT.

## CONCLUSIONS

In this paper we have presented a comparison of both single point measurements of plasma and magnetic field as well as global estimates of energy storage and release to outputs of a simulation driven with actual solar wind data. There is remarkable degree of correspondence between the observations and the simulation. The single-point measurements show qualitative, and sometimes quantitative agreement with the simulation.. Both the data and the simulation show the two substorm onsets, and the energy budgets are consistent with each other once one makes allowances for the unphysical eastward electrojet that develops in the simulation.. There are also areas of disagreement, such as the exact values of the various quantities compared at the spacecraft locations, the exact levels of auroral dissipation in the simulation versus the observations (especially the spurious eastward electrojet), and the exact times for substorm onset and intensification, that need to be addressed. We have some indications that increasing the spatial resolution in the code will help with many of these issues. Nonetheless, the results are very encouraging, since the ability to simulate the global flow of energy in the magnetosphere-ionosphere system as well as the environment in specific regions of space is a crucial step in creating a viable space weather system

*Acknowledgments.* The authors would like to thank T. Mukai and S. Kokubun for Geotail plasma and magnetic field data, respectively, R. Lepping for the Wind and IMP 8 magnetic field data, A. J. Lazarus for the Wind plasma data, and J. C. Samson for providing the CANOPUS data and for stimulating discussions and comments. We also wish to thank WDC-C2 for Geomagnetism, Kyoto University, for providing provisional Dst on-line. This work was supported by NASA grants NAG-56256, NAG-54662, and NAGW-3222, and by NSF grant ATM-9527055.

# REFERENCES

Akasofu, S.-I., The development of the auroral Substorm, *Planet. Space Sci.*, 12, 273-282, 1964.

Akasofu, S.-I., Energy coupling between the solar wind and the magnetosphere, *Space Sci. Rev.*, 28, 121, 1981.

Baker, D. N., Particle and field signatures of substorms in the near magnetotail, in *Magnetic Reconnection in Space and Laboratory Plasmas*, edited by E. W. Hones, Jr., 193-202, AGU, Washington, D. C., 1984.

Baker, D. N., T. A. Fritz, R. L. McPherron, D. H. Fairfield, Y. Kamide, and W. Baumjohann, Magnetotail energy storage and release during the CDAW 6 substorm analysis intervals, *J. Geophys. Res.*, 90, 1205-1216, 1986.

Baumjohann, W., and Y. Kamide, Hemispherical joule heating and the AE indices, *J. Geophys. Res.*, 89, 383, 1984.

Fedder, J. A. and J. G. Lyon, The solar wind-magnetosphere-ionosphere current-voltage relationship, *Geophys. Res. Lett.*, 14, 880-883, 1987.

Holzer, R. E., R. L. McPherron, and D. H. Hardy, A quantitative empirical model of the magnetospheric flux transfer process, *J. Geophys. Res.*, 91, 3287-3293, 1986.

Hones, E. W., Jr., Plasma sheet behavior during substorms, in *Magnetic Reconnection in Space and Laboratory Plasmas*, edited by E. W. Hones, Jr., 178-184, AGU, Washington, D. C., 1984.

Lopez, R. E., On the role of reconnection during substorms, *Proc. International Conference on Substorms-2*, 175-182, 1994

Lopez, R. E., C. C. Goodrich, G. D. Reeves, R. D. Belian, and A. Taktakishvili, Mid-tail plasma flows and the relationship to near-Earth substorm activity: A case study, *J. Geophys. Res.*, 99, 23561-23569, 1994.

Lopez, R. E., and D. N. Baker, Evidence for particle acceleration during magnetospheric substorms, *Ap. J. S.*, 90, no. 2, 531-539, 1994.

Lui, A. T. Y., A synthesis of magnetospheric substorm models, *J. Geophys. Res.*, 96, 1849, 1991.

McPherron, R. L., C. T. Russell, and M. P. Aubry, Satellite studies of magnetospheric substorms on August 15, 1968: 9. Phenomenological model for substorms, *J. Geophys. Res.*, 78, 3131-3149, 1973.

Siscoe, G. E. Hilner. T. L. Killeen, L. J. Lanzerotti, and W. Lotko, Developing Service Promises Accurate Space Weather Forecasts in the Future, *EOS*, 75, no 31, 365-366, August 2, 1994.

Zwickl, R. D., L. F. Bargatze, D. N. Baker, C. R. Clauer, and R. L. McPherron, An evaluation of the total magnetospheric energy output parameter, Ut, in *Magnetotail Physics*, edited by A. T. Y. Lui, 155-159, Johns Hopkins University Press, 1987.

---

C. C. Goodrich, R. Lopez, and K. Papadopoulos, all at Department of Astronomy, University of Maryland, College Park, MD, 20742

J. G. Lyon at Department of Physics, Dartmouth College, Hannover, NH, 03755

M. Wiltberger, Department of Physics, University of Maryland, College Park, MD, 20742

# Large-Scale Dynamics of the Magnetospheric Boundary: Comparisons between Global MHD Simulation Results and ISTP Observations

J. Berchem[1], J. Raeder[1], M. Ashour-Abdalla[1], L. A. Frank[2], W. R. Paterson[2], K. L. Ackerson[2], S. Kokubun[3], T. Yamamoto[4], and R. P. Lepping[5]

Understanding the large-scale dynamics of the magnetospheric boundary is an important step towards achieving the ISTP mission's broad objective of assessing the global transport of plasma and energy through the geospace environment. Our approach is based on three-dimensional global magnetohydrodynamic (MHD) simulations of the solar wind-magnetosphere-ionosphere system, and consists of using interplanetary magnetic field (IMF) and plasma parameters measured by solar wind monitors upstream of the bow shock as input to the simulations for predicting the large-scale dynamics of the magnetospheric boundary. The validity of these predictions is tested by comparing local data streams with time series measured by downstream spacecraft crossing the magnetospheric boundary. In this paper, we review results from several case studies which confirm that our MHD model reproduces very well the large-scale motion of the magnetospheric boundary. The first case illustrates the complexity of the magnetic field topology that can occur at the dayside magnetospheric boundary for periods of northward IMF with strong $B_X$ and $B_Y$ components. The second comparison reviewed combines dynamic and topological aspects in an investigation of the evolution of the distant tail at 200 $R_E$ from the Earth.

## 1. INTRODUCTION

Global magnetohydrodynamic (MHD) simulations have been used for about 20 years to model the time-dependent interaction of the solar wind with the Earth's magnetosphere [*Leboeuf et al.*, 1978; 1981; *Lyon et al.*, 1980; 1981]. However, it is only recently that results from global MHD simulation models and observations have been compared directly. Most of this modelling has been motivated by the International Solar Terrestrial Physics (ISTP) program. Indeed, since the inception of the program, it has been realized that numerical simulations, in particular global MHD models, can be used to link local measurements made by the different spacecraft of the mission, and thus that they have the potential to provide valuable theoretical tools for pursuing the ISTP mission's broad objective of assessing the global transport of plasma and energy through the geospace environment. In response, global models were considerably improved [see review by *Walker and Ashour-Abdalla*, 1994]. In preparation for Geotail, global MHD simulations were extended to model the distant magnetotail structure and convection processes well beyond 100 $R_E$ [*Fedder et al.*, 1995a; *Fedder and Lyon*, 1995; *Usadi et al.*, 1993; *Raeder et al.*, 1995; *Mobarry et al.*, 1996]. Resolution of the computational mesh used in the simulations was also improved to allow the investigation of magnetospheric boundaries and mesoscale structures such as magnetic flux ropes in the magnetotail [*Walker and Ogino*,

[1]Institute of Geophysics and Planetary Physics, University of California, Los Angeles, California 90095-1567
[2]Department of Physics and Astronomy, The University of Iowa, Iowa City, Iowa 52242
[3]Solar-Terrestrial Environment Laboratory, Nagoya University, Toyokawa, Aichi 442, Japan
[4]Institute of Space and Astronautical Science, Sagamihara, Kanagawa 229, Japan
[5]NASA/Goddard Space Flight Center, Greenbelt, Maryland 20771

Geospace Mass and Energy Flow: Results From the International Solar-Terrestrial Physics Program
Geophysical Monograph 104
Copyright 1998 by the American Geophysical Union

1996] and at the high-latitude magnetopause [*Berchem et al.*, 1995a, b].

New refinements in the models and computational advances have allowed us to undertake the type of studies that were envisioned at the inception of the ISTP mission. The principle of these studies is to use solar wind plasma and magnetic field measurements as input parameters to drive the three-dimensional global MHD simulations and then to compare the simulation results with observations. Several approaches have been taken in carrying out these comparisons. One approach is to produce synthetic auroral emissions derived from the simulation results and to compare them with actual images from auroral spacecraft [*Fedder et al.*, 1995b]. Another approach is to produce time series from the simulations and compare them with those measured by instruments onboard equatorial spacecraft located at a downstream location [*Frank et al.*, 1995; *Berchem et al.*, 1997; *Raeder et al.*, 1997].

In this paper, we review results from two studies based on time series comparisons. The first of these studies illustrates the complexity of the magnetic field topology and the convection patterns that can occur at the dayside magnetospheric boundary for periods of northward interplanetary magnetic field (IMF) with strong $B_X$ and $B_Y$ components. This study is based on events observed on December 27, 1994, while Geotail was skimming the dayside magnetopause and the Wind spacecraft was monitoring the solar wind. The second comparison reviewed in this paper focuses on the dynamic and topological aspects of the magnetotail boundary by investigating the time evolution of the distant tail cross section at 200 $R_E$ from the Earth. The events used in that study were observed by Geotail on July 7, 1993, when the direction of the IMF was predominantly northward and marked by a slow rotation of its clock angle component. Throughout this paper, we use the Earth-centered solar ecliptic (GSE) coordinate system to indicate the locations of the spacecraft and to display the observations and simulation results.

## 2. SIMULATION MODEL

The magnetospheric part of our simulation model is based on a single fluid MHD description. The code solves the normalized resistive MHD equations as an initial value problem, using an explicit conservative predictor-corrector scheme for time stepping and hybridized numerical fluxes for spatial finite differencing [*Raeder et al.*, 1995; *Berchem et al.*, 1995a, b]. A nonlinear function of the current density is used to model the explicit local resistivity. This model includes a threshold that is a function of the local normalized current density and that has been calibrated to avoid spurious dissipation [*Raeder et al.*, 1996a]. Similar phenomenological resistivity models have been used in local MHD simulation models [e.g. *Sato and Hayashi*, 1979; *Ugai*, 1985] and are based on the assumption that current driven instabilities are responsible for the anomalous resistivity that produces reconnection.

A spherical shell with a radius of 3.7 $R_E$ is placed around the Earth to exclude the region where the Alfvén velocity becomes too large. Inside the shell, the MHD equations are not solved and a static dipole magnetic field is assumed. To close the field aligned currents we assume a two-dimensional ionosphere and self-consistently determine the electrostatic potential by solving the ionospheric potential equation. Our current model takes three ionization sources (solar EUV, precipitating electrons, diffuse electron precipitation) into account to compute the ionospheric Hall and Pedersen conductances that are needed to solve the potential equation. The ionospheric potential is then mapped to the 3.7 $R_E$ shell where it is used as a boundary condition for the magnetospheric flow velocity. A detailed description of the model can be found in *Raeder et al.* [1996a].

## 3. DAYSIDE MAGNETOSPHERIC BOUNDARY

This study focused on a time interval on December 27, 1994, during which Geotail was skimming the dayside magnetospheric boundary while Wind was monitoring the solar wind conditions. Coordinated data sets from the two spacecraft provided a unique opportunity to study the structure and dynamics of the dayside magnetospheric boundary [*Berchem et al.*, 1996]. The Wind measurements that were used as the driving input to our MHD simulation model are displayed in Figure 1. The time interval starts at 1700 UT when the Wind spacecraft was located at $R_W$ = (22.2, -53.8, -1.3) $R_E$, and ends at 2100 UT. It is marked by a slow and gradual transition from a northward to a southward IMF orientation. In addition, several strong enhancements are observed in the number density and the thermal pressure of the solar wind flow. An interesting feature is the large and almost steady X and Y components of the IMF during the period before 1900 UT. We show below that this IMF orientation created a complicated magnetic field topology in the dusk magnetosheath.

Geotail measurements and time series from the simulation are shown in Figure 2. Dotted traces represent the plasma parameters derived from the response of the Comprehensive Plasma Instrumentation (CPI) [*Frank et al.*, 1994] and the measurements from the Magnetic Field (MGF) experiment [*Kokubun et al.*, 1994], while solid traces show the computed values along the Geotail trajectory. We can follow the motion of spacecraft relative to the magnetospheric boundary by looking at the density and bulk velocity profiles measured by Geotail. From 1700 UT until about 1800 UT, the Geotail spacecraft is for the most part skimming the magnetopause on its magnetospheric side. This is clearly indicated by the low densities and strong northward $B_Z$ component seen during that period. Small density peaks and bulk flow velocity fluctuations observed during that period suggest that the spacecraft is indeed very close to the boundary. The density jump observed by Wind around 1735 UT reaches the magnetopause at about 1800 UT, pushing it inward. From that time until the end of the interval shown, Geotail stays mostly on the magnetosheath side of the magnetopause, except for two short and one long incursions into the magnetosphere. These reentries are observed around 1830 UT and 1855 UT, and from about 1910 UT until 1930 UT. They match fairly well the density depressions observed in the upstream solar wind around 1810 UT and 1835 UT, and from about 1850 UT until 1910 UT, respectively. At the time of its last outbound crossing of the magnetopause, about 1940 UT, Geotail was located very close to the dusk terminator at $R_G$ = (1.2, 14.9, 0.1) $R_E$. An important feature to consider is that the X-component of the magnetic field measured by Geotail in the magnetosheath is negative ($B_X < 0$), and thus has a direction opposite to that observed in the solar wind ($B_X > 0$).

In comparing the time series from the simulation with those measured by the Geotail spacecraft, we find good agreement between simulated and measured magnetic fields; in particular the simulation traces are a very good match for the measured

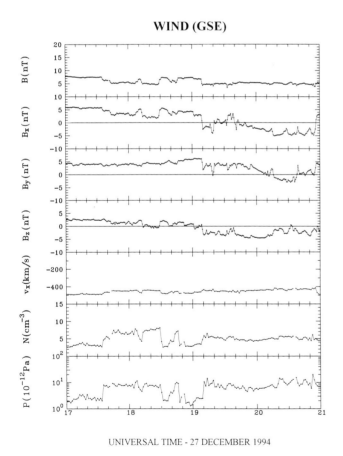

**Figure 1.** Plasma and field parameters measured by the Wind spacecraft during the period 1700-2100 UT on December 27, 1994 and plotted using the GSE system of coordinates. The Wind location is $R_{IMP}$ = (22.2, -53.8, -1.3) $R_E$ in GSE coordinates.

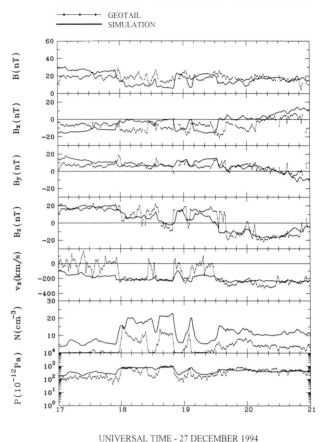

**Figure 2.** Time series from the global MHD simulation (dotted traces) and measurements (solid traces) of the magnetic field and ion plasma parameters from the MGF and CPI experiments onboard the Geotail spacecraft during the period 1430-1730 UT on July, 7, 1993. The data are plotted using the GSE system of coordinates. From 1700 UT until about 1800 UT, the Geotail spacecraft is for the most part skimming the magnetopause from its magnetospheric side. The density jump observed by Wind around 1735 UT reaches the magnetopause at about 1800 UT, pushing it inward. From that time until the end of the interval shown, Geotail stays mostly on the magnetosheath side of the magnetopause, except for two short and one long incursions into the magnetosphere. The Geotail location at 1940 UT was $R_G$ = (1.2, 14.9, 0.1) $R_E$ in GSE coordinates.

Z-component of the magnetic field. The simulation tracks are also in good accord with the plasma parameters, though the computed densities are somewhat higher than the actual values. Discrepancies observed in the X-component of the bulk flow velocity prior to 1800 UT are not of too much concern. Indeed, the large fluctuations in the flow velocities observed by the spacecraft during that period do not correlate with any feature in the ion density or magnetic field observations. It is thus unlikely that they result from the large scale motion of the boundary and are probably due to local turbulence on scale lengths much smaller than the simulation grid size in that region (0.28 $R_E$). Furthermore, the very low plasma densities in the boundary layer may have affected some of the measurements of the bulk flow.

Plate 1a shows color coded contour plots of the plasma density in the equatorial plane that were computed for time 1836 UT. To emphasize details in the distribution of densities, the maximum range of these contours was constrained to 12 cm$^{-3}$. The simulation results unambiguously show the formation and the convection of plasma inhomogeneities into the magnetosheath. Comparison with the time series from the Wind spacecraft indicates that these magnetosheath fluctuations are created by variations in the solar wind density. For example, the low density plasma shown by the green and yellow contours observed in the magnetosheath corresponds to the density drop observed around 1810 UT by the Wind spacecraft, while high densities shown by the red contours in the flank and the subsolar magnetosheath are associated with the high densities observed before and after the drop. Similar contours plots (not shown here) obtained at different times in the simulation show that the response of the boundary to these density fluctuations is consistent with Geotail time series. In addition, as is also discernable in Plate 1a, the tailward convection of the gusts of plasma in the magnetosheath appears as a large amplitude wave propagating on the surface of the magnetospheric boundary. The detailed study of the boundary response to these pressure pulse-like events [e.g. *Sibeck*, 1992] is currently under study and will be reported elsewhere.

Plate 1b displays a top view of the equatorial plane on which we have plotted color coded contours of the magnetic field topology (MFT) deduced from simulation results for time 1836 UT. The Geotail trajectory during the 1700-2100 UT interval is shown by the white trace with red spheres (the black trace represents the trajectory during the 17 hours prior to the interval), and the circular red area around the Earth represents the spherical shell used as the inner boundary of the model. MFT contours are calculated by tracing the field line passing through each grid point of the simulation domain and determining whether or not the field line is connected to the Earth. We also resolve the issue of whether or not a connected field line is connected at both ends, i.e. whether the field line is closed or open. Results are then mapped back to each grid point using color coding. Areas shown in blue, green and yellow in Plate 1b thus indicate regions threaded by closed, open and unconnected field lines respectively. In addition, the intersection of the three colors observed in the dawn sector showing that the three type of topologies (open, close and unconnected) are present, implies that reconnection is occurring in that region. The peculiar diagonal shape of the unconnected (yellow) field lines results from the relatively large X and Y-components of the IMF during the time interval.

Another interesting feature revealed by the MFT contours shown in Plate 1b is the thin layer of open field lines (green area) in front of the magnetopause. Panels (a) and (b) of Plate 2 show respectively a front view (looking towards the tail) and a top view of some of the open field lines (in yellow) threading that region. Red spheres indicate the location of Geotail, and closed field lines are shown in blue. These plots clearly show that the formation of the open layer identified in Plate 1b results from the reconnection of closed field lines with magnetosheath field lines draping over the dayside magnetopause. Characteristic bends in the field lines traced in the morning sector indicate unambiguously that the southern merging line comes near the ecliptic plane in the dawn sector, in agreement with the diagnostics from the MFT plots. Such a merging site is consistent with patterns proposed for antiparallel merging at the dayside magnetopause for northward IMF with a strong $B_X$ component [*Crooker*, 1979; *Luhmann et al.*, 1984]. Indeed, correlation of the features observed by Wind and Geotail indicates that the time delay between the two spacecraft is about 26 minutes. Hence Plate 1 and 2 correspond to the interaction of the IMF measured around 1810 UT, i.e. $B_{IMF}$ = (5, 4, 2) nT.

The top view (Plate 2b) also shows that the draping of the magnetosheath field over the afternoon and dusk sectors accounts for the negative X component of the magnetic field observed by Geotail. This is expected to occur because the IMF has a large Y component. Something less expected, though, is the warping of the field lines in the dusk magnetosheath because of the X component of the IMF. An immediate consequence of the large reversal of the field is the creation of a high-density region with a low magnetic field, i.e. high beta plasma, in the dusk flank of the magnetosheath. Since the field lines are open, such a configuration can explain the presence of a relatively steady boundary layer populated by a mixture of magnetospheric and solar wind plasmas. Furthermore, because the IMF is mostly steady during the time interval considered, the open field lines shown in Plate 2 can been viewed as the time history of the motion of the field lines. They show that a field line that is reconnected on the dawnside convects over the entire dayside boundary because of the field tension and the tailward motion of its open end in the dusk flank of the magnetosheath. An interesting consequence of this transport of plasma from dawn to dusk is that it suggests sunward flows in the dawn and pre-noon regions of the boundary layer identified above. Further studies involving spacecraft in both dawn and dusk regions will be necessary to confirm such a convection pattern during steady solar wind conditions marked by large X and Y IMF components.

## 4. DISTANT MAGNETOTAIL BOUNDARY

The period of interest in this comparative study is 1330-1730 UT on July 7, 1993. On that day at 1500 UT, IMP-8 and Geotail were positioned at $R_{IMP}$ = (18.0, -27.9, 15.2) $R_E$

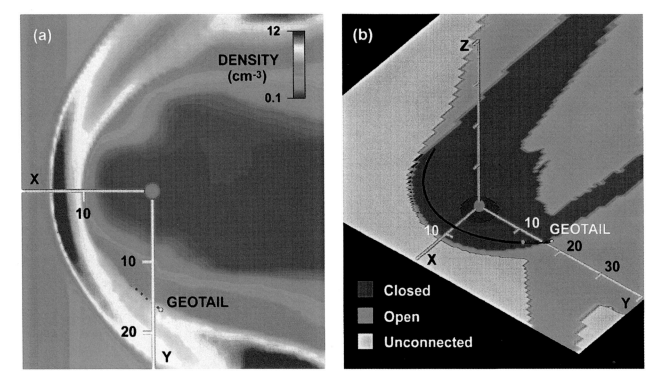

**Plate 1.** Panel (a) shows color coded contour plots of the plasma density in the equatorial plane that were computed using the simulation results for time 1836 UT. To emphasize details in the distribution of densities, the maximum range of these contours was constrained to 12 per cm$^3$. Comparison with the Wind time series shown in Figure 1 indicates that the plasma inhomogeneities seen in the magnetosheath correspond to solar wind fluctuations. Panel (b) shows a top view of the equatorial plane in which we have plotted color coded contours of the magnetic field topology (MFT) derived from the simulation for time 1836 UT. The Geotail trajectory during the 1700-2100 UT interval is shown by the white trace with red spheres (the black trace represents the trajectory during the 17 hours prior to the interval); the circular red area around the Earth indicates the spherical shell used as the inner boundary of the model. MFT contours are calculated by tracing the field line passing through each grid point of the simulation domain. Areas shown in blue, green and yellow indicate regions threaded by closed, open and unconnected field lines respectively. The intersection of the three colors observed in the dawn sector, indicates that reconnection is occurring in that region. The peculiar diagonal shape of the unconnected (yellow) field lines results from the relatively large X and Y-components of the IMF during the time interval.

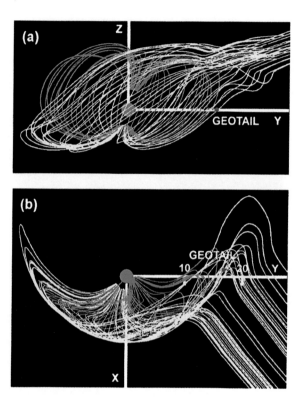

**Plate 2.** Panel (a) shows a front view (looking towards the tail) and Panel (b) a top view of some of the open field lines (in yellow) and closed field lines (in blue) threading the dayside magnetosphere for time 1836 UT. The Geotail trajectory during the 1700-2100 UT interval is shown by the white trace with red spheres. These plots clearly show the formation of an open layer resulting from the reconnection of closed field lines with magnetosheath field lines draped over the dayside magnetopause. Characteristic bends in the field lines traced in the morning sector indicate that reconnection occurs in the dawn region. Because of the X component of the IMF the field lines are considerably warped in the dusk magnetosheath.

and $R_G = (-196.1, -0.5, 8.2)$ $R_E$ respectively. Figure 3 displays IMP-8 ion and magnetic field observations for July 7, 1993 from 1200 UT to 2400 UT. The solar wind plasma parameters and the field intensity remain fairly steady throughout the 1330-1630 UT time period that was used as input to our global MHD-ionosphere model. The most important feature of this time period is the large rotations of the IMF clock angle. From being almost due north at around 1330 UT, the IMF rotates slowly towards dusk ($B_Y > 0$) until 1420 UT and then oscillates back toward a northward direction, with the result that the field is again almost due north around 1502 UT and 1540 UT. Notice also that the $B_X$ component remains relatively constant until 1540 UT.

Figure 4 shows time series extracted from the simulation superimposed over Geotail measurements. The time period

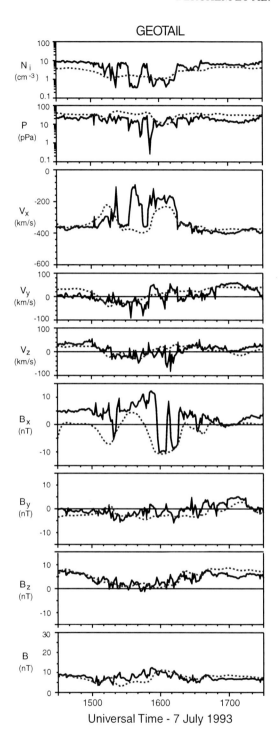

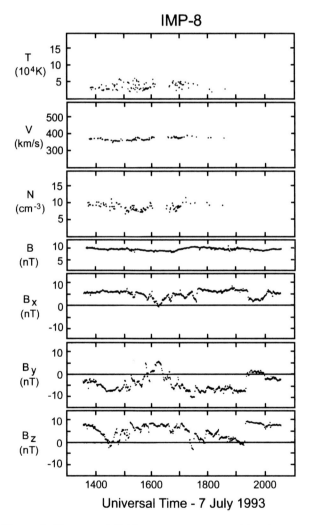

**Figure 3.** Plasma and field parameters measured by the IMP-8 spacecraft during the period 1300-2100 UT on July 7, 1993, and plotted using the GSE system of coordinates. The IMP-8 location is $R_{IMP} = (18.0, -27.9, 15.2)$ $R_E$ in GSE coordinates.

**Figure 4.** Times series from the global MHD simulation (dotted traces) and Geotail measurements (solid traces) during the period 1430-1730 UT on July 7, 1993. The observation of a tailward magnetic field component ($B_X < 0$) is quite surprising when it is considered that the Geotail spacecraft is located well above the equatorial plane. $R_G = (-196.1, -0.5, 8.2)$ $R_E$ in GSE coordinates.

256 GLOBAL MHD SIMULATION OF THE MAGNETOSPHERIC BOUNDARY

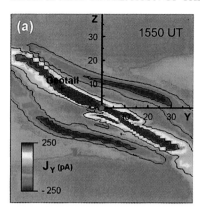 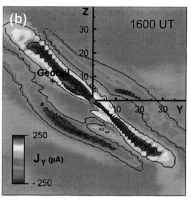 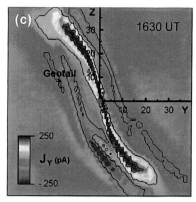

**Plate 3.** Cross-sectional cuts of the Y component of the current density ($J_Y$) taken at the spacecraft X location ($X_S = -196\ R_E$). The black cross marks the position of the Geotail spacecraft. Note that this position has been shifted about 16 $R_E$ toward dawn from the actual YS GSE location to allow for the aberration resulting from the Earth's orbital motion. Color coded contours are calculated for time 1550 UT (a), 1600 UT (b), and 1630 UT (c) and are displayed using the GSE system of coordinates. We have used a low value for the upper cut off of the current density ($J_Y$) to reveal the current structure of the tail. Red contours ($J_Y > 0$) give us a good indication of the location of the current sheet while blue contours ($J_Y < 0$) show the magnetopause current layer.

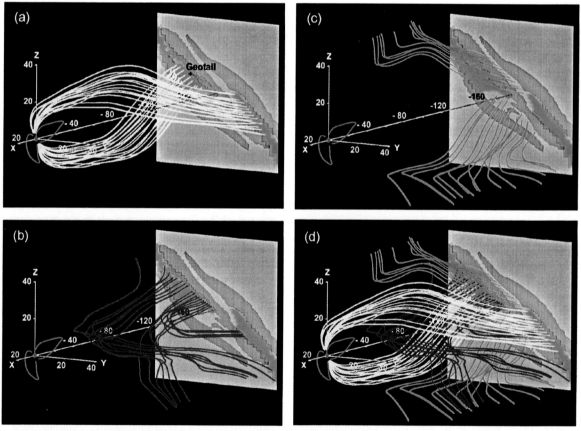

**Plate 4.** Three-dimensional renderings of the magnetosphere using magnetic field lines calculated for time 1600 UT. The magneto-sphere is viewed from the dusk side, and MFT contours are displayed as a cross section at $X = -196\ R_E$. Semi-opaque layers of the current density contours displayed in Plate 3(b) have been superimposed over the MFT plot to show areas of the cross section where the current density, $J_Y$, is non-zero. Each of the 4 groups of field lines displayed has been color coded to help distinguish the different types of topology encountered. The simplest group, shown in blue, consists of closed field lines (Panel a). Yellow field lines represent open field lines (Panel a) while the red (Panel b) and green (Panel c) field lines indicate newly unconnected solar wind field lines that have been reconnected in the center of the tail and at the high-latitude magnetopause respectively. Panel (d) is a combination of Panels (a), (b), and (c). It shows the flat braiding of the lobe field lines along the axis of the tail, with the plane of the braid lying in the direction of the IMF.

faster oscillations of the IMF clock angle. Another interesting feature not visible here but easily deduced from the IMF orientation observed by IMP-8 (see *Berchem et al.* [1997]) is that the east-west axis of the current sheet remains roughly aligned along the direction of the clock angle component of the IMF during its rotation.

Plate 4 shows three-dimensional renderings of the magnetosphere using magnetic field lines calculated for time 1600 UT. The plate has four panels labeled (a) to (d). Panels (a), (b) and (c) show individual groups of field lines cutting the cross-sectional plane passing by the location of Geotail, while Panel (d) is a composite of Panels (a), (b) and (c). Before discussing the group of field lines displayed in each panel, we use Panel (a) to present the common format used to display them. The magnetosphere is viewed from the dusk side, and contours of the magnetic field topology (MFT) are displayed as a cross section at $X = -196$ $R_E$. In addition, we have superimposed over the MFT plot semi-opaque layers that show areas of the cross section where the current density, $J_Y$, is non-zero. These layers correspond to the current density contours displayed in Plate 3(b). Although they are wider than the actual current layers, they give us a good indication of the locations of the magnetopause and the current sheet. The composite plot allows us to delineate more precisely the different areas of the tail cross section. In particular, it reveals that about half of the tail lobes are in fact threaded by field lines that are reconnected to the solar wind (yellow areas). This important feature will become more evident when the subsequent panels are considered. Note that Geotail's location, marked by a black cross, has been indicated in Panel (a) for clarity.

There are 4 groups of field lines displayed in the panels of Plate 4. Each group has been color coded to help distinguish the different types of topology encountered. Starting from the top of Plate 4, Panel (a) shows two groups of field lines. The simplest group, shown in blue, consists of closed field lines. These field lines delineate clearly the near-Earth plasma sheet and the dayside magnetosphere. They also reveal the high elevation of the plasma sheet above the ecliptic plane. This is due to the large inclination of the magnetic field dipole axis (22°) that is expected for a time period near the summer solstice. Above and below the closed field lines lies the second group of field lines, shown in yellow. These field lines are obviously open, as indicated by their threading of the green areas of the MFT cross section, and thus belong to the tail lobes. It is readily seen on this three-dimensional rendering that field lines originating from the upper left region of the green area are connected to the southern polar cap, whereas those from the lower right are connected to the southern polar cap. This indicates that the open field lines were formed by reconnection of dayside closed field lines with northward directed IMF field lines at the high-latitude magnetopause and have been convected along the flanks of the magnetosphere [*Russell*, 1972; *Maezawa*, 1976; *Crooker*, 1979; *Cowley*, 1982; *Reiff and Burch*, 1985]. The global picture shows that the two bundles of field lines are twisted around the X-axis. This configuration explains the inversion of the lobes that we noted earlier when discussing the two-dimensional cross section plots. The third group of field lines is shown in red in Panel (b). The yellow color coding of the areas threaded by these field lines and the field lines' curliness indicate that they are "unconnected" field lines resulting from reconnection at the center of the tail. A closer look at Panel (b) shows that there are two distinct sets. One set is composed of field lines threading the central part of the current sheet; the second set consists of field lines that emanate from some of the small yellow islands seen inside the green areas of the MFT cross section. It is interesting to note that the kinks observed in these field lines are sharper and closer to Earth than those of the first set, suggesting that they reconnected more recently than the first set.

The field lines shown in green on Panel (c) make up the fourth and last group. As are those in the previous group, they are unconnected field lines. However, these field lines are former lobe field lines that have been reconnected with solar wind field lines at the high-latitude magnetopause. They are the complement of the lobe field lines shown in yellow in Panel (a) and thus have the same twist. An interesting point is that these unconnected field lines cross the 200 $R_E$ plane in a region that one might expect to be filled by open field lines from the lobes. Nevertheless, since it takes at least 40 minutes for the information to propagate from the reconnection region to the location where the MFT cross section is computed, the plasma observed in that cross section still has the characteristics of the tail lobes. Such a configuration seems to be the counterpart of the magnetospheric asymmetries associated with the Y-component of a southward directed IMF. When this occurs, flux is added preferentially to the dusk side of the northern lobe and to the dawn side of the southern lobe for IMF $B_Y < 0$ [*Cowley*, 1981]. Results from the simulation suggest that for a northward IMF with $B_Y < 0$, flux is removed preferentially from the dawn side of the northern lobe and from the dusk side of the southern lobe.

The final global picture composed of the four groups of field lines is quite remarkable (Panel d). It shows how the combination of reconnection, both in the center of the tail and at the high-latitude magnetospheric boundary, and the slow rotation of the IMF east-west component result in an intricate enmeshment of the field lines of the distant tail. As we noted when presenting the two-dimensional results, though, the resulting structure is not one given by a simple rotation or twist of the distant tail along its axis. The seemingly complex configuration is in reality quite simple compared to the three-dimensional topology of the lobes shown in Panel (a). Panel (a) clearly shows that open field lines, initially created by the high-latitude reconnection of closed field lines with northward IMF, cross each other in the distant tail. Obviously this is because the ends of the field lines connected to the northern hemisphere are located south of the equatorial plane and vice versa for the southern lobe. As their open ends start convecting tailward, the dipole field in the near-Earth region prevents the field lines from penetrating, and they simply drape over the magnetosphere to form the tail flank boundary layer [*Song and Russell*, 1992; *Raeder et al.*, 1997]. Further down the distant tail, as the magnetospheric field weakens, the strong draping of the magnetosheath field lines pushes these field lines towards the center in the tail. Eventually they become lobe field lines, and the most central ones reconnect in the cross-over region to form the new interplanetary field lines shown in red in Panel (b). In the process, adjacent lobe field lines also reconnect to the IMF at the high-latitude magnetopause (green field lines of Panel c) to form new interplanetary field lines which also cross the center of the tail. The final picture of the distant tail thus looks like a flat braiding of the lobe field lines along the axis of the tail, with the plane of the braid lying in the direction of the IMF.

## 5. DISCUSSION AND SUMMARY

The two studies reviewed in this paper reveal several interesting points. First, they show very good agreement between time series extracted from the simulations and measurements from the Geotail spacecraft. Although a few discrepancies were found, the simulations correctly reproduced most of the observed features of the large scale motion of the dayside magnetospheric boundary, as well as the complex structure and dynamics of the distant tail 200 $R_E$ down stream from the Earth. The very good agreement found in the initial study of the large scale motion of the magnetotail boundary at a downstream distance of about 81 $R_E$ [*Frank et al.*, 1995] and in the more recent investigation of the formation of a broad boundary layer in the middle tail (X = -46 $R_E$) during a period of northward IMF [*Raeder et al.*, 1997] have already confirmed the validity and the power of the approach. The two comparisons presented here reinforce our confidence that the simulation model is able to reproduce the large scale motion of the magnetospheric boundary from the subsolar magnetopause to 200 $R_E$ down stream from the Earth. These studies also demonstrate the complexity of the magnetospheric configurations driven by simple changes in the solar wind. Analysis of the simulation results for Geotail's pass on December 27 1994, established that the spacecraft was skimming a region of the dusk boundary layer that contained mostly open field lines which had been reconnected in the dawn region. The simulation also revealed that, because of the relatively large and steady $B_X$ and $B_Y$ components of the IMF during the interval investigated, the draped field lines were considerably warped in the dusk region, creating a region of high beta plasma (weak magnetic field) in the surrounding magnetosheath. The study of the structure and dynamics of the distant magnetotail boundary at the Geotail apogee on July 7 1993, showed that the distant tail was severely flattened by the pressure of the magnetosheath field. It also demonstrated that for predominantly northward IMF the combination of reconnection and the torque imposed by slow rotations in the IMF clock angle result in a complicated braiding of the magnetotail field lines. In particular, the study predicted accurately the occurrence of the rapid crossing of the southern tail lobe at an unexpected location.

As we mentioned in the introduction, using solar wind measurements as input parameters to three-dimensional global MHD simulation models and making direct comparisons between simulated time series and observations constitute a new approach to global modeling. At the beginning of the mission it was thought that running simulations for generic steady solar wind conditions would be enough to link the measurements from the ISTP spacecraft that are needed to measure the transport of mass, momentum and energy through the magnetosphere. However, as we pointed out above, even simple changes in the solar wind conditions produce dramatic changes in the magnetic field configuration of the magnetotail. An obvious reason for such a response is that the magnetosphere has a memory, i.e. the state of the magnetosphere at a given time depends on its states prior to that time. For example, the twist of the lobe field lines observed by Geotail on July 7 1993, is related to the reconnection that occurred at the high-latitude magnetopause more than 60 minutes prior to the spacecraft observations. It is thus not surprising that actual solar wind input to the magnetosphere is needed to model accurately its time evolution.

Despite the very good results already obtained, considerably more must be achieved before we will be able to predict the state of the magnetosphere for any solar wind conditions. To date most of the comparative studies that we have carried out involve time periods when the IMF was predominantly northward. Although studies including time periods with southward IMF conditions are more challenging, preliminary results of the global modeling substorm events are encouraging [*Raeder et al.*, 1996b]. Furthermore, global simulations have limitations inherent in the single fluid MHD description and thus preclude the description of phenomena that result from the identity of individual particles. The self-consistent global particle models needed to resolve these microscopic processes and produce quantitative predictions are far beyond our present computing capabilities. Nevertheless, the recent large scale kinetics modeling that uses time varying electromagnetic fields from global MHD simulations [*Ashour-Abdalla et al.*, 1997a, b; *El-Alaoui et al.*, 1997] have proved that such simulations can provide the basic topology required to gain significant insights into the history and behavior of the plasma distributions through the magnetosphere. Although much work remains to attain a satisfactory representation of all the physical processes involved, it is obvious that the progress in global modeling accomplished during the ISTP program has contributed tremendously to the development of accurate predictions of magnetospheric phenomena.

*Acknowledgments.* The IMP-8 and WIND plasma measurements in the solar wind were kindly supplied by A. J. Lazarus and K. I.

Paularena (CSR, Massachusetts Institute of Technology) and by K. Ogilvie (NASA/Goddard Space Flight Center) respectively. Computations were performed at the Cornell Supercomputer Center. This research was supported in part at the University of California, Los Angeles, under NASA grants NAGW-1100, NAGW-4541, and NAGW-4543. At the University of Iowa this work was supported under NASA grant NAG5-2371. UCLA-IGPP publication #4877.

## REFERENCES

Ashour-Abdalla, M., M. El-Alaoui, V. Peroomian, J. Raeder, L. A. Frank, W. R. Paterson, J. M. Bosqued, Determination of particle sources from observed distribution functions, this volume, 1997.

Ashour-Abdalla, M., M. El-Alaoui, V. Peroomian, J. Raeder, R. J. Walker, R. L. Richard, L. M. Zelenyi, L. A. Frank, W. R. Paterson, J. M. Bosqued, R. P. Lepping, K. Ogilvie, S. Kokubun, T. Yamamoto, Ion sources and acceleration mechanisms inferred from local distribution functions, *Geophys. Res. Lett., in press*, 1997.

Berchem, J., J. Raeder, and M. Ashour-Abdalla, Reconnection at the magnetospheric boundary: Results from global magnetohydrodynamic simulation, in *Physics of the Magnetopause*, edited by P. Song, B. U. Ö. Sonnerup and M. F. Thomsen, Geophys. Monograph, p. 205, American Geophys. Union, Washington, D. C., 1995a.

Berchem, J., J. Raeder, and M. Ashour-Abdalla, Magnetic flux ropes at the high-latitude magnetopause, *Geophys. Res. Lett., 22*, 1189, 1995b.

Berchem, J., J. Raeder, M. Ashour-Abdalla, L. A. Frank, W. R. Paterson, K. L. Ackerson, S. Kokubun, T. Yamamoto, R. P. Lepping, and K. Ogilvie, WIND/GEOTAIL comparative studies of the dayside magnetospheric boundary: Initial results from global MHD modeling, *EOS, 77*, 241, 1996.

Berchem, J., J. Raeder, M. Ashour-Abdalla, L. A. Frank, W. R. Paterson, K. L. Ackerson, S. Kokubun, T. Yamamoto, and R. P. Lepping, The distant tail at 200 $R_E$: Comparison between Geotail observations and the results from a global magnetohydrodynamic simulation, *J. Geophys. Res., in press*, 1997.

Crooker, N. U., Dayside merging and cusp geometry, *J. Geophys. Res., 84*, 951, 1979.

Cowley, S. W. H., Magnetospheric and ionospheric flow and the interplanetary magnetic field, in *The Physical Basis of the Ionosphere in the Solar-Terrestrial System*, pp 4-11, AGARD, Neuilly, France, 1982.

Cowley, S. W. H., Magnetospheric asymmetries associated with the Y-component of the IMF, *Planet. Space Sci., 29*, 79, 1981.

El-Alaoui, M., M. Ashour-Abdalla, J. Raeder, V. Peroomian, L. A. Frank, W. R. Paterson, and J. M. Bosqued, Understanding the structure in the magnetotail ion distribution by using global MHD simulation and ion trajectory calculations, this volume, 1997.

Fairfield, D. H., Waves in the vicinity of the magnetopause, Average and unusual locations of the earth's magnetopause and bow shock, in *Magnetospheric Particles and Fields*, edited by B. M. McCormac, pp 67-77, D. Reidel, Boston, MA, 1976.

Fedder, J. A., and J. G. Lyon, The Earth's magnetosphere is 165 $R_E$ long: Self-consistent currents, convection, magnetospheric structure and processes for northward interplanetary magnetic field, J. Geophys. Res., 100, 3623, 1995.

Fedder, J. A., J. G. Lyon, S. P. Slinker, and C. M. Mobarry, Topological structure of the magnetotail as a function of the interplanetary field, *J. Geophys. Res., 100*, 3613, 1995a.

Fedder, J. A., S. P. Slinker, J. G. Lyon, and R. D. Elphinstone, Global numerical simulation of the growth phase and the expansion onset for a substorm observed by Viking, *J. Geophys. Res., 100*, 19083, 1995b.

Frank, L. A., K. L. Ackerson, W. R. Paterson, J. A. Lee, M. R. English and G. L. Pickett, The Comprehensive Plasma Instrument (CPI) for the GEOTAIL spacecraft, *J. Geomag. Geoelectr., 46*, 23, 1994.

Frank, L. A., M. Ashour-Abdalla, J. Berchem, J. Raeder, W. R. Paterson, S. Kokubun, T. Yamamoto, R. P. Lepping, F. V. Coroniti, D. H. Fairfield, and K. L. Ackerson, Observations of plasmas and magnetic fields in Earth's distant magnetotail: Comparison with a global MHD model, *J. Geophys. Res., 100*, 19,177, 1995.

Gosling, J. T., D. N. Baker, S. J. Bame, E. W. Hones Jr., D. J. McComas, R. D. Zwickl, J. A. Slavin, and E. J. Smith, Plasma entry in the distant tail lobes: ISEE-3, *Geophys. Res. Lett., 11*, 1078, 1984.

Kokobun, S., T. Yamamoto, M. H. Acuña, K. Hayashi, K. Shiokawa, and H. Kawano, The GEOTAIL magnetic field experiment, *J. Geomag. Geoelectr., 46*, 7, 1994.

Leboeuf, J. N., T. Tajima, C. F. Kennel, and J. M. Dawson, Global simulation of the time dependent magnetosphere, *Geophys. Res. Lett., 5*, 609, 1978.

Leboeuf, J.N., T. Tajima, C. F. Kennel, and J. M. Dawson, Global simulation of the three-dimensional magnetosphere, *Geophys. Res. Lett., 8*, 257, 1981.

Lyon, J. G., S. H. Brecht, J. A. Fedder, and P. J. Palmadesso, The effects on the Earth's magnetotail from a shock in the solar wind, *Geophys. Res. Lett., 7*, 721, 1980.

Lyon, J. G., S. H. Brecht, J. D. Huba, J. A. Fedder, and P. J. Palmadesso, Computer simulation of a geomagnetic substorm, *Phys. Rev. Lett., 46*, 1038, 1981.

Luhmann, J., G., R. J. Walker, C. T. Russell, N. U. Crooker, J. R. Spreiter, and S. S. Stahara, Patterns of potential magnetic field merging sites on the dayside magnetopause, *J. Geophys. Res., 89*, 19, 1739, 1984.

Maezawa, K., Magnetic convection induced by the positive and negative z-component of the interplanetary magnetic field: Quantitative analysis using polar cap magnetic records, *J. Geophys. Res., 81*, 2289, 1976.

Mobarry, C. M., J. A. Fedder, and J. G. Lyon, Equatorial plane convection from simulations of the Earth's magnetosphere, *J. Geophys. Res., 101*, 7859, 1996.

Paterson, W., and L. A. Frank, Survey of plasma parameters in Earth's distant magnetotail with the GEOTAIL, *Geophys. Res. Lett., 21*, 2971, 1994.

Raeder, J., R. J. Walker, and M. Ashour-Abdalla, The structure of the distant geomagnetic tail during long periods of northward IMF, *Geophys. Res. Lett., 22*, 349, 1995.

Raeder, J., J. Berchem, and M. Ashour-Abdalla, The importance of small scale processes in global MHD simulations: Some numerical experiments, in *Physics of Space Plasmas*, edited by T. Chang and J. R. Jasperse, MIT Center for Theoretical Geo/Cosmo Plasma Physics, MIT, Massachusetts, 14, p. 403, 1996a.

Raeder, J., J. Berchem, M. Ashour-Abdalla, Global MHD simulation of the May 19/20 1996 substorm event and comparisons with ISTP/GGS observations, EOS, 77, 639, 1996b.

Raeder, J., J. Berchem, M. Ashour-Abdalla, L. A. Frank, W. R. Paterson, K. L., Ackerson, S. Kokubun, T. Yamamoto, and J. A. Slavin, Boundary layer formation in the magnetotail: Geotail observations and comparisons with a global MHD simulation, *Geophys. Res. Lett., in press*, 1997.

Reiff, P. H., and J. L. Burch, IMF $B_y$-dependent plasma flow and

Birkeland currents in the dayside magnetosphere 2. A global model for northward and southward IMF, *J. Geophys. Res., 90*, 1595, 1985.

Russell, C. T., The configuration of the magnetosphere, in *Critical Problems of Magnetospheric Physics*, edited by E. R. Dyer, pp. 1-16, IUCSTP Secretariat, National Academy of Sciences, Washington, D. C., 1972.

Sato, T., and T. Hayashi, Externally driven magnetic reconnection and a powerful magnetic energy converter, *Phys. Fluids, 22*, 1189, 1979.

Sibeck, D. G., Transient events in the outer magnetosphere: Boundary wave or flux transfer events, *J. Geophys. Res., 97*, 4009, 1992.

Sibeck, D. G., G. L. Siscoe, J. A. Slavin, E. J. Smith, B. T. Tsurutani, and R. P. Lepping, The distant magnetotail's response to a strong interplanetary magnetic field By: twisting, flattening, and field line bending, *J. Geophys. Res., 90*, 4011, 1985.

Sibeck, D. G., and J. T. Gosling, Magnetosheath density fluctuations and magnetopause motion, *J. Geophys. Res., 101*, 31, 1996.

Siscoe, G. L., L. A. Frank, K. L. Ackerson, W. R. Paterson, Properties of the mantle like magnetotail boundary layer: GEOTAIL data compared with a mantle model, *Geophys. Res. Lett., 21*, 2975, 1994a.

Siscoe, G. L., L. A. Frank, K. L. Ackerson, W. R. Paterson, Irregular, long-period boundary oscillations beyond ~100 $R_E$: GEOTAIL plasma observations, *Geophys. Res. Lett., 21*, 2979, 1994b.

Slavin, J. A., B. T. Tsurutani, E. J. Smith, D. E. Jones, and D. G. Sibeck Average configuration of the distant tail (<220 $R_E$) magnetotail: ISEE-3 initial magnetic field results, *Geophys. Res. Lett., 10*, 973, 1983.

Slavin, J. A., E. J. Smith, D. G. Sibeck, D. N. Baker, R. D. Zwickl, and S.-I. Akasofu, An ISEE 3 study of average and substorm conditions in the distant tail, *J. Geophys. Res., 90*, 10,875, 1985.

Song, P., and C. T. Russell, Model of the formation of the low-latitude boundary layer for strongly northward interplanetary field, *J. Geophys. Res., 97*, 1411, 1992.

Tsurutani, B. T., D. E. Jones, and D. G. Sibeck, The two-lobe structure of the distant tail, *Geophys. Res. Lett., 11*, 1066, 1984.

Ugai, M., Temporal evolution and propagation of a plasmoid associated with asymmetric fast reconnection, *J. Geophys. Res., 90*, 9576, 1985.

Usadi, A., A. Kageyama, K. Watanabe, and T. Sato, A global simulation of the magnetosphere with a long tail; Southward and northward interplanetary magnetic field, *J. Geophys. Res., 98*, 7503, 1993.

Walker, R. J., and M. Ashour-Abdalla, The magnetosphere in the machine, large scale theoretical mode of the magnetosphere, IUGG Quadrennial Report, *Rev. of Geophys.*, supplement, 639, 1994.

Walker, R. J., and T. Ogino, A global magnetohydrodynamic simulation of the origin and evolution of magnetic flux ropes in the magnetotail, *J. Geomag. Geoelectr., 48*, 765, 1996.

---

J. Berchem, J. Raeder, and M. Ashour-Abdalla, Institute of Geophysics and Planetary Physics, University of California, Los Angeles, California 90095-1567.

L. A. Frank, W. R. Paterson, and K. L. Ackerson, Department of Physics and Astronomy, The University of Iowa, Iowa City, Iowa 52242.

S. Kokubun, Solar-Terrestrial Environment Laboratory, Nagoya University, 3-13 Honohara, Toyokawa, Aichi 442, Japan.

R. P. Lepping, NASA/Goddard Space Flight Center, Greenbelt, Maryland 20771.

T. Yamamoto, Institute of Space and Astronautical Science, Sagamihara, Kanagawa 229, Japan.

# Study of an Isolated Substorm with ISTP Data

A.T.Y. Lui[1], D.J. Williams[1], R.W. McEntire[1], S. Ohtani[1], L.J. Zanetti[1], W.A. Bristow[1], R.A. Greenwald[1], P.T. Newell[1], S.P. Christon[2], T. Mukai[3], K. Tsurada[3], T. Yamamoto[3], S. Kokubun[4], H. Matsumoto[5], H. Kojima[5], T. Murata[5], D.H. Fairfield[6], R.P. Lepping[6], J.C. Samson[7], G. Rostoker[7], G.D. Reeves[8], A.L. Rodger[9], H.J. Singer[10]

An isolated substorm of moderate intensity (AE index ~500 nT), which occurred on February 9, 1995 after an extended interval of strong northward interplanetary magnetic field was examined with data from eleven spacecraft in space (Wind, IMP 8, Geotail, six geostationary satellites, one DMSP satellite, and Freja) and two networks of ground stations (Canopus and SuperDARN) covering both the northern and southern hemispheres. The extensive coverage of this event provides us with results (1) confirming some expected substorm phenomena, (2) bringing out some unusual characteristics possibly related to the isolated nature of the substorm, and (3) revealing some surprising features difficult to reconcile with the traditional substorm model. In the first category are several features in the substorm growth phase (equatorward movement of the auroral ovals and intensification of the large-scale field-aligned current and the cross-tail current systems) and in the substorm expansion phase (breakup arc located well within the closed field line region, dipolarization and particle injection, strong plasma flow, complex magnetic field structure, enhanced dawn–dusk electric field, and plasma sheet thinning in the mid-tail region). In the second category are the unusually long duration of the growth phase and the long time delay between substorm expansion onset and particle injection onset at the geostationary altitude. In the third category are the new evidence for multiple particle acceleration sites during substorm expansion, for sunward plasma flow during the late expansion phase of a substorm being not related to a single acceleration site (X-line) moving from the near-Earth tail to the more distant tail, and for possible identification of the optical signature of a bursty bulk flow event observed in the mid-tail region.

[1]The Johns Hopkins University Applied Physics Laboratory, Laurel, MD 20723, USA
[2]The University of Maryland, College Park, MD 20742, USA
[3]Institute of Space and Astronautical Science, Sagamihara, Kanagawa 229, Japan
[4]Solar Terrestrial Environment Laboratory, Nagoya University, Toyokawa, Aichi 442, Japan
[5]RASC, Kyoto University, Kyoto, Japan
[6]NASA/GSFC, Greenbelt, MD, USA
[7]University of Alberta, Edmonton, Canada
[8]Los Alamos National Laboratory, Los Alamos, NM 87545, USA
[9]British Antarctic Survey, Madingley Rd., Cambridge CB3 0ET, United Kingdom
[10]NOAA Space Environment Center, 325 Broadway, Boulder, CO 80303, USA

## INTRODUCTION

One of the main objectives of the ISTP (International Solar Terrestrial Physics) program is to investigate the flow of energy, momentum, and mass from the Sun through the magnetosphere to the ionosphere and the atmosphere. Achieving this objective in the ISTP era has the distinct advantage over previous attempts because of the unprecedented multi-point measurements available and planned for ISTP activities. In this paper, we address this ISTP task using eleven spacecraft and two networks of ground stations from the ISTP data base.

Studying an isolated substorm is particularly appropriate to address this ISTP task because a large amount of energy, momentum, and mass from the solar wind pass through the magnetosphere during a substorm episode. Furthermore, studying an isolated substorm instead of a substorm embedded within a sequence of substorm disturbances allows one to eliminate the possible interference from preceding substorm activity. Consequently, a clearer identification of features genuinely associated with the substorm growth phase or the expansion onset can be made without the ambiguity introduced by the remnants of activity related to the recovery of a previous substorm. Before the onset of this substorm under study, a magnetic cloud passed over the Earth, engulfing the Earth's magnetosphere with a prolonged period of steady northward interplanetary magnetic field (IMF). This sets up an ideal situation for our study since the magnetosphere was then brought to a state with very little, if not the minimum, activity present. The extensive coverage of substorm activity during this event provides us with results which range from verifying some of our conventional expectations about substorm phenomena as well as revealing some surprises which are different from the anticipation of the traditional model for the substorm expansion and recovery development.

## OBSERVATIONS

The suite of data available for this event may be divided into four categories according to the region where measurements were taken. These categories are solar-wind/magnetosheath, ground-based/auroral-region, geostationary altitude, and magnetotail. Figure 1 shows the locations of the high-altitude spacecraft Wind, IMP 8, Geotail, and six geostationary satellites during this event. Wind was in the far upstream region at ~200 $R_E$ in front of the magnetosphere. IMP 8 was near the dusk meridian, initially in the solar wind but later entered the magnetosheath before the substorm onset. Geotail was

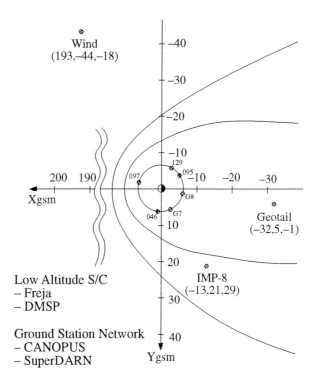

**Figure 1.** Spacecraft locations at the time of the substorm expansion onset (~0431 UT on February 9, 1995) for this study.

near the midnight meridian at X ~ −32 $R_E$. In the geostationary orbit were four LANL satellites (1984–129, 1987–097, 1989–046, and 1990–095), GOES-7, and GOES-8 distributed at various local times. Not shown in the sketch are two low-altitude satellites DSMP F11 and Freja. This suite of satellites makes a total of eleven spacecraft in space gathering data for this event. Supplementing these data are observations from two networks of ground stations, Canopus and SuperDARN, which cover both the northern and southern hemispheres. These constitute a rich data set, truly unprecedented in terms of coordinated observations.

### Solar Wind/Magnetosheath Observations

A good starting point to examine this event is from the Wind observation. The IMF and the solar wind plasma parameters from Wind are displayed in Figure 2 for February 8–9, 1995. The IMF turned northward at ~14 UT on February 8 and remained strongly northward

($B_z$ in the range of 2–10 nT) for an extended period with a brief ~2-hr break at ~02–04 UT on February 9 when $B_z$ was slightly negative. The IMF $B_y$ component was quite strong, in the neighborhood of ~4–12 nT throughout most of this period. The solar wind velocity was quite steady at the nominal speed of ~400 km/s. The solar wind dynamic pressure varied somewhat from ~1 to 4 nPa, mostly due to the variation in the solar wind number density. This is part of an interplanetary magnetic cloud studied earlier by *Lepping et al.* [1996].

IMP 8 observations showed similar characteristics of the solar wind as that seen by the Wind spacecraft in spite of the fact that IMP 8 was separated from Wind by ~65 $R_E$ in GSM Y-coordinate. There is, however, one major difference on the duration and magnitude of the southward IMF seen by the two satellites during this time interval. While Wind observed the brief southward excursion of IMF for less than 2 hrs, IMP 8 detected the southward IMF for a duration of ~4 hr, as shown in the bottom panel of Figure 2, which is about twice as long as that seen by Wind. This difference may be related to the bow shock crossings of IMP 8 as indicated by the abrupt changes in the $B_z$ component and in the other two magnetic field components (not shown) as well. IMP 8 entered the magnetotail during the later part of the day.

The time difference on the detection of southward IMF turning between Wind and IMP 8 is ~66 min. This is a bit longer than expected based solely on the solar wind flow speed and the separation distance of Wind and IMP 8 (206 $R_E$/(4 $R_E$/min) ≈ 51 min). However, this time delay would be consistent if the orientation of the solar wind IMF structure was aligned with the magnetic field orientation as pointed out by *Winglee et al.* [1997].

*Ground-Based/Auroral-Region Observations*

The brief southward turning of IMF led to an isolated substorm, as indicated in Figure 3. The IMF $B_z$ component from Wind and the solar wind energy transfer function, Akasofu's $\epsilon$ parameter [*Akasofu*, 1981], based on measurements from Wind are reproduced in the top two panels. These two parameters are time-shifted to reflect the estimated propagation time between Wind and the subsolar magnetopause. The slant lines below the second panel provide a guide to the time shift applied. We have adopted the solar wind parameter measured by Wind to calculate the $\epsilon$ parameter because (a) IMP 8 passed into the magnetotail during the later part of the day and provided no solar wind information, (b) IMP 8 crossed the bow shock several times and thus the solar wind parameters were modified by the bow shock, and (c) IMP 8

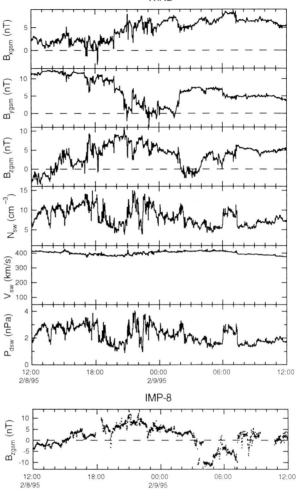

**Figure 2.** The solar wind parameters as monitored by WIND and IMP 8.

data have several data gaps while Wind data have none. The $\epsilon$ parameter, which measures the solar wind input energy to the magnetosphere, was quite small (below $10^4$ MW) before the southward turning of the IMF. Afterwards, the $\epsilon$ parameter reached to large values (above $10^5$ MW) where substorm activity is anticipated. The auroral kilometric radiation (AKR) index based on the observed kilometric radiation from Geotail is shown in the third panel. This index is defined as the logarithm in decibel of the AKR power flux in 50–800 kHz frequency range normalized at a fixed geocentric distance (25 $R_E$) [*Murata et al.*, 1997]. For this period, the AKR index at 64s resolution indicates an intensification of AKR emission starting at ~0431 UT, followed by another one at

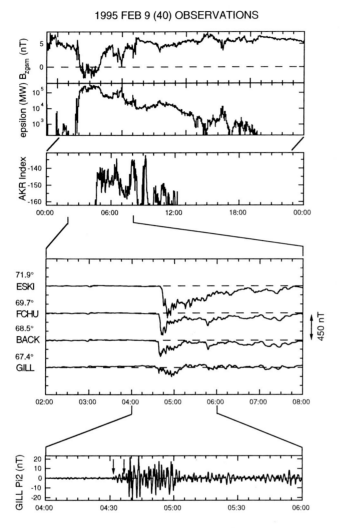

**Figure 3.** Shown here are the $B_z$ component in the solar wind (measured by WIND), the computed Akasofu's $\epsilon$ parameter, and the AKR index (in units of V/m/√(kHz) observed by GEOTAIL), magnetograms from Canopus, and Pi2 pulsation from Gillam on February 9, 1995. The two Pi2 onsets are indicated by arrows in the bottom panel.

~0435 UT. The onset time closely matched to the time of northward turning of IMF and thus the substorm might have been triggered by northward turning of IMF [*Rostoker*, 1983; *Lyons*, 1996]. The magnetic stations from Canopus, which were in the pre-midnight sector (~23.5 MLT at the substorm onset time), registered this isolated substorm activity. Shown in the middle of the figure are the magnetograms from Eskimos, Fort Churchill, Back, and Gillam, illustrating a negative bay of ~500 nT started near the time of the increase in the AKR index. There were two Pi2 micropulsation onsets detected at Gillam, the first one commencing at ~0431 UT and the next one at ~0436 UT, similar to the onset times indicated by the AKR index.

Observations from Meridian Scanning Photometer (MSP) chain in the Canopus network were assembled from Gillam and Rankine for the 5577 Å auroral emission in the top panel of Plate 1. The latitude refers to the Pace latitude [*Baker and Wing*, 1989]. At the beginning of the interval, the auroral luminosity was very low but was still noticeable at ~66°–75°. Starting at ~0300 UT, the auroral emission became enhanced and simultaneously moved equatorward. This equatorward motion is consistent with expected expansion of the auroral oval during the substorm growth phase in response to southward IMF. The brightening of auroral arc at substorm onset and the subsequent poleward expansion of auroral activity to form the auroral bulge are quite evident.

At the substorm onset, the auroral emission intensified tremendously at 66.5° and migrated rapidly poleward. Based on the 6300 Å emission (not shown here but is given in Figure 5 in *Winglee et al.*, 1997), the polar cap boundary was at least 72°, indicating that the substorm expansion onset took place well within the closed field line region, consistent with results from other earlier studies [*Lui and Burrows*, 1978; *Samson et al.*, 1992]. Noticeable also from the MSP data is the equatorward movement and detachment of some auroral luminosity from the main poleward edge of the auroral bulge, which started at ~0535 UT. It moved from ~71° to ~67.5° at ~0550 UT. The magnetograms from the north–south chain of Canopus indicated a similar and simultaneous equatorward movement of the westward substorm auroral electrojet.

The bottom panel of Plate 1 shows the ionospheric plasma convection speeds along the line of sight at F-region altitudes determined by radar at Goose Bay. Positive (negative) speed represents convection toward (away from) the radar station. At the start of the interval shown, the backscattered signal from the auroral electrojet indicated the auroral electrojet to be at ~68.5° Pace latitude. The equatorward movement of the auroral electrojet commenced at ~0300 UT, like the auroral luminosity seen by the Canopus MSP. By the end of the substorm growth phase, the auroral electrojet had moved ~3° equatorward, in agreement with the equatorward displacement of the 5577 Å auroral emission. The convection pattern within the radar field of view showed a development from one with mainly westward convection electric field to one with mainly northward convection

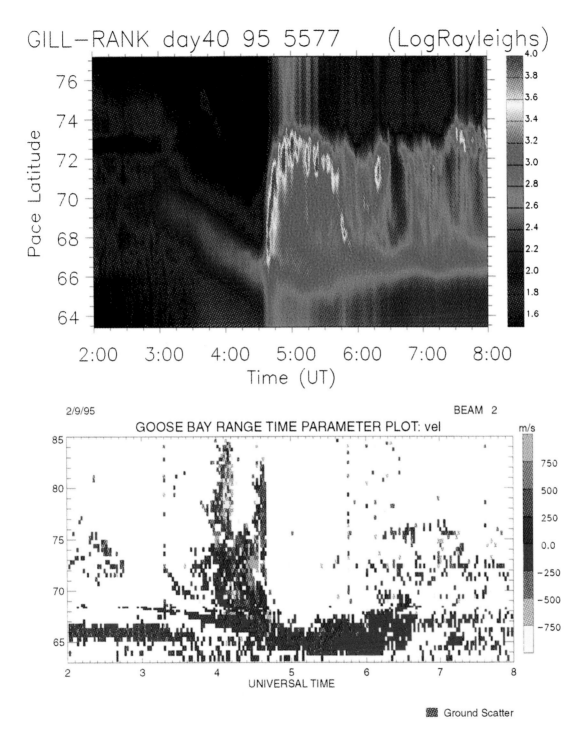

Plate 1. The top panel is constructed from Canopus stations Gillam and Rankine to show the time variation in the latitudinal profile of the 5577 Å auroral emission on February 9, 1995. The bottom panel shows the convection speeds along the line of sight from Goose Bay radar.

electric field. Although not shown here, the radar observation from Halley station (conjugate to Goose Bay) indicated equatorward movement of the backscattered signal (interpreted here as the auroral electrojet) from about −68° at 0200 UT to about −65° (also 3° equatorward displacement) at 0430 UT just before the substorm onset. These observations show the auroral electrojet in both hemispheres responded to southward turning of the IMF in a very similar manner.

DMSP F11 crossed the auroral oval in the dusk-midnight sector. Using the identification scheme of *Newell et al.* [1996], particle precipitation from this satellite indicated that the plasma sheet boundary (defined as the latitude where the electron average energy is neither increasing or decreasing with latitude) expanded from −73.8° Pace latitude (22.6 MLT) at 0008 UT to −67.7° (23.4 MLT) at 0334 UT. This lowering in latitude of the plasma sheet boundary is quite consistent with the equatorward movement of the auroral electrojets in both hemispheres.

Two passes of Freja over the auroral region are relevant to this study. The first one took place at 0224–0254 UT in which the satellite crossed the midnight meridian at 71° Pace latitude. The auroral activity was so low that no large-scale field-aligned current (FAC) was discernible. In the second pass at 0414–0444 UT, Freja crossed the auroral oval at ~03 MLT between 0433:55 and 0435 UT, just after substorm onset. An east-west magnetic field perturbation of ~120 nT at ~67–69° PACE latitude was observed. This low value implies the large-scale FACs to be still rather weak (but enhanced relative to the first pass) in the early morning local time a few minutes after the substorm onset. Thus, the substorm disturbance was still localized in MLT at this stage.

*Geostationary Altitude Observations*

At the geostationary altitude, there were six satellites distributed in local times of 3.4 hr (1990–095), 5.0 hr (1984–129), 11.4 hr (1987–097), 17.5 hr (1989–046), 19.5 hr (GOES-7), and 23.5 hr (GOES-8) at the time of substorm onset. As shown in Figure 4, dispersive energetic electron injections were seen by three of the four LANL satellites, namely, 1984–129 (LT = UT + 0.5), 1987–097 (LT = UT + 6.9), and 1989–046 (LT = UT − 11.0). For the fourth LANL satellite (1990–095), there was a data gap during this time. The onset time of electron injection based on the dispersive feature (i.e., extrapolating injection time to infinite energy particles) is found to be ~0444 UT, which is ~12–13 min delay with respect to the first Pi2 onset or AKR index

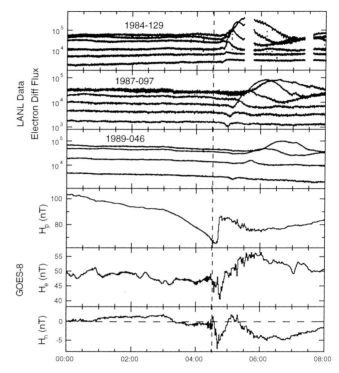

**Figure 4.** The observations at the geostationary altitude on February 9, 1995. The top three panels are from the LANL satellites. Energetic electron fluxes are stacked up (30–45 keV, 45–65 keV, 65–95 keV, 95–140 keV, 140–200 keV, and 200–300 keV for the top two panels; 50–75 keV, 75–105 keV, 105–150 keV, 150–225 keV for the third panel). The bottom three panels are magnetic field components measured by GOES-8. The $H_p$ component is positive northward parallel to the Earth's spin axis, $H_e$ is positive radially inward, and $H_n$ is positive eastward. The vertical dashed line marks the expansion onset time.

increase, and ~8 min delay with respect to the second Pi2 onset and AKR index increase.

The magnetic field measurements from GOES-8 (LT = UT − 5.0) are given in Figure 4 as well. GOES-8 showed a gradual depression of the $H_p$ component starting at ~03 UT, indicating tail field stretching near midnight coincident with the beginning of the equatorward expansion of the auroral oval. The magnetic field dipolarized at ~0440 UT, which is ~9 min delay from the first Pi2 onset and ~4 min delay from the second one. For GOES-7 (LT = UT − 9.0), only the $H_p$ component (not shown) remained useful. Nevertheless, it provided the information that GOES-7, which was 4 hr in LT to the west of GOES-8, also detected a substantial decrease in the $H_p$ component just like GOES-8 during the growth phase.

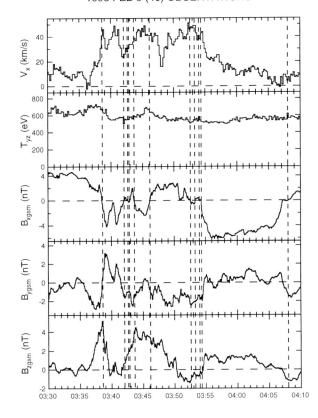

**Figure 5.** Geotail observations during the early part of the substorm growth phase on February 9, 1995. Vertical lines mark the neutral sheet crossings.

## Magnetotail Observations

For this event, Geotail was located slightly in the pre-midnight mid-tail region where magnetic reconnection is anticipated to occur at substorm expansion onset. During the early part of the growth phase (0330–0410 UT), Geotail was in the plasma sheet at ~(−32, 6, −1) $R_E$ in GSM coordinates and crossed the neutral sheet twelve times as shown in Figure 5. For the first six crossings (0335–0350 UT), the $B_z$ component was all positive. For the last six crossings (0350–0408 UT), the $B_z$ component was all negative, ranging in values from −0.25 to −1.2 nT. These last six crossings were associated with slow earthward plasma flow (< 50 km/s). The occurrence of negative $B_z$ in the mid-tail neutral sheet during the substorm growth phase may imply a mid-tail X-line formed earthward of Geotail location. If this is the case, then the X-line was formed ~40 min before expansion onset since the first negative $B_z$ in the neutral sheet was seen at ~0350 UT and the expansion onset was at 0431 UT. Furthermore, this X-line did not produce significant plasma flow at Geotail since very slow earthward (not tailward) plasma flow was seen during the negative $B_z$ interval.

The plasma, magnetic field, and electric field measurements from Geotail for a more extended period for the substorm are displayed in Figure 6. A noticeable sudden decrease in plasma density occurred at ~0415 UT, accompanied by a temperature increase and small disturbances in the electric and magnetic fields. These are probably the signatures of Geotail's entry into the plasma sheet boundary layer. At ~0425 UT, Geotail entered a low density region representative of the tail lobe. At ~0435 UT, ~4 min after the substorm onset, Geotail re-entered the plasma sheet boundary layer where a brief

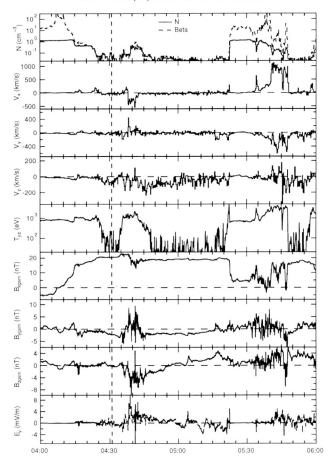

**Figure 6.** Geotail observations showing the plasma moments, magnetic and electric field measurements on February 9, 1995. Plasma beta is plotted in the same panel as the plasma number density. The vertical dashed line marks the expansion onset time.

interval of weak (~100 km/s) sunward flow was observed. Moderate (~300–500 km/s) tailward flow occurred immediately afterwards but it did not last for the entire period of Geotail's residence in the plasma sheet boundary layer (0435–0445 UT). Therefore, the tailward flow burst was extremely transient and/or localized. Plasma flow reversal occurred not just in the $x$-component but also in the $y$-component with a smaller magnitude and a shorter duration. The local magnetic field turned southward at ~0435 UT before the start of the weak sunward flow. The negative $B_z$ was accompanied by significant changes in the $B_y$ component, indicating that the field change was due to a three-dimensional field structure. Also, the negative $B_z$ was associated with enhancements of the dawn–dusk electric field to an average of several mV/m (up to 10 mV/m) in the duskward direction. The computed total magnetic and plasma pressure (not shown here but is given in Figure 1 in *Nagai et al.*, 1997) indicated a relatively steady increase from ~0.13 nPa at ~0335 UT to ~0.22 nPa at 0438 UT. This steady increase in total pressure is consistent with the expected increase in the tail lobe field strength during the growth phase. Since expansion onset occurred at 0431 UT, the pressure increase in the mid-tail as monitored by Geotail continued for ~7 min after the first Pi2 (substorm expansion) onset and for ~2 min after the second Pi2 onset in this event.

Plate 2 shows the ion and electron energy spectrograms from the Low Energy Plasma (LEP) instrument on Geotail [*Mukai et al.*, 1994]. Measurements are sorted for plasma coming from the four directions of dawnward, sunward, duskward, and antisunward. A careful examination shows that sunward–dawnward flow occurred at the ~0435 UT re-entry of the plasma sheet boundary. Interestingly, the strong tailward flow seen after re-entry was mainly from the high energy (> ~4 keV) portion of the ion population. The lower energy ion population was rather isotropic, much like what was observed prior to the plasma dropout at ~0425 UT. This clearly indicates that in spite of the large southward field (−10 nT) observed locally at this time, the plasma in the plasma sheet boundary layer remained rather undisturbed, just like before the plasma dropout at ~0425 UT. The tailward flow was mainly due to a high-energy tailward flowing population moving past the satellite. This is verified by an examination of the ion velocity distribution (not shown) which indicates distinctly the presence of two ion populations. The transient passing of this tailward flowing population is further indicated by its disappearance well before Geotail exited the plasma sheet boundary layer and while $B_z$ was still strongly negative.

The measurements of energetic ions and electrons from the Energetic Particles and Ion Composition (EPIC) instrument on Geotail [*Williams et al.*, 1994] shown in Plate 3 substantiate the above interpretation of transient energetic particle population passing through an ambient plasma. The anisotropy spectrograms indicate that tailward streaming of energetic ions (~67 keV) and electrons (> 38 keV) were both quite intermittent. At ~0522 UT, Geotail re-entered the plasma sheet once again and detected rather isotropic energetic electrons. However, tailward streaming of energetic ions was still quite prominent and lasted till ~0541 UT. This implies that the process leading to acceleration of energetic particles and their tailward streaming during substorm was still active and lingered on in the region earthward of Geotail. In addition, intermittent sunward streaming of energetic ions was simultaneously seen to mingle with the tailward streaming activity. The co-existence of intermittent tailward and sunward streaming of energetic particles indicates that there were more than one particle source responsible for the streaming activity seen. These activities form a sharp contrast to the rather stagnant (i.e. no significant bulk flow) of the plasma population during ~0522–0534 UT (Figure 6). From the energy spectrograms in Plate 2 and the plasma moment plots in Figure 6, it can be seen that the plasma at this time had characteristics quite similar to the pre-substorm activity at ~0400 UT.

At ~0535 UT, strong (>1000 km/s) sunward-dawnward flow began and lasted for ~10 min. This transient flow may be classified as a bursty bulk flow event [*Angelopoulos et al.*, 1992]. There was no noticeable tailward plasma flow preceding the start of the strong sunward–dawnward plasma flow even though Geotail was well within the plasma sheet during this time (plasma beta being in the range of 1–100). Furthermore, the strong sunward plasma flow arose at the time when tailward streaming of energetic ions was still detectable. This observation forms a powerful evidence that the process which produces tailward streaming of energetic ions did not move down the tail past Geotail to give rise to the observed strong sunward plasma flow. Note that the occurrence of strong sunward–dawnward plasma flow at Geotail coincided with the start of equatorward detachment of auroral luminosity from the main poleward edge of the auroral bulge as seen by Canopus MSP. This suggests that the detached auroral feature may be the optical signature of the bursty bulk flow observed by Geotail.

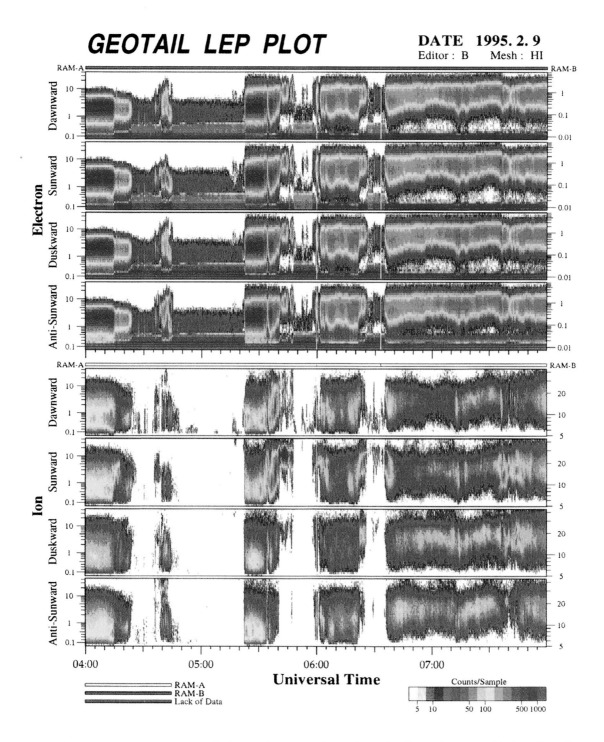

**Plate 2.** The energy spectrograms of plasma electrons and ions from the plasma instrument LEP on Geotail on February 9, 1995. Energies in keV are given on the left side of each panel.

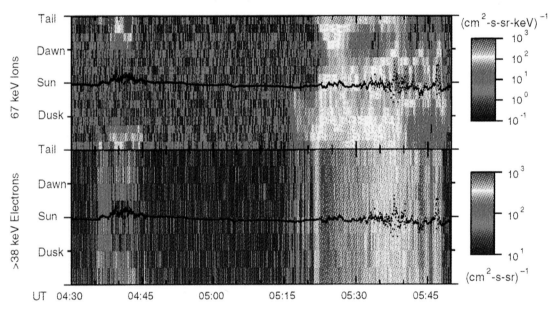

**Plate 3.** The anisotropy spectrogram of 67 keV ions and energetic electrons > 38 keV measured by EPIC instrument on Geotail on February 9, 1995.

## SUMMARY AND DISCUSSION

An isolated substorm is investigated with ISTP data from eleven spacecraft and two networks of ground stations covering both the northern and the southern hemispheres. These observations are unprecedented in providing global monitors of the substorm event, extending to all regions in the near-Earth space, namely, the solar wind, the magnetosheath, the magnetotail, the inner magnetosphere, the auroral altitude, and the ground. Analysis of this comprehensive set of measurements leads to confirmation of some expected substorm phenomena, to awareness of some unusual characteristics for this isolated substorm, and revelation of some surprising features which are difficult to reconcile with the traditional substorm model.

The ionospheric and magnetospheric phenomena resulting from the interaction of geomagnetic field with southward IMF, the substorm growth phase [*McPherron et al.*, 1973], are well substantiated in this isolated substorm. These phenomena include (a) the equatorward expansion of the polar cap (observed in DMSP and Canopus MSP data), (b) the equatorward shift of the auroral oval/electrojet in both hemispheres (observed in DMSP, Canopus MSP, Goose Bay and Halley data), (c) the enhancement of the large-scale magnetic-field-aligned current system (observed in Freja data), (d) the increase in the lobe magnetic field strength (and thus the cross-tail current intensity) as monitored in the mid-tail region (observed in Geotail data), and (e) the decrease in the poloidal component of the magnetic field at the geostationary orbit (observed in GOES-7 and GOES-8 data). The features (a) and (b) imply indeed energy storage of the magnetosphere during this interaction. Features (c), (d) and (e) indicate that the energy storage phase is associated with intensification of the large-scale magnetospheric current system which encompasses not just the cross-tail current but also regions 1 and 2 field-aligned currents. Feature (e) further indicates the earthward approach of the cross-tail current in this storage phase, as theorized by *Siscoe and Cummings* [1969]. For substorm expansion, expected features seen in this study include breakup arc location being well within the closed field line region, dipolarization and particle injection at the geostationary altitude, strong plasma flow, complex magnetic field structure, enhanced dawn-dusk electric field, and plasma sheet thinning in the mid-tail region.

One unusual aspect in the growth phase of this substorm is its long duration of ~90 min instead of the typical duration of ~40-60 min. This long duration is quite comparable with that for another isolated substorm on March 9, 1995 studied by *Rodger et al.* [1997] in which the growth phase lasted for ~80 min. Therefore, long duration growth phase may be common among isolated substorms but not for substorms inside a multiple-substorm sequence. Another unexpected feature found in this study is the difference in the duration of southward IMF between Wind and IMP 8 observations. One possibility is that there were spatial structures of the IMF in the solar wind. Solar wind observations taken at locations far from the Sun–Earth line should be viewed with caution. Another possibility is that the difference may arise as a result of the IMF being modified after passing through the bow shock. This is an important possibility in much need for clarification because after all it is the magnetic field in the magnetosheath that interacts with the geomagnetic field.

The second unusual aspect of this substorm is the time delay between ground indications of substorm onset and substorm injection onset inferred from observations at the geostationary orbit. Although some time delay (about a few min) is expected between these onsets as indicated by a recent study [*Reeves et al.*, 1996], the delay of ~8–13 min is unusually long. It would be interesting to examine in the future whether this long delay may be common for isolated substorms just like the long duration of growth phase. There are at least two possible interpretations for the long delay. Since there were two Pi2 onsets relating to the expansion onset, one may suggest that there were two particle injections, with the first one not reaching the geostationary altitude and the second one delayed significantly due to its initial injection location being further down the tail than usual. Another possibility is that there was only one particle injection with its initial location quite far down the tail. That substorm injection may not reach geostationary altitude is indicated by previous observations of an isolated substorm after a prolonged quiet period reported by *Lui et al.* [1975]. They showed that in spite of a lack of injection at the geostationary altitude for an isolated substorm, earthward plasma injection was observed at a downstream distance of 11 $R_E$. That a substantial delay can be expected if the initial injection location is far down the tail is supported by a detailed examination on the radial propagation of substorm injections by *Reeves et al.* [1996] who noted that the radial propagation speed is typically very slow (~24 km/s on the average).

The most surprising observed features of this substorm were noted in the mid-tail region. One of them is the occurrence of negative $B_z$ in the mid-tail neutral sheet starting from ~40 min before the substorm expansion onset. This may imply mid-tail magnetic reconnection occurring well before expansion onset to form an X-line earthward of $X \approx -32\ R_E$. If this is the case, this X-line did not produce significant plasma flow or particle energization since very slow earthward (not tailward) plasma flow and no significant particle energization were seen with the negative $B_z$ occurrence. In other words, magnetic reconnection in this case is rather benign with no significant energy conversion from magnetic field to particles. This interpretation is consistent with the MHD simulation of this event by *Winglee et al.* [1997]. They suggest that a mid-tail X-line can form without triggering a substorm expansion onset and that additional processes appear to be necessary to disrupt the near-Earth current sheet for substorm expansion. Another possible interpretation for the occurrence of negative $B_z$ in the mid-tail neutral sheet is that it is a spatially localized feature due to current filamentation of the cross-tail current sheet. No dynamic process such as plasma jetting or particle energization is associated with such a change in magnetic field topology. It is important to conduct future studies of isolated substorms to discriminate between these different possibilities.

Another surprising feature in mid-tail region is the continual buildup of total pressure (and thus tail lobe magnetic field) for ~7 min after the ground expansion onset. This is unexpected if expansion onset is initiated by magnetic reconnection in the tail to release its stored energy. Since this expansion onset is consistent with being triggered by a northward turning of IMF, the dayside reconnection rate is not anticipated to exceed the nightside reconnection rate to allow further tail energy buildup after expansion onset. Furthermore, about 4 min after the ground expansion onset, large southward magnetic field and tailward plasma flow occurred in the mid-tail region. The traditional interpretation would be that these are signatures of an X-line. However, a detail examination shows that tailward flow arises from a transient energetic ion population streaming down the tail while the ambient plasma appears to be undisturbed by the activity. Moreover, the ambient plasma remained undisturbed after the passage of the tailward streaming population. The lack of response from the ambient plasma is quite contrary to the expectation of magnetic reconnection occurring nearby as implied by the observed magnitude of southward magnetic field. The variation in the dawn–dusk component of the magnetic field simultaneous with the occurrence of southward magnetic field further suggests that the magnetic field structure at that time is complicated and three-dimensional, unlike the relatively simple two-dimensional nature for an X-line. In particular, the magnetic structure is not consistent with a two-dimensional X-line superposed on a uniform $B_y$ component. A number of three-dimensional MHD simulations of substorm activity in the tail predicts the formation of an X-line and a plasmoid over a large-scale. The X-line and plasmoid in these simulations typically extend ~20 $R_E$ in the dawn–dusk direction (e.g., *Birn and Hesse*, 1991). Therefore, for an observer close to the tail axis, the magnetic field geometry resembles a two-dimensional X-line superposed on a relatively uniform $B_y$ component. The magnetic field structure observed by Geotail is more complex than that from such simulations. Another possible interpretation is that the complex magnetic field structure represents the diffusion region of the magnetic reconnection process. However, counter to this interpretation is the undisturbed (and unenergized) character of the ambient plasma after the passing of the tailward streaming population when the magnetic field was still strongly southward and complex. Another possibility is to relate the complex magnetic field structure to a magnetic flux rope. However, since it was observed in the plasma sheet boundary, the magnetic flux rope is inevitably linked to an X-line reconnecting open magnetic field lines. Large particle energization and tailward ejection of the flux rope (thus tailward convection of the local plasma) are anticipated, in sharp contrast to the observation of undisturbed ambient plasma in this structure.

One more surprising feature in the mid-tail region is the occurrence of strong sunward plasma flow in the late expansion phase when Geotail was in the plasma sheet prior to its appearance. This is an extremely fortuitous situation because almost all previous studies of satellite re-entry to the plasma sheet during the late expansion phase showed an expanding plasma sheet with strong sunward plasma flow. The traditional interpretation of these events in previous studies is that an X-line moves past the satellite location to further downstream, allowing the plasma sheet to expand and engulf the satellite and sunward plasma flow arises from magnetic reconnection at the X-line. This interpretation can be verified or refuted if the satellite were within the plasma sheet before the commencement of strong sunward plasma flow in the late expansion phase. The observations reported here provides one such case and they are inconsistent with such a traditional interpretation. Similar findings were reported by *Mukai et al.* [1996]. These observations, however, are quite consistent with the substorm synthesis model in which multiple sites of current disruption occur during the substorm expansion which subsequently lead to magnetic reconnection further downstream at the late-expansion/early-recovery phase of a substorm [*Lui*, 1991]. In this scenario, the process responsible for substorm expansion onset is different from the process which generates strong sunward plasma flow in association with plasma sheet recovery during the late expansion or early recovery phase. Current disruption can generate earthward as well as tailward plasma flow depending on the modification of the local Lorentz force. One distinction between current disruption and the traditional magnetic reconnection model is that no slow mode shock needs to be formed in current disruption for fast conversion of magnetic field energy into particle energy [*Lui*, 1996]. Another distinction is that current density is reduced at the current disruption region whereas current density is enhanced at the X-line or diffusion region in the magnetic reconnection picture. That the current density is enhanced and not reduced at the X-line is not well aware by most advocates of magnetic reconnection but is shown to be the case by the MHD simulation of *Birn and Hesse* [1991]. Note that based on their simulation (their Figure 8), the overall north–south integrated current at the X-line region is reduced due to the thinning of the plasma sheet (thus less contribution to the integrated current from the reduced thickness) and the reduction of current density at the outer edges of the plasma sheet.

Finally, another important feature emerging from this study is that the transient strong sunward plasma flow (bursty bulk flow) observed in the mid-tail region at the late-expansion/early-recovery phase of the substorm appears to be associated with a visible optical feature moving equatorward on the ground as seen in the Canopus MSP observation. This optical feature has similarities to the north–south aligned auroral fingers observed by the Viking UV imager as reported by *Rostoker et al.* [1987]. They have also related these optical features to longitudinally confined earthward injection of plasma in the tail. The confinement in local time compares well with the expected narrow dawn–dusk extent of bursty bulk flows [*Angelopoulos et al.*, 1997].

This study represents an attempt to use the comprehensive ISTP data set to address several outstanding issues on substorms. It is important to conduct future studies for more isolated substorms to ascertain the general features exhibited by these substorms and to help resolving the various substorm theories currently under consideration.

*Acknowledgments.* This work was supported by funding from the Space Physics Division of the National Aeronautics and Space Administration (under the Department of Navy Task IAF; contract N00039-91-C-001) and the Atmospheric Sciences Division of the National Science Foundation (Grant ATM-9622080) to the Johns Hopkins University Applied Physics Laboratory.

# REFERENCES

Akasofu, S.-I., Energy coupling between the solar wind and the magnetosphere, *Space Sci. Rev.*, 28, 121, 1981.

Angelopoulos, V., et al., Bursty bulk flows in the inner central plasma sheet, *J. Geophys. Res.*, 97, 4027, 1992.

Angelopoulos, V., et al., Magnetotail flow bursts: association to global magnetospheric circulation, relationship to ionospheric activity and direct evidence for localization, *Geophys. Res. Lett.*, 24, 2271, 1997.

Baker, K. B., and S. Wing, A new magnetic coordinate system for conjugate studies at high latitudes, *J. Geophys. Res.*, 94, 9139, 1989.

Birn, J., and M. Hesse, The substorm current wedge and field-aligned currents in MHD simulations of magnetotail reconnection, *J. Geophys. Res.*, 96, 1611, 1991.

Lepping, R. P., et al., Magnetic cloud-bow shock interaction: WIND and IMP-8 observations, *Geophys. Res. Lett.*, 23, 1195, 1996.

Lui, A. T. Y., A synthesis of magnetospheric substorm models, *J. Geophys. Res.*, 96, 1849, 1991.

Lui, A. T. Y., Current disruption in the Earth's magnetosphere: Observations and models, *J. Geophys. Res.*, 101, 13067, 1996.

Lui, A. T. Y., and J. R. Burrows, On the location of auroral arcs near substorm onset, *J. Geophys. Res.*, 83, 3342, 1978.

Lui, A. T. Y., et al., Observation of the plasma sheet during a contracted oval substorm in a prolonged quiet period, *J. Geophys. Res.*, 81, 1415, 1976.

Lyons, L. R., Substorms: Fundamental observational features, distinction form other disturbances, and external triggering, *J. Geophys. Res.*, 101, 13011, 1996.

McPherron, R. L., C. T. Russell, and M. P. Aubry, Satellite studies of magnetospheric substorms on August 15, 1968, 9, Phenomenological model for substorms, *J. Geophys. Res.*, 78, 3131, 1973.

Mukai, T., et al., GEOTAIL and INTERBALL-TAIL correlative observations of magnetotail dynamics and substorms, *EOS* Supplement, F611, 1996.

Nagai, T., et al., Plasma sheet pressure changes during the substorm growth phase, *Geophys. Res. Lett.*, 24, 963, 1997.

Newell, P. T., et al., Morphology of nightside precipitation, *J. Geophys. Res.*, 101, 10737, 1976.

Rodger, A. S., et al., 9 March 1995—a simple isolated substorm? *Geophys. Res. Lett.*, submitted 1997.

Rostoker, G., Triggering of expansive phase intensifications of magnetospheric substorms by northward turnings of the interplanetary magnetic field, *J. Geophys. Res.*, 88, 69812, 1983.

Rostoker, G., et al., North–south structure in the midnight sector auroras as viewed by the VIKING imager, *Geophys. Res. Lett.*, 14, 407, 1987.

Samson, J. C., et al., Substorm intensifications and field line resonances in the nightside magnetosphere, *J. Geophys. Res.*, 97, 8495, 1992.

Siscoe, G. L., and W. D. Cummings, On the cause of geomagnetic bays, *Planet. Space Sci.*, 17, 1795, 1969.

Winglee, R., et al., Magnetospheric/Ionospheric activity during an isolated substorm: A comparison between WIND/Geotail/IMP 8/CANOPUS observations and modeling, this monograph, 1997.

A. T. Y. Lui, D. J. Williams, R. W. McEntire, S. Ohtani, L. J. Zanetti, W. A. Bristow, R. A. Greenwald, and P. T. Newell, Applied Physics Laboratory, Johns Hopkins University, Laurel, MD 20723.

S. P. Christon, The University of Maryland, College Park, MD 20742.

T. Mukai, K. Tsuruda, and T. Yamamoto, Institute of Space and Astronautical Science, Sagamihara, Kanagawa 229, Japan.

S. Kokubun, Solar Terrestrial Environment Laboratory, Nagoya University, Toyokawa, Aichi 442, Japan.

H. Matsumoto, H. Kojima, and T. Murata, RASC, Kyoto University, Kyoto, Japan.

D. H. Fairfield and R. P. Lepping, NASA/GSFC, Greenbelt, MD, USA.

J. C. Samson and G. Rostoker, University of Alberta, Edmonton, Canada.

G. D. Reeves, Los Alamos National Laboratory, Los Alamos, NM 87545, USA.

A. L. Rodger, British Antarctic Survey, Madingley Rd., Cambridge CB3 OET, United Kingdom.

H. J. Singer, NOAA Space Environment Center, 325 Broadway, Boulder, CO 80303, USA.

# Structure of the Magnetotail Reconnection Layer in 2-D Ideal MHD Model

Y. Lin and X. X. Zhang

*Physics Department, Auburn University, Auburn, Alabama*

A two-dimensional ideal MHD model is developed to calculate the structure of the reconnection layer, which is assumed to have evolved from an initial current sheet. In particular, the symmetric reconnection layers, which may be present in the distant magnetotail, has been calculated for various guide field $B_{y1}$ and normal component of magnetic field $B_{z1}$ in the lobes. In the cases with $B_y = 0$, a pair of slow shocks are present in the reconnection layer, as in the *Petschek's* [1964] model. The strength of the slow shock and the angle of the slow shock front change with $B_{z1}$ and the lobe plasma beta $\beta_1$. In the cases with $B_{y1} \neq 0$, two rotational discontinuities are present upstream of two slow shocks, respectively. The slow shocks are weaker than those in the cases with $B_{y1} = 0$. The acceleration of plasma flow in the reconnection layer is through both the rotational discontinuities and slow shocks. The application of the present model to the magnetotail reconnection is discussed.

## INTRODUCTION

The existence and the role of magnetohydrodynamic (MHD) waves and discontinuities in the outflow region of magnetic reconnection were first proposed by *Petschek* [1964]. The layered structure in the outflow region which contains several MHD discontinuities is now referred to as the reconnection layer [e.g., *Heyn et al.*, 1988; *Shi and Lee*, 1990; *Lin and Lee*, 1994]. *Petschek's* [1964] steady-state model describes the symmetric case with equal plasma density, equal magnetic field strength, and antiparallel magnetic fields on the two sides of current layer. Such symmetric case may be applied to the distant magnetotail, where the field and plasma in the two lobes are highly symmetric. The reconnection layer consists of a pair of slow shocks emanating from a small central diffusion region. The energy conversion in the magnetic reconnection is mainly through the slow shocks. Symmetric reconnection models have also been proposed by *Sonnerup* [1970], *Yeh and Axford* [1970], *Priest and Forbes* [1986], and *Priest and Lee* [1991]. On the other hand, *Levy et al.* [1964] provided an extremely asymmetric reconnection model for the dayside magnetopause, in which the slow shocks in Petschek's [1964] model are replaced by a rotational discontinuity and a slow expansion fan. Satellite observations of plasma and field signatures have shown the existence of rotational discontinuities at the magnetopause [e.g., *Paschmann et al.*, 1979; *Sonnerup et al.*, 1981] and slow shocks in the distant magnetotail [e.g., *Feldman et al.*, 1984].

The study for the structure of reconnection layers in the magnetosphere is important. It not only provides signatures of magnetic field and plasma as needed to identify a magnetic reconnection event, but also predicts possible MHD waves that would propagate locally and globally from the X line. The structure of reconnection layer has been studied by solving the one-dimensional (1-D) ideal MHD Riemann problem under various magnetic field and plasma conditions [e.g., *Heyn et al.*, 1988; *Lin and Lee*, 1994]. The transition from *Petschek's* [1964] symmetric model to *Levy et al.*'s [1964] extremely asymmetric model has been studied. It is found that in general cases rotational discontinuities, slow shocks, slow expansion waves, and contact discontinuity may be present in the reconnection layer.

In the presence of a finite resistivity, 1-D MHD simulations of the Riemann problem associated with the evolution of an initial current sheet after the onset of magnetic reconnection show that the rotational discontinuities are replaced by intermediate shocks and time-dependent intermediate shocks [*Lin et al.*, 1992]. The MHD discontinuities that are predicted from the 1-D simulations have also been found in two-dimensional (2-D) resistive MHD simulations of quasi-steady reconnections [*Sato*, 1979; *Ugai*, 1984; *Scholer*, 1989; *Shi and Lee*, 1990; *Yan et al.*, 1992]. Recently, the reconnection layer has been further studied by 3-D MHD simulations [*Ugai and Shimizu*, 1996]. In addition, the evolution of the outflow region of magnetic reconnection in transient stages has been studied by theoretical approaches [e.g., *Semenov et al.*, 1992] and MHD simulations [e.g., *Birn*, 1984; *Ogino and Walker*, 1984; *Ugai*, 1995].

Nevertheless, the resistivity imposed in the MHD simulations may not be true in collisionless plasmas. In 1-D hybrid simulations for the structure of the reconnection layer in collisionless plasmas, time-dependent intermediate shocks are found to quickly evolve to a steady rotational discontinuity [*Lin and Lee*, 1994]. The existence of the rotational discontinuity is consistent with the ideal MHD solution. In order to investigate the large-scale structure of quasi-steady reconnection in the distant magnetotail, 2-D hybrid simulations have been performed by *Krauss-Varban and Omidi*, [1995]; *Lin and Swift*[1996], *Nakamura et al.* [1997] for the *Petschek*-type symmetric cases. The 2-D simulation by *Lin and Swift* [1996] indicates that in order for the two slow shocks in the reconnection layer to clearly separate a long simulation system is needed. A kinetic simulation with a very large simulation domain, however, has not been performed because it is highly time consuming.

The purpose of this paper is to present a 2-D ideal MHD model for the structure of the reconnection layer. The calculation and a parameter search are conducted for the symmetric reconnection layers in the distant magnetotail. The advantage of using the ideal MHD formulation is that it constructs the reconnection layer by directly calculating the jumps of physical quantities across the discontinuities while ignoring their internal structure. The ideal MHD can help us to understand the evolution of the 2-D configuration of magnetic reconnection in numerical simulations and predict the structure at distances far from the X line. The structure of steady-state reconnection layer in 2-D reconnection configuration has been calculated by *Heyn et al.* [1988] using an approximation based on the 1-D MHD results, and the existence of fast mode waves is ignored. In our study, the main reconnection layer is calculated directly from the 2-D model, and the fast waves are also considered in the calculation. The dependence of the boundary layer structure on the reconnection rate is discussed. Cases with a zero guide field ($B_y = 0$) as well as with a finite guide field ($B_y \neq 0$) are investigated. The 2-D ideal MHD model is given in section 2. The calculation results for cases with $B_y = 0$ are shown in section 3, and the results for cases with $B_y \neq 0$ are shown in section 4. A summary is given in section 5.

## 2-D MHD FORMULATION

In a magnetic reconnection, the reconnection layer develops from the X line in an initial current sheet. Assume that the initial current sheet separates two uniform plasma regions with antiparallel magnetic field components in the $x$ direction. The current sheet normal is assumed to be in the $z$ direction, and the X line of the reconnection is located at $(x, z) = (0, 0)$. The total pressure $(P + P_B)$ is constant across the initial current sheet, where $P$ is the thermal pressure and $P_B = B^2/2\mu_0$ is the magnetic pressure.

In general, the solution of the 1-D Riemann problem, which concerns the evolution of an initial current sheet associated with a magnetic reconnection, consists of seven discontinuities or waves. These discontinuities and waves includes two fast mode shocks or waves, two rotational discontinuities, two slow mode shocks or slow expansion waves, and a contact discontinuity[*Jeffrey and Taniuti*, 1964]. Among these MHD discontinuities, the fast and slow shocks are corresponding to the fast and slow MHD linear modes, respectively, and the rotational discontinuity is an Alfven mode discontinuity. The contact discontinuity is an entropy mode structure. At the fast shock, the normal component of inflow velocity is greater than the linear fast mode speed in upstream, and the magnetic field plasma density, and temperature increase from upstream to downstream. At the slow shock, the normal inflow speed is greater than the upstream slow mode speed. The plasma density and temperature increase but the magnetic field decreases across the slow shock. At the rotational discontinuity, the normal inflow speed equals the normal component of

Alfven velocity in upstream. The field and plasma density conserve across the rotational discontinuity. The normal flow velocity relative to the contact discontinuity is zero. Across the contact discontinuity, the field, pressure, and flow velocity do not change, whereas the plasma density can change arbitrarily.

In general cases with asymmetric field and plasma quantities on the two sides of the current sheet, the reconnection layer consists of a fast expansion wave, a rotational discontinuity, and a slow shock or slow expansion wave on one side from the contact discontinuity and another set of fast mode, rotational discontinuity, and slow mode on the other side [e.g., *Lin and Lee*, 1994]. The fast expansion waves are very weak because of the total pressure balance in the initial current sheet. Among these wave modes, the fast wave has the fastest propagation speed and quickly propagates away. In the main reconnection layer, the rotational discontinuity is followed by the slow mode. The contact discontinuity has a zero wave speed in plasma and thus stays at the center of the reconnection layer. The role of the rotational discontinuity is to change the direction of magnetic field, while the fast and slow modes change the magnitude of magnetic field and plasma density. The contact discontinuity is required to match the difference in the plasma density on the two sides.

For a symmetric current sheet which separates two plasma regions with equal magnetic field strength, equal plasma density, antiparallel $B_x$ component, and a common guide field $B_y$, the solution of the Riemann problem consists of a pair of fast expansion waves, rotational discontinuities and slow shocks. For the symmetric case with $B_y = 0$, the reconnection layer consists of two fast expansion waves and two slow shocks. In this paper, we solve for the structure of the reconnection layer in symmetric cases with various $B_y$. The slow expansion waves do not appear in the symmetric cases.

Figure 1 shows the coordinate system and the 2-D model used in this paper. Since the two weak fast expansion waves quickly propagate away from the main reconnection layer [e.g., *Lin and Lee*, 1994] and the group velocity and thus front of the fast waves are nearly isotropic relative to the X line, we assume that the front of the two identical fast expansion waves FE and FE', which propagate with a velocity $V_F$, are parallel to the initial current sheet, as shown in Figure 1. On the other hand, the fronts of rotational discontinuities and slow shocks emanate from the X line in the quasi-steady reconnection. Two symmetric rotational discontinuities RD and RD' are located ahead of two identical slow shocks SS and SS', respectively. The angles $\alpha_r$ and $\alpha_s$ represent, respectively, the angles of RD and SS fronts relative to the $x$ axis. The discontinuities and waves divide the $xy$ plane into 8 regions, as indicated in Figure 1.

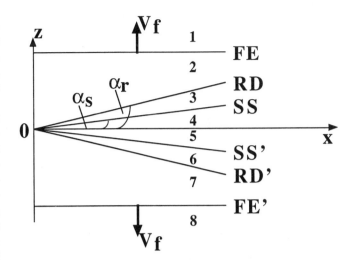

**Figure 1.** A sketch of the 2-D coordinate system used in this study. In general, the symmetric reconnection layer may consist of two rotational discontinuities (RD and RD') and two slow shocks (SS and SS'), whose fronts emanate from the X line with angles $\alpha_r$ and $\alpha_s$, respectively.

Note that in the symmetric cases region 4 and region 5 contain the same magnetic field and plasma, whereas in asymmetric cases a contact discontinuity exists nearly at the center. Region 1 and region 8 contain unperturbed field and plasma on the two sides of the initial current sheet. We only consider the situation that FE and FE' are far from the center line $z = 0$ and calculate the structure of the reconnection layer in which the fronts of rotational discontinuities and slow shocks are behind the front of fast waves.

The ideal MHD equations are used to calculate the variations of physical quantities across the infinitely thin discontinuities RD, RD', SS, and SS'. The set of ideal MHD equations can be written as

$$\frac{\partial \rho}{\partial t} + \nabla \cdot (\rho \mathbf{V}) = 0 \quad (1)$$

$$\frac{\partial \rho \mathbf{V}}{\partial t} + \nabla \cdot [(P + \frac{B^2}{2\mu_0})\mathbf{I} + \rho \mathbf{V}\mathbf{V} - \frac{\mathbf{B}\mathbf{B}}{\mu_0}] = 0 \quad (2)$$

$$\frac{\partial \epsilon_T}{\partial t} + \nabla \cdot [(\frac{\rho V^2}{2} + \frac{P}{\gamma - 1} + P)\mathbf{V} + \mathbf{E} \times \mathbf{B}] = 0 \quad (3)$$

$$\frac{\partial \mathbf{B}}{\partial t} - \nabla \times \mathbf{E} = 0 \quad (4)$$

$$\nabla \cdot \mathbf{B} = 0 \quad (5)$$

with

$$\epsilon_T = \frac{1}{2}\rho V^2 + \frac{P}{\gamma - 1} + \frac{B^2}{2\mu_0}$$

and
$$\mathbf{E} = -\mathbf{V} \times \mathbf{B}$$
where $\rho$ is the mass density of plasma, $\mathbf{V}$ is the plasma flow velocity, $P$ is the thermal pressure, $\mathbf{B}$ is the magnetic field, $\mathbf{E}$ is the electric field, $t$ is time, and $\mu_0$ is the permeability of free space.

To obtain the 2-D steady-state structure, the plasma and field are assumed to be uniform in each of the 8 regions, and the fronts of discontinuities are located along radial direction. The physical quantities are assumed to be only functions of polar angle $\alpha$ and to change only across the discontinuities. Such an assumption has also been used by *Neubauer* [1975] for oblique interaction between MHD discontinuities. The jumps of physical quantities across each discontinuity are determined by the Rankine-Hugoniot jump conditions of the discontinuity [e.g., *Landau and Lifshitz*, 1960], which are obtained from the conservation of the flux of mass, momentum, and energy across the discontinuity and the Maxwell equations in steady state. Let $\mathbf{z}'$ be along the normal $\mathbf{n}$ of the RD front, $\mathbf{y}' = \mathbf{y}$, and $\mathbf{x}'$ form a right-hand coordinate system with $\mathbf{y}'$ and $\mathbf{z}'$. For $B_{x1} > 0$ and $B_{z1} > 0$, the jump conditions for RD can be written as

$$\rho_2 = \rho_3 \tag{6}$$

$$P_2 = P_3 \tag{7}$$

$$B_2 = B_3 \tag{8}$$

$$V_{z'2} = V_{z'3} \tag{9}$$

$$B_{z'2} = B_{z'3} \tag{10}$$

$$V_{x'2} + \frac{B_{x'2}}{\sqrt{\mu_0 \rho_2}} = V_{x'3} + \frac{B_{x'3}}{\sqrt{\mu_0 \rho_3}} \tag{11}$$

$$V_{y'2} + \frac{B_{y'2}}{\sqrt{\mu_0 \rho_2}} = V_{y'3} + \frac{B_{y'3}}{\sqrt{\mu_0 \rho_3}} \tag{12}$$

$$V_{z'2} = -\frac{B_{z'2}}{\sqrt{\mu_0 \rho_2}} \tag{13}$$

where

$$\begin{bmatrix} V_{z'} \\ V_{x'} \end{bmatrix} = \begin{bmatrix} \cos \alpha_r & -\sin \alpha_r \\ \sin \alpha_r & \cos \alpha_r \end{bmatrix} \begin{bmatrix} V_z \\ V_x \end{bmatrix} \tag{14}$$

$$\begin{bmatrix} B_{z'} \\ B_{x'} \end{bmatrix} = \begin{bmatrix} \cos \alpha_r & -\sin \alpha_r \\ \sin \alpha_r & \cos \alpha_r \end{bmatrix} \begin{bmatrix} B_z \\ B_x \end{bmatrix} \tag{15}$$

The subscripts 2 and 3 represent the physical quantities in regions 2 and 3, respectively.

The jump conditions for SS are given as

$$B_{z'3} = B_{z'4} \tag{16}$$

$$\rho_3 V_{z'3} = \rho_4 V_{z'4} \tag{17}$$

$$B_{z'3} V_{x'3} - V_{z'3} B_{x'4} = B_{z'4} V_{x'4} - V_{z'4} B_{x'4} \tag{18}$$

$$B_{z'3} V_{y'3} - V_{z'3} B_{y'4} = B_{z'4} V_{y'4} - V_{z'4} B_{y'4} \tag{19}$$

$$\rho_3 V_{z'3} V_{x'3} - \frac{B_{z'3} B_{x'3}}{\mu_0} = \rho_4 V_{z'4} V_{x'4} - \frac{B_{z'4} B_{x'4}}{\mu_0} \tag{20}$$

$$\rho_3 V_{z'3} V_{y'3} - \frac{B_{z'3} B_{y'3}}{\mu_0} = \rho_4 V_{z'4} V_{y'4} - \frac{B_{z'4} B_{y'4}}{\mu_0} \tag{21}$$

$$\rho_3 V_{z'3}^2 + P_3 + \frac{B_{z'3}^2}{2\mu_0} = \rho_4 V_{z'4}^2 + P_4 + \frac{B_{z'4}^2}{2\mu_0} \tag{22}$$

$$\left( \frac{1}{2} \rho_3 V_3^2 + \frac{\gamma P_3}{\gamma - 1} + \frac{B_3^2}{2\mu_0} - \frac{B_{z'3}^2}{\mu_0} \right) V_{z'3}$$
$$- \frac{B_{x'3} V_{x'3} + B_{y'3} V_{y'3}}{\mu_0} B_{z'3} =$$
$$\left( \frac{1}{2} \rho_4 V_4^2 + \frac{\gamma P_4}{\gamma - 1} + \frac{B_4^2}{2\mu_0} - \frac{B_{z'4}^2}{\mu_0} \right) V_{z'4}$$
$$- \frac{B_{x'4} V_{4'3} + B_{y'4} V_{y'4}}{\mu_0} B_{z'4} \tag{23}$$

where the $x'y'z'$ system is the local coordinate system in the shock normal frame of SS, and the subscript 4 represents the quantities in region 4. The transformation from the $x$ and $z$ components to the $x'$ and $z'$ components is given by (14) and (15), with $\alpha_r$ being replaced by angle $\alpha_s$ of the SS front.

For the weak fast expansion waves FE, we assume that they also satisfy the conservation of mass, momentum, and energy, with the energy equation being replaced by the conservation of entropy. The set of jump conditions for FE can be written as

$$a_6 R_\rho^{\gamma+3} + a_5 R_\rho^{\gamma+2} + a_4 R_\rho^{\gamma+1} + a_3 R_\rho^3$$
$$+ a_2 R_\rho^2 + a_1 R_\rho + a_0 = 0 \tag{24}$$

$$P_2 = P_1 R_\rho^\gamma \quad (25)$$

$$V_{z'2} = M_{IF} B_{z'1}(1 - R_\rho)/R_\rho \quad (26)$$

$$V_{x'2} = M_{IF} B_{x'1}(1 - R_\rho)/(R_\rho - M_{IF}^2) \quad (27)$$

$$V_{y'2} = M_{IF} B_{y'1}(1 - R_\rho)/(R_\rho - M_{IF}^2) \quad (28)$$

$$B_{z'2} = B_{z'1} \quad (29)$$

$$B_{x'2} = R_\rho B_{x'1}(1 - R_\rho^2)/(R_\rho - M_{IF}^2) \quad (30)$$

$$B_{y'2} = R_\rho B_{y'1}(1 - R_\rho^2)/(R_\rho - M_{IF}^2) \quad (31)$$

with

$$\begin{aligned}
R_\rho &\equiv \rho_2/\rho_1 \\
a_6 &= \beta_1/(2M_{IF}^2 \cos^2 \theta_{nB1}) \\
a_5 &= -\beta_1/\cos^2 \theta_{nB1} \\
a_4 &= \beta_1 M_{IF}^2/(2\cos^2 \theta_{nB1}) \\
a_3 &= -1/\cos^2 \theta_{nB1} - \beta_1/(2M_{IF} \cos^2 \theta_{nB1}) \\
    &\quad + \tan^2 \theta_{nB1} M_{IF}^2/2 \\
a_2 &= 1 + 2M_{IF}^2 + \beta_1/\cos^2 \theta_{nB1} + \tan^2 \theta_{nB1} \\
a_1 &= -[M_{IF}^4 + 2M_{IF}^2 + \beta_1 M_{IF}^2/(2\cos^2 \theta_{nB1}) \\
    &\quad + \tan^2 \theta_{nB1} M_{IF}^2/2] \\
a_0 &= M_{IF}^2
\end{aligned}$$

where $M_{IF} \equiv v_{n1}/C_{I1}$ is the intermediate Mach number of the fast mode wave, $v_{n1}$ is the upstream normal flow speed relative to a center point of the propagating fast expansion wave, and $C_{I1}$ is the normal component of Alfven velocity in region 1. The above equations for the quasi-steady fast wave have also been used by *Lin and Lee* [1994] in the calculation of the 1-D Riemann problem associated with the reconnection layer. Their ideal MHD results for the cases with $B_y = 0$ agree very well with the corresponding MHD simulations. It should be mentioned that in the 2-D cases, the above assumptions of the fast wave with a straight wave front may not be very accurate. In some cases it is found that apparent errors are present in the $x$ component of flow velocity, which is usually very small.

In general cases, a contact discontinuity exists at the center of the reconnection layer, across which

$$P_4 = P_5 \quad (32)$$

$$V_{z'4} = V_{z'5} \quad (33)$$

$$B_{z'4} = B_{z'5} \quad (34)$$

$$V_{x'4} = V_{x'5} \quad (35)$$

$$B_{x'4} = B_{x'5} \quad (36)$$

$$V_{y'4} = V_{y'5} \quad (37)$$

$$B_{y'4} = B_{y'5} \quad (38)$$

The equations for FE', RD', and SS' are the same as those for FE, RD, and SS, respectively. In principle, a total of fifty-five equations can be written for the seven discontinuities in general asymmetric cases. In the calculation, the quantities in regions 1 and 8 are given. The total number of unknowns is also fifty-five, including eight physical quantities in each region for regions 2 through 7, five angles of the discontinuity fronts, and two propagation speeds or intermediate Mach numbers of the fast expansion waves. In the symmetric cases, the equations are greatly simplified because of the presence of identical discontinuities on the two sides of the reconnection layer. The contact discontinuity disappears due to the symmetry.

The normalization of physical quantities is given below. The plasma density is normalized to the density $\rho_1$ in region 1, the magnetic field is normalized to $B_{x1}$, and the pressure is normalized to the magnetic pressure $B_{x1}^2/\mu_0$. The flow velocity is in units of the Alfven speed $B_{x1}/\sqrt{\mu_0 \rho_1}$. The spatial coordinates are normalized to the domain length $L_z$ in the $z$ direction. The simplified set of equations are solved by minimization of functions using the direction set (Powell's) method with MATHEMATICA software. The solution is accurate to $10^{-3}$.

## RESULTS FOR CASES WITH $B_y = 0$

In case 1, the guide field $B_y$ is assumed to be zero in region 1 and region 8, and the antiparallel magnetic field component $B_{x1} = -B_{x8} = 1$. The $z$ component of magnetic field, which is equivalent to the reconnection rate, is chosen to be $B_{z1} = B_{z8} = 0.1$ and the plasma beta $\beta_1 = \beta_8 = 0.2$, the density $\rho_1 = \rho_8 = 1$. The plasma flow speed $V_1 = V_8 = 0$, which corresponds to a non-driven, spontaneous reconnection.

The top panel of Figure 2 sketches the structure of the main reconnection layer in the $xz$ plane. A pair of slow shocks SS and SS' are present. Two fast expansion waves have quickly propagated away and are not shown in the figure. It is found from the MHD equations that $B_y = 0$ everywhere in the 2-D system if $B_y = 0$ in regions 1 and 8. Rotational discontinuities are not present in the case with $B_y = 0$, and thus regions 3 and 6 disappear. The angle $\alpha_s$ of the SS front is calculated to be 2.9°. The magnetic field lines are greatly bent across the slow shock, as shown in Figure 2, and the field strength decreases. Note that the flaring angles of the shock fronts are much smaller than they appear in the top panel of Figure 2.

In order to see the change of physical quantities across the slow shocks, we plot in the middle and bottom panels of Figure 2 the profiles of $B_x$, $B_z$, $V_x$, $\rho$, $P$, and $B$ along $x = 4.8$ as a function of $z$. Across the slow shock, the $B_x$ component decreases to zero, as shown in Figure 2. The plasma density $\rho$ increases by a factor of 2.10, thermal pressure increases by a factor of 6.06, and the $V_x$ component of flow velocity increases from nearly zero to 0.903 because of the strong field tension force. The magnetic energy is converted to the flow and thermal energy across the slow shock. The complete set of solution for the quantities in regions 2 and 4 is given in Table 1.

**Table 1.** Results of Case 1

| Region | 1 | 2 | 4 |
|---|---|---|---|
| $\rho$ | 1 | 0.988 | 2.100 |
| $P$ | 0.1 | 0.099 | 0.606 |
| $B_x$ | 1 | 0.988 | 0.000 |
| $B_y$ | 0 | 0.000 | 0.000 |
| $B_z$ | 0.1 | 0.100 | 0.044 |
| $V_x$ | 0 | 0.001 | 0.899 |
| $V_y$ | 0 | 0.000 | 0.000 |
| $V_z$ | 0 | -0.040 | 0.027 |

A parameter search has been conducted for the dependence of the structure of reconnection layer on the normal field component $B_{z1}$, or roughly the reconnection rate. Figure 3 shows the density $\rho_4$ in region 4 downstream of SS, velocity $V_{x4}$, pressure ratio $P_4/P_1$, and the angle $\alpha_s$ of the SS front as a function of $B_{z1}$ for $\beta_1 = 0.2$. It is seen that the shock angle $\alpha_s$ increases with $B_{z1}$. The density $\rho_4$ slightly decreases with $B_{z1}$. This behavior of $\rho_4$ is different from that obtained from the 1-D ideal MHD model of the Riemann problem. In the 1-D model, the increase of $B_z$ leads to the decrease of the shock normal angle of slow shock, which in turn causes the increase of compressibility of shock. In the 2-D model, however, the angle $\alpha_s$ of the SS front increases with $B_{z1}$, and thus the shock normal angle of SS does not decrease.

The flow velocity $V_{x4}$ also decreases with $B_{z1}$, as seen in Figure 3. In the 2-D model of the symmetric reconnection layer, $B_{x4} = 0$, but the tangential magnetic field is non-zero downstream of SS because of the obliquity of the SS front. The slow shocks are thus nonswitch-off shocks, whereas in the 1-D model they are switch-off shocks. The change in tangential velocity across the slow shock becomes smaller as the shock becomes nonswitch-off. In addition, the ratio $P_4/P_1$ decreases with $B_{z1}$ as the slow shock becomes more nonswitch-off.

Figure 4 presents the dependence of the quantities $\rho_4$, $V_{x4}$, $P_4/P_1$, and $\alpha_s$ on $\beta_1$ for $B_{z1} = 0.1$. The slow shock is weaker as $\beta_1$ increases. Note that $\beta_1 \simeq \beta_2$. As a result, the density, flow, and the pressure jump in region 4 decrease. On the other hand, the shock angle $\alpha_s$ is found to increase with $\beta_1$. This is due to the increase of the slow mode speed with plasma beta.

## RESULTS FOR CASES WITH $B_y \neq 0$

In case 2, the guide field are non-zero, with $B_{y1} = B_{y8} = 0.5$. Other quantities are the same as in case 1, with $B_{x1} = -B_{x8} = 1$, $B_{z1} = B_{z8} = 0.1$, $\rho_1 = \rho_8 = 1$, $\beta_1 = \beta_8 = 0.2$, and $V_1 = V_8 = 0$.

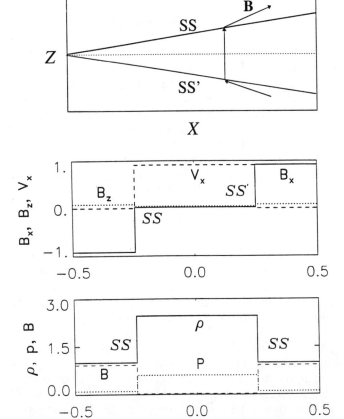

**Figure 2.** Results of case 1 with $B_y = 0$: the fronts of slow shocks and a sketch of field line in the reconnection layer (top panel) and the profiles of various physical quantities as a function of $z$ along $x = 4.6$ (middle and bottom panels).

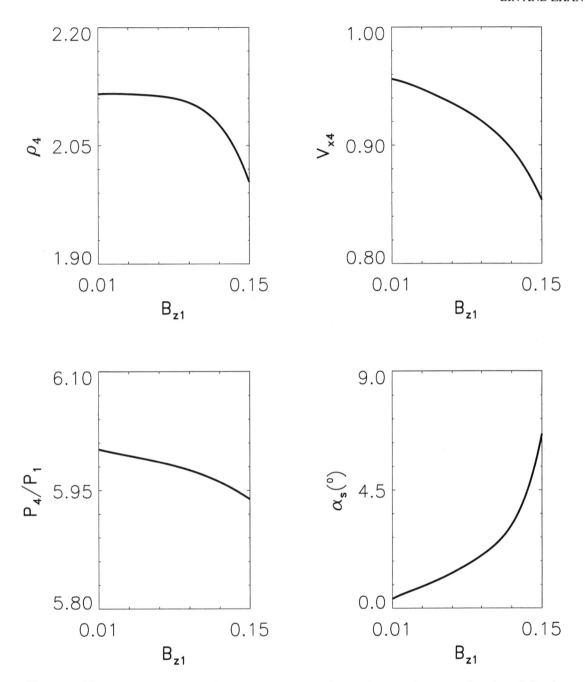

**Figure 3.** The density $\rho_4$, velocity $V_{x4}$, pressure ratio $P_4/P_1$, and the angle $\alpha_s$ as a function of $B_{z1}$ for the cases with $B_y = 0$ and $\beta_1 = 0.2$.

Figure 5 shows the structure of the resulting reconnection layer. In addition to the two slow shocks SS and SS', two rotational discontinuities RD and RD' are also present in the reconnection layer if $B_y \neq 0$. The fronts of discontinuities in the main reconnection layer are shown in the left column of Figure 5. The rotational discontinuities RD and RD' are located ahead of the slow shocks SS and SS', respectively. Similar to case 1, two weak fast expansion waves have propagated away from the main reconnection layer.

The right column of Figure 5 shows the spatial profiles of $B_x$, $B_y$, $B$, $\rho$, $P$, $V_x$, and $V_y$. Across RD, the magnetic field changes direction but the field strength, plasma density, and pressure are conserved. Note that

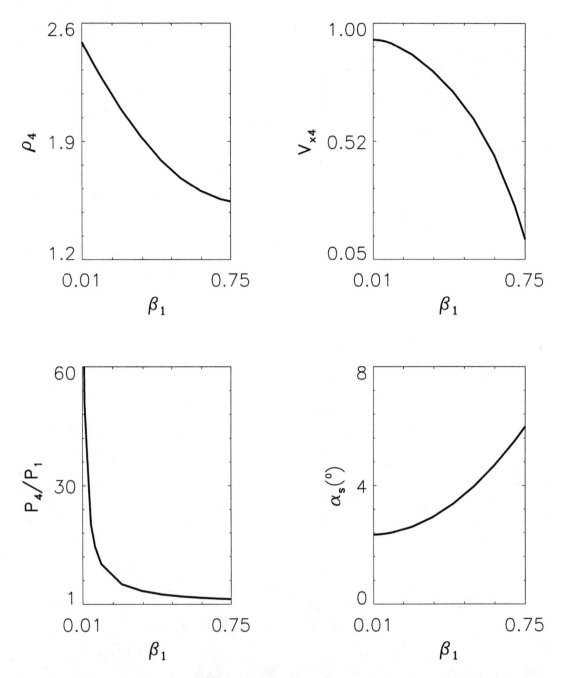

**Figure 4.** Various quantities as a function of $\beta_1$ for the cases with $B_y = 0$ and $B_{z1} = 0.1$.

the $B_x$ component of magnetic field is not equal to zero in region 3 downstream of RD, as is different from the results from the 1-D ideal MHD model for the Riemann problem [e.g., *Lin and Lee*, 1994]. This is due to the obliquity of the RD front in the 2-D configuration. The flow is accelerated across RD, and the flow speed in region 3 is nearly equal to the Alfven speed in region 1. Across the slow shock SS, $B_x$ decreases to zero at the center of the reconnection layer, while $\rho$ and $P$ increase. A considerable tangential magnetic field, however, exists in region 4, and the nonswitch-off slow shock is much weaker than that in case 1. The magnetic field and plasma quantities in regions 2, 3, and 4 are given in Table 2. The angles of RD front and SS front relative to the $x$ axis are $\alpha_r = 2.6°$ and $\alpha_s = 1°$.

Figure 6 shows the dependence of $\rho_4$, $P_4/P_1$, $B_{y3}$, and $B_{y4}$ on the guide field $B_{y1}$. The density $\rho_4$ is found to decrease with $B_{y1}$ because the slow shock is getting weaker as $B_{y1}$ increases. Likewise, the ratio $P_4/P_1$ also decreases. The angle between RD and SS fronts (not shown) increases with the guide field $B_{y1}$ because of the slow down of SS as it is getting weaker. On the other hand, since across RD the tangential field rotates and becomes a downstream along $B_y$ while the field strength does not change, $B_{y3}$ is larger for a larger $B_{y1}$. This increase in $B_{y3}$, together with the fact that SS is weaker for a larger $B_{y1}$, then results in an increase in $B_{y4}$, as plotted in the lower right panel of Figure 6. Note that $B_{y1} = 0$ is a singular case, in which $B_{y3} = 0$ and $B_{y4} = 0$.

## SUMMARY AND DISCUSSION

In summary, we have developed a 2-D ideal MHD model to calculate the structure of steady-state reconnection layer, which is assumed to have developed from

**Table 2.** Results of Case 2

| Region | 1 | 2 | 3 | 4 |
|---|---|---|---|---|
| $\rho$ | 1 | 0.955 | 0.955 | 1.993 |
| $P$ | 0.125 | 0.117 | 0.117 | 0.474 |
| $B_x$ | 1 | 0.954 | 0.010 | 0.000 |
| $B_y$ | 0.5 | 0.477 | 1.070 | 0.656 |
| $B_z$ | 0.1 | 0.094 | 0.048 | 0.050 |
| $V_x$ | 0 | 0.009 | 0.992 | 0.968 |
| $V_y$ | 0 | 0.005 | -0.601 | 0.000 |
| $V_z$ | 0 | -0.056 | -0.013 | 0.000 |

an initial current sheet in a spontaneous reconnection. The calculation has been performed for symmetric cases, which may be applied to the distant magnetotail, with a given normal component of magnetic field $B_{z1}$ corresponding to a certain reconnection rate.

In the cases with the guide field $B_y = 0$, a pair of slow shocks are present in the reconnection layer, as in the *Petschek's* [1964] model. The plasma is accelerated by the slow shocks in the outflow region of magnetic reconnection. The strength of the slow shock and the angle of the slow shock front change with $B_{z1}$ and the lobe plasma beta $\beta_1$. The presence of the slow shocks in the magnetotail reconnection layer has also been obtained in the 2-D hybrid simulations [e.g., *Lin and Swift*, 1996].

In the cases with $B_y \neq 0$, two rotational discontinuities are present upstream of two slow shocks, respectively. The tangential magnetic field rotates across the rotational discontinuity to a direction nearly along the $y$ axis but with a finite $B_x$ component. This result is different from that obtained from the 1-D MHD model of the Riemann problem, in which $B_x = 0$ behind the rotational discontinuity. The slow shocks are weaker than those in the cases with $B_y = 0$. The acceleration of plasma flow in the reconnection layer is through both the rotational discontinuities and slow shocks. The structure of reconnection layer as a function of $B_{y1}$ has been searched.

Slow shocks have been observed by satellites in the plasma sheet boundary layer associated with a distant-tail X line [*Feldman et al.*, 1984; *Smith et al.*, 1984; *Ho et al.*, 1994; *Saito et al.*, 1996]. The magnetic field decreases from lobe to the central plasma sheet across the slow shocks, the plasma density increases, and the ions are heated. The plasma is accelerated earthward or further tailward. The MHD jump conditions for slow shocks are found to be nearly satisfied. On the other hand, satellite observations from magnetotail neutral sheet crossings often indicate that the dawn-dusk field component $B_y$ is non-zero and maintains the same sign throughout the crossing [e.g., *Lui*, 1984]. It has been suggested that the $B_y$ component of the interplanetary

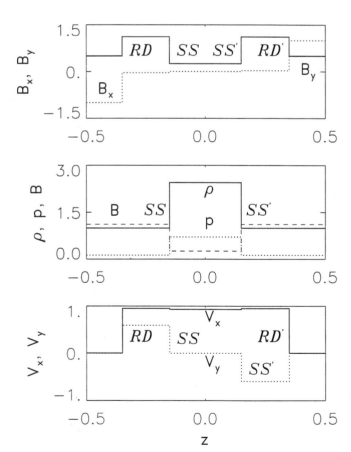

**Figure 5.** Results of case 2 with $B_{y1} = 0.5$: spatial profiles of physical quantities.

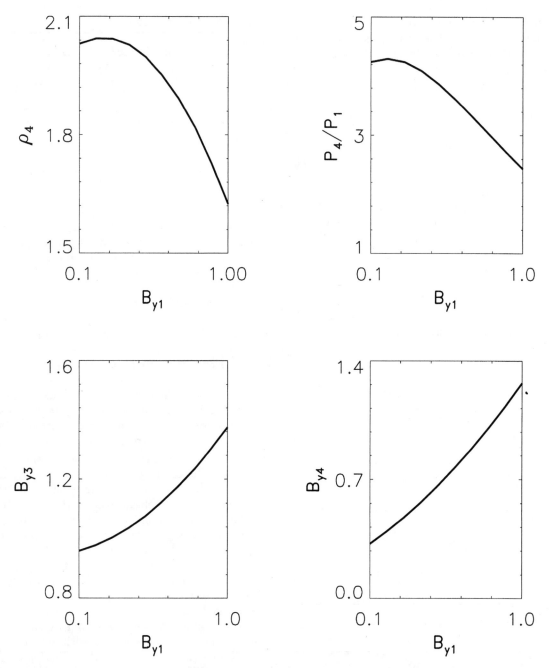

**Figure 6.** The dependence of $\rho_4$, $P_4/P_1$, $B_{y3}$, and $B_{y4}$ on the guide field $B_{y1}$ for the cases with $\beta_1 = 0.2$ and $B_{z1} = 0.1$.

magnetic field (IMF) can penetrate partially into the tail lobe. The effects of a finite lobe $B_y$ on the magnetic field and plasma signatures in the magnetotail reconnection layer have been studied by using 1-D MHD and hybrid simulations [*Lin and Lee*, 1995]. The results in this paper provide the exact solution of the structure of reconnection layer in the 2-D ideal MHD model.

Previous 2-D simulations of magnetic reconnection have shown that in the outflow region a steady-state reconnection layer can develop behind a leading, transient bulge of magnetic field [e.g., *Scholer*, 1989; *Shi and Lee*, 1990; *Ugai*, 1992; *Lin and Swift*, 1996]. If the duration of the reconnection is long (short) the steady-state reconnection layer (transient bulge) may dominate the reconnection configuration. In the distant tail, where the lobe $\beta_1$ is low, strong slow shocks are expected to be present as a result of quasi-steady reconnection if the guide field $B_y$ is very small. Nevertheless, for a typical reconnection rate around 0.1 and a small lobe $B_{y1}$, the flaring angle of the slow shock front is very close to that of the rotational discontinuity ahead of the slow shock. Consequently, the slow shock and the rotational discontinuity may be hard to distinguish.

In the ideal MHD, the discontinuities do not have a width or transition region. Our calculation is only for the jump of field and plasma quantities across the discontinuities. Nevertheless, in the real plasma sheet boundary layer, the internal structure of RD and SS is determined by ion dynamics in collisionless plasmas. The RD and SS have a finite thickness, which may be on the order of a few to ten's of the lobe ion inertial length. In the 2-D hybrid simulation of *Lin and Swift* [1996], it is shown that the discontinuities may not be well separated in the regions not far enough from the X line, and thus the jumps in field and plasma quantities across the discontinuities change with the $x$-distance from the X line.

The flaring angles of the discontinuity fronts are determined by the reconnection rate $B_{z1}$, or the reconnection electric field. They can also be estimated by measuring the flow speeds near the discontinuities. Considering $V_{x2} \ll V_{x3}$, from (9) and (14) the angle of the RD front can be written as

$$\alpha_r \simeq \frac{V_{z3} - V_{z2}}{V_{x3}} \qquad (39)$$

The shock angle of SS can be written as

$$\alpha_s \simeq \frac{V_{z3}}{V_{x3} - V_{x4}\frac{\rho_4}{\rho_3}} \qquad (40)$$

Note that $V_{z4} = 0$ has been considered in obtaining (40). In general the flow speed $V_{x3} \simeq V_{x4}$ in the reconnection layer, and the shock angle $\alpha_s$ is thus proportional to $|V_{z3}/V_{x4}|$. This result indicates that the steady-state shock front is formed by the propagation of the shock wave in the normal direction of the current sheet as well as the convection of field lines with accelerated plasmas in the tangential direction, as discussed by *Lin et al.* [1992] for the relation between the 1-D and 2-D simulation results. In the incompressible case with $B_y = 0$, the shock front angle becomes $|V_{z2}/V_{x4}|$, consistent with the shock characteristics obtained in the 2-D MHD simulation by *Yan et al.* [1992].

Finally, it should be mentioned that in collisionless plasmas the strength of the discontinuities may be modified by the ion kinetic effects. Ion beams are found upstream of the slow shock in hybrid simulations of the reconnection layer [e.g., *Lin and Lee*, 1995; *Krauss-Varban and Omidi*, 1995; *Lin and Swift*, 1996]. A temperature anisotropy is present in the reconnection layer. In the cases with a finite guide field $B_{y1}$, the slow shock behind the rotational discontinuity may become much weaker than that obtained in the MHD model.

*Acknowledgments.* This work was supported by the ONR grant NAVY-N00014-951-0839 and NSF grant ATM-9507993 to the Auburn University.

## REFERENCES

Birn, J., in E. W. Hones, Jr. (ed.), *Magnetic Reconnection in Space and Laboratory Plasmas*, p.264, AGU, Washington, D. C, 1984.

Dungey, J. W., Interplanetary magnetic field and the auroral zones, *Phys. Rev. Lett.*, 6, 47, 1961.

Feldman, W. C., et al., Evidence for slow-mode shock in the deep geomagnetic tail, *Geophys. Res. Lett.*, 11, 599, 1984.

Heyn, M. F., H. K. Biernat, R. P. Rijnbeek, and V. S. Semenov, The structure of reconnection layer, *J. Plasma Phys.*, 40, 235, 1988.

Ho, C. M., B. T. Tsurutani, E. J. Smith, and W. C. Feldman, A detailed examination of an X-line region in the distant tail: ISEE-3 observations of jet flow and $B_z$ reversals and a pair of slow shocks, *Geophys. Res. Lett.*, 21, 3031, 1994.

Krauss-Varban, D., and N. Omidi, Large-scale hybrid simulations of the magnetotail during reconnection, *Geophys. Res. Lett.*, 22, 3271, 1995.

Jeffrey, A. and T. Taniuti, *Non-Linear Wave Propagation*, Academic, Orlando, Fla., 1964.

Landau, L. D. and E. M. Lifshitz, *Electrodynamics of Continuous Media*, Pergamon Press, 1960.

Levy, R. H., H. E. Petschek, and G. L. Siscoe, Aerodynamic aspects of the magnetospheric flow, *AIAA J.*, 2, 2065, 1964.

Lin, Y., L. C. Lee, and C. F. Kennel, The role of intermediate shocks in magnetic reconnection, *Geophys. Res. Lett.*, 19, 229, 1992.

Lin, Y. and L. C. Lee, Structure of reconnection layers in the magnetosphere, *Space Sci. Rev.*, 65, 59, 1994.

Lin, Y., and L. C. Lee, A simulation study of the Riemann problem associated with the magnetotail reconnection, *J. Geophys. Res.*, 100, 19,227, 1995.

Lin, Y., and D. W. Swift, A two-dimensional hybrid simulation of the magnetotail reconnection layer, *J. Geophys. Res.*, 101, 19,859, 1996.

Lui, A. T. Y., Characteristics of the cross-tail current in the Earth's magnetotail, in *Magnetospheric Currents*, Geophys. Monogr. Ser., Vol. 28, edited by T. A. Potemra, p.158, AGU, Washington D. C., 1984.

Nakamura, M. S., M. Fujimoto, and K. Maezawa, Ion dynamics and resultant velocity space distributions in the course of magnetic reconnection, *J. Geophys. Res.*, in press, 1997.

Neubauer, F. M., Nonlinear oblique interaction of interplanetary tangential discontinuities with magnetogasdynamic shocks, *J. Geophys. Res., 80*, 1213, 1975.

Ogino, T. and R. J. Walker, *Geophys. Res. Lett., 11*, 1018, 1984.

Petschek, H. E., Magnetic field annihilation, in AAS-NASA Symposium on the Physics of Solar Flares, *NASA Spec. Publ., SP-50*, 425-439, 1964.

Priest, E. R., and T. G. Forbes, *J. Geophys. Res., 91*, 5579, 1986.

Priest, E. R., and L. C. Lee, *J. Plasma Phys., 44*, 337, 1991.

Saito, Y., T. Mukai, T. Terasawa, A. Nishida, S. Machida, S. Kokubun, and T. Yamamoto, Foreshock structure of the slow-mode shocks in the Earth's magnetotail, *J. Geophys. Res., 101*, 13,267, 1996.

Sato, T., Strong plasma acceleration by slow shocks resulting from magnetic reconnection, *J. Geophys. Res., 84*, 7177, 1979.

Scholer, M., Undriven magnetic reconnection in an isolated current sheet, *J. Geophys. Res., 94*, 8805, 1989.

Semenov, V. S., I. V. Kubyshkin, V. V. Lebedera, M. V. Sidneva, R. P. Rijnbeek, M. F. Heyn, H. K. Biernat, and C. J. Farrugia, A comparison and review of steady-state reconnection and time-varying reconnection, *Planet. Space Sci., 40*, 63, 1992.

Shi, Y., and L. C. Lee, Structure of the reconnection layer at the dayside magnetopause, *Planet. Space. Sci., 38*, 437, 1990.

Smith, E. J., J. A. Slavin, B. T. Tsurutani, W. C. Feldman, and S. J. Bame, Slow mode shocks in the Earth's magnetotail: ISEE-3, *Geophys. Res. Lett., 11*, 1054, 1984.

Sonnerup, B. U. O., Magnetic field reconnection, in *Solar System Plasma Physics, vol. III*, edited by C. F. Kennel, L. T. Lanzerotti, and E. N. Parker, p.45, North Holland, New York, 1979.

Sonnerup, B. U. O., G. Paschmann, I. Papamastorakis, N. Sckopke, G. Haerendel, S. J. Bame, J. R. Asbridge, J. T. Gosling, and C. T. Russell, Evidence for magnetic reconnection at the Earth's magnetopause, *J. Geophys., Res., 86*, 10,049, 1981.

Ugai, M., *Plasma Phys. Contr. Fusion, 26*, 1549, 1984.

Ugai, M., Computer studies on development of the fast reconnection mechanism for different resistivity models, *Phys. Fluids. B, 4*, 2953, 1992.

Ugai, M., Computer studies on plasmoid dynamics associated with the spontaneous fast reconnection mechanism, *Phys. plasmas., 2*, 3320, 1995.

Ugai, M., and T. Shimizu, Computer studies on the spontaneous fast reconnection mechanism in three dimensions, *Phys. Plasmas, 3*, 853, 1996.

Yan, M., L. C. Lee, and E. R. Priest, Fast magnetic reconnection with small shock angles, *J. Geophys. Res., 97*, 8277, 1992.

Yeh, T., and W. Axford, *J. Plasma Phys., 4*, 207, 1970.

---

Y. Lin and X. X. Zhang, 206 Allison Lab, Physics Department, Auburn University, Auburn, AL 36849-5311

# Particle Acceleration Due to Magnetic Field Reconnection in a Model Current Sheet.

Victor M. Vazquez, Maha Ashour-Abdalla,[1]

*Department of Physics and Astronomy, UCLA, Los Angeles, California*

Robert L. Richard

*Institute of Geophysics and Planetary Physics, UCLA, Los Angeles, California*

We have investigated the acceleration of particles to energies up to the order of 1 MeV in the magnetotail by using the method of large scale kinetics (LSK). We used a Harris neutral sheet model combined with an explosively growing tearing mode perturbation to model a magnetotail magnetic field undergoing rapid reconnection. We calculated the trajectories of protons as well as electrons. We found that protons and electrons can be accelerated by the inductive electric field to energies above 1 MeV. We found that a proton's final energy depends strongly on its initial position, but not on its initial velocity. In addition, we observed both regular and inverse temporal dispersion in our model.

## 1. INTRODUCTION

The electrostatic potential drop across the Earth's magnetotail is rarely above 100 kV. However, particle bursts with energies above 1 MeV have been observed in the Earth's magnetotail. It has been shown that explosive tearing mode reconnection in the magnetotail is capable of accelerating particles to high energies [*Galeev et al.*, 1978; *Galeev*, 1979; *Zelenyi et al.*, 1990]. Observations by spacecraft imply that these energetic particle bursts are related to magnetic field reconnection events [*Krimigis and Sarris*, 1980; *Sarris et al.*, 1996]. We calculated the trajectories of protons and electrons numerically to simulate the acceleration process by using the large scale kinetic (LSK) technique [*Ashour-Abdalla et al.*, 1993, 1995].

## 2. SIMULATION MODEL

Our magnetic field model consists of a static Harris neutral sheet configuration with a time-varying component which models the explosive growth of the tearing mode [*Zelenyi et al.*, 1990]:

$$\vec{B} = B_{ox} tanh(\frac{z}{L})\hat{e}_x + \frac{B_{oz}}{(1-\frac{t}{\tau_r})} sin(kx)\hat{e}_z.$$

The tearing mode instability model leads to the formation of multiple magnetic islands centered around the current sheet. Our coordinate system is such that there is an X-line at the origin with additional X-lines at $x = \pm 10$ and $\pm 20 R_E$, and $z = 0$ corresponds to the current sheet plane (See Figure 1a).

Our magnetic field parameters are: $B_{ox} = 2 \times 10^{-8} T$, $B_{oz} = 2 \times 10^{-9} T$, $\tau_r = 10 sec$, $L = \frac{1}{6}R_E$, $k = \frac{2\pi}{10 R_E}$. The time-varying magnetic field induces an electric field,

---

[1]Also at Institute of Geophysics and Planetary Physics, UCLA, Los Angeles, California.

Geospace Mass and Energy Flow: Results From the International Solar-Terrestrial Physics Program
Geophysical Monograph 104
Copyright 1998 by the American Geophysical Union

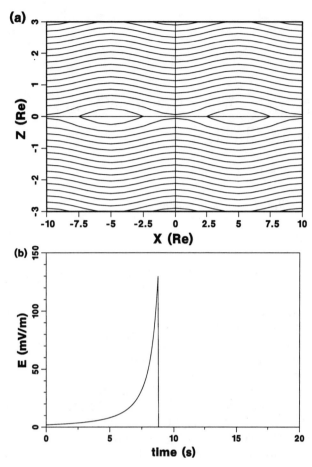

**Figure 1.**: Magnetic and electric field models: (a) Magnetic field topology at $t_s$, (b) Electric field profile at $x = 0$ in mV/m as a function of time.

$$\vec{E} = \frac{B_{oz}}{k\tau_r} \frac{cos(kx)}{(1-\frac{t}{\tau_r})^2} \hat{e}_y,$$

which is calculated from Faraday's Law. Near the X and O-lines the magnetic field is weak and it is possible for the electric field to accelerate charged particles. Instability saturation was set to occur when $B_z = 0.8 B_{ox}$. At that time, $t_s = 8.75$ seconds, the electric field was turned off ($\vec{E} = 0$, see Figure 1b). The electric field's maximum value was $\sim 130$ mV/m, and there was no cross-tail electrostatic potential. The magnetic field is held fixed after the instability saturation, $t_s$.

The simulation system size was $-20.0 < x < 20.0 R_E$, $-50.0 < y < 50.0 R_E$, $-10.0 < z < 10.0 R_E$. Particles were launched with a Maxwellian velocity distribution with a temperature of 1.0 keV. The initial particle density had the form $n = n_o sech^2(\frac{z}{L})$. Particles were uniformly distributed in x. We calculated the trajectories by solving the Lorentz force equation, $\vec{F} = q(\vec{E} + \vec{v} \times \vec{B})$, using a fourth-order Runge-Kutta routine.

## 3. SIMULATION RESULTS

### 3.1. Protons

In our system, protons gained as much as 1 MeV. Particles with energies in this range have been observed by *Krimigis and Sarris,* [1980], and to energies in the order of 1 MeV/q by *Kirsch et al.,* [1981]. The amount of energy which a proton gained depended on its initial position relative to the magnetic field's X and O-lines. The greatest energization occurred near the X-line because the magnetic field was weak and the particles could cover a large distance across the electric potential. Protons reaching energies of 1 MeV at saturation were those near an X-line initially. What we call "midenergy" protons were produced near the O-lines. These were protons that were launched near $x = \pm 5 R_E$, and had final energies in the range of 100 to 150 keV. A proton that starts near an O-line will $\vec{E} \times \vec{B}$ drift away from the current and that O-line. A particle initially above and below the $z = 0$ plane near an X-line will $\vec{E} \times \vec{B}$ drift towards that X-line. Near $x = \pm 2.5 R_E$ are what we call the stagnation regions. In these regions the electric field is weak, and the magnetic field (at z=0) is at a local maximum. The $\vec{E} \times \vec{B}$ drift is toward the stagnation region on either side of it because the electric field changes sign at $x = \pm 2.5 R_E$. Protons which had initial positions near $x \simeq \pm 2.5 R_E$ as well as protons that started elsewhere but drifted toward $x = \pm 2.5 R_E$ remained at low energies.

Two different types of velocity dispersion are evident in our results: normal dispersion, when high energy particles arrive at the detector before lower energy particles, and inverse dispersion, when low energy particles arrive at a detector before higher energy particles. Inverse velocity dispersion has been observed by various spacecraft (IMP7 and 8 [*Krimigis and Sarris,* 1980], and more recently, GEOTAIL (See Figure 3 in *Sarris et al.,* [1996])).

Figure 2(a) and figure 2(b) display the energies of protons that cross a planar virtual detector, as a function of time. Each plot displays the protons that crossed at a different detector along the x-axis. Early in the run, all particles have low energies, but as the electric field increases, higher energy particles are seen.

### 3.2. Electrons

We followed electrons in the same model fields that were used for the protons. Because electrons could reach relativistic speeds, the particle push was modified to include relativistic effects. The highest energy electron reached an energy near 10 MeV, an order of magnitude greater than the highest energy protons.

Because they are moving at speeds comparable to

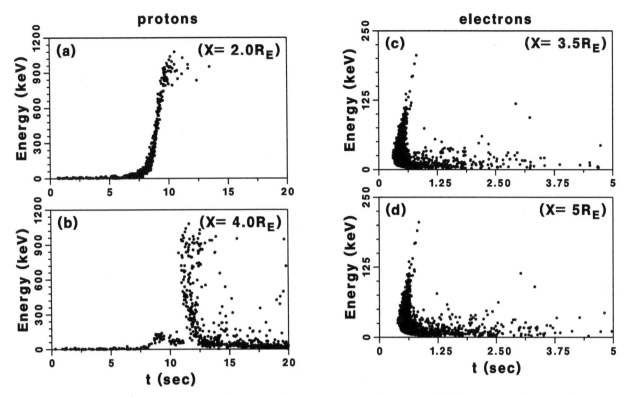

**Figure 2.:** Particle energies as a function of time at virtual detectors: (a) Protons at the $x = 2R_E$ detector and (b) at the $x = 4R_E$ detector. (c) Electrons at the $x = 3.5R_E$ detector and (d) at the $x = 5R_E$ detector. All plots show the energy in keV as a function of the time in seconds.

the speed of light, electrons travel very far in a short period of time. Figures 2(c) and (d) show the dispersion plots for electrons. Inverse dispersion dominates the dispersion plots for electrons, but normal dispersion is also visible.

## 4. CONCLUSIONS

In summary, we have reproduced two of the features of the observations by the IMP7 and 8 spacecraft: particles with energies above 1 MeV and normal and inverse velocity dispersion [*Krimigis and Sarris*, 1980].

Both protons and electrons can be accelerated to energies greater than 1 MeV in a tail-like magnetic field configuration undergoing explosive reconnection. In our model, the final energy of a proton depended strongly on its initial position at the onset of reconnection, but not on its initial velocity. For protons, there are three different spatial regions, relative to magnetic field structures, of distinct particle behavior. Electron energies were not strongly dependent on either initial position or velocity. For electrons, strong acceleration occurred only near X-lines, and very high energy electrons resulted from these runs. Although electrons have been observed at energies above 1 MeV [*Sarris et al.*, 1976; *Krimigis and Sarris*, 1980], higher energies are not expected to be observed due to low electron fluxes at higher energies and the sensitivity threshold of instruments aboard spacecraft [*Zelenyi et al.*, 1990]. Both species displayed inverse as well as regular velocity dispersion.

*Acknowledgments.* We would like to thank L. M. Zelenyi and A. L. Taktakishvili for useful discussions. This work was supported by NASA International Solar Terrestrial Physics program grant NAG5-1100. Computing support was provided by the Pittsburgh Supercomputing Center.

## REFERENCES

Ashour-Abdalla, M., J. Berchem, J. Büchner and L. M. Zelenyi, Shaping of the magnetotail from the mantle: global and local structuring, *J. Geophys. Res.*, 98, 5651, 1993.

Ashour-Abdalla, M., L. M. Zelenyi, V. Peroomian, R. L. Richard, and J. M. Bosqued, The mosaic structure of plasma bulk flows in the Earth's magnetotail, *J. Geophys. Res.*, 100, 19,191, 1995.

Galeev, A. A., Reconnection in the Magnetotail, *Space Sci. Rev.*, 23, 411-425, 1979.

Galeev, A. A., F. V. Coroniti, M. Ashour-Abdalla, Explosive Tearing Mode Reconnection in the Magnetospheric Tail, *Geophys. Res. Lett.*, 5, 707-711, 1978.

Kirsch, E., S. M. Krimigis, E. T. Sarris, and R. P. Lepping, Detailed study on acceleration and propagation of energetic protons and electron in the magnetotail during substorm activity, *J. Geophys. Res., 86*, 6727-6738, 1981.

Krimigis, S. M., and E. T. Sarris, Energetic Particle Bursts in the Earth's magnetotail, in *Dynamics of the Magnetosphere*, edited by S.-I. Akasofu, pp. 599-630, D. Reidel, Hingham, Mass., 1980.

Sarris, E. T., S. M. Krimigis, and T. P. Armstrong, Observations of magnetospheric bursts of high-energy protons and electrons at 35 $R_E$ with IMP 7, *J. Geophys. Res., 81*, 2341-2355, 1976.

Sarris, E. T., V. Angelopoulos, R. W. McEntire, D. J. Williams, S. M. Krimigis, A. T. Lui, E. C. Roelof, and S. Kokubun, Detailed Observations of a Burst of Energetic Particles in the Deep Magnetotail by Geotail, *J. Geomag. Geoelectr., 48*, 649-656, 1996.

Zelenyi, L. M., A. S. Lipatov, J. G. Lominadze, and A. L. Taktakishvili, The dynamics of the energetic proton bursts in the course of the magnetic field topology reconstruction, *Planet. Space Sci., 32*, 312-324, 1984.

Zelenyi, L. M., J. G. Lominadze, and A. L. Taktakishvili, Generation of the Energetic Proton and Electron Bursts in Planetary Magnetotails, *J. Geophys. Res., 95*, 3883-3891, 1990.

---

V. M. Vazquez, M. Ashour-Abdalla, R. Richard, Department of Physics and Astronomy and Institute of Geophysics and Planetary Physics, UCLA, Los Angeles, CA. (e-mail: victor@igpp.ucla.edu; mabdalla@igpp.ucla.edu; rrichard@igpp.ucla.edu)

# Modeling Magnetotail Ion Distributions with Global Magnetohydrodynamic and Ion Trajectory Calculations

M. El-Alaoui, M. Ashour-Abdalla, J. Raeder, V. Peroomian

*Institute of Geophysics and Planetary Physics, University of California, Los Angeles, CA 90095-1567*

L. A. Frank, W. R. Paterson

*Department of Physics and Astronomy, The University of Iowa, Iowa City*

J. M. Bosqued

*Centre d'Etude Spatiale des Rayonnements, CNRS, Toulouse, France*

On February 9, 1995, the Comprehensive Plasma Instrumentation (CPI) on the Geotail spacecraft observed a complex, structured ion distribution function near the magnetotail midplane at $x \sim -30\ R_E$. On this same day the Wind spacecraft observed a quiet solar wind and an interplanetary magnetic field (IMF) that was northward for more than five hours, and an IMF $B_y$ component with a magnitude comparable to that of the IMF $B_z$ component. In this study, we determined the sources of the ions in this distribution function by following approximately 90,000 ion trajectories backward in time, using the time-dependent electric and magnetic fields obtained from a global MHD simulation. The Wind observations were used as input for the MHD model. The ion distribution function observed by Geotail at 1347 UT was found to consist primarily of particles from the dawn side low latitude boundary layer (LLBL) and from the dusk side LLBL; fewer than 2% of the particles originated in the ionosphere.

## 1. INTRODUCTION

A fundamental goal of magnetospheric physics has been to determine the mechanism for transport of plasmas through the solar wind-magnetosphere-ionosphere system. Growing evidence indicates that the solar wind and the ionosphere, the main sources of hot plasma in the magnetosphere, contribute roughly comparable amounts [*Shelley et al.*, 1982; *Lennartsson and Shelley*, 1986]. The most definitive information on relative contributions of the ionosphere and the solar wind has been deduced from ion composition measurements, with $O^+$ ions used as tracers of ionospheric plasma and $He^{++}$ ions used as solar wind tracer ions. The $H^+$ ions that make up a large fraction of the plasma, however, may come from either source. This study focuses only on $H^+$.

To understand the transport of plasma through the solar wind-magnetosphere-ionosphere system, it is necessary to determine not only the sources of the plasma, but also its transport mechanisms throughout the magnetosphere and the geomagnetic tail. The MHD models yield a picture of the overall configuration of the magnetosphere and bulk transport, but the MHD paradigm neglects important physics such as particle drift motion. To address this deficiency, we used time-dependent Large-Scale Kinetics (LSK) to determine the history of the particles in the measured distribution function and then used this information to extend the capability of the MHD model. In this paper we apply an approach in which we use both numerical simulations and observations to study the origins and transport of magnetotail plasma. We begin with a global magnetohydrodynamic (MHD) simulation of the interaction between the solar wind and the magnetosphere that employs observed solar wind and interplanetary magnetic field (IMF) parameters for the solar wind boundary conditions of the simulation [*Ashour-Abdalla et al.*, this issue]. The MHD code is run to model the time dependent response of the magnetosphere to the changing solar wind and to obtain the global magnetic and electric fields [*Raeder et al.*, 1997]. Next we launch ions in accordance with an observed distribution function and follow the trajectories backwards in time in the MHD electric and magnetic fields [*El-Alaoui et al.*, 1995; *Ashour-Abdalla et al.*, 1997a]. We chose to apply this approach first to a study of the "quiet" magnetosphere when the IMF is northward for an extended time. Although the magnetosphere is less dynamic during intervals of northward IMF, the magnetic configuration is still very complex.

## 2. OBSERVATIONS

On February 9, 1995, between 0800 UT and 1400 UT the Wind spacecraft was located near (190, 40, 2) $R_E$ in GSE coordinates. The observed IMF and the solar wind parameters were quite stable for the entire 6 hours. During this time interval a northward IMF ($B_z \sim 5$ nT) with a significant $B_y$ component was observed. The $B_y$ component was directed duskward and was comparable in magnitude to the $B_z$ component. The solar wind speed and density averaged roughly 400 km/s and 5 cm$^{-3}$ respectively. During this same time interval the Geotail satellite was located in the tail, 30 $R_E$ from Earth. Plate 1 shows $v_y$ - $v_z$, $v_x$ - $v_z$ and $v_x$ - $v_y$ cuts of the 3-D ion distribution function measured at 1347 UT by the Hot Plasma Analyzer (HP) of the Comprehensive Plasma Instrumentation (CPI) [*Frank et al.*, 1994]. The arrow (in the middle panel) represents the projection of the magnetic field vector onto the $z$ - $x$ plane. This distribution function is non-Maxwellian and has additional structures in the direction perpendicular to the magnetic field. The distribution is also warmer than in the previous time interval (not shown here) [*Ashour-Abdalla et al.*, 1997c].

## 3. METHODOLOGY

The IMF and solar wind data from the Wind spacecraft were used as input to a global MHD simulation. The simulation was validated by comparing its magnetic field and flow velocity time series at the nominal Geotail location (30 $R_E$ downtail) with those observed by Geotail from the time interval 0800 UT to 1400 UT (see *Ashour-Abdalla et al.* [1997c] for a complete description). The comparison showed reasonable agreement between the model's results and the observations for the period when Geotail was in the boundary of the plasma sheet. However, the MHD simulation did not reproduce the multiple penetrations of the central plasma sheet made by the Geotail spacecraft, though the agreement was better when the "virtual satellite" in the MHD model was moved upward in $z$ by 1.5 $R_E$. This lack of agreement was evidently a result of the fact that Geotail was in a region of steeper gradients than those reproducible within the spatial resolution of the MHD model.

The 3-D distribution function shown in Plate 1 was partitioned into (75 km/s)$^3$ bins with the number of model ions in each $v_x$ - $v_y$ - $v_z$ bin chosen to be proportional to the phase space density measured by Geotail for each bin. The fine-scale structure of the observed distribution was preserved in this partitioned distribution function. We followed the trajectories of 90,000 noninteracting ions in the time dependent MHD electric and magnetic fields backward in time until they reached their origin, i.e. the magnetopause (as determined from the MHD total current) or the ionosphere (taken to be those particles approaching to within 3.7 $R_E$ of the Earth). Magnetic and electric fields were obtained from the global MHD calculations at four minute intervals. We used linear interpolation in space and time to obtain the field values between time and spatial grid points for the trajectory calculations. In these calculations the electric field was taken to be

$$\vec{E} = -\vec{U} \times \vec{B} + \eta \vec{J}$$

where $\vec{U}$ is the bulk flow velocity, $\vec{B}$ is the magnetic field, $\vec{J}$ is the current density, and $\eta$ is the resistivity. Resistive ($\eta \vec{J}$) contributions to $\vec{E}$ become significant only near the magnetopause and near the x-lines. A 4th order Runge-Kutta method was used to integrate the ion trajectories with respect to time. One time step in the

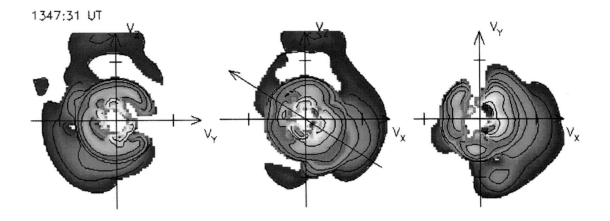

**Plate 1.** Velocity distribution function measured by Geotail on February 9, 1995, at 1347 UT.

particle trajectory calculation was 0.002 of the local ion gyro-period, with an upper limit applied to prevent the time step's becoming too large in weak field regions.

## 4. RESULTS

The calculated points of origin for all of the simulation's ions are shown in Plate 2. The upper left panel shows these points projected onto the noon-midnight meridian plane ($y = 0$), and in the lower left panel they are projected onto the equatorial plane ($z = 0$). Solid black curves indicate the intersections of the magnetopause with the $y = 0$ and $z = 0$ planes. The color scale to the right of each panel denotes the number of particles in each $1 R_E \times 1 R_E$ bin. The particles come from both the dawn and the dusk flanks of the magnetosphere, and the sources form broad bands in the $y$ and $z$ directions along the magnetopause.

For the dusk side magnetopause, the upper left panel shows that the source region is localized around $z = 0$ at $x = -30 R_E$. Further downtail, the particles enter at larger values of $z$, reaching as high as the vicinity of $z = 10 R_E$ at $x = -120 R_E$. Most of the particles from the dusk side magnetopause come from 0 to $30 R_E$ down the tail. Further down the tail the density of the entering particles is very low, with only a few particles per bin. The dawn side magnetopause particle entry is concentrated much further downtail. The density of the particles per bin increases downtail on the dawn side magnetopause, with the peak density occurring between $-90 R_E$ and $-120 R_E$. The entry region is near $z = -10 R_E$ at $x = -120 R_E$. The ionospheric contribution (labeled "ionosphere" in Plate 2) is small, providing only 2% of the ions in the observed Geotail distribution. This is because Geotail was relatively far from the Earth ($\sim 30 R_E$). Also, the IMF was steady and northward for more than 6 hours on this day and had a large $B_y$ component. The ionosphere is therefore not a viable source of plasma for this time period. The right panel of Plate 2 shows the MHD current density as a function of $y$ and $z$ at $x = -100 R_E$, and reveals the twisting of the magnetotail. The black dots are projections of all of the ion entry points onto this plane ($x = -100 R_E$). It can be seen that the entry regions correspond to the LLBL. The twisting of the tail explains the location of the LLBL above (below) $z = 0$ on the dusk (dawn) side.

In Plate 3 the $v_y$ - $v_z$, $v_x$ - $v_z$ and $v_x$ - $v_y$ cuts of the initial 3-D Geotail ion distribution function are color coded according to the source regions of the ions. If more than 75% of the particles in a bin have been supplied by a single source, the bin is given the color indicating that source; otherwise the bin is black. No bin is dominated by the ionospheric source. The low energy particles streaming tailward along the $x$ axis are mainly from the dusk side LLBL (red dots). The particles streaming earthward are from the dawn side LLBL (green dots). The higher energy part of the tailward streaming distribution is also populated by the dawnside LLBL (Plate 3a, 3b).

Figure 1 and Figure 2 show two typical test particle trajectories from our calculations. The top panel of each figure displays an individual trajectory in a three-dimensional perspective, and the lower panel of each figure shows the particle energy versus time with the parameter of adiabaticity $\kappa$ (shown with gray dots) superimposed [*Büchner and Zelenyi*, 1986, 1989]. $\kappa$ is defined as

294 MODELING MAGNETOTAIL ION DISTRIBUTIONS

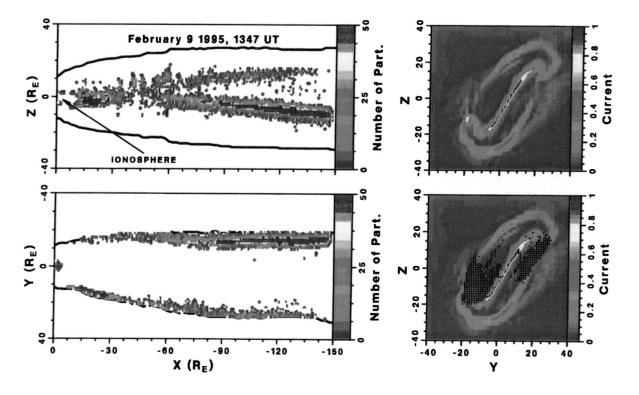

**Plate 2.** Magnetospheric entry points for all particles projected onto the x-y (lower left-hand panel) and x-z planes (upper left-hand panel). The upper right panel shows a cross section o the MHD total current density at the Geotail location (x = - 30 $R_E$) at 1347 UT. This panel shows the twisting of the geomagnetic tail around the Sun-Earth axis toward the dusk side. The lower right-hand panel shows a projection of all entry points onto this plane (black dots).

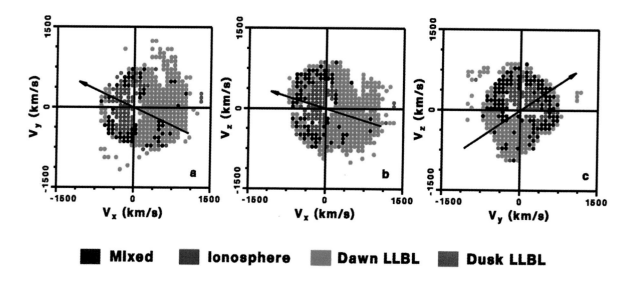

**Plate 3.** 3-D cuts of the velocity distribution function color-coded according to dominant particle source. The color table is shown below the figure.

$$\kappa \equiv \sqrt{R_c/\rho_L}$$

where $R_c$ is the local magnetic field radius of curvature and $\rho_L$ is the ion Larmor radius. The upper panel of Figure 1a depicts an ion that originated in the dawn side magnetopause at $x \sim -90\ R_E$, where it has a kinetic energy of $\sim 2$ keV. The particle crosses the magnetotail from dawn to dusk, executes several nonadiabatic ($\kappa \leq 1$) interactions with the current sheet (gray dots in the lower panel) during which it gains energy, and finally arrives at Geotail (green dot) with an energy of $\sim 4$ keV. This type of particle makes up the bulk of the distribution measured in the direction perpendicular to $\vec{B}$.

Figure 2 shows a particle that enters the magnetosphere at $x \sim -20\ R_E$ on the dusk side of the magnetosphere. This ion becomes trapped on near-Earth closed field lines, the particle then executes several adiabatic ($\kappa \gg 1$) bounces while drifting dawnward until it encounters Geotail. This orbit is representative of the ions entering the magnetosphere on the dusk flank between $x = -10\ R_E$ and $-30\ R_E$ (see Plate 2). The particle's energy does not change significantly during its journey and the particle arrives at Geotail (shown by the green dot) with an energy of 2 keV.

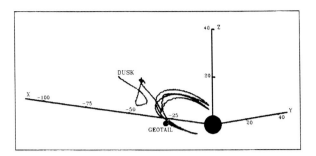

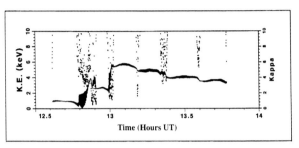

**Figure 2.** Trajectory of a typical ion from the dusk side LLBL measured by Geotail at 1347 UT. The figure is in the same format as Fig. 1.

## 5. SUMMARY

The calculations presented in this paper were based on observations made on February 9, 1995, by the Wind spacecraft, positioned upstream of the magnetosphere in the solar wind, and by the Geotail satellite, located 30 $R_E$ downtail, while the IMF was northward and all solar wind parameters were steady. A relatively large $B_y$ component led to a magnetospheric configuration unlike any that had previously been produced by analytical models or idealized simulations. The complexity of the magnetosphere during this time period resulted from the twisting of the tail around the Sun-Earth axis toward the dusk side. Test particles were launched at the Geotail location $(-29., 1.53, -2.58)\ R_E$ and were followed backward in time through an MHD field model, which was also regressed through its previous states. This time-dependent large scale kinetic study identified the origins of different ion subsets of the ion velocity distribution function measured on February 9, 1995, at 1347 UT by Geotail. The results can be summarized as follows:

1. The IMF $B_y$ component caused a twisting of the magnetotail current sheet. Because the IMF $B_y$ was steady for more than 6 hours, the magnetotail cross section showed little change as a function of time.

2. Two sources, the dawn side LLBL and the dusk side LLBL, contributed most of the observed ion population.

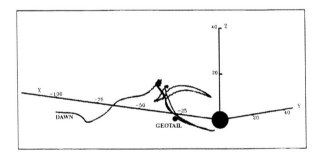

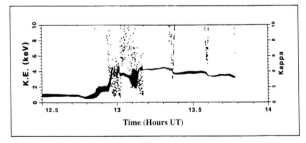

**Figure 1.** Trajectory of a typical ion from the dawn side LLBL measured by Geotail at 1347 UT. The upper panel shows a three-dimensional perspective of the particle's trajectory. The lower panel shows the particle's kinetic energy (scale on the left), and $\kappa$ (gray dots, scale on the right).

3. The ionospheric contribution was minor, making up only 2% of the ions observed by Geotail. The smallness of this contribution could be a result of the following factors: 1) the IMF was northward for an interval of more than 6 hours; and 2) Geotail was postitioned beyond $x \sim -20\ R_E$, a region where the ionospheric contribution to the magnetotail plasma population may be small [e.g. *Lennartsson and Shelley*, 1986].

4. The dawn side LLBL ions interacted with the current sheet tailward of Geotail. These particles experienced nonadiabatic ($\kappa < 1$) behavior and were substantially energized during their neutral sheet crossing. They accounted for the earthward streaming portion of the distribution function and the outer edge of the tailward streaming portion. The source density of these particles increased downtail.

5. The dusk side LLBL ions reached the Geotail location mainly by convecting adiabatically earthward from the duskward flank of the magnetosphere to form the low energy part of the tailward streaming portion of the distribution function. In contrast to the dawn side entry, the peak in the dusk side entry region occurred earthward of Geotail. The dusk side ions became trapped in adiabatic trajectories as soon as they entered the magnetosphere and arrived at Geotail while streaming tailward.

*Acknowledgments.* The authors thank K. Ogilvie for providing the Wind plasma data and R. P. Lepping for providing the Wind IMF data. We also thank R. L. Richard for helpful discussions. This work was supported by NASA grants NAG5-1100 and NAGW-4553. Computing support was provided by the Cornell Theory Center, the UCLA Office of Academic Computing, the Maui High Performance Computing Center and the San Diego Supercomputer Center. UCLA/IGPP publication number 4937.

## REFERENCES

Ashour-Abdalla, M., et al., Ion sources and acceleration mechanisms inferred from local distribution functions, *Geophys. Res. Lett.*, 24, 955, 1997a.

Ashour-Abdalla, M., et al., Determination of particle sources for a Geotail distribution funcion observed on May 23, 1995, in *Geophysical Monograph Series*, *Global Observations and Models in the ISTP Era*, edited by J. Horwitz, G. L. Gallagher, and W. K. Peterson, this issue, 1997b.

Ashour-Abdalla, M., J. Raeder, M. El-Alaoui, and V. Peroomian, Magnetotail structure and its internal particle dynamics during northward IMF, in *Geophysical Monograph Series*, *The Earth's Magnetotail: New Perspectives*, edited by A. Nishida, submitted, 1997c.

Büchner, J., and L. M. Zelenyi, Deterministic chaos in the dynamics of charged particles near a magnetic field reversal, *Phys. Lett. A, 118*, 395, 1986.

Büchner, J., and L. M. Zelenyi, Regular and chaotic charged particle motion in magnetotail-like field reversals, 1. Basic theory of trapped motion, *J. Geophys. Res.*, 94, 11821, 1989.

El-Alaoui, M., M. Ashour-Abdalla, J. Raeder, and J. M. Bosqued, Simulation of ion trajectories in the magnetotail using time-dependent electromagnetic fields, in AGU Spring Meeting, Baltimore, (*EOS*, vol. 76, no. 16), 1995.

Frank, L. A., K. L. Ackerson, W. R. Paterson, J. A. Lee, M. R. English, and G. L. Pickett, The Comprehensive Plasma Instrumentation (CPI) for the GEOTAIL spacecraft, *J. Geomagn. Geoelec.*, 46, 23, 1994.

Lennartsson, W., and E. G. Shelley, Survey of 0.1- to 16- keV/e plasma sheet ion composition, *J. Geophys. Res.*, 91, 3061, 1986.

Raeder, J., J. Berchem, M. Ashour-Abdalla, L. A. Frank, W. R. Paterson, K. L. Ackerson, S. Kokubun, and T. Yamamoto, and J. A. Slavin, Boundary layer formation in the magnetotail: Geotail observations and comparisons with a global MHD model, *Geophys. Res. Lett.*, 24, 951, 1997.

Shelley, E. G., W. K. Peterson, A. G. Ghielmetti, and J. Geiss, The polar ionosphere as a source of energetic magnetospheric plasma, *Geophys. Res. Lett.*, 9, 941, 1982.

---

Mostafa El-Alaoui, Maha Ashour-Abdalla, Joachim Raeder, and Vahé Peroomian, Institute of Geophysics and Planetary Physics, University of California, Los Angeles, CA 90095-1567 (e-mail mostafa@igpp.ucla.edu).

L. A. Frank, and W. R. Paterson, Department of Physics and Astronomy, The University of Iowa, Iowa City, IA 52242.

J. M. Bosqued, Centre d'Etude Spatiale des Rayonnements, CNRS, BP 4346, 31029 Toulouse Cedex, France.

# Determination of Particle Sources for a Geotail Distribution Function Observed on May 23, 1995

M. Ashour-Abdalla[1], M. El-Alaoui[1], V. Peroomian[1], J. Raeder[1], R. L. Richard[1], R. J. Walker[1], L. M. Zelenyi[2], L. A. Frank[3], W. R. Paterson[3], J. M. Bosqued[4], R. P. Lepping[5], K. Ogilvie[5], S. Kokubun[6], and T. Yamamoto[7]

On May 23, 1995, the Comprehensive Plasma Instrumentation (CPI) onboard the Geotail spacecraft observed a complex and structured ion distribution function near the magnetotail midplane at $x \sim -10\ R_E$. On the same day, the Wind spacecraft observed a very high density ($\sim 40\ cm^{-3}$) solar wind and an interplanetary magnetic field (IMF) that was predominantly northward but had several southward turnings. We have inferred the sources of the ions in this distribution function by following approximately 90,000 ion trajectories backward in time using time-dependent electric and magnetic fields obtained from a global MHD simulation. Wind data were used as input for the MHD model. We found that three sources contributed to this distribution: the ionosphere, the plasma mantle which had near-Earth and distant tail components, and the low latitude boundary layer (LLBL). Moreover, distinct structures in the low energy part of the distribution function were found to be associated with individual sources. Structures near 0° pitch angle were made up of either ionospheric or plasma mantle ions, while structures near 90° pitch angle were dominated by ions from the LLBL source. Particles that underwent nonadiabatic acceleration were numerous in the higher energy part of the ion distribution function, whereas ionospheric and LLBL ions were mostly adiabatic. A large proportion of the near-Earth mantle ions underwent adiabatic acceleration, while most of the distant mantle ions experienced nonadiabatic acceleration.

[1]Institute of Geophysics and Planetary Physics, University of California, Los Angeles, California
[2]Space Research Institute, Russian Academy of Sciences, Moscow, Russia
[3]Department of Physics and Astronomy, The University of Iowa, Iowa City, Iowa
[4]Centre d' Etude Spatiale des Rayonnements, CNRS, Toulouse, France
[5]NASA/Goddard Space Flight Center, Greenbelt, Maryland
[6]STELAB, Nagoya University, Toyokawa, Aichi, Japan
[7]ISAS, Sagamihara, Kanagawa, Japan

Geospace Mass and Energy Flow: Results From the International Solar-Terrestrial Physics Program
Geophysical Monograph 104
Copyright 1998 by the American Geophysical Union

## 1. INTRODUCTION

A fundamental issue of magnetospheric physics is determining the origin of magnetospheric plasma (see review by *Hill* [1974]). Early investigations considered likely plasma sources to be solar wind plasma diffusing across the magnetopause [*Eastman and Hones*, 1979] and the polar wind. Observations revealed a layer of enhanced density adjacent to the magnetopause, which was termed the low latitude boundary layer (LLBL); the properties of this region indicated that it was a site of solar wind mass, momentum and energy transfer to the magnetosphere [*Hones et al.*, 1972a; *Eastman et al.*, 1976]. *Hones et al.* [1972b] also identified another potential source, the high latitude plasma mantle. *Pilipp and*

*Morfill* [1978] postulated that the mantle source could supply all of the plasma observed in the magnetotail. The plasma mantle and the LLBL are likely entry points for solar wind plasma. The first indication that ions of terrestrial origin could also constitute a substantial component of the energetic ion population in the magnetosphere was given by ion composition measurements [*Shelley et al.*, 1972] which showed fluxes of $O^+$, evidently of ionospheric origin, to be more common than $H^+$ fluxes during magnetic storms. Because there is a negligible amount of $O^+$ in the solar wind, this component was used as a tracer of ionospheric plasma. Since then, composition measurements from different satellites have established the presence of substantial $O^+$ fluxes everywhere in the magnetosphere at various times [*Geiss et al.*, 1978; *Lennartsson et al.*, 1979, 1981; *Lundin et al.*, 1980; *Balsiger et al.*, 1980; *Sharp et al.*, 1981]. Exactly how the ions reach a given region of the magnetotail from a plasma source, as well as the precise locations of the source regions, remains an open question.

Numerous studies have investigated plasma transport through the magnetotail; these include studies of ion beam acceleration and velocity dispersion in the plasma sheet boundary layer (PSBL) [*Jaeger and Speiser*, 1974; *Cowley*, 1980; *Williams*, 1981; *Green and Horwitz*, 1986; *Schindler and Birn*, 1987; *Curran and Goertz*, 1989], processes occurring near magnetic x-lines [*Speiser*, 1965; *Amano and Tsuda*, 1978; *Wagner et al.*, 1981; *Martin*, 1986; *Martin and Speiser*, 1988; *Savenkov et al.*, 1991], and studies of ion dynamics when there is a normal component of the magnetic field that is constant or slowly varying downtail [*Lyons and Speiser*, 1982; *Zelenyi et al.*, 1988; *Büchner and Zelenyi*, 1989]. A major result of particle trajectory studies was a recognition of the importance of the chaotic nature of ions [*Büchner and Zelenyi*, 1986, 1989; *Chen and Palmadesso*, 1986]. These studies used idealized electric and magnetic field models to show that particles could experience nonadiabatic acceleration in regions of highly curved magnetic field lines. *Ashour-Abdalla et al.* [1993] developed large-scale kinetics (LSK) in which the trajectories of a large number of noninteracting ions were calculated in model magnetic and electric fields to investigate the importance of nonadiabatic motion in determining the magnetotail plasma population. By using a uniform electric field and a 2D reduction of the *Tsyganenko* [1989] magnetic field model, they showed that the magnetotail could be populated from a plasma mantle source [*Ashour-Abdalla et al.*, 1993]. Moreover, they found that highly structured magnetotail ion distributions could result from nonadiabatic ion behavior [*Ashour-Abdalla et al.*, 1994, 1995]. A comparison of LSK distribution functions with those measured by the Galileo spacecraft showed substantial qualitative agreement between the observed and calculated distribution functions [*Ashour-Abdalla et al.*, 1993].

The Geotail spacecraft has given us new insight into the plasma distributions in the magnetotail. The advanced Geotail instruments allow a full 3-D coverage of velocity space with high time resolution. The detailed velocity distribution functions reveal puzzling details: bumps, beams and holes which presumably are signatures of the ions' histories. In data collected from the distant tail, *Frank et al.* [1996] observed nongyrotropic distribution functions near the equatorial current sheet. Closer to Earth, the ion distribution functions were complex and anisotropic, containing small-scale structures which *Frank et al.* [1996] attributed to the particles' "memory" of remote tail acceleration and transport processes.

The objective of this research is to use LSK to determine the ion sources and transport processes that led to the distribution functions observed in the *Frank et al.* [1996] studies. In section 2 we introduce Geotail and Wind observations relevant to this study, and in section 3 we outline the methodology used to calculate the magnetic and electric field models and to perform the LSK calculation. In section 4 we discuss the MHD results, compare them to Geotail observations, and show the results of the LSK calculation. We summarize the results and their implications in section 5.

## 2. OBSERVATIONS

On May 23, 1995, between 0900 UT and 1100 UT the Geotail satellite was located in the near-Earth, near equatorial, magnetotail region ($x = -9.5\ R_E$, $y = -3.3\ R_E$, $z = 1.3\ R_E$ in the Earth centered solar-ecliptic (GSE) coordinates we use in this paper). During this time period, the Comprehensive Plasma Instrumentation (CPI) [*Frank et al.*, 1994] onboard the spacecraft observed complex and highly structured ion distribution functions. For the observations presented here the Hot Plasma (HP) analyzer was operated in a mode by which the velocity distributions of ions were determined from more than 3000 samples over the energy-per-unit charge (E/Q) range of 22 V to 48 kV with a repetition rate of 22 s. Plate 1 displays $v_\parallel$ versus $v_\perp$ for six distribution functions measured by Geotail between 1030 UT and 1036 UT. *Frank et al.* [1996] also published a series of ion velocity distributions measured between 1400 UT and 1406 UT on this day. They concluded from their results that the distribution functions were nonisotropic. Plate 1 shows the departure of the pitch angle distributions from isotropy with structuring occurring at both high (≥ 500 km/s) and low velocities. Distinct structures in the low energy part of the distribution function can be seen along the $v_\perp = 0$ axis, at $v_\parallel = 0$, and at more oblique angles.

During this day the Wind spacecraft was located at $x = 246\,R_E$, $y = 15\,R_E$, and $z = -9\,R_E$. Figure 1 shows the interplanetary magnetic field components, the solar wind velocity, and density measured by Wind and used in the MHD simulation. The input differs from the measurements only in that the values are 3 min averages and that the IMF $B_x$ was held constant at 4 nT. Between 0700 and 1100 UT the IMF $B_z$ component was predominately northward but was interrupted by several small southward turnings. During this time interval, the Wind spacecraft observed a solar wind of unusually high density (~40 cm$^{-3}$); the solar wind speed and thermal pressure had average values of 350 km/s and 18 pPa, respectively.

## 3. METHODOLOGY

The understanding of the structures in the chosen distribution function requires an accurate three-dimensional, time-dependent global scale model for the magnetospheric magnetic and electric fields. Currently, such self-consistent fields can be obtained from global MHD simulations alone.

While the three-dimensional, time-dependent magnetic field can be obtained directly from the global MHD simulation, the electric field is calculated from the simulation parameters as:

$$\vec{E} = -\vec{v} \times \vec{B} + \eta \vec{J}$$

where $\vec{v}$ is the bulk flow velocity, $\vec{B}$ is the magnetic field, $\vec{J}$ is the current density, and $\eta$ is the resistivity. Resistive $\eta\vec{J}$ contributions to $\vec{E}$ become significant only near the magnetopause and near x-lines. The accurate calculation of particle trajectories requires that we use the full electric field, including the resistive term, from the MHD simulation. To investigate the origin of structuring occurring in the distribution functions and the acceleration mechanisms that produced them, the velocity distribution function chosen (lower left panel of Plate 1) is a highly structured one measured by Geotail at 1033:34 UT, when the spacecraft was in the near-Earth plasma sheet ($x = -9.5\,R_E$, $y = -3.3\,R_E$, $z = 1.3\,R_E$). With this distribution function as the starting point for our calculations, we created a numerical distribution function

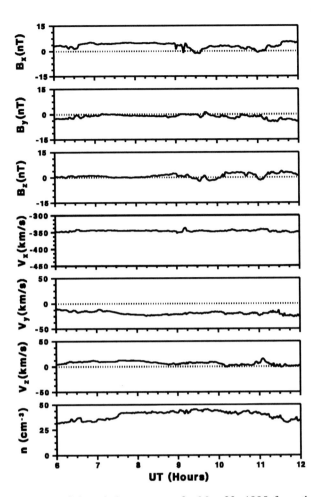

**Figure 1.** Solar wind parameters for May 23, 1995 from the Wind spacecraft. The figure shows, from top to bottom: the magnetic field components, the velocity components, and the proton density.

in which f(v) measured by Geotail in each $v_\parallel$ - $v_\perp$ bin (50 km/s × 50 km/s) was multiplied by a constant (1.65 × 10$^{27}$ in this case) to obtain the number of particles in the corresponding bin of the numerical distribution. The particles in each bin are distributed randomly in phase angle; this phase angle randomization was deemed necessary because, in the case of non-adiabatic acceleration, different phase angles can result in different particle trajectories. The resulting distribution function is given in Plate 2, where the color scale denotes the number of particles assigned to a bin. The structuring of the observed distribution is preserved in the numerical distribution function, and this numerical distribution function was assumed in following the trajectories of approximately 90,000 H$^+$ ions backwards in time from the Geotail location until they reached their "origin," i.e. ei-

300 DETERMINATION OF PARTICLE SOURCES

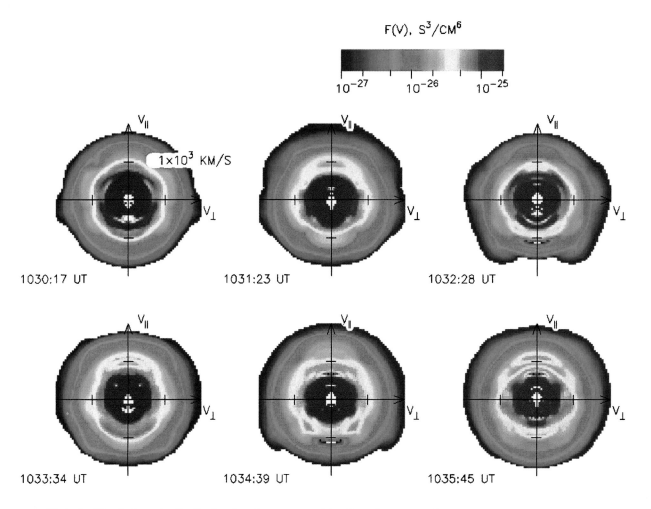

**Plate 1.** Six pitch angle distribution functions measured by Geotail on May 23, 1995. The phase space densities are color coded according to the bar on top.

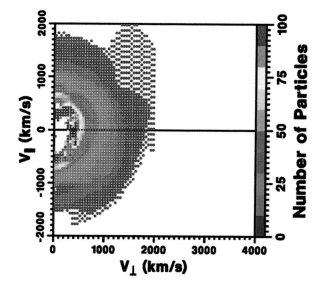

**Plate 2.** Discretized form of the distribution function measured by Geotail at 1033:34 UT. The color scale on the right gives the number of particles per bin.

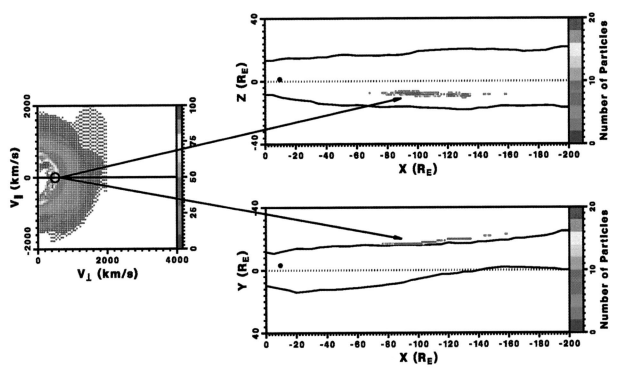

**Plate 8.** (Left-hand panel) Numerical distribution function with circle showing location of structure at $v_\perp \sim 500$ km/s and $v_\parallel \sim 0$ km/s. Entry points for all ions in circled structure are projected onto (upper right-hand panel) the noon-midnight meridional plane and (lower right-hand panel) the equatorial plane, in a format similar to Plate 5. Geotail's location is shown with a large black dot.

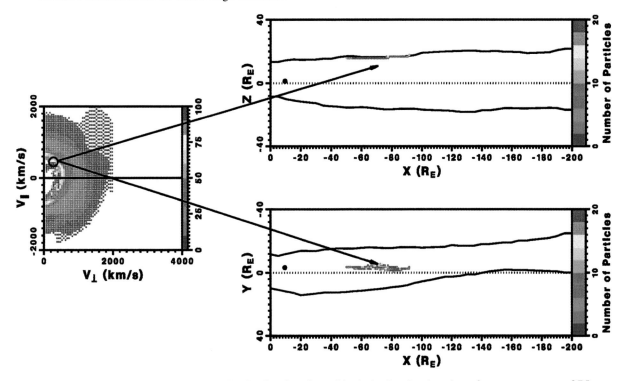

**Plate 9.** (Left-hand panel) Numerical distribution function with circle showing location of structure at $v_\perp \sim 275$ km/s and $v_\parallel \sim 450$ km/s. Entry points for all ions in circled structure are projected onto (upper right-hand panel) the noon-midnight meridional plane and (lower right-hand panel) the equatorial plane, in a format similar to Plate 5. Geotail's location is shown with a large black dot.

308 DETERMINATION OF PARTICLE SOURCES

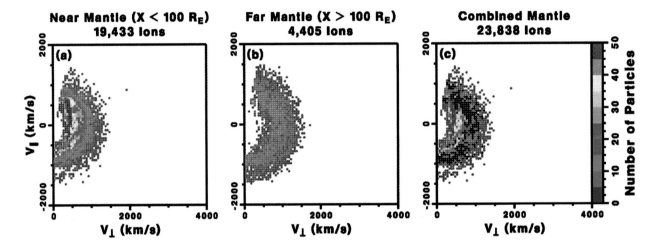

**Plate 10.** Division of the contribution of mantle ions to the numerical distribution function (total 23,838 ions) into (a) near-Earth mantle ions (19,433 ions), (b) distant mantle ions (4,405 ions), and (c) all mantle ions, with particles from the high-energy structure in the distant mantle source overlaid using black dots.

Plate 9 show that the ions in this structure originated in a narrow high latitude region ($y \sim 0$, $z \geq 18\ R_E$) and a region in $x$ where $-85\ R_E < x < -55\ R_E$.

Figure 5 shows a particle trajectory typical of the near-Earth ($x \geq -90\ R_E$) portion of the plasma mantle. The ion depicted enters the magnetotail at $x \sim -80\ R_E$, convects earthward and is trapped in the near-Earth region without interacting with the current sheet. The $\kappa$ for this particle remains high ($\kappa > 4$, with $\kappa \gg 10$ for most of the orbit) and the ion is only weakly energized during its orbit. Plate 9 indicates that only a small portion of the near-Earth plasma mantle source has contributed ions to the structure at $v_\perp \sim 275$ km/s and $v_\parallel \sim 450$ km/s. To reach a better understanding of the role of mantle ions arriving from the distant tail, we plot in Figure 6 the trajectory of a mantle ion originating from $x \sim -110\ R_E$. In contrast to the ion from the near-Earth mantle source, this particle undergoes several nonadiabatic interactions with the current sheet at $x \sim -25\ R_E$ and is quickly energized to $\sim 10$ keV. Note that $\kappa < 1$ for the period when the ion gains most of its energy (gray dots in panel (d) of Figure 6). Thus, ions from the distant tail mantle source arrive at Geotail with higher energies than particles originating from the near-Earth mantle source. This point is illustrated in Plate 10, in which we have separated the contribution of the plasma mantle source to the Geotail distribution into two portions: that from the near-Earth mantle ($x \geq -100\ R_E$, Plate 10a), and that from the distant mantle ($x < -100\ R_E$, Plate 10b). The peak of the distribution from the near-Earth mantle occurs at $v_\parallel = 500$ km/s, while the much smaller peak in the distant tail mantle ion distribution occurs at $v \sim 1000$ km/s. This is also illustrated in Plate 10c, where we have superposed the distant tail mantle peak (shown with black dots) onto the combined mantle distribution. Plate 10c unambiguously illustrates that the two mantle sources are responsible for populating distinct regions of phase space.

## 5. DISCUSSION AND CONCLUSIONS

In this paper we have presented a detailed analysis of an observed Geotail distribution function based on a combination of MHD and LSK techniques. A virtual satellite was placed in an MHD electric and magnetic field model at the position of the Geotail satellite at the time of the observation. The MHD simulation was based on solar wind conditions observed by the Wind spacecraft. A structured distribution function observed at Geotail provided the velocity space distribution for launching time-reversed ions in the MHD model. Plasma man-

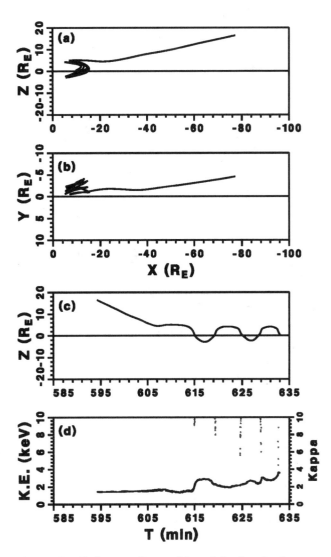

**Figure 5.** Trajectory of a particle originating in the near-Earth plasma mantle, in a format similar to Fig. 4.

tle, LLBL and ionospheric sources all contributed to this distribution function. While the majority of the particles arrived at the Geotail position through adiabatic processes the highest energy ions were influenced by nonadiabatic acceleration at the current sheet. The specific sources and transport histories of individual structures were identified.

Just as we checked on the validity of the MHD model it is advisable to consider the reliability of the particle trajectory results. Because the initial distribution function was based on the Geotail distribution we clearly must reproduce this distribution. Unfortunately, the particle trajectory results cannot be compared with observations as easily as the MHD result could. However, there

310 DETERMINATION OF PARTICLE SOURCES

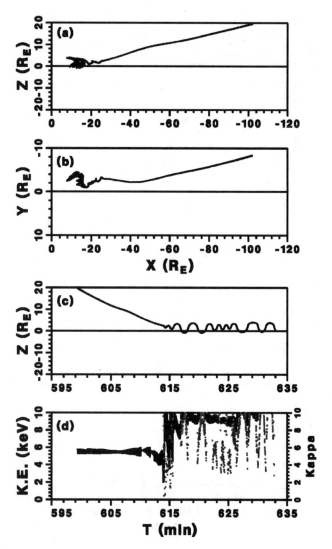

**Figure 6.** Trajectory of a particle originating in the distant plasma mantle, in a format similar to Fig. 4.

are two features of the results that give us confidence in them. One is that the inferred sources are consistent with well established sources of magnetospheric protons (ionosphere, LLBL and plasma mantle) that could all reasonably be active under the observed IMF conditions. A second is the strong correspondence of individual structures in the distribution with individual sources. While any point in phase space could be mapped to a source location by the methods we employed and any distribution could than be divided up among various sources, a "bump" on the observed distribution in a region of phase space mapping to an actual magnetospheric source, say, the plasma mantle would not be expected if this association were random.

Another interesting feature of the results is that, in spite of non-adiabatic or multiple encounters with the current sheet, the particles are seen to originate from relatively narrow regions on the magnetopause, indicating that the transport process is not diffusive in the sense that it does not seem to destroy information about the source locations of the particles' origin, a promising result for this type of calculation. This result also confirms previous model calculations [*Ashour-Abdalla et al.*, 1996] that indicate non-adiabatic processes contribute to the production of higher energy ions in the plasma sheet, especially for a plasma mantle source whose ions interact with the weak field in the distant tail. Furthermore, note that while MHD simulations would trace the origin of a parcel of fluid at a location within the magnetosphere to a single source location, our calculation shows that multiple sources can contribute to a given observed distribution. This shows that, though the validity of MHD results have been confirmed in general, LSK calculations are required to explain distribution functions in detail, at least for certain conditions. The ions reflect the past history of the magnetotail configuration that in turn reflects the past history of the solar wind.

The backwards tracing of particles in the magnetosphere is currently limited by the failure of the model in the extremely sharp gradients near the magnetopause (which is the reason for stopping the particles before they reach this region). The sources that we determine in this calculation may be influenced by processes occurring prior to entry that we cannot capture. For example, *Richard et al.* [1994] showed that LLBL particles could originate at the cusp reconnection region. Ultimately it would be a great benefit to space weather prediction (as well as to a general understanding of magnetospheric physics) if particles could be run forward in time from the solar wind and ionosphere and distribution functions obtained everywhere in the magnetosphere, but this would require vast numbers of particles. Currently our method seems to be the only tractable one that can resolve the full details of a local distribution function.

*Acknowledgments.* This work was supported by NASA grants NAG5-1100, NAGW-4553. Computing support was provided by the Maui High Performance Computing Center, the Office of Academic Computing at UCLA, the San Diego Supercomputer Center, and the Cornell Theory Center. UCLA/IGPP publication number 4939.

## REFERENCES

Amano, K., and T. Tsuda, Particle trajectories at a neutral point, *J. Geomag. Geoelect.*, *30*, 7, 1978.

Ashour-Abdalla, M., J. Berchem, J. Büchner, and L. Zelenyi, Shaping of the magnetotail from the mantle: Global and local structuring, *J. Geophys. Res.*, *98*, 5651, 1993.

Ashour-Abdalla, M., L. M. Zelenyi, V. Peroomian, and R. L. Richard, Consequences of magnetotail ion dynamics, *J. Geophys. Res.*, *99*, 14,891, 1994.

Ashour-Abdalla, M., L. M. Zelenyi, V. Peroomian, R. L. Richard, and J. M. Bosqued, The mosaic structure of plasma bulk flows in the Earth's magnetotail, *J. Geophys. Res.*, *100*, 19,191, 1995.

Ashour-Abdalla, M., L. A. Frank, W. R. Paterson, V. Peroomian, and L. M. Zelenyi, Proton velocity distributions in the magnetotail: Theory and observations, *J. Geophys. Res.*, *101*, 2587, 1996.

Balsiger, H., P. Eberhardt, J. Geiss, and D. T. Young, Magnetic storm injection of 0.9 - 16 keV/e solar and terrestrial ions into the high-altitude magnetosphere, *J. Geophys. Res.*, *85*, 1645, 1980.

Büchner, J., and L. M. Zelenyi, Deterministic chaos in the dynamics of charged particles near a magnetic field reversal, *Phys. Lett. A*, *118*, 395, 1986.

Büchner, J., and L. M. Zelenyi, Regular and chaotic charged particle motion in magnetotail-like field reversals, 1. Basic theory of trapped motion, *J. Geophys. Res.*, *94*, 11821, 1989.

Chen, J., and P. J. Palmadesso, Chaos and nonlinear dynamics of single particle orbits in a magnetotail-like magnetic field, *J. Geophys. Res.*, *91*, 1499, 1986.

Cowley, S. W. H., Plasma populations in a simple open model magnetosphere, *Space Sci. Rev.*, *26*, 217, 1980.

Curran, D., and C. K. Goertz, Particle distributions in a two-dimensional reconnection field geometry, *J. Geophys. Res.*, *94*, 272, 1989.

Eastman, T. E., and E. W. Hones, Characteristics of the magnetospheric boundary layer and magnetopause layer as observed by IMP 6, *J. Geophys. Res.*, *84*, 2019, 1979.

Eastman, T. E., E. W. Hones, S. J. Bame, and J. R. Asbridge, The magnetospheric boundary layer: Site of plasma, momentum and energy transfer from the magnetosheath into the magnetosphere, *Geophys. Res. Lett.*, *3*, 685, 1976.

Frank, L. A., K. L. Ackerson, W. R. Paterson, J. A. Lee, M. R. English, and G. L. Pickett, The Comprehensive Plasma Instrumentation (CPI) for the GEOTAIL spacecraft, *J. Geomagn. Geoelec.*, *46*, 23, 1994.

Frank, L. A., M. Ashour-Abdalla, J. Berchem, J. Raeder, W. R. Paterson, S. Kokubun, T. Yamamoto, R. P. Lepping, F. V. Coroniti, D. H. Fairfield, and K. L. Ackerson, Observations of plasmas and magnetic fields in Earth's distant magnetotail: Comparison with a global MHD model, *J. Geophys. Res.*, *100*, 19177, 1995.

Frank, L. A., W. R. Paterson, K. L. Ackerson, S. Kokubun, and T. Yamamoto, The Plasma velocity distributions in the near-Earth plasma sheet: A first look with the Geotail spacecraft, *J. Geophys. Res.*, *101*, 10627, 1996.

Geiss, J., H. Balsiger, P. Eberhardt, H. P. Walker, L. Weber, D. T. Young, and H. Rosenbauer, Dynamics of magnetospheric ion composition as observed by the Geos mass spectrometer, *Space Sci. Rev.*, *22*, 537, 1978.

Green, J. L., and J. L. Horwitz, Destiny of earthward streaming plasma in the plasma sheet boundary layer, *Geophys. Res. Lett.*, *13*, 76, 1986.

Hill, T. W., Origin of the plasma sheet, *Rev. of Geophys. and Space Phys.*, *12*, 379, 1974.

Hones, E. W., S.-I. Akasofu, S. J. Bame, M. D. Montgomery, S. Singer, and S.-I. Akasofu, Measurements of magnetotail plasma flow made with Vela 4B, *J. Geophys. Res.*, *77*, 5503, 1972a.

Hones, E. W., S.-I. Akasofu, S. J. Bame, and S. Singer, Outflow of plasma from magnetotail into the magnetosheath, *J. Geophys. Res.*, *77*, 6688, 1972b.

Jaeger, E. F., and T. W. Speiser, Energy and pitch angle distributions for auroral ions using the current sheet acceleration model, *Adv. Space Res.*, *28*, 129, 1974.

Kokubun, S., T. Yamamoto, M. H. Acuna, K. Hayashi, K. Shiokawa, and H. Kawano, The GEOTAIL magnetic field experiment, *J. Geomag. Geoelec.*, *46*, 7, 1994.

Lennartsson, W., E. G. Shelley, R. D. Sharp, R. G. Johnson, and H. Balsiger, Some initial ISEE-1 results on the ring composition and dynamics during the magnetic storm of December 11, 1977, *Geophys. Res. Lett.*, *6*, 483, 1979.

Lennartsson, W., R. D. Sharp, E. G. Shelley, R. G. Johnson, and H. Balsiger, Ion composition and energy distribution during 10 magnetic storms, *J. Geophys. Res.*, *86*, 4678, 1981.

Lepping, R. P., et al., The Wind Magnetic Field Investigation, *Space Sci. Rev.*, *71*, 207, 1995.

Lundin, R., L. R. Lyons, and N. Pissarenko, Observations of the ring current composition at L < 4, *Geophys. Res. Lett.*, *7*, 425, 1980.

Lyons, L. R., and T. W. Speiser, Evidence for current-sheet acceleration in the geomagnetic tail, *J. Geophys. Res.*, *87*, 2276, 1982.

Martin, R. F., Chaotic particle dynamics near a two-dimensional magnetic neutral point with application to the geomagnetic tail, *J. Geophys. Res.*, *91*, 985, 1986.

Martin, R. F., and T. W. Speiser, A predicted energetic ions signature of a neutral line in the geomagnetic tail, *J. Geophys. Res.*, *93*, 11,521, 1988.

Ogilvie, K. W., et al., SWE, A comprehensive plasma instrument for the Wind spacecraft, *Space Sci. Rev.*, *71*, 55, 1995.

Pilipp, W. G., and G. Morfill, The formation of the plasma sheet resulting from plasma mantle dynamics, *J. Geophys. Res.*, *83*, 5670, 1978.

Raeder, J., R. J. Walker, and M. Ashour-Abdalla, The structure of the distant geomagnetic tail during long periods of northward IMF, *Geophys. Res. Lett.*, *22*, 349, 1995.

Raeder, J., J. Berchem, and M. Ashour-Abdalla, The importance of small scale processes in global MHD simulations: Some numerical experiments, in *The Physics of Space Plasmas*, edited by T. Chang and J. R. Jasperse, *14*, MIT Center for Theoretical Geo/Cosmo Plasma Physics, Cambridge, MA, 403, 1996.

Raeder, J., J. Berchem, M. Ashour-Abdalla, L. A. Frank, W. R. Paterson, K. L. Ackerson, R. P. Lepping, K. Ogilvie, S. Kokubun, T. Yamamoto, and D. Fairfield, The distant tail under strong northward IMF conditions: Global MHD results for the Geotail March 29, 1993 observations, *J. Geophys. Res.*, submitted, 1997.

Richard, R. L., R. J. Walker, and M. Ashour-Abdalla, The population of the magnetosphere by solar wind ions when the interplanetary magnetic field is northward, *Geophys. Res. Lett.*, *21*, 2455, 1994

Savenkov, B. V., L. M. Zelenyi, M. Ashour-Abdalla, and J. Büchner, Regular and chaotic aspects of charged particle motion in a magnetotail-like field with a neutral line, *Geophys. Res. Lett.*, *18*, 1587, 1991.

Schindler, K., and J. Birn, On the generation of field-aligned plasma flow at the boundary of the plasma sheet, *J. Geophys. Res.*, *92*, 95, 1987.

Sharp, R. D., D. L. Carr, W. K. Peterson, and E. G. Shelley, Ion streams in the magnetotail, *J. Geophys. Res.*, *86*, 4639, 1981.

Shelley, E. G., R. G. Johnson, and R. D. Sharp, Satellite observations of energetic heavy ions during a geomagnetic storm, *J. Geophys. Res.*, *77*, 6104, 1972.

Speiser, T. W., Particle trajectories in a model current sheet based on the open model of magnetosphere, with applications to auroral particles, *J. Geophys. Res.*, *70*, 4216, 1965.

Tsyganenko, N. A., A magnetospheric magnetic field model with a warped tail current sheet, *Planet. Space Sci.*, *37*, 5, 1989.

Wagner, J. S., P. C. Gray, J. R. Kan, T. Tajima, and S.-I. Akasofu, Particle dynamics in reconnection field configurations, *Planet. Space Sci*, *29*, 391, 1981.

Williams, D. J., Energetic ion beams at the edge of the plasma sheet: ISEE 1 observations plus a simple explanatory model, *J. Geophys. Res.*, *86*, 5501, 1981.

Zelenyi, L. M., J. Büchner, and D. V. Zogin, Quasiadiabatic ion acceleration in the Earth's magnetotail, *Proceedings of Varenne-Abustuman Workshop on Plasma Astrophysics*, Eur. Space Agency Spec. Publ., ESA-SP285, 1988.

---

Maha Ashour-Abdalla, Mostafa El-Alaoui, Vahé Peroomian, Joachim Raeder, Robert L. Richard, and Ray J. Walker, Institute of Geophysics and Planetary Physics, University of California, Los Angeles, CA 90095-1567 (e-mail mabdalla@igpp.ucla.edu).

Lev M. Zelenyi, Space Research Institute, Russian Academy of Sciences, 117810 Moscow, Russia.

L. A. Frank, and W. R. Paterson, Department of Physics and Astronomy, The University of Iowa, Iowa City, IA 52242.

J. M. Bosqued, Centre d'Etude Spatiale des Rayonnements, CNRS, BP 4346, 31029 Toulouse Cedex, France.

R. P. Lepping, and K. Ogilvie, NASA/Goddard Space Flight Center, Greenbelt, MD 20771.

S. Kokubun, STELAB, Nagoya University, Toyokawa, Aichi 442, Japan.

T. Yamamoto, ISAS, Sagamihara, Kanagawa 229, Japan.

# Three-Dimensional Reconnection in the Earth's Magnetotail: Simulations and Observations

J. Büchner[1], J.-P. Kuska[1], B. Nikutowski[1], H. Wiechen[1], J. Rustenbach[2], U. Auster[3], K. H. Fornacon[3], S. Klimov[4], A. Petrukovich[4], S. Savin[4]

In the course of magnetospheric substorms the reconnection process is supposed to release energy, previously stored in the Earth's magnetotail. The causal relationship between substorms and reconnection is, however, still open. In particular the route to reconnection via a collisionless tearing mode instability has become questionable. All attempts have failed to relate the onset of magnetotail reconnection to two-dimensional collisionless sheet tearing. Hence, new kinetic approaches are under development to understand the stability of the magnetotail current sheet and the unstable decay of thin sheets in three dimensions. In order to derive typical signatures for comparison with *in situ* satellite measurements we have simulated three-dimensional reconnection in the framework of resistive magnetohydrodynamics (MHD) as well as by a fully kinetic particle-in-cell (PIC) plasma code. In the MHD approach the magnetic signatures of reconnection are closely related to the macroscopic ideal plasma flow pattern. In the kinetic treatment the decay of thin current sheets starts with a sausage mode bulk current instability. The inertia of the different mass particles causes a Hall effect and currents with typical magnetic signatures. The simulated features of three-dimensional reconnection are then compared with INTERBALL-1 and GEOTAIL magnetic field observations during substorm-related plasma sheet encounters.

## 1. INTRODUCTION

Reconnection is a plasma process able to convert most efficiently the energy of flows and currents into heat and particle acceleration [Vasyliunas, 1975; Axford, 1984]. It plays a key role in the near-Earth neutral line model of magnetospheric substorms, which is still the best available framework for ordering the complex, global manifestations of substorms [Baker et al., 1996]. In a resistive magnetohydrodynamic (MHD) description reconnection can be caused by a tearing instability of current sheets [Furth et al., 1963]. In collisionless plasmas Landau damping on electrons can drive a similar instability [Coppi et al., 1966]. In fact, numerical simulations have verified the collisionless tearing instability of Harris [1962] current sheet sheet equilibria in two dimensions [Zwingmann, 1990; Pritchett, 1991]. Any normal magnetic field component $B_n$ perpendicular to a current sheet inhibits, however, Landau damping on electrons. Although in this case Landau damping on ions might provide the dissipation [Schindler, 1974], theo-

---

[1]Max-Planck-Institut für Aeronomie, Katlenburg-Lindau, Germany
[2]Max-Planck-Institut für extraterrestrische Physik, Außenstelle Berlin, Germany
[3]Technische Universität Braunschweig, Germany
[4]Space Research Institute, Russian Academy of Sciences, Moscow, Russia

retical considerations have revealed an apparent stability of current sheets against tearing if $B_n \neq 0$ [Galeev and Zelenyi, 1976; Lembége and Pellat, 1982]. The reason is the large demand of free energy which is needed to compress plasma particles on their flux tubes before the sheet can tear off [Pellat et al., 1991; Brittnacher et al., 1994]. The stabilizing effect of the normal magnetic field was confirmed for two-dimensional configurations by numerical plasma simulations [Pritchett, 1994; Pritchett and Büchner, 1995].

Recently we have suggested that the two-dimensionality of previous tearing instability theories is the reason for their failure and that the current sheet decay is, instead, a three-dimensional process [Büchner, 1995, 1996]. In order to verify, whether reconnection in the Earth's magnetotail in the context of substorms is three-dimensional, it is appropriate to look for its typical signatures. For comparison with measurements we have carried out both resistive MHD as well as collisionless kinetic plasma simulations of non-stationary three-dimensional reconnection.

In section 2. we report the results of three-dimensional resistive MHD simulations. We derive the typical magnetic field signatures, which characterize three-dimensional reconnection in the fluid approach. In section 3. we consider signatures which arise beyond the limits of single-fluid MHD. For this purpose we use kinetic plasma simulations. In section 4. we look at INTERBALL-1 and GEOTAIL magnetic field observations in the plasma sheet during substorms to find out, whether the simulated signatures of three-dimensional reconnection can be found.

## 2. THREE-DIMENSIONAL RESISTIVE MHD SIMULATIONS OF MAGNETOTAIL RECONNECTION

First we looked for the magnetic field signatures of three-dimensional magnetotail reconnection in the single-fluid MHD approach. We have carried out MHD simulations in a model, where the plasma sheet is initially embedded in a two-dimensional tail-like equilibrium. Such model is valid near the midnight meridian plane where reconnection is supposed to get off. The coordinates $X, Y, Z$ are chosen in accordance with the direction of the GSM coordinate system. The initially invariant direction is $y$ (dawn-dusk). We normalize spatial variables to the the length scale sheet half-thickness $L_z$ (about 1 $R_E$) and temporal variables to the time scale $\tau_A$, the Alfven transit time through the sheet. The initial equilibrium includes a distant

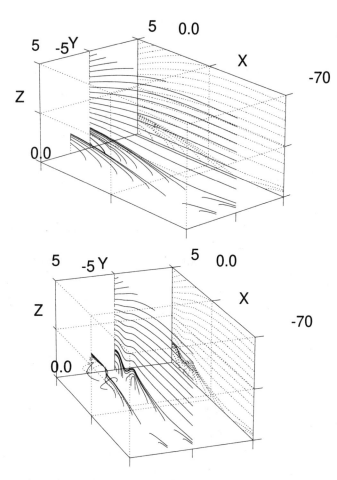

**Figure 1.** Model magnetic field lines at $t = 4\tau_A$ (1a), upper panel, and at $t = 45\tau_A$ (1b), lower panel.

neutral line near $x = -50$. The upper panel of Figure 1 depicts the magnetic field line configuration 4 $\tau_A$ after simulation has been started, i.e. almost of the initial equilibrium state. Reconnection is launched by switching on a finite resistivity, localized near the center of the model plasma sheet, with the profile function $\eta = \eta_o \cdot \exp\{-(x+10)^2\} \cdot \exp\{-y^2\}$. As usual in MHD simulations with high Reynolds numbers, i.e. negligible numerical dissipation, one has to impose a sufficiently high resistivity in order to achieve reconnection at a reasonable time scale. In our calculations we added to the practically ideal, high numerical-method based magnetic Reynolds-number (typically reaching tens of thousands) plasma, a localized finite resistivity $\eta_o = 0.01$. This corresponds to a local magnetic Reynolds-number of the order of 100. Reconnection starts after about 22.5 $\tau_A$, which correspond to about 15 minutes in real time. The lower panel of Figure 1 depicts the magnetic field line configuration after 45 $\tau_A$, when three-dimensional reconnection is well developed. In the symmetry plane $y = 0$ one sees the reconnection region with an X-point evolving near $x = -10$. Field lines starting in the plane $x = -2$, however, are wrapped around following the typical flow pattern of three-dimensional reconnection [Wiechen et al., 1996]. For comparison with observations we have derived time series of the variation of magnetic fields components at different locations relatively to the reconnection site. We show here only the results for satellites located in north of the magnetical equator $z = 0$ from which, due to the symmetry of the problem, the field values in the southern hemisphere follow immediately. Figures 2 and 3 depict time series of the magnetic field variations seen dawnward of the midnight meridian, earthward (Figure 2) and tailward (Figure 3) of a three-dimensional reconnection region. One can see that $B_y$ follows the $B_z$ variation. Figures 4 and 5 depict time series of the magnetic field variations, obtained for satellites located duskward of the midnight meridian, earthward (Figure 4) and tailward (Figure 5) of a three-dimensional reconnection region. At this side

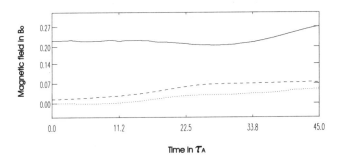

**Figure 2.** A sample time series of magnetic field components (the solid, dotted, dashed lines represent $B_x, B_y, B_z$, respectively) seen by a satellite dawnward and earthward of a three-dimensional reconnection site.

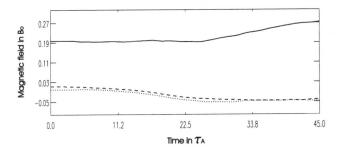

**Figure 3.** A sample time series of magnetic field components (the solid, dotted, dashed lines represent $B_x, B_y, B_z$, respectively) seen by a satellite dawnward and tailward of a three-dimensional reconnection site.

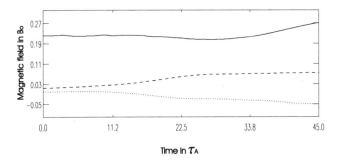

**Figure 4.** Magnetic field components (the solid, dotted, dashed lines represent $B_x, B_y, B_z$, respectively) seen by a satellite duskward and earthward of a three-dimensional reconnection site.

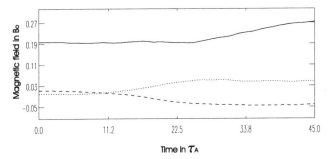

**Figure 5.** Magnetic field components (the solid, dotted, dashed lines represent $B_x, B_y, B_z$, respectively) seen by a satellite duskward and tailward of a three-dimensional reconnection site.

of the reconnection region $B_y$ changes in a sense opposite to the $B_z$ variation.

These signatures of the magnetic field variation around a three-dimensional reconnection region in the MHD approach are due to the ideal plasma conductivity outside, which lets the magnetic field move together with the plasma heading for the finite resistivity zone.

## 3. THREE-DIMENSIONAL KINETIC SIMULATIONS OF RECONNECTION

Another factor influencing the magnetic field variation around a reconnection region is due to the Hall effect of the different mass charged particles. The Hall effect scales with the sheet thickness as mentioned by Vasyliunas [1975] and becomes important in thin sheets. It was discussed for two-dimensional reconnection geometries by Sonnerup [1979], its consequences for the two-dimensional tearing instability were presented by Terasawa [1983]. But no analysis exists for three-dimensional reconnection, whose kinetic properties become investigated just now. Since single fluid MHD simulations do not distinguish between the reaction of different mass charged particles we have carried out kinetic simulations using the three-dimensional electromagnetic particle-in-cell (PIC) code GISMO [Büchner and Kuska, 1996]. The code explicitely integrates the complete set of Maxwell's equations for scalar ($\phi$) and vector potentials ($\vec{A}$). For highest accuracy GISMO solves also the Poisson equations. The field values are calculated on a three-dimensional equidistant rectangular spatial mesh. The distances between the grid planes are $\Delta_x = \Delta_y = \Delta_z = \Delta$ in the $X$, $Y$ and $Z$ directions, respectively. The GISMO code integrates the relativistic equations of particle motion using a Runge Kutta scheme with stepsize control. The sheet is initialized by loading a particle distribution in accordance with

$$f_j(\vec{r},\vec{v}) = \frac{n_j(Z)}{\pi^{3/2} \cdot v_{tj}^3} \cdot$$

$$\cdot \exp\left\{\frac{m_j}{2k_B T_j}\left(v_x^2 + (v_y - u_{dyj})^2 + v_z^2\right)\right\}, \quad (1)$$

where

$$n_j(Z) = n_e(Z) = n_i(Z) = n_o \cdot \cosh^{-2}\left(\frac{Z}{L_z}\right) \quad (2)$$

and

$$L_z = \frac{\sqrt{k_B \cdot (T_e + T_i)}}{\sqrt{2\pi \, n_o \, e^2}} \cdot \frac{c}{u_{di} - u_{de}}$$

$j = e, i$ denote electrons and ions, respectively. $T_e$ and $T_i$ are the electron and ion temperatures and $u_{dj}$ the drift velocities of the particles. $k_B$ is the Boltzmann constant. We abbreviate the thermal Larmor radius $\rho_{oj} = v_{tj}/\Omega_{oj}$ for $v_{tj} = \sqrt{k_B T_j/m_j}$, the thermal velocity of the $j$- particle species. $\Omega_{oi} = eB_o/m_i c$ is the gyro-frequency in the asymptotic field, $e$ is the elementary charge, $m_i$ the ion mass and $c$ the speed of light. The current in a Harris-equilibrium is diamagnetic. It is closely related with the density gradient in the $Z$ direction and causes a magnetic field

$$\vec{B} = B_o \cdot \tanh\left(\frac{Z}{L_z}\right) \cdot \vec{e}_x$$

where

$$B_o = \sqrt{8\pi \cdot n_o \cdot k_B \cdot (T_e + T_i)}. \quad (3)$$

In the simulations we prefer the frame, in which the equilibrium electric field vanishes at $t = 0$. This can be done by relating the electron and ion drift velocities $u_{dye}$ and $u_{dyi}$ as $u_{dye}/u_{dyi} = -T_e/T_i$. In order to avoid additional numerical instabilities due to unequal electron and ion temperatures (Pritchett, 1994) we assume $T_e = T_i = T$ at $t = 0$, so that $u_{dyi} = -u_{dye} = u_d$. The sheet half-width is given by

$$L_z = \frac{\sqrt{k_B \cdot T}}{\sqrt{4\pi \, n_o \, e^2}} \cdot \frac{c}{u_d} = \frac{v_{ti}}{u_d} \cdot \lambda_i = 2 \cdot \rho_{oi} \cdot \frac{v_{ti}}{u_d} \quad (4)$$

Equation (4) relates the half-thickness $L_z$ via the drift and ion thermal velocity ($u_d/v_{ti}$) to the ion inertial lengths $\lambda_i$ as well as to the Larmor-radius in the external field $\rho_{oi}$.

In two dimension the Harris equilibrium is known to be unstable to a tearing instability, characterised by a growing perturbation $\propto \exp\{\gamma t + ik_x X\}$. The tearing growth rate is small in thick sheets $L_z \gg \rho_{oi}$, for which the theory originally was developed [Coppi et al., 1966]. The plasma- and current sheet thinning down to $L_z \to \rho_{oi}$, observed before substorms, drives the tail current sheet, however, to a thickness comparable with the ion inertial length $\lambda_i$. We therefore start our simulations with such thin sheets. Technically this can be done by chosing a thermal velocity over drift speed $u_d \to v_{ti}$.

The results, presented here, are based on simulation runs incorporating more than 3,000,000 particles. For fields and particles periodic boundary conditions are applied in the $X$ and $Y$ directions. The distance between current sheet and upper (lower) boundaries (in the $Z$ direction) is chosen in a way that almost no particles reach the boundaries at $Z = 0$ and $\Delta Z = \Delta X/2$, where the potentials are put zero. The field equations are solved on a $64 \times 128 \times 64$ as well as on a $128 \times 128 \times 64$ mesh in the $X$, $Y$ and $Z$ directions, respectively. High frequency waves are damped away before reaching a boundary. In order to avoid grid-induced instabilities it is necessary to resolve the Debye length $\lambda_D = \sqrt{k_B T/4\pi n_o e^2}$. As a rough rule of the thumb the grid distance $\Delta$ should resolve $\lambda_D$ about three times. Since the main sheet-reconnecting interaction is supposed to take place inside the current sheet, we tried to resolve the whole sheet by twenty grid spaces, i.e. $L_z \approx 10 \, \Delta$. Hence, in order to fulfill the numerical stability condition, the sheet half-thickness $L_z$ should not exceed 30 $\lambda_D$. From this condition one obtains a lower limit of $u_{dj} > 0.03 \, c$ for the drift velocity. Since we aim at the investigation of thin sheets $\rho_{oj}/L_z \to 1$, according to equation (4) one obtains a condition on the thermal speed $v_{tj} > 0.015 \, c$ or $k_B T_i > 2.3 \, MeV$ and $k_B T_e > 1.3 \, keV$! Since we choose $T_i = T_e$ but want to exclude relativistic effects, we start the simulations with a mass-ratio-unity electron-positron like plasma and decrease the electron mass later. All particles are distributed in accordance to an initial temperature of $k_B T_j = 40\ keV$. By the choice of the density $n_o$ at $Z = 0$ ($1 cm^{-3}$), one controls the asymptotic magnetic field strength $B_o$ and, therefore, the time scaling gyrofrequency $\Omega_o$.

For comparison we have first carried out simulations with a mass ratio $M = M_i/m_e = 1$. In the three-dimensional simulations a drift instability occurs at the edges of the current sheet. This is in agreement with the predictions of Huba et al. [1977], the simulations of Winske and Liever [1978] and the discussion of Huba et al. [1980]. As Zhu and Winglee [1996] we see a drift wave developing at the edges of the sheet. It becomes damped away at time scales much shorter than that of the macroscopic current sheet decay. Then, however, in contrast to Zhu and Winglee [1996], a whole-sheet sausage mode current instability develops. It structures the sheet in the $Y$, the current direction, and enhances the growth of macroscopic magnetic reconnection. The sausage mode current instability can be identified in the density structure at least after $t\,\Omega_o = 5$. Its structure can clearly be seen in Figure 6, comparing density isosurfaces at $t\,\Omega_o = 9.3$ (lower panel) with the initial one at $t = 0$. To demonstrate the transition to the sausage mode instability we show in Figure 7 a sequence of density isosurface cuts through the plane $X = 0.5$ at $t = 0$, $t\,\Omega_o = 0.7, 5.8$ and $6.5$, subsequently. Figure 8 depicts the sequence of density isosurface cuts through the plane $X = 0.5$ between $t\,\Omega_o = 7.2$ and $9.3$, allowing to determine wavelength, frequency and propagation speed of the instability driven wave, propagating in the positive $Y$-, the current flow direction according to $\propto \exp\{-i\omega t + i k_y Y\}$. The simulations reveal for $M = 1$ a wave frequency $\omega \approx \Omega_o$ a wave number $k_y L_z \approx 1$, i.e. a propagation speed $v_w = \omega/k_y \approx \Omega_o L_z$. This is, due to our choice of $u_d/v_t = 2$ and equation (4), $v_w \approx 2\,v_t = u_d$. Soon after the sausage instability has started, $B_z$ becomes strongly modulated and macroscopic reconnection starts. However, in the mass ratio unity case the field lines stay in a plane $Y = const.$ (not shown here), i.e. in contrast to the simulations of Zhu and Winglee [1996] no significant shear is generated. We think that the strong shear, they have seen, is due to their simulation setup with boundaries, which are reached by their relativistic particles almost immediately. We also found field shear for a mass ratio $M = 1$ but only when running the simulations in smaller boxes and with higher initial temperatures.

A characteristic field shear and $B_y$ excursion arises, however, as soon as a finite mass ratio is considered. In order to simulate charge separation effects in three-

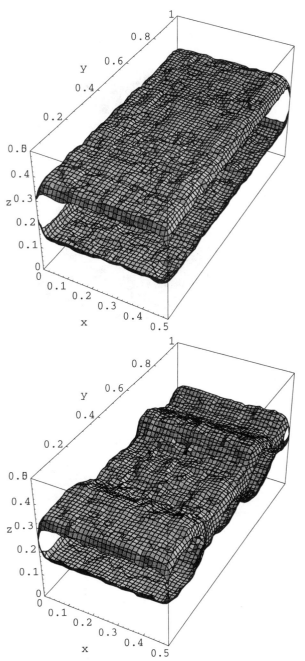

**Figure 6.** Density isosurfaces at $t = 0$ (a) and $t\,\Omega_o = 9.3$ (b).

dimensional spontaneous reconnection through thin current sheets we have carried out calculations with negatively charged particles ("electrons"), which by technical reasons are only 64 times lighter than the positively charged ions. In case of a mass ratio $M = 64$ the sequence of events initially appeared to be the same as in the mass ratio unity case that first a sausage mode

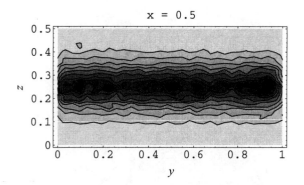
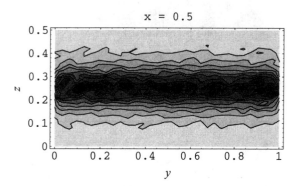
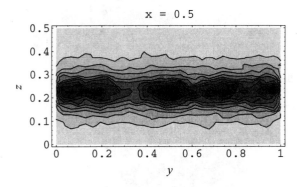
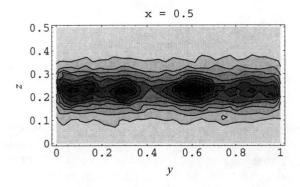

**Figure 7.** Sequence of density isosurface cuts through the plane $X = 0.5$ at $t = 0$ (a), $t\,\Omega_o = 0.7$ (b), $t\,\Omega_o = 5.8$ (c) and $t\,\Omega_o = 6.5$ (d).

current instability starts. It grows, however, faster than in the mass ratio unity case and has a shorter wavelength in the $Y$ direction. The current instability causes dissipation and, as a result, macroscopic reconnection quickly evolves out of the microscopic fluctuations, much faster than in the mass ratio unity case. A major difference compared with the $M = 1$ case is the appearance of a modulated shear field component $B_y$. In order to understand the reason for the appearance of $B_y \neq 0$ we checked its consistency with a Hall effect acting in the sheet due to the different masses of the charged particles. A Hall currents arises in reconnection regions of thin sheets due to the different inertial response of ions and electrons to the reconnection electric field [Vasyliunas, 1975; Sonnerup, 1979]. For two-dimensional sheet tearing Terasawa [1983] has analyzed the Hall effect. Figure 9 depicts his prediction of a two-dimensional Hall current system in relation to reconnection $X$- and $O$-lines. In three dimensions, however, the excursion of the particle motion is limited in the $Y$ direction. As a result the pattern of the shear magnetic field $B_y$ differs from the two-dimensional case. This can be seen in Figure 10 which depicts both the magnetic field lines in a plane of a sausage (upper panel) and the $B_y$ component pointing out of this plane (lower panel). The snapshot corresponds to $t\,\Omega_0 = 20$, when the instability went into saturation. The corresponding three-dimensional plasmoid magnetic field structure is depicted in Figures 11 - 13, corresponding to early moments when reconnection just has started. Figures 11 - 13 depict the magnetic field structure which is determined by the analogous deviation of $B_y$ and $B_z$ from the average level. This is in contrast to the axial $B_y$ fields of flux ropes, generated by reconnection with the interplanetary magnetic field (IMF) at the flanks of the magnetosphere and co-related with the $B_y$ component of the IMF [Hughes et al., 1987]. The $B_y$ field perturbation caused by the 3D Hall effect changes its sign inside a plasmoid, but only once, in contrast to the two-dimensional Hall effect in a tearing mode [Terasawa, 1983].

## 4. OBSERVATIONS

In an attempt to verify the magnetic signatures discussed in sections 2. and 3. we looked for appropriate INTERBALL-1 and GEOTAIL observations. In Fall 1995 the main INTERBALL-1 spacecraft encountered the central plasma sheet several times at distances between 10 and 15 $R_E$ (= Earth radii) from Earth. We present two substorm-related FGM-I mag-

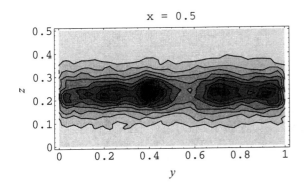

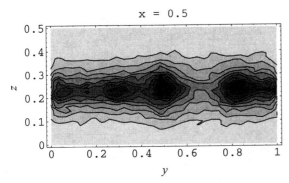

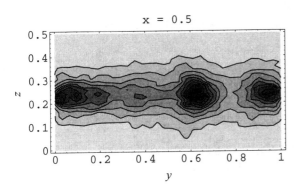

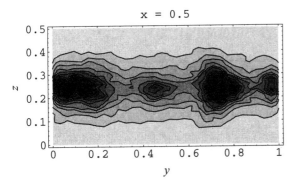

**Figure 8.** Sequence of density isosurface cuts through $X = 0.5$ at $t\,\Omega_o = 7.2$ (a), $t\,\Omega_o = 7.9$ (b), $t\,\Omega_o = 8.6$ (c) and $t\,\Omega_o = 9.3$ (d).

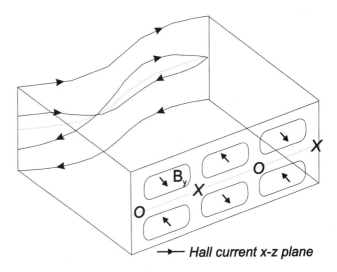

**Figure 9.** Hall currents and resulting magnetic fields in two-dimensional tearing mode reconnection (after [Terasawa, 1983]).

netic field observations in the plasma sheet. One took place away (October 28 - 29, 1995) and the other well inside the tail current sheet (November 28, 1995). More details about these events are described in [Büchner et al., 1997; Petrukovich et al., 1997]. Here we focus on the comparison of the substorm-related magnetic field signatures with the predicted signatures of three-dimensional reconnection. In fact, we have found corresponding variations the $B_y$ magnetic field component during substorm-related dipolarisation events in the INTERBALL-1 data set. In the following subsections we give several examples.

### 4.1 INTERBALL - The October 29, 1995 Event

Magnetograms of auroral stations indicate substorm activity starting at 00:20 UT on October 29. During this period INTERBALL-1 was located in the tail, close to the neutral plane ($Z_{GSM} \approx -0.1 R_E$) on the duskside ($Y_{GSM} \approx -8 R_E$), at a distance of about $X_{GSM} = -12 R_E$ from the Earth. Figure 14 depicts the magnetic field variations in GSM coordinates, measured by the FGM-I magnetometer [Klimov et al., 1997]. Starting at 23:00 UT the total magnetic field strength has increased from about 15 $nT$ up to 25 $nT$ (not shown here). This indicates a typical growth phase related energy loading of the tail. The magnetic field dropped to 15 $nT$ first time at 23:30 UT. It went up again to 25 $nT$ at about 23:45 UT. Then it started to decrease. Our Figure depicts the magnetic field variation after 23:50 UT. From this moment the magnetic field stagnates at

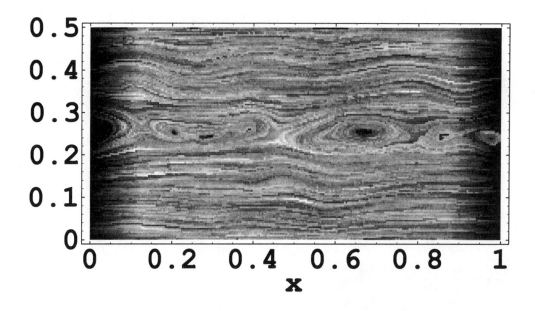

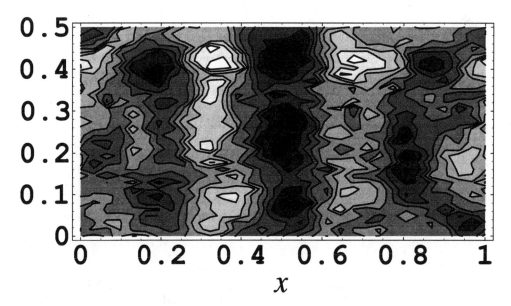

**Figure 10.** Magnetic field lines (upper panel) and $B_y$ component (11 step greyscale with white corresponding to maximum positive, dark gray to zero and black to maximum negative values $B_y$) in a plane cutting through a sausage mode density maximum at $t\,\Omega = 20$ (saturation).

a level of about 20 nT. The field was at that time concentrated mainly in the $X$-component. The shear field component $B_y$ was about $4-5\,nT$ and $B_z$ was less than $2\,nT$. A $B_x$ component larger than $10\,nT$ indicates that the spacecraft is sliding along the plasma sheet, staying north of it, not crossing the neutral plane during the time interval considered here. At about 00:06 UT, however, both the $B_x$ and the $B_y$ components start to decrease. They reach the levels $15\,nT$ and $2\,nT$, respectively, at 00:11 UT. Then $B_z$ starts to increase up to 5 nt in about 10 minutes while $B_x$ decreases to 10 nT – like in a dipolarisation event which ends shortly after 00:20 UT, the moment of ground substorm onset. Three minutes after the ground onset $B_z$ reaches a maximum of $5\,nT$. At the same time $B_y$ reverses its sign from plus to minus. In the October 29th event the pe-

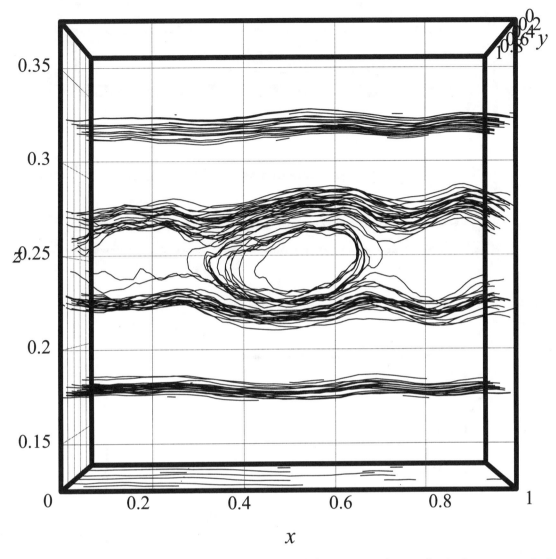

**Figure 11.** Magnetic field lines, seen from the negative $Y$ direction at $t\,\Omega = 10$ (beginning reconnection).

riod of dawnward directed shear field $B_y x$ lasted about 7 - 8 minutes. At the same time $B_z$ exhibited a clear dipolarisation-like increase up to 8 $nT$. The enhancement of $B_z$ is clearly substorm-related.

*4.2 INTERBALL and GEOTAIL - the November 28, 1995*

On November 28, 1995 both the INTERBALL-1 and GEOTAIL spacecraft together were in a favourable position to study the substorm related plasma sheet activity. Since a more detailed discussion of the configuration and obervations can be found in the literature [Büchner et al., 1997; Petrukovich et al., 1997], we concentrate here on the magnetic field observations of INTERBALL-1 and GEOTAIL spacecraft. At 11:30 UT, when INTERBALL-1 was located at $X_{GSM} = -11.5 R_E$, $Y_{GSM} = -1.9 R_E$, $Z_{GSM} = -2.5 R_E$, GEOTAIL was located at $X_{GSM} = -27.8 R_E$, $Y_{GSM} = -6.4 R_E$ and $Z_{GSM} = -1.5 R_E$. Hence the two spacecraft appeared close to the midnight meridian, where at 11:27 UT an auroral breakup was observed at Kotsebue and at 11:28 UT a Pi2 onset took place at low latitudes [Petrukovich et al., 1997]. Magnetic field observations, performed onboard INTERBALL by FGM-I and onboard GEOTAIL by MGF [Kokubun et al., 1994] are depicted in Figures 15 and Figures 16, respectively. Just

322 THREE-DIMENSIONAL MAGNETOTAIL RECONNECTION

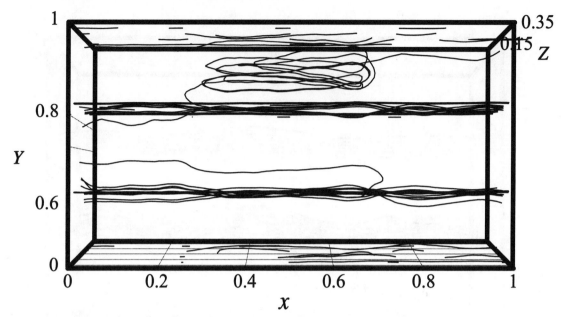

**Figure 12.** Zoomed magnetic field lines, seen from the negative $Z$ direction at $t\,\Omega = 10$ (beginning reconnection).

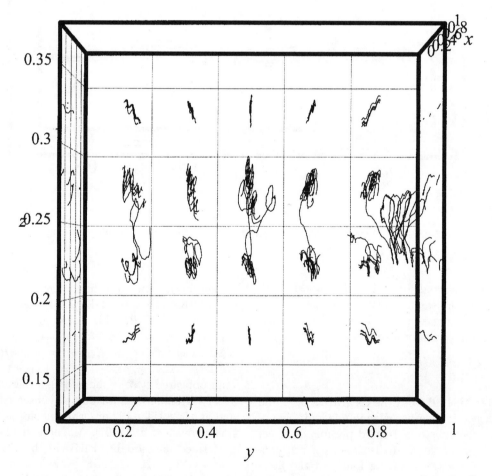

**Figure 13.** Zoomed magnetic field lines, seen from the negative $Z$ direction at $t\,\Omega = 10$ (beginning reconnection).

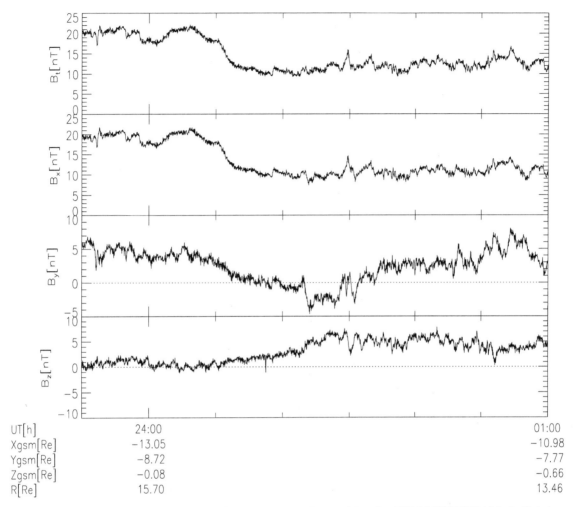

**Figure 14.** Total field and magnetic field components observed by the FGM-I INTERBALL-1, October 28, 1995, 23:50 UT - October 29, 1995, 01:00 UT.

before the ground onset, starting at about 10:26 UT, the INTERBALL spacecraft observed the same dawnward directed magnetic shear ($B_y$) burst, as in the October 29 case, discussed in section 4.1. This time it lasts about 5 minutes. During this period $B_x$ was already negative, i.e. the satellite has crossed the neutral plane southbound between 10:10 UT and 10:15 UT. Hence, it was already southward of the neutral plane when the substorm activation occured. This time, however, the $B_z$ component exhibited an excursion in the negative direction during the $B_y$ - shear burst. In contrast to the observation of October 29th, when $B_z$ increased during the shear-burst in accordance to a field-dipolarisation, the reconnection perturbation now moved earthward over the spacecraft, as flow observations show. Instead of being in the dipolarisation region, however, a reconnection induced perturbation has traveled earthward over the INTERBALL spacecraft. At the same time GEOTAIL was located northward of the neutral plane (positive $B_x$), $15R_E$ behind INTERBALL-1. Like INTERBALL-1, GEOTAIL saw a negative excursion of the $B_z$- field component, indicating the traversal of a plasmoid. The $B_y$ (shear-) burst, however, points now in the dawnward direction. This result agrees well with the prediction from the Hall-current, generated near a reconnection event and demonstrated by kinetic simulations in section 3..

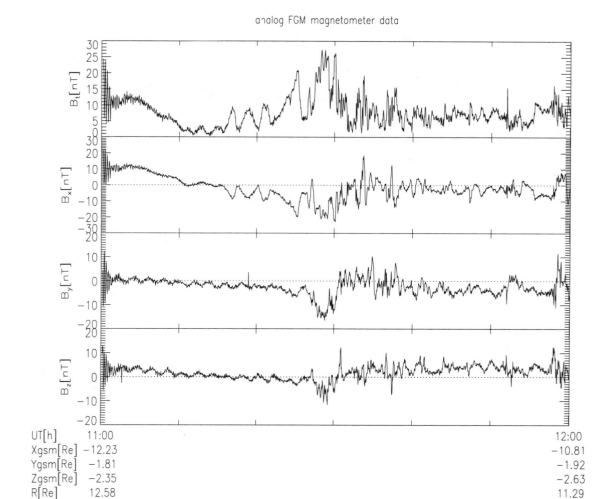

**Figure 15.** Total field and magnetic field components observed by the FGM-I INTERBALL-1, November 28, 1995, 11:00 UT - 12:00 UT.

## 5. DISCUSSION AND SUMMARY

We have derived characteristic magnetic field signatures of three-dimensional magnetotail reconnection using MHD as well as kinetic plasma simulations. As a result we have presented signatures as they would seen by satellites from different positions relatively to the reconnection site. In the MHD limit the characteristic magnetic field signature follows from the draoing of the magnetic field around the reconnection region. This is in accordance with the three-dimensional flow pattern of ideal plasma toward the reconnection site. The kinetic simulations reveal three-dimensional reconnection through thin current sheets which grows faster than the two-dimensional tearing mode instability. It is characterized by an unstable sausage mode wave which quickly

provides the necessary dissipation while anomalous resistivity would lead to a much slower resistive tearing instability [Furth et al., 1963]. Neither in our more restricted initial simulations [Büchner and Kuska, 1996], nor in the extended ones [Büchner and Kuska, 1997] we saw a kink mode instability as Zhu and Wingle [1996] or Pritchett et al. [1996] did. Comparing numerical experiments on collisionless reconnection for a particle mass ratio $M = M_i/m_e = 1$ with $M = M_i/m_e = 64$ we have demonstrated the consequences of charge separation between lighter (electrons) and heavier (ions) particles. It causes a Hall current and a shear field component $B_y$. It appeared that the resulting magnetic field shear structure is different compared to two dimensions: In three dimensions the shear field component does not change its sign accross the sheet, in the $Z$ direction, but only

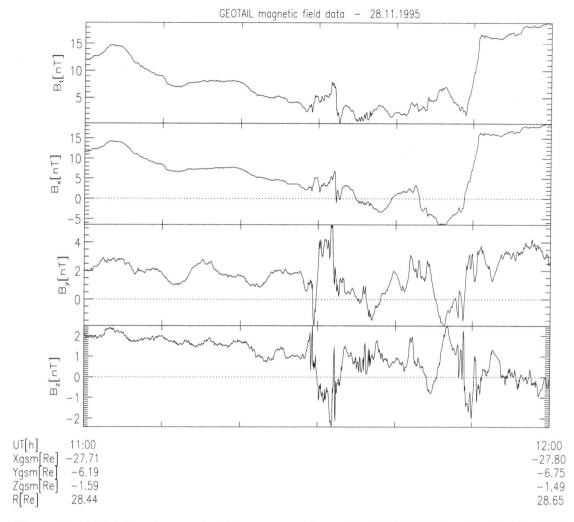

**Figure 16.** Total field and magnetic field components observed by GEOTAIL, November 28, 1995, 11:00 UT - 12:00 UT.

across a reconnection region and the plasmoid center, i.e. along the $X$ direction.

Looking at the ISTP-satellite data of INTERBALL-1 and GEOTAIL we think we have seen the magnetic field signatures discussed here in this paper. We show only a few examples out of a larger data set. In the first example INTERBALL-1 was located outside the current sheet during a substorm activation tailward of the spacecraft position (October 29, 1995). In this case dipolarisation was seen in $B_x$ and $B_z$. But there is also the $B_y$-signature which can be interpreted in accordance with the three-dimensional Hall effect. In the second example, demonstrated here in this paper (the November 28, 1995 case), INTERBALL-1 was located right inside the current sheet when a substorm related reconnection perturbation moved earthward. A transient, lasting only 5 minutes, dawnward directed $B_y$ excursion took place. At the same time GEOTAIL was located along the same magnetic meridian but tailward of the reconnection site. It, therefore, observed the consequences of reconnection from the tail side of the event and exhibited a duskward $B_y$ excursion typical for changing to the other side of the reconnection region in the three- dimensional Hall effect.

In this paper we have demonstrated the principal observability of the signatures of three-dimensional reconnection, arising at different levels of sophistication of the treatment of reconnection. Of course, the characteristic magnetic field signatures of three-dimensional reconnection will be used best by multi-satellite observations, e.g. in the CLUSTER project.

*Acknowledgments.* The Lindau authors thank Prof. Sir W.I. Axford FRS for his support of their work in the MPAe. The work of J.-P. Kuska on the GISMO code was financed by a grant of the German Research Society DFG, the work of B. Nikutowski and H. Wiechen was supported by the German Space Agency DARA, contract 50 QN 9202 / 9707. The Editor thanks two referees for evaluating the manuscript.

## REFERENCES

Axford, I., Magnetic field reconnection, in *Reconnection in space and Laboratory Plasma,* edited by E.W. Hones Jr., pp. 4-14, Geophysical Monograph 30, AGU, Washington D.C., 1984.

Baker, D.N., T.I. Pulkkinen, V. Angeleopoulus, W. Baumjohann, and R.L. McPherron, Neutral line model of substorms: Past results and present view *J. Geophys. Res., 101,* 12,975, 1996.

Birdsall, C.K., and A.B. Langdon, *Plasma Physics via Computer Simulation,* 423 pp., IOP Publishing, Plasma Physics Series, 1991.

Brittnacher, M., K.B. Quest, and H. Karimabadi, On the energy principle and ion tearing in the magnetotail, *J. Geophys. Res. Lett., 21,* 1591, 1994.

Büchner, J., Multiscale coupling in reconnection – Three-dimensional current sheet tearing, in *Multiscale coupling in space plasmas,* edited by T. Chang et al., MIT scientific publishers, 79, 1995.

Büchner, J., Three-dimensional current sheet tearing in the earth's magnetotail, *Adv. Space Res., 18,* 267, 1996.

Büchner, J., and J.-P. Kuska, Three-dimensional collisionless reconnection through thin current sheets, Theory and self-consistent simulation, *ESA-SP 389,* 373, 1996.

Büchner, J., and J.-P. Kuska, Numerical simulation of three-dimensional reconnection due to the instability of collisionless current sheets, *Adv. Space Res., 19,* 1817, 1997.

Büchner, J., H. Wiechen, B. Nikutowski, U. Auster, K.H. Fornacon, J. Rustenbach, S. Klimov, S. Romanov, S. Savin, and A. Otto, INTERBALL-1 plasma sheet encounters and three-dimensional MHD modeling results, *Adv. Space Res.,* 1997, in press.

Coppi, B, G. Laval, and R. Pellat, A model for the influence of the Earth magnetic tail on geomagnetic phenomena, *Phys. Rev. Lett., 16,* 1207, 1966.

Furth H.P., J. Killeen, and M.N. Rosenbluth, *Phys. Fluids, 6,* 459, 1963.

Galeev, A.A., and L.M. Zelenyi, Tearing instability in plasma configuration, *Sov. Phys. JETP, Engl. Transl., 43,* 1113, 1976.

Harris, E.G., On a plasma sheath separating regions of oppositely directed magnetic field, *Nuovo Cimento, 23,* 115, 1962.

Hockney, R.W., and J.W. Eastwood, *Computer Simulation Using Particles,* 433 pp., Adam-Hilger, 1988.

Huba,J.D., N.T. Gladd, and K. Papadopoulus, The lower hybrid drift instability as a source of anomalous resistivity for magnetic reconnection, *Geophys. Res. Lett., 4,* 125, 1977.

Huba,J.D., J.F. Drake, and N.T. Gladd, Lower hybrid drift instability in field reversed plasmas, *Phys. Fluids, 23,* 552, 1980.

Hughes, W.J., and D.G. Sibeck, On the 3-dimensional structure of plasmoids, *Geophys. Res. Lett., 14,* 636, 1987.

Klimov, S., S. Romanov, E. Amata, J. Blecki, J. Büchner, B. Nikutowski, and 32 more authors, ASPI experiment: Measurements of fields and waves on board the INTERBALL-1 spacecraft, *Ann. Geophysicae, 15,* 514, 1997.

Kokubun, S., T. Yamamoto, M. Acuna, K. Hayashi, K. Shiokawa, and H. Kawano, The GEOTAIL magnetic field experiment, *J. Geomag. Geoelectr., 46,* 7, 1994.

Lembége B., and R. Pellat, Stability of a thick two-dimensional quasi-neutral sheet, *Phys. Fluids, 25,* 1995, 1982.

Matsumoto, H., and T. Sato, *Computer Simulation of Space plasma,* 723 pp., D. Reidel Publ. Comp., Dordrecht, Boston, Lancester, 1985.

Pellat, R., F.V. Coroniti, and P.L. Pritchett, Does ion tearing exist? *Geophys. Res. Lett., 18,* 143, 1991.

Priest, E.R., and P. Demoulin, Three-dimensional magnetic reconnection without null points, *J. Geophys. Res., 100,* 23,443, 1995.

Pritchett, P.L., Effect of electron dynamics on collisionless reconnection in two-dimensional magnetotail equilibria, *J. Geophys. Res., 99,* 9935, 1994.

Pritchett, P.L., and J. Büchner, Collisionless reconnection in configurations with a minimum in the equatorial magnetic field and with magnetic shear, *J. Geophys. Res., 100,* 3601, 1995.

Petrukovich, A.A., V.A. Sergeev, L.M. Zelenyi, T.Mukai, S.Kokubun, J. Büchner, E.Yu. Budnick, C.S.Deehr, A.O. Fedorov, V.Grigor'eva, T.J.Hughes, N.F. Pissarenko, S.A. Romanov, I.Sandahl, Two spacecraft observations of reconnection pulse during the auroral breakup, *J. Geophys. Res., 102,* 1997, in press.

Pritchett, P.L., F.V. Coroniti, and V.K. Decyk, Three-dimensional stability of thin quasi-neutral current sheets, *J. Geophys. Res., 101,* 27,413, 1996.

Schindler, K., A theory of substorm mechanisms, *J. Geophys. Res., 79,* 2803, 1974.

Sonnerup, B.U.Ö, Magnetic field reconnection, in *Solar System Plasma Physics,* ed. L.T. Lanzerotti, C.F. Kennel, and E.N. Parker, 3, 47, 1979.

Terasawa, T., Hall current effect on tearing mode instability, *Geophys. Res. Lett., 10,* 475, 1983.

Vasyliunas, V.M., Theoretical models of magnetic field line merging, 1, *Rev. Geophys. Space Phys., 13,* 303, 1975.

Wiechen, H., J. Büchner, and A. Otto, Reconnection in the near-Earth plasma sheet: A three-dimensional model, *J. Geophys. Res., 101,* 24,911, 1996.

Winske, D., Particle simulation studies of the lower hybrid drift instability, *Phys. Fluids, 24,* 1069, 1981.

Zhu, Z, and R.M. Winglee, Tearing instability, flux ropes, and kinetic current sheet kink instability in the Earth's magnetotail: A three-dimensional perspective from particle simulations, *J. Geophys. Res., 101,* 4885, 1996.

Zwingmann, W., J. Wallace, K. Schindler, and J. Birn, Particle simulation of magnetic reconnection in the magnetotail configuration, *J. Geophys. Res., 95,* 20,877, 1990.

---

J. Büchner, J.-P. Kuska, B. Nikutowski, H. Wiechen, Max-Planck-Institut für Aeronomie, Postfach 20, Katlenburg-Lindau, D-37191, Germany. (e-mail: buechner@linmpi.mpg.de)

# Drift-Shell Splitting in an Asymmetric Magnetic Field

Mei-Ching Fok

*Universities Space Research Association, NASA Goddard Space Flight Center, Greenbelt, Maryland*

Thomas E. Moore

*Laboratory for Extraterrestrial Physics, NASA Goddard Space Flight Center, Greenbelt, Maryland*

In a day-night asymmetric magnetic configuration, charged particles with different pitch angles gyrating along the same field line at one local time will be mirroring on different field lines as they drift to other local times. This drift-shell splitting effect is one of the important factors in determining the particle pitch angle distribution at distances greater than 5 earth radii. In this paper, the ring current ion pitch angle distribution is calculated in an activity-dependent Tsyganenko model, during the magnetic storm on May 2, 1986. Our simulation shows that, near the storm maximum, the observed low-energy (< 10 keV) pitch angle distributions on the dayside are strongly affected by drift-shell splitting. We also found that a fluctuating magnetic configuration produces diffusion in pitch angle, and that this effect is pronounced only for high-energy ions. Excellent agreement with the measured pitch angle distributions is obtained when a time-dependent realistic field model is employed.

## 1. INTRODUCTION

In a longitudinally uniform magnetic field (i.e., a pure dipole) and in the absence of electric field, charged particles with different equatorial pitch angles, originally mirroring along the same field line at one local time, will find themselves bouncing along the same field line when they drift to other local times. However, in reality, especially beyond 5 earth radii, the magnetic field is not day-night symmetric. In an asymmetric configuration, particles initially on the same field line but with different mirroring points will drift on different drift shells and spread over a region of field lines [*Roederer*, 1967]. Equivalently, particles with different pitch angles seen at one observation point may come from different locations. As a result, any pitch angle diffusion will lead to a radial diffusion owing to drift-shell splitting and vice versa [*Roederer*, 1967]. The effect of drift-shell splitting is an important factor in determination of the pitch angle distribution (PAD) of charged particles. *Sibeck et al.* [1987] showed qualitatively that drift-shell splitting in the distorted nondipolar magnetic field during disturbed periods can generate butterfly (relative minimum at 90° pitch angle) and head-and-shoulder (excess of near 90° pitch angle particles) energetic particle PADs which appear in the dayside magnetosphere. Drift-shell splitting separates the high and low pitch angle particles from nightside injections as they move to the dayside magnetosphere. The higher pitch angle particles move radially away from the earth and particles with lower pitch angles follow more circular drift paths. They argued, consequently, that butterfly PADs form at the same geocentric distance as the injection and head-and-shoulder PADs at a point slightly further radially outward.

Geospace Mass and Energy Flow: Results From the International Solar-Terrestrial Physics Program
Geophysical Monograph 104
Copyright 1998 by the American Geophysical Union

We have developed a kinetic model to solve the temporal variation of the ring current energy and pitch angle distributions in a magnetic dipole field [*Fok et al.*, 1995, 1996]. In the simulation of the main phase of the storm on May 2, 1986 [*Fok et al.*, 1996], we found, near the inner edge of the ring current, the observed perpendicular PAD is a result of strong charge exchange loss of the field-aligned ions. For energies greater than tens of keV, however, the observed round-shape PAD cannot be solely explained by the charge exchange loss. An additional pitch angle diffusion process with diffusion coefficient about $5 \times 10^{-6}$ s$^{-1}$ is required in order to match the data from AMPTE/CCE (Active Magnetospheric Particle Tracer Explorers / Charge Composition Explorer) satellite. Recently, our model has been extended to include a realistic magnetic field model [*Tsyganenko*, 1989]. This improved ring current model was used to simulate the sudden enhancements in the equatorial ion fluxes and the corresponding ionospheric precipitation during a substorm expansion [*Fok and Moore*, 1997]. In the present paper, the effects of drift-shell splitting on the ring current pitch angle distribution are studied using our model with an asymmetric magnetic field. The main phase of the storm on May 2, 1986 is simulated. The calculated PADs are compared with those results generated in a dipole field to distinguish the shell-splitting effects. Observed ring current H$^+$ PADs by the AMPTE/CCE satellite will be used to evaluate the success of the modeling approach.

## 2. PREVIOUS CALCULATIONS IN MAGNETIC DIPOLE FIELD

We have previously modeled the H$^+$ pitch angle distribution in a magnetic dipole field during the main phase of the storm on May 2, 1986 [*Fok et al.*, 1996]. The initial condition is a quiet time distribution. The H$^+$ differential flux averaged over pitch angle compiled by *Sheldon and Hamilton* [1993] during the quietest days in 1985–1987, seen by CCE is used as initial distribution at 0200 UT, May 2, 1986. Another input to the model is the boundary condition near geosynchronous orbit. The instantaneous energy and pitch angle distributions on the nightside boundary ($r_o \sim 6.75$) are obtained by interpolation in time of measurements from two CCE passes at 0800 and 2400 UT, where $r_o$ is the equatorial distance in earth radius ($R_E$).

The initial ion PADs are estimated by radial distance and the charge exchange cross sections with the neutral hydrogen, assuming that the quiet time PAD is mainly shaped by the charge exchange loss [*Fok et al.*, 1996]. Figure 1 plots the initial PAD of H$^+$ (dotted lines) of 4

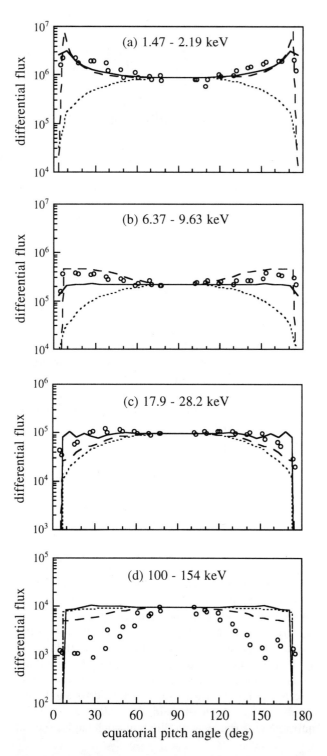

**Figure 1.** Calculated pitch angle distributions in a dipole field (solid lines) and in a Tsyganenko model (dashed lines) at ~ 1800 UT, 5.7 $R_E$, 0800 LT. Open circles are observations from AMPTE/CCE at the same location and time, on May 2, 1986. Dotted lines are the initial distributions at 0200 UT. All data sets are scaled so that they match at 90° pitch angle.

different energies at $r_0 \sim 5.7$. The round shape PADs of the low energy H$^+$ (Figure 1a, b, and c) represent relatively strong charge exchange loss of field aligned ions at these energies. The charge exchange cross section of H$^+$ decreases with increasing energy, so the initial PAD of the 100–154 keV ions are fairly isotropic. After 16 hours of simulation, the calculated dayside PADs at 0800 UT in a dipole field at the same radial distance are overlaid in Figure 1 in solid lines. The CCE measurements at the same time and location are also shown in open circles. The representation of the dashed lines will be discussed in the later sections. Charge exchange loss is not strong at this distance. If the magnetic field is a pure dipole and thus there is no drift-shell splitting, the shape of the dayside PAD should be similar to that on the nightside. As shown in the Figure 1 (solid lines), we predict a field-aligned distribution in the lowest energy range and flat distributions for higher energies. These simulated PADs on the dayside strongly depend on the corresponding distributions measured on the nightside boundary. However, observations on the dayside (open circles) show a field-aligned distribution in the energy range of 6.37–9.63 keV and perpendicular distributions for energies $\gtrsim$ 20 keV. As we have suggested, additional processes must be taken into account to determine the ion pitch angle distributions.

## 3. RING CURRENT MODELING IN REALISTIC MAGNETIC CONFIGURATION

Our ring current model has been extended to adapt any magnetic field configuration [*Fok and Moore*, 1997]. In the new version of the model, particle drifts are calculated in the coordinates spanned by two Euler-potentials: $(\alpha, \beta)$. We have chosen $\beta = \phi_i$ and $\alpha = -M_E \cos 2\lambda_i / 2 r_i$, where $M_E$ is the Earth's magnetic dipole moment; $\lambda_i$, $\phi_i$ and $r_i$ are the magnetic latitude, magnetic local time and radial distance, respectively, of the ionospheric foot point of the field line. We have used the ionospheric foot point to label a field line since magnetic variation at the ionosphere is very small even when the magnetosphere is compressed or expanded due to substorm activities. Field lines thus can be regarded as rooted at the ionosphere.

The bounce-averaged drift of a charge particle can be represented by [*Northrop*, 1963]:

$$\langle \dot{\alpha} \rangle = -\frac{1}{q} \frac{\partial H}{\partial \beta} \quad , \quad \langle \dot{\beta} \rangle = \frac{1}{q} \frac{\partial H}{\partial \alpha} \qquad (1)$$

where $H$ has the form:

$$H = \sqrt{p^2 c^2 + m_0^2 c^4} + q\Phi + q\alpha \,\partial\beta/\partial t$$

$$= \sqrt{p^2 c^2 + m_0^2 c^4} + q\Phi + q\alpha\Omega \qquad (2)$$

$\Phi$ is the cross-tail potential and $\Omega$ is the angular velocity of the rotation of the Earth. $\langle \dot{\alpha} \rangle$ and $\langle \dot{\beta} \rangle$ include gradient-curvature drift, electric drift due to convection, and corotation. The convection model employed is that of Volland-Stern [*Volland*, 1973; *Stern*, 1975], with the field strength parameterized by the $Kp$ index [*Maynard and Chen*, 1975]. The changing of the magnetospheric configuration during substorms do not yield any non-zero value of $\partial\beta/\partial t$ $(= \partial\phi_i/\partial t)$ because we have assumed that the ionospheric foot points of field lines are unchanged due to substorm activities. The substorm induced electric field and the resulting bounce-averaged drift are treated implicitly by continuously changing the gradient-curvature drift according to the instantaneous magnetic configuration.

It is convenient to express the bounce-averaged velocity in terms of $\lambda_i$ and $\phi_i$,

$$\langle \dot{\lambda}_i \rangle = -\frac{1}{q\xi} \frac{\partial H}{\partial \phi_i} \quad , \quad \langle \dot{\phi}_i \rangle = \frac{1}{q\xi} \frac{\partial H}{\partial \lambda_i} \qquad (3)$$

where $\xi$ is equal to $M_E \sin 2\lambda_i / r_i$. The bounce-averaged kinetic equation of ring current ion species $s$ is given by [*Fok and Moore*, 1997]:

$$\frac{\partial \bar{f}_s}{\partial t} + \langle \dot{\lambda}_i \rangle \frac{\partial \bar{f}_s}{\partial \lambda_i} + \langle \dot{\phi}_i \rangle \frac{\partial \bar{f}_s}{\partial \phi_i}$$

$$= -v\sigma_s \langle n_H \rangle \bar{f}_s - \left( \frac{\bar{f}_s}{0.5\tau_b} \right)_{loss\ cone} \qquad (4)$$

where $\bar{f}_s = \bar{f}_s(t, \lambda_i, \phi_i, M, K)$, is the average distribution function on the field line between mirror points. $M$ is the relativistic magnetic moment and $K = J/\sqrt{8 m_0 M}$. $\sigma_s$ is the cross section for charge exchange of species $s$ with the neutral hydrogen and $n_H$ is the hydrogen density. $\tau_b$ is the bounce period. The second term on the right hand side of (4) is applied only to particles with pitch angle inside the loss cone, which is defined at 800 km. In this study, only loss due to charge exchange with the neutral hydrogen and loss at the loss cone are considered.

## 4. DRIFT-SHELL SPLITTING IN ASYMMETRIC MAGNETIC FIELD

The H$^+$ pitch angle distributions at various energies are re-calculated in a realistic magnetic field model with the

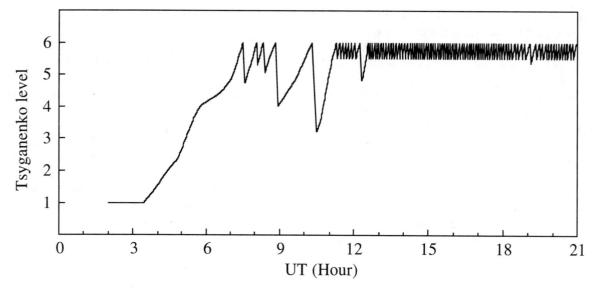

**Figure 2.** Simulated levels (IOPT) of the Tsyganenko model as a function of UT on May 2, 1986.

same initial and boundary conditions used in *Fok et al.* [1996] for the May 2, 1986 storm. The magnetic field configuration is represented by the Tsyganenko model at various activity levels (EXT89AE, IOPT = 1–6). An algorithm has been established to simulate the instantaneous magnetic field configuration driven by the AE index [*Fok et al.*, 1996]. The simulated Tsyganenko level (IOPT) as a function of UT on May 2, 1986 are plotted in Figure 2. Simulation starts when the magnetosphere is in quiet condition (level = 1). The magnetosphere then slowly evolves to the higher levels (increased stretching and inflation). When it reaches the level of 6, it collapses to lower levels. At the early main phase (before 1100 UT), the rate of stretching is small and the amplitude of relaxation is large. However, during the highly disturbed period, the magnetosphere is rapidly stretched and only small relaxations occur. The bounce-averaged drift velocities $(\langle \dot{\lambda}_i \rangle$ and $\langle \dot{\phi}_i \rangle)$ are calculated according to the instantaneous magnetic field configuration. Velocities at continuous levels are obtained by interpolation between the 6 discrete levels.

We overlay on Figure 1 the calculated PADs (dashed lines) in the Tsyganenko field at the same time and location (1800 UT, 5.7 $R_E$, and 0800 LT) as those previously obtained in a dipole field. Both sets of calculated fluxes are scaled to match the measurements at 90° pitch angle. As shown in the figure, for ion energy below 10 keV (Figure 1a, b), more field-aligned distributions are predicted in an asymmetric field than in a dipole field. We have traced backward along the drift paths of particles with different equatorial pitch angles. We found these low-energy ions drift along open trajectories and reach the observation point from dawn. However, field-aligned particles come from the local times farther toward nightside than field-perpendicular particles. The new simulation predicts more field-aligned ions at this observation point because they come from closer to the nightside injection region. With the consideration of an asymmetric field, great improvement has been made in agreement with the observation, especially in the energy range of 6.37–9.63 keV.

Figures 1c and 1d show the results in energy ranges of 17.9–28.2 and 100–154 keV. In these two cases, including a realistic magnetic field model gives rounder pitch angle distributions. It seems that a diffusive process in pitch angle is taking place to smooth out the distribution. Ions at these energies exhibit closed drift paths and diffusive transport dominates their motions [*Chen et al.*, 1994]. We suspect that the fluctuation of the magnetosphere (see Figure 2) during the late main phase produces particle radial diffusion, as well as pitch angle diffusion. Our suggestion is confirmed when we perform a test run with Tsyganenko level kept at a constant of 5.5 after first reaching this value at ~ 0700 UT. The test simulation gives relatively flat (with respect to the dashed line in Figure 1c, d) distributions for both 17.9–28.2 keV and 100–154 keV ions. In other words, without the high-frequency (~ 10 hr$^{-1}$) fluctuating magnetic field, the round shape PAD for high-energy ions cannot be reproduced. Of course, even with the time-varying magnetic field, the observed strong perpendicular PAD for 100–154 keV ions is not accurately predicted (Figure 1d). Other processes that cause pitch angle diffusion, such as wave-particle interactions and electric fluctuation, should also be considered. For ions below 10 keV, there is nearly no difference between the calculated PADs from the test simulation and those from the fluctuating magnetic field. This indicates that diffusive transport of low-energy ions is unimportant.

## 5. DISCUSSION AND SUMMARY

The effect of drift-shell splitting on the ring current ion pitch angle distribution has been modeled in an activity-dependent Tsyganenko magnetic field model. We compare the calculated PADs in a realistic field at 5.7 $R_E$ with those previously calculated in a dipole configuration and find significant differences. Moreover, great improvement is obtained in agreement with the CCE particle observations with the asymmetric magnetic field.

We have quantitatively shown that magnetic fluctuations result in pitch angle scattering. We plan to simulate the diffusive effect from electric variations as well. In future work, a realistic electric field model based on interplanetary magnetic field and solar wind conditions [e.g., *Weimer*, 1995; *Papitashvili*, 1994] will be incorporated in our model. The role of wave-particle interactions on ring current distribution will also be investigated.

In conclusion,
(1) We have calculated the ring current ion pitch angle distributions in a day-night asymmetric magnetic field model during the main phase of the storm on May 2, 1986.
(2) We found drift-shell splitting is responsible for the observed dayside field-aligned PADs for low-energy (< 10 keV) ions.
(3) We have shown that drift-shell splitting in a fluctuating magnetic configuration results in pitch angle diffusion for high-energy (> 20 keV) ions.

## REFERENCES

Chen, M. W., L. R. Lyons, and M. Schulz, Simulations of phase space distributions of storm time proton ring current, *J. Geophys. Res.*, 99, 5745-5759, 1994.

Fok, M.-C., and T. E. Moore, Ring current modeling in a realistic magnetic field configuration, *Geophys. Res. Lett.*, 24, 1775-1778, 1997.

Fok, M.-C., T. E. Moore, J. U. Kozyra, G. C. Ho, and D. C. Hamilton, Three-dimensional ring current decay model, *J. Geophys. Res.*, 9619-9632, 1995.

Fok, M.-C., T. E. Moore, and M. E. Greenspan, Ring current development during storm main phase, *J. Geophys. Res.*, 101, 15,311-15,322, 1996.

Maynard, N. C., and A. J. Chen, Isolated cold plasma regions: Observations and their relation to possible production mechanisms, *J. Geophys. Res.*, 80, 1009-1013, 1975.

Papitashvili, V. O., B. A. Belov, D. S. Faermark, Ya. I. Feldstein, S. A. Golyshev, L. I. Gromova, and A. E. Levitin, Electric potential patterns in the Northern and Southern polar regions parameterized by the interplanetary magnetic field, *J. Geophys. Res.*, 99, 13,251-13,262, 1994.

Roederer, J. G., On the adiabatic motion of energetic particles in a model magnetosphere, *J. Geophys. Res.*, 72, 981-992, 1967.

Sheldon, R. B., and D. C. Hamilton, Ion transport and loss in the earth's quiet ring current 1. Data and standard model, *J. Geophys. Res.*, 98, 13,491-13,508, 1993.

Sibeck, D. G., R. W. McEntire, A. T. Y. Lui, R. E. Lopez, and S. M. Krimigis, Magnetic field drift shell splitting: cause of unusual dayside particle pitch angle distributions during storms and substorms, *J. Geophys. Res.*, 92, 13,485-13,497, 1987.

Stern, D. P., The motion of a proton in the equatorial magnetosphere, *J. Geophys. Res.*, 80, 595-599, 1975.

Tsyganenko, N. A., A magnetospheric magnetic field model with a warped tail current sheet, *Plant. Space Sci.*, 37, 5-20, 1989.

Volland, H., A semiempirical model of large-scale magnetospheric electric fields, *J. Geophys. Res.*, 78, 171-180, 1973.

Weimer, D. R., Models of high-latitude electric potentials derived with a least error fit of spherical harmonic coefficients, *J. Geophys. Res.*, 100, 19,595-19,607, 1995.

---

M.-C. Fok, and T. E. Moore, Lab. for Extraterrestrial Physics, NASA/GSFC, Code 692, Greenbelt, MD 20771. (e-mail: fok@gsfc.nasa.gov, thomas.e.moore@gsfc.nasa.gov)

# Comparison of Photoelectron Theory Against Observations

G. V. Khazanov and M. W. Liemohn

*Space Sciences Laboratory, NASA Marshall Space Flight Center, ES-83, Huntsville, Alabama*

Presented here are comparisons of a superthermal electron interhemispheric transport model with satellite data and previous transport model results. Good agreement is shown with the results of the two-stream superthermal electron transport model, as well as with other ionospheric calculations when the same assumptions have been applied. The concept of plasmaspheric transparency is considered, and a method of applying transparencies calculated by this model to ionospheric transport models is presented. Explicit comparisons are made with energy spectra from the Atmospheric Explorer E satellite and pitch angle distributions from the Dynamic Explorer 2 satellite. Implicit comparisons are made by examining the influence of superthermal electrons in the formation of observed thermal plasma quantities. This type of comparison is performed with Dynamic Explorer 1 and Akebono observations.

## INTRODUCTION

This study tests the results of a time-dependent, interhemispheric transport model with observations and other superthermal electron transport models to see how well our results compare. This model [*Khazanov et al.*, 1993; *Khazanov and Liemohn*, 1995; *Liemohn and Khazanov*, 1995; *Liemohn et al.*, 1997] unifies the spatial regions of the ionosphere and magnetosphere into a single calculation scheme. The algorithm was developed to model superthermal electron fluxes without any restrictions on space or time in the calculation. It handles photoionization and impact ionization sources, elastic and inelastic scattering with atmospheric neutral particles, interactions with the thermal plasma, inhomogeneities in the geomagnetic field, and internally and externally imposed forces, such as the ambipolar electric field. This non-steady-state kinetic model has been used to investigate the role of superthermal electrons on plasmaspheric refilling, especially during the initial transient stages of the refilling process.

Recently, a driver program has been created that can combine this model with any thermal plasma model, and this approach was used by *Liemohn et al.* [1997] to self-consistently coupling collisional and electrodynamical interactions between the superthermal electrons and thermal plasma species. Collisional coupling between plasma species includes Coulomb interactions, and plays a dominant role in the ionosphere and along filled flux tubes. Electrodynamic coupling encompasses electric fields and potentials, as well as inclusion of the superthermal electron population in the quasineutrality current balance conditions, and often dominates in the collisionless regime, especially along depleted or open flux tubes. In that study, the kinetic model was coupled with the time-dependent, field-aligned, hydrodynamic thermal plasma model of *Guiter et al.* [1995] to examine the influence of electrodynamic coupling on plasmaspheric refilling.

The comparisons with models is limited to the ionosphere, because that is where most of the modeling efforts have been focused. The interhemispheric transport calculation of the present model, however, allows for the calculation of plasmaspheric transparencies. This is the fraction of particles that reach the topside of the conjugate ionosphere after transversing the plasmasphere, and was shown in *Khazanov and Liemohn* [1995] to be a complicated function of the atmospheric sources and scattering processes as

Geospace Mass and Energy Flow: Results From the International Solar-Terrestrial Physics Program
Geophysical Monograph 104
Copyright 1998 by the American Geophysical Union

well as the flux tube thermal plasma content. A method for using plasmaspheric transparencies calculated by the present model in ionospheric transport calculations is discussed.

Explicit data comparisons with Atmospheric Explorer E (AE-E) and Dynamics Explorer 2 (DE 2) data are limited to several cases: the local equilibrium region, the transition region, and the transport dominated region. The local equilibrium region contains those altitudes dominated by collisions, where local production and loss chemistry is the dominant factor in determining the distribution function. This region is typically located below about 200 km in the ionosphere. Above this is a transition region where plasma transport begins to have a significant role in the development of the distribution, while local effects are also still important. The third region is above this transition region, where the collisional scale lengths become larger than the transport scale lengths. In this region, motion of the particles along the field line is the major contributor to the formation of the distribution function. The transport-dominated region extends out into the magnetosphere, and, along closed field lines, connects to the ionospheric transition region of the conjugate hemisphere.

Implicit comparisons can also be made with satellite data, comparing observations of thermal plasma parameters with results from this model. The first is with DE 1 data during quiet times, when collisional effects should dominate, and the second is with Akebono data in the high-latitude region, where electrodynamic coupling should be the major interaction mechanism between the superthermal electrons and the thermal plasma.

## COMPARISONS WITH OTHER MODELS

It is also useful to compare the results of this model with other commonly used methods of superthermal electron transport modeling for cases when certain restrictions to this model can be applied to match its results with previous calculations. *Cicerone et al.* [1973] performed a comparison among three quite different approaches: a diffusion equation formulation [*Nisbet*, 1968; *Swartz*, 1972]; a two-stream approximation of the general transport equation [*Nagy and Banks*, 1970]; and a Monte Carlo technique to simulate trajectories [*Cicerone and Bowhill*, 1970, 1971]. An updated version of the two-stream model is available and will be discussed later. Here, however, a comparison with published results is desired to show the closeness of the ionospheric models when they use identical input parameters. They give the atmospheric and ionospheric density distributions, so it is relatively easy to compare with results from this model. Figure 1 shows this comparison for the 0° solar zenith angle case of *Cicerone et al.* [1973]. The "net flux" quantity shown here is the net directional flux defined as

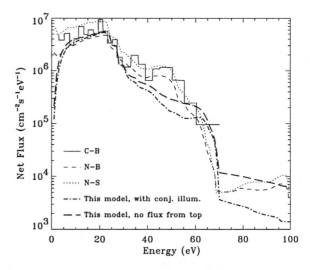

**Figure 1.** Comparison of net directional fluxes for this model and three others from *Cicerone et al.* [1973]. Results are for the 0° solar zenith angle case at the top of the ionosphere (580 km) of the previous study, with no downward flux above this altitude (except for one case for this model, which includes plasmaspheric and conjugate effects).

$$\phi_{NET}(s, E) = \int_{\varphi=0}^{2\pi} \left[ \int_{\theta=0}^{\pi} \phi(s, E, \theta) \cos\theta \sin\theta d\theta \right] d\varphi \\ = 2\pi \int_{\mu=-1}^{1} \phi(s, E, \mu) \mu d\mu \quad (1)$$

and represents a differential bulk flow of the photoelectrons. Here, $\phi$ is the differential flux of superthermal electrons, $s$ is distance along the field line, $E$ is kinetic energy, $\theta$ is pitch angle, $\varphi$ is velocity azimuthal angle, and $\mu=\cos\theta$. As in the earlier comparison, "N-S" refers to the diffusion equation formulation, "N-B" to the two-stream model, and "C-B" to the Monte Carlo technique. Two curves from our model are shown: one includes the plasmasphere and conjugate ionosphere; while the other has no downward flux from above 580 km. The latter case should be consistent with the assumptions of the other three models. While the results of the present model are close to the earlier model results below about 30 eV, they are somewhat different above this energy, although never by more than a factor of three. This is not unexpected, however, because the new model uses updated solar spectra and cross section information [from Dr. Solomon, private communication, 1994]. Without rewriting the cross section calculation of our code to match theirs (or vice versa), this is the closest comparison that can be presented with the *Cicerone et al.* [1973] results. A comparison with the multistream model of *Oran and Strickland* [1978] (such as that done by *Winningham et al.* [1989]) is underway and will be the subject of a future publication.

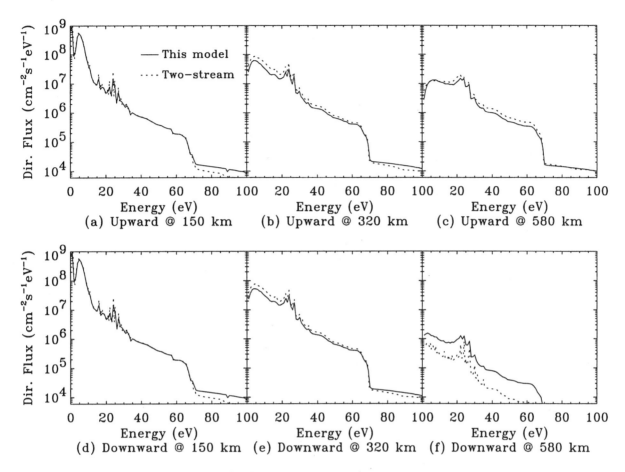

**Figure 2.** Upward and downward fluxes for this model (solid lines) and an updated two-stream model (dotted lines) [S. C. Solomon, private communication, 1994] at various ionospheric altitudes. Results are for noon on day 178 of 1983 at 41° invariant latitude, with no downward fluxes above 580 km.

There is, however, another alternative for comparison with the two-stream model of *Nagy and Banks* [1970]. Our model uses the same solar spectrum and cross section calculations as Dr. Solomon's two-stream model, and so a direct comparison can be made between these models. Such a comparison is shown in Figure 2 for conditions similar to that of the *Cicerone et al.* [1973] study. Here, the IRI model [*Bilitza*, 1990] is used for the thermal plasma density, and the MSIS model [*Hedin*, 1991] is used for the neutral atmosphere profiles for day 178 of 1983 at $L=2$ (41° invariant latitude), which is a similar location in the solar and annual cycle to the previous comparison. The plots on the top are upward directional fluxes, and the plots on the bottom are downward directional fluxes,

$$\phi_{UP} = 2\pi \int_{\mu=0}^{1} \phi\mu d\mu \quad \text{and} \quad \phi_{DOWN} = 2\pi \int_{\mu=0}^{-1} \phi\mu d\mu \quad (2)$$

which, when subtracted, would yield the net flux quantity shown in Figure 1 ($\phi_{NET}=\phi_{UP}-\phi_{DOWN}$). The results for both models are for no downward flux above 580 km, making them single-ionosphere calculations. There appears to be close agreement between the two models, especially at low altitudes. Even at 580 km, the upward fluxes never differ by more than 20%. The downward results in Figure 2f arise from two sources: local photoionization and local backscattering, since there are no particles precipitating from above. Thus, this discrepancy is due to differences in the collision operators: the two-stream model uses backscatter probabilities; while the present model uses the Boltzmann equation for scattering with neutrals and the Fokker-Planck equation for scattering with the thermal plasma.

The absence of any downflowing superthermal electron flux, however, is highly unlikely on the dayside. Even if the conjugate ionosphere is in darkness, backscattering due to Coulomb scattering and elastic collisions with atmospheric neutrals will provide some amount of downflowing flux [*Khazanov and Liemohn*, 1995]. However, the two-stream model cannot account for the changing magnetic field and the trapping of particles along the flux tube, so simply extending this ionospheric model into the plasmasphere will produce unphysical results.

Plasmaspheric transport has been included in two-stream calculations through the use of a plasmaspheric transparency, $T(E)$ [e.g., *Lejeune*, 1979]. There are several ways to define this quantity, so the first definition will be that of *Takahashi* [1973] as the ratio of particle fluxes precipitating into one ionosphere divided by the particle fluxes flowing out of the conjugate ionosphere,

$$T(E) = \frac{\int_0^1 \mu \phi(s_{2,top}, E, \mu) d\mu}{\int_0^1 \mu \phi(s_{1,top}, E, \mu) d\mu} \quad (3)$$

where $s_{1,top}$ is the location of the "top" of one ionosphere and $s_{2,top}$ is the altitude of the location of the "top" of the conjugate ionosphere. Note that this is not the probability of a single particle reaching the conjugate hemisphere, but rather is an attenuation factor such that the downward flux entering the conjugate ionosphere is equal to the upward flux leaving the first ionosphere multiplied by $T(E)$. Such a quantity can be used by an ionospheric transport model to parameterize the plasmaspheric processes, and a calculation in the conjugate ionosphere could then be conducted with a precipitating flux at the upper boundary.

Another definition for plasmaspheric transparency that could be used with ionospheric models is the ratio of the flux flowing down from the plasmasphere to the flux flowing up into the plasmasphere,

$$T^*(E) = \frac{\int_{-1}^0 \mu \phi(s_{1,top}, E, \mu) d\mu}{\int_0^1 \mu \phi(s_{1,top}, E, \mu) d\mu} \quad (4)$$

Notice that both integrals are at the same spatial location: the top of one of the ionospheres. $T^*(E)$ is an attenuation factor to obtain the flux entering an ionosphere given the flux leaving that same ionosphere. With this quantity, an ionospheric model can iterate to a solution in one ionosphere, without having to perform a calculation in the conjugate ionosphere.

Both of these quantities can be calculated with our interhemispheric model. Figure 3 shows two plasmaspheric transparencies for the same time as Figure 2, except the calculation is extended along the entire tilted dipole field line. The results are for a "filled" flux tube, when the thermal plasma density along the field line can be assumed to be proportional to the magnetic field [*Newberry et al.*, 1989]. Also, $s_{1,top}$ is taken in the northern ionosphere and $s_{2,top}$ in the southern ionosphere. The solid line shows $T(E)$ with both ionospheres illuminated (S. I. stands for

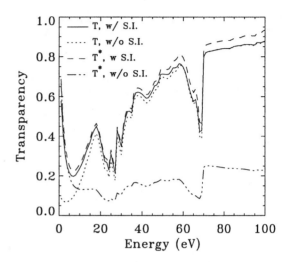

**Figure 3.** Plasmaspheric transparency versus energy from equations (3) and (4) for similar conditions to Figure 2, with $n_e \propto B$ in the plasmasphere, with and without conjugate hemisphere illumination.

"southern illumination"), while the dotted line shows $T(E)$ with the southern hemisphere source term artificially omitted. There is a difference between these two results at low energies. The features of these curves are due to many things, including the features of the photoelectron production spectra of both ionospheres as well as the scattering processes included in the plasmasphere and both ionospheres/thermospheres.

The transparency from (4) is also shown in Figure 3. The dashed line is $T^*(E)$ with southern hemisphere illumination, and the dash-dot-dot-dot line is $T^*(E)$ without this source included. There is quite a difference between these two lines. The only difference between $T(E)$ and $T^*(E)$ with southern illumination is because the photoelectron sources are not exactly symmetric due to the tilt of the dipole. The difference between these quantities without the southern source is drastic, since the numerator in (4) is produced only by backscattered electrons that started in the northern ionosphere. From these results, it can be concluded that the high-energy particles leaving the plasmasphere are primarily unhindered through the plasmasphere, with backscattering occurring in the conjugate ionosphere, while the low-energy electrons that leave the plasmasphere are mostly those backscattered in the plasmasphere and not from the conjugate ionosphere. Keep in mind that these results are for a filled $L=2$ flux tube, and will be different for other conditions.

These transparency results can now be used in the two-stream model. As mentioned above, these transparencies are dependent on the ionospheric and thermospheric processes, and so using these quantities with the two-stream model is not self-consistent. However, as seen in Figure 2c, the upward fluxes at 580 km for the two models are quite close, and so it is expected that the results will not be far from a self-consistent calculation.

Figure 4 shows results from such a calculation. Shown here are downward directional fluxes, similar to those shown in Figure 2f, using $T^*(E)$ from Figure 3 in the two-stream model to attenuate the outflowing fluxes for use as a precipitating upper boundary condition, and then iterating to a converged solution. Notice that the results of the two models are quite close, never differing by more than 30%. Also notice that all of these results are greater than either of the results plotted in Figure 2f.

This shows that a physical calculation of plasmaspheric and conjugate effects can be included in a single-ionosphere transport model by using an interhemispheric transport model to calculate the necessary parameters. Although the calculation is not self-consistent between the spatial regions, it can improve the accuracy of the results from the ionospheric model, especially in the transition and transport-dominated regions of the upper ionosphere. If only one calculation is required for a given set of ionospheric conditions, then the potential gain is not that big. However, if many calculations are needed for similar conditions, then a few results from a spatially-unified model could be used to enhance the accuracy of the ionospheric model's results.

## COMPARISON WITH OBSERVATIONS

The true test of any numerical model, however, is how well it can reproduce observations and explain the processes occurring in the plasma responsible for the forma-

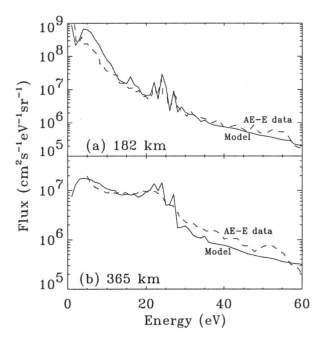

**Figure 5.** Comparison of model results (solid lines) with AE-E data (dashed lines) at (a) 182 km and (b) 365 km altitude on day 355 of 1975. The satellite data is reproduced from *Doering et al.* [1976].

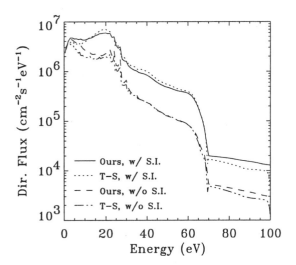

**Figure 4.** Downward fluxes for this model and an updated two-stream model at 580 km, with and without conjugate hemisphere illumination. Results are analogous to Figure 2f, except plasmaspheric and conjugate effects are included in our model, and the transparencies in Figure 3 have been used in the two-stream results.

tion of the distribution functions. Since *in situ* measurements of superthermal electron distributions have primarily been conducted in the ionosphere, we will limit our presentation to the three ionospheric spatial regions mentioned above: the collision-dominated region, the transition region, and the transport-dominated region.

The Atmospheric Explorer satellites offer a plentiful supply of superthermal electron energy spectra. The electron spectrometers on the AE satellites are ideal for comparison because of the fine spectral resolution achieved in the low-energy range. Although the data is spin-averaged, this is not a big problem since the satellites flew through the low to middle ionosphere, where the distribution function is nearly isotropic. Figure 5 shows omnidirectional fluxes from the AE-E satellite and model results for similar conditions. The data in this plot is reproduced from *Doering et al.* [1976], for day 355 of 1975 at 182 and 365 km altitude. This first altitude is in the region where collisions with neutrals dominate the formation of the distribution function, and local production and loss mechanisms are the major processes in the calculation. The second altitude is in the transition region, where transport is starting to have a significant influence on the distribution. The solar zenith angles for the two spectra are 50° and 37°, respectively. Since AE-E flew in a nearly equatorial orbit, the model comparisons are made at 0° geographic latitude, choosing an appropriate morningside local

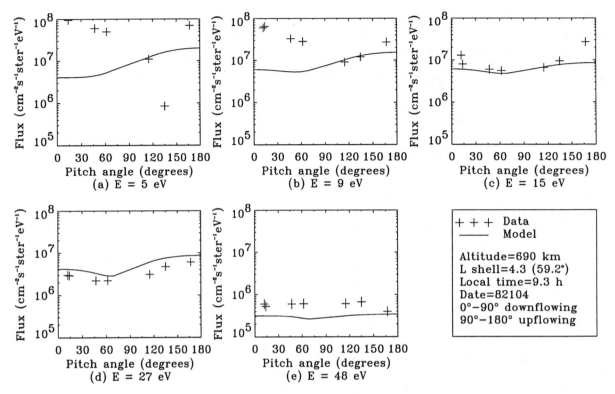

**Figure 6.** Comparison of model results with DE 2 LAPI pitch angle distributions. Data courtesy of R. E. Erlandson [private communication, 1994].

time with the given solar zenith (because the data collection occurred there, *Doering et al.* [1976]). As above, initial profiles for the thermal plasma are taken from the IRI model and atmospheric parameters are taken from the MSIS model.

In Figure 5a, the spectra agree closely for most of the energy range. The model predicts a slightly higher flux in the 5-15 eV range, but this difference is less than a factor of two. Figure 5b also shows good agreement, with the model predicting more definition in the 20-30 eV range and lower fluxes above 30 eV by a factor of less than two. These differences could be explained by uncertainties in the experimental data, differences in the neutral atmosphere or ionospheric plasma profiles, or uncertainties in the collisional cross sections used in the model. The larger fluxes at low energy and the increased definition of the production peaks in the model results indicate that the thermal plasma density from IRI is probably lower than the actual densities; a higher plasma density would act to smooth out these features of the distribution function. It is thought that this difference is not due to detector resolution, because $\Delta E/E$ was 2.5% and the production peaks clearly appear in the low altitude measurements. The comparison does show, however, that the model accurately calculates the main features of the photoelectron spectrum in the local equilibrium and transition regions of the ionosphere.

At higher altitudes, the pitch angle dependence of the distribution function becomes more important. This makes a comparison with AE data less informative in the upper ionosphere and plasmasphere. Therefore, a direct comparison with data for this region should be made with DE 2 results from the Low Altitude Plasma Instrument (LAPI). LAPI had less energy resolution than the AE electron spectrometers (with a $\Delta E/E$ of 32%), but had a much narrower field of view and allows for pitch angle distribution comparisons. Figure 6 shows this comparison for the 104th day of 1982 at a local time of 9.3 h and altitude of 690 km [data from R. E. Erlandson, private communication, 1994; error bars from *Winningham et al.*, 1989]. Note that the pitch angle distribution is defined by the data, with 0° being downstreaming particles and 180° being the upflowing electron fluxes. The distributions are shown at energies of (a) 5 eV, (b) 9 eV, (c) 15 eV, (d) 27 eV, and (e) 48 eV. Notice that the model compares reasonably well for most of the cases shown. There is quite a bit of disagreement in the 5 eV results, as well as part of the 9 eV results, but this could be due to spacecraft charging effects or other processes not included in the model. The trends in the data distributions of the other energies are reflected in the model results, and the magnitudes of the model results are not far from the measured values (within a factor of two).

The availability of superthermal electron velocity-space observations is quite limited outside the ionosphere, and so other methods of comparing results with data will be considered. These are indirect comparisons, where the influence of superthermal electrons on the thermal plasma is used to compare with measurements. Two examples of this will be presented here for the two limits of interaction between the superthermal electrons and the thermal plasma: Coulomb collisions only and electrodynamic coupling only. The topic of when to use either of these limits, or a combination of the two, is discussed in detail in *Liemohn and Khazanov* [1997].

One such comparison is shown through the calculations of *Newberry et al.* [1989]. In that study, a comparison was made between data from the retarding ion mass spectrometer (RIMS) on the DE 1 satellite and the Field Line Interhemispheric Plasma (FLIP) model during quiet times (when geomagnetic activity levels were low). The FLIP model solves hydrodynamic equations for the thermal plasma along a flux tube, combined with a superthermal electron two-stream transport model to calculate heating rates in the thermal plasma energy equations. This model includes a phenomenological factor (trapping factor) to represent the amount of energy lost to the plasmasphere from the photoelectrons. This factor is analogous to the transparencies discussed in the previous section. Without this trapping factor, the observed ion temperatures could not be reproduced, and it was concluded that good agreement is achieved between the calculated and measured ion temperatures when ~55% of the total photoelectron flux is trapped in the plasmasphere.

We conducted a similar study with our model and used the thermal electron profile in the ionosphere and plasmasphere from *Newberry et al.* [1989], and we found that the portion of energy absorbed in the plasmasphere due to Coulomb losses with the thermal plasma is 0.53. This shows that our calculations are in agreement with phenomenological modeling and measurements of the thermal structure of the plasmasphere during quiet times. The accuracy of the comparison also indicates that Coulomb interaction with the thermal plasma is the dominant process acting on the superthermal electrons in the plasmasphere where the data was collected.

Another indirect comparison can be made for the other coupling limit, when Coulomb collisions are expected to play a secondary role to electrodynamic interactions with the thermal plasma. This situation is expected in the plasmasphere during and after a geomagnetic disturbance, when the thermal plasma is depleted and the superthermal electrons will be a significant population. Another scenario is along open field lines, when the flow of superthermal electrons is unbalanced by flows from a conjugate source region and the thermal plasma cannot build up to the "filled" levels that were possible at lower latitudes. Data from the polar-orbiting satellite Akebono offers the opportunity for an indirect comparison at high-latitudes above the collisional ionosphere, where electrodynamic coupling would be expected to play a major role in the interactions between the thermal and superthermal plasma populations.

To compare model results with electron temperature measurements in the limit of electrodynamic coupling only, a slightly different method must be used: the model results are a two-part calculation. First, superthermal electron densities at 500 km are obtained from the numerical model used for the previous comparisons, assuming a polar cap latitude (75° up to 90°) with photoionization as the primary source of superthermal electrons (i.e., no precipitating energetic electrons creating secondary superthermal electrons). Then an analytical solution to the collisionless, steady-state kinetic equation is used for the superthermal electrons and thermal ion calculation, which is combined with a fluid treatment for the thermal electrons. This yields a self-consistent calculation of the plasma along a field line that includes the electrostatic potential in the equations of all the plasma species. For details on the second part of this calculation, please see *Khazanov et al.* [1997]. The relative concentration of photoelectrons at the 500 km interface level, $n_{p0}$, varies with solar zenith angle and is thereforeanalogous to latitude (decreasing with increasing latitude).

Figure 7 shows such a comparison of this type of calculation with Akebono data from *Abe et al.* [1993]. The data is from April 28 and May 10, 1991, as the satellite passed from the dayside to the nightside at progressively higher altitudes. These two polar passes represent high and low geomagnetic activity (April and May, respectively). Data fluctuations were presumed to be latitudinal or local time variations in the ionospheric conditions. Shown with the data are three curves at different levels of $n_{p0}$ (0.02%, 0.03%, and 0.06%), that were calculated using the *Khazanov and Liemohn* [1995] model for the conditions of an illuminated polar cap at various latitudes. Only those results are shown that cover the extent of the data, because the data was taken as the satellite crossed through the polar cap at a constantly changing altitude, latitude, and local time. This shows that the interaction between the superthermal electrons and thermal plasma through the electrostatic potential can produce electron temperatures comparable to observed values in the polar cap, where this is expected to be the dominant coupling mechanism.

## DISCUSSION

We have shown that our kinetic, time-dependent, spatially-unified superthermal electron transport numerical model compares reasonably well with previous ionospheric transport models. When identical input parameters are used, this model compares very well with the two-stream kinetic model of *Nagy and Banks* [1970], even though our model includes pitch angle diffusion for the scattering terms and N-B uses backscatter probabilities. When plas-

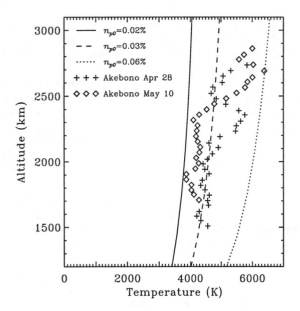

**Figure 7.** Comparison of results for several values of $n_{p0}$ with data from *Abe et al.* [1993], from April 28 and May 10, 1991.

maspheric transparencies are used to obtain a realistic down-flowing flux at the upper boundary of the two-stream model, the two models are in even better agreement, especially in the upper ionosphere.

Plasmaspheric transparency can be defined several ways, and two such definitions have been shown and used in this study. With a plasmaspheric transparency calculated from a spatially-unified model such as ours, the accuracy of the results from a single-ionosphere model can be improved. It will not be a spatially self-consistent model, but if the boundary conditions do not change much between calculations, this type of calculation could be beneficial.

We also illustrated the ability of this model to reproduce superthermal electron observations. Comparisons with Atmospheric Explorer energy spectra were shown for the local equilibrium and transition regions of the ionosphere, and comparisons with Dynamic Explorer pitch angle distributions were shown in the upper ionosphere. There is good agreement is in each of these comparisons.

Finally, we showed that this model can be used to reproduce thermal plasma data, measuring the influence of the superthermal electrons on the thermal plasma through collisional and electrodynamic processes. Dynamic Explorer data was used to show this during low geomagnetic activity, and it was determined that Coulomb collisions are the primary interaction mechanism in the plasmasphere during quiescent times. A comparison with Akebono electron temperature data illustrated that electrodynamic coupling processes are dominant along polar cap field lines. The determination of when electrodynamic coupling plays an important role in the interaction between superthermal electrons and thermal plasma is discussed in *Liemohn and Khazanov* [1997].

*Acknowledgments.* The authors would like to thank the conveners of the Huntsville 96 Workshop for the invitation to present our results at the meeting and to contribute this manuscript to the monograph. We would also like to thank Dr. Robert Erlandson for the DE 2 LAPI data, and Dr. Stan Solomon for providing the two-stream model for direct model comparisons. M. W. Liemohn was supported at the University of Michigan by NASA GSRP grant NGT-51335 and by the National Research Council through a Marshall Space Flight Center Postdoctoral Research Associateship. G. V. Khazanov was funded at the University of Alabama in Huntsville by the National Science Foundation under grant ATM-9523699, and also held a National Research Council-Marshall Space Flight Center Senior Research Associateship while this work was performed.

## REFERENCES

Abe, T., B. A. Whalen, A. W. Yau, S. Watanabe, E. Sagawa, and K. I. Oyama, Altitude profile of the polar wind velocity and its relationship to ionospheric conditions, *Geophys. Res. Lett.*, 20, 2825, 1993.

Bilitza, D., Progress report on IRI status, *Adv. Space Res.*, 10(11), 3, 1990.

Cicerone, R. J., and S. A. Bowhill, Photoelectron escape fluxes obtained by a Monte Carlo technique, *Radio Sci.*, 5, 49, 1970.

Cicerone, R. J., and S. A. Bowhill, Photoelectron fluxes in the ionosphere computed by a Monte Carlo method, *J. Geophys. Res.*, 76, 8299, 1971.

Cicerone, R. J., W. E. Swartz, R. S. Stolarski, A. F. Nagy, and J. S. Nisbet, Thermalization and transport of photoelectrons: A comparison of theoretical approaches, *J. Geophys. Res.*, 78, 6709, 1973.

Doering, J. P., W. K. Peterson, C. O. Bostrom, and T. A. Potemra, High resolution daytime photoelectron energy spectra from AE-E, *Geophys. Res. Lett.*, 3, 129, 1976.

Guiter, S. M., T. I. Gombosi, and C. E. Rasmussen, Two-stream modeling of plasmaspheric refilling, *J. Geophys. Res.*, 100, 9519, 1995.

Hedin, A. E., Extension of the MSIS thermospheric model into the middle and lower atmosphere, *J. Geophys. Res.*, 96, 1159, 1991.

Khazanov, G. V., and M. W. Liemohn, Non-steady-state ionosphere-plasmasphere coupling of superthermal electrons, *J. Geophys. Res.*, 100, 9669, 1995.

Khazanov, G. V., M. W. Liemohn, T. I. Gombosi, and A. F. Nagy, Non-steady-state transport of superthermal electrons in the plasmasphere, Geophys. Res. Lett., 20, 2821, 1993.

Khazanov G. V., M. W. Liemohn, and T. E. Moore, Photoelectron effects on the self-consistent potential in the collisionless polar wind, *J. Geophys. Res.*, 102, in press, 1997.

Lejeune, G., "Two stream" photoelectron distributions with interhemispheric coupling: A mixing of analytical and numerical methods, *Planet. Space Sci.*, 27, 561, 1979.

Liemohn, M. W. and G. V. Khazanov, Non-steady-state coupling processes in superthermal electron transport, in

*Cross-Scale Coupling in Space Plasmas, Geophysical Monograph 93*, AGU, Washington, D. C., 181, 1995.

Liemohn, M. W., and G. V. Khazanov, Determining the significance of electrodynamic coupling between superthermal electrons and thermal plasma, submitted to *Encounter Between Global Observations and Models in the ISTP Era,* AGU Monograph, 1997.

Liemohn. M. W., G. V. Khazanov, T. E. Moore, and S. M. Guiter, Self-consistent superthermal electron effects on plasmaspheric refilling, *J. Geophys. Res., 102,* in press, 1997.

Nagy, A. F., and P. M. Banks, Photoelectron fluxes in the ionosphere, *J. Geophys. Res., 75,* 6260, 1970.

Newberry, I. T., R. H. Comfort, P. G. Richards, and C. R. Chappell, Thermal $He^+$ in the plasmasphere: Comparison of observations with numerical calculations, *J. Geophys. Res., 94,* 15265, 1989.

Oran, E. S., and D. J. Strickland, Photoelectron flux in the Earth's ionosphere, *Planet. Space Sci., 26,* 1161, 1978.

Nisbet, J. W., Photoelectron escape from the ionosphere, *J. Atmos. Terr. Phys., 30,* 1257, 1968.

Swartz, W. E., *Electron production recombination and heating in the F region of the ionosphere, Sci. Rep. 381,* Ionosphere Res. Lab., Penn. State Univ., University Park, 1972.

Winningham, J. D., D. T. Decker, J. U. Kozyra, J. R. Jasperse, and A. F. Nagy, Energetic (>60 eV) atmospheric photoelectrons, *J. Geophys. Res., 94,* 15,335, 1989.

---

G. V. Khazanov and M. W. Liemohn: Space Sciences Laboratory, NASA/MSFC ES-83, Huntsville, AL 35812.

G. V. Khazanov is also at: Center for Space Physics, Aeronomy, and Astrophysics Research, Department of Physics, The University of Alabama in Huntsville, Huntsville, AL 35899.

# Determining the Significance of Electrodynamic Coupling Between Superthermal Electrons and Thermal Plasma

M. W. Liemohn and G. V. Khazanov

*Space Sciences Laboratory, NASA Marshall Space Flight Center, ES-83, Huntsville, Alabama*

The necessity of electrodynamic coupling in addition to collisional coupling between superthermal electrons (SEs) and thermal plasma is discussed. Collisional coupling typically involves Coulomb collision terms in the SE kinetic equation and heating rate terms in the thermal plasma energy equations. Electrodynamic coupling encompasses the inclusion of superthermal electron terms in the quasineutrality condition, flux balance equation, and electric field formulation, as well as the inclusion of the self-consistent electric field in the SE kinetic equation. The case of plasmaspheric refilling is investigated, and two methods of quantitatively determining when electrodynamic effects are important and should be included in the calculation are presented. Using the methods determined from the previous case, results from a polar wind numerical calculation are investigated and it is determined that even a very small amount of photoelectrons are significant in the polar cap.

## INTRODUCTION

Superthermal electrons (SEs) are an important part of near-Earth space plasma, often traveling vast distances before depositing their energy through various mechanisms. The effects of SEs on the thermal plasma can be significant, sometimes even the dominant process in the formation of the thermal plasma distribution, and so it is often necessary to include these effects in thermal plasma calculations.

The influence of superthermal electrons has been known for many decades beginning with studies of ionospheric temperatures [e.g., *Hanson*, 1963]. These effects were determined to be caused by both local SE production [*Hoegy et al.*, 1965] and by nonlocal production, specifically from the conjugate ionosphere [*Fontheim et al.*, 1968]. *Nagy and Banks* [1970] developed the first two-stream kinetic model of SE transport in the ionosphere, and many other SE transport models soon followed [Cf. the review sections of *Cicerone et al.*, 1973; *Winningham et al.*, 1989;

*Khazanov and Liemohn*, 1995]. These models have been used for many purposes, most notably to calculate heating rates into the thermal plasma and excitation rates for atmospheric emissions.

There have been several models that combine the superthermal calculation with that of the thermal plasma. The earliest is *Lemaire* [1972], who developed a self-consistent collisionless kinetic description of the polar wind that included a SE population. More recently, the model of *Min et al.* [1993] calculates $E_{\parallel}$ from the thermal plasma equations and then uses this electric field in the steady-state kinetic equation for the superthermal electrons in the aurora and at midlatitudes. *Tam et al.* [1995] treat the ions and superthermal electrons kinetically along with a fluid approach for the thermal electrons, and obtain steady-state polar wind results that are collisionally and electrodynamically self-consistent. *Khazanov et al.* [1997] reexamined the collisionless kinetic description of the polar wind to determine the maximum influence of photoelectrons on ion outflows. The first time-dependent model of self-consistent coupling is that of *Liemohn et al.* [1997], where SE effects on thermal plasma refilling were investigated.

In *Liemohn et al.* [1997], a driver program was introduced to couple the time-dependent, spatially-unified, kinetic model of SE transport [*Khazanov and Liemohn*, 1995; *Liemohn and Khazanov*, 1995] with the time-dependent, field-aligned, hydrodynamic model of the thermal plasma [*Guiter et al.*, 1995]. This study self-consistently coupled the two models, both collisionally and electrodynamically. Collisional coupling involves Coulomb collision terms in the SE kinetic equation and heating rate terms in the thermal plasma energy equations. Electrodynamic (collisionless) coupling encompasses the inclusion of superthermal electron terms in the quasineutrality condition, flux balance equation, and electric field formulation, as well as the inclusion of the self-consistent electric field in the SE kinetic equation. In that study, both of these coupling mechanisms were included all of the time, but the computational cost of electrodynamically coupling the models is high, and the question should be asked: when is it necessary to electrodynamically couple the SE and thermal plasma calculations?

This study presents a quantitative method of determining when electrodynamic effects are significant and should be included in the calculation. This analysis is based on results from *Liemohn et al.* [1997], which focused on plasma flows along closed field lines. A case study of results from *Khazanov et al.* [1997] is also conducted, to determine the necessity of self-consistent coupling in the polar cap region in the presence of photoelectrons.

## RESULTS

In *Liemohn et al.* [1997], it was determined that the primary electrodynamic coupling mechanism is the inclusion of a SE term in the flux balance equation,

$$\sum_i n_i u_i - n_e u_e - n_s u_s = \frac{j}{e} \quad (1)$$

where the subscripts $e$, $s$, and $i$ represent thermal electrons, superthermal electrons, and ion species; $n$ is density; $u$ is bulk flow; and $j/e$ is the current density. This equation was used by setting $j=0$ and solving for the thermal electron flux, with the other fluxes calculated from the hydrodynamic (ions) and kinetic (SE) equations. In that study, photoionization was started in the ionospheres connected to a greatly depleted plasmaspheric flux tube ($L=4$), and the flows of superthermal and thermal plasma along the field line were modeled and analyzed. It was seen that, during the first few minutes of refilling, the thermal electron velocity reverses and flows down from the equatorial plane in the plasmasphere into the ionospheres, exactly opposite to the case when the SE flux is not included in this equation (see Figure 6 of *Liemohn et al.* [1997]). However, by 15 min the thermal electron velocity had returned to its value without electrodynamic coupling for most of the field line, because the SEs from each ionosphere had balanced each other to the point of reducing the SE flux to an insignificant level in the plasmasphere. Only near the ionospheres was there a difference, because the counterflowing SE streams are not balanced and the SE flux is still quite strong.

Because it is the fluxes ($nu$) that appear in (1), this quantity should be examined during the early stage time period to determine the conditions leading to the strong electrodynamic coupling between the SEs and the thermal plasma. Figure 1 shows these fluxes for the SEs, thermal electrons, and the sum of the thermal ions at various times after beginning the refilling process, with (solid lines) and without (dotted lines) the SE flux included in (1) (as well as other less significant electrodynamical coupling terms). Note that because it is assumed that there is no current, the thermal electron flux is found by subtracting the SE flux from the total ion flux. In Figure 1, the SE fluxes with and without electrodynamic coupling are quite close to each other, but the fluxes without are slightly bigger. This is because the electric field in the SE kinetic equation decelerates the electrons as they move from the ionospheres towards the equatorial plane. At 10 sec, the SEs dominate equation (1), causing the fluid thermal electrons to compensate with a nearly equal but oppositely-directed particle flux. After 1 min, the thermal ions are developing a stronger presence in (1) near the equatorial plane, and so the thermal electron flux is beginning to change from the SE-dominated case. By 3 min, the ions have become a major term in (1), and the thermal electrons have started flowing back towards the equatorial plane for most of the plasmasphere. After 5 min, the thermal electron flow has almost returned to its "no coupling" levels, and by 15 min the contribution of the superthermal flux term in (1) is significant only in small regions near the ionospheres. Notice that the ion fluxes are nearly identical with and without electrodynamic coupling. The reader is referred to *Liemohn et al.* [1997] for further discussion of the simulation.

## ANALYSIS

Qualitatively, it appears that the SE flux term is important only for the first few minutes of refilling. How does this translate into a quantitative algorithm for determining when this form of coupling is necessary in the equations?

First, criteria must be established for setting limits on what significant means in terms of influencing the results. Two possibilities will be discussed: one based on the maximum ratio of SE flux to total ion flux; and another based on the average ratio at any point along the simulation range. For this analysis, the results along the field above 1 $R_E$ from the surface in each of the conjugate ionospheres will be used. This is because the plasmasphere is the region of interest, and so attention will be restricted to the SE contribution in this region.

### Method 1: Maximum ratio

The ratio of SE flux to the total ion flux at any point along the field line is given by

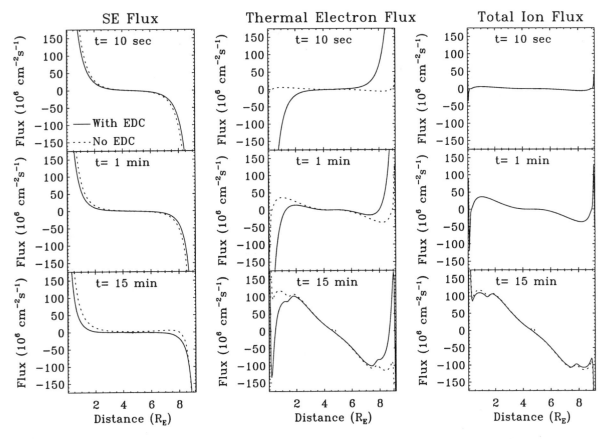

**Figure 1.** Particle fluxes from *Liemohn et al.* [1997] of superthermal electrons (left column), thermal electrons (center column), and the total ion flux (right column) during the early stages of plasmaspheric refilling with (solid lines) and without (dotted lines) electrodynamic coupling effects. Results are along an $L=4$ flux tube, with distances measured from the base of the northern ionosphere in Earth radii.

$$R(s) = |n_s(s)u_s(s)| \Big/ \left|\sum_i n_i(s)u_i(s)\right| \qquad (2)$$

and represents the relative contribution of each of these terms to (1). A value greater than unity means that SEs are dominant at this spatial point, while a value less than unity means that the ion flux is dominant. So, one indication that the SE flux is no longer a critical factor in (1) would be to define a threshold value for including or omitting the SE terms. What is needed, then, is the maximum value of this ratio, $R_{max}=\max[R(s), s=s_{low}...s_{high}]$, where $s_{low}$ and $s_{high}$ are the limits of the simulation range or the region of interest.

As was mentioned above, though, the SEs were decelerated due to the ambipolar electric field, and so the ratio will be smaller with coupling than without coupling. It is necessary, then, to have two thresholds, one for turning on electrodynamic coupling and the other for turning it off, with some overlap in the values so the processes are not flickering on and off when the results are near the threshold. The following thresholds are proposed: turn this extra coupling on when $R_{max}$ exceeds 1.0, and turn it off when it falls below 0.5.

Figure 2a shows $R_{max}$ for the results with and without electrodynamic coupling from Figure 1. The times presented are minutes after the sources are turned on in the ionospheres, as in Figure 1, and are shown out to 30 min of simulation time. Using the above-mentioned thresholds, electrodynamic coupling it important for the first 11 min of refilling, after which time its influence is minimal. The terms would not need to be included during the times shown, because $R_{max}$ at 30 min without coupling is only 0.7. It should be noted that the results beyond 30 min are reaching the limits of validity for the hydrodynamic model, when unphysical shocks develop and taint the results.

*Method 2: Average ratio*

Another method of determining the significance of the electrodynamic coupling terms is to use the average ratio of SE flux to total ion flux. The average value for $R$ is defined as

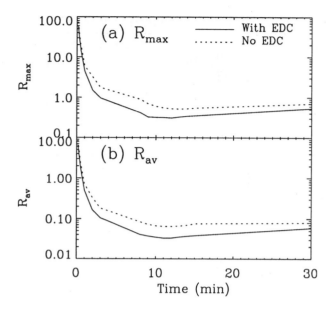

**Figure 2.** Flux ration calculations showing (a) $R_{max}$ and (b) $R_{av}$ with electrodynamic coupling (solid lines) and without (dotted lines) based on the simulations from *Liemohn et al.* [1997].

$$R_{av} = \int_{s_{low}}^{s_{high}} R(s)ds \bigg/ \int_{s_{low}}^{s_{high}} ds \qquad (3)$$

As with Method 1, two thresholds for $R_{av}$ need to be defined, one for turning coupling on (compared to $R_{av}$ without coupling) and one for turning coupling off (compared to $R_{av}$ with coupling). These thresholds should be much tighter than those for Method 1, because $R_{av}$ is a measure of the overall influence on the electrodynamic coupling terms. Therefore, turn on and turn off thresholds for $R_{av}$ of 0.10 and 0.05, respectively, will be chosen.

Figure 2b presents $R_{av}$ with and without electrodynamic coupling. As you can see, $R_{av}$ is quite high for the first few minutes of refilling, confirming that the SEs are indeed the major term (1) and that electrodynamic coupling should be included. However, after 11 min, $R_{av}$ drops below 0.05, and the electrodynamic coupling terms no longer have a strong influence on the thermal plasma calculation. Again, $R_{av}$ in Figure 2b does not exceed the turn on threshold, and the influence remains small.

## CASE STUDY

It is possible to apply these methods to other cases in order to determine the significance of electrodynamic coupling between superthermal electrons and the thermal plasma. One region of interest is the dayside polar cap, where photoelectrons are thought to play a role in the escape of upflowing ions. At high latitudes, the flows are not balanced by particles streaming down the field line from the conjugate ionosphere, because these field lines are either open or so long that the conjugate point is unimportant. Therefore, the influence of the superthermal electrons in this region should be greater than that at midlatitudes, as was the case for the plasmaspheric refilling study above.

Let us examine the results of *Khazanov et al.* [1997], who presented a collisionless, steady-state kinetic model of the polar wind in the presence of photoelectrons. Their simulation domain is from 500 km out to 5 $R_E$ along a high-latitude, open field line. Examining the Plates in this paper reveals that the photoelectrons have quite an influence on the outflowing ions and thermal electrons, especially for a photoelectron concentration at the base greater than 0.1%. Let us examine these results based on the analysis above.

The flux calculation in *Khazanov et al.* [1997] for the kinetic populations is given by

$$n_\alpha(s)u_\alpha(s) = n_{0\alpha}u_{0\alpha}\left[1 - y_{s^{ub}}^2 \exp\left(-\frac{B_{s^{ub}}}{B_0}\frac{z_\alpha(s^{ub})}{y_{s^{ub}}^2}\right)\right]$$
$$\times \frac{B(s)}{B_{s^{ub}}}e^{-z_\alpha(s^{ub})} = \frac{B(s)}{B_{s^{ub}}}n_\alpha(s^{ub})u_\alpha(s^{ub}) \qquad (4)$$

Here $z_\alpha(s)$ depends on the difference of the electrostatic and gravitational potentials, $y$ is a function of the magnetic field strength $B(s)$, the "0" subscript refers to the boundary conditions at the base altitude, and $s^{ub}$ is the upper level of the simulation domain. This shows that the fluxes of photoelectrons, oxygen ions, and hydrogen ions are only dependent on the flux at $s^{ub}$ and the magnetic field strength. Therefore, $R(s)$ from (2) will be constant and $R_{max}$ will be equal to $R_{av}$. In this case, the thresholds for turning on and off electrodynamic coupling should be taken from method 2 above.

Results from this study are presented in Figure 3. It should be noted that the boundary conditions (at 500 $R_E$) are $6\times10^4$ cm$^{-3}$ for O$^+$, $1\times10^3$ cm$^{-3}$ for H$^+$, ion temperatures of 2000 K, a thermal electron temperature of 2500 K, and a 20 eV characteristic energy for the photoelectrons. The fluxes at the upper boundary (5 $R_E$) for the three populations are given in Figure 3a. Notice that the proton fluxes are unaffected by the presence of photoelectrons in these results. This is because the protons are escaping along the field line even when there are no photoelectrons, and thus the flux of these particles is determined entirely by the boundary condition at the base. The oxygen ion fluxes, however, are greatly dependent on the photoelectrons, and even surpass the proton flux for $n_{p0} \geq 0.5\%$.

The flux ratio is shown in Figure 3b. The curve has reached an asymptote from roughly $n_{p0} \sim 0.1\%$ and below. The curvature of the result at the upper end of the scale indicates that there could be another high concentration asymptotic limit. The ratio crosses unity near $n_{p0}$=0.003%, and the trend towards $R$=0 as $n_{p0}$ goes to zero is seen. Because these results are all produced with electrodynamic coupling included, the turn-off threshold is $R$=0.05, which is the case for $n_{p0}$≤0.0002%. This is a very small concentration of photoelectrons.

Another indicator of superthermal-thermal plasma coupling is the oxygen density profile, plotted in Figure 4. Results are shown up to 4000 km, below which the distributions for widely different photoelectron contents are similar enough to show on the same scale. Notice that the change in the O$^+$ density is getting bigger with increasing photoelectron concentration. This is also seen in the O$^+$ flux at 5 $R_E$ presented in Figure 3. At low $n_{p0}$ values, oxygen has little response, then it goes through a rapid response with a maximum near $n_{p0}$=0.05%, and then begins to asymptote again at very high photoelectron concentrations.

The electrodynamic coupling between photoelectrons and the thermal plasma in the polar cap region was originally discussed by *Lemaire* [1972]. Using the collisionless, steady-state kinetic equation for all species (H$^+$, O$^+$, thermal electrons, and photoelectrons), he also showed increased ion densities at high altitudes when photoelectrons were included. It is therefore useful to show a comparison of his results with results from the *Khazanov et al.* [1997] model for similar boundary conditions. The boundary conditions, taken at a low altitude limit of 950 km, are $7\times10^3$ cm$^{-3}$ for O$^+$, 320 cm$^{-3}$ for H$^+$, thermal plasma species temperatures of 3000 K, and a 10 eV photoelectron

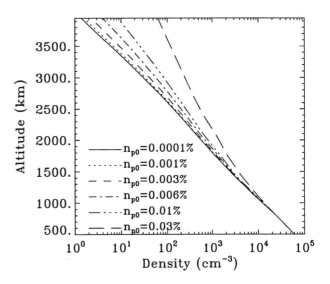

**Figure 4.** Oxygen ion density profiles for several photoelectron concentrations at the base for the polar wind results of *Khazanov et al.* [1997].

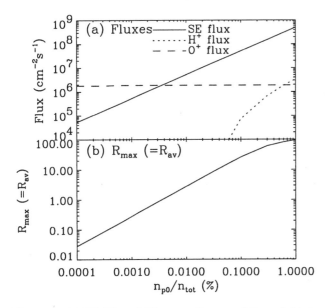

**Figure 3.** (a) SE, H$^+$, and O$^+$ escape fluxes at 5 $R_E$ and (b) the flux ratio $R$ (=$R_{max}$=$R_{av}$) as a function of photoelectron concentration at the base (500 km) for the polar wind results of *Khazanov et al.* [1997].

characteristic energy. It is expected that the lower densities and higher temperatures for the thermal plasma, combined with a lower characteristic energy for the photoelectrons, should decrease the photoelectron influence from that calculated above.

A comparison of the results of *Lemaire* [1972] with that of *Khazanov et al.* [1997] is shown in Figure 5. Both models predict that $n_{H^+}$ is not affected by the presence of photoelectrons at the concentrations taken for these calculations (0.0% and 0.0094% at 950 km). Also, both models predict a crossover of the O$^+$ and H$^+$ densities at an altitude of less than an $R_E$. The O$^+$ densities increase very slightly with the inclusion of the photoelectrons. This concentration of 0.0094% is quite low, so it is not surprising that the influence is small, similar to Figure 4. In Figure 4, however, the influence at $n_{p0}$=0.01% is greater than the influence here in Figure 5, as expected from the differing boundary conditions. For $n_{p0}$=0.0094%, $R$=0.439, indicating these results are close to $n_{p0}$~0.002% in the case study above. However, the thermal electron fluxes are still influenced by the presence of the photoelectrons, and inclusion of self-consistent electrodynamic coupling is advised. A similar situation arose in the plasmaspheric refilling results, where huge changes in the thermal electron distribution were seen without a strong signature in the ion results. Larger changes in the ion results would be expected for $n_{p0}$>0.1%, as seen in the results of *Khazanov et al.* [1997].

## CONCLUSION

Based on the need for fast yet accurate calculations of superthermal and thermal plasma populations, a system of

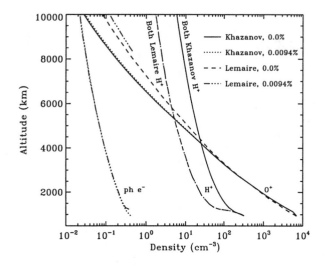

**Figure 5.** Polar wind densities from the *Lemaire* [1972] model and the *Khazanov et al.* [1997] model, with and without a photoelectron population included. Note that the boundary conditions are different than those for Figures 3 and 4.

determining when it is necessary to include electrodynamic (collisionless) coupling terms in addition to collisional coupling terms has been presented. Two methods of determining the level of significance the superthermal electron flux has in the flux balance equation are discussed, and, for the plasmaspheric refilling study of *Liemohn et al.* [1997], it was concluded that electrodynamic coupling terms were significant only for the first 11 minutes of the calculation (determined, coincidentally, by both methods). For a case study of the polar wind results of *Khazanov et al.* [1997], it was determined that the self-consistent coupling is unimportant when the photoelectron concentration at the base (500 km) is below 0.0002% of the total plasma density. This very small threshold density indicates that self-consistent coupling should always be taken into account on the dayside polar cap where the upflowing particles are not balanced by flows from the conjugate hemisphere. This conclusion is supported by a comparison with the original photoelectron-polar wind results of *Lemaire* [1972].

*Acknowledgments.* MWL would like to thank the conveners of the Huntsville 96 Workshop for travel support to attend the meeting. This work was supported at the University of Michigan by NASA GSRP grant NGT-51335. MWL held a National Research Council-Marshall Space Flight Center Postdoctoral Research Associateship while this work was performed. GVK was funded at the University of Alabama in Huntsville by the National Science Foundation under grant ATM-9523699, and also held a National Research Council-Marshall Space Flight Center Senior Research Associateship during part of this work.

## REFERENCES

Cicerone, R. J., W. E. Swartz, R. S. Stolarski, A. F. Nagy, and J. S. Nisbet, Thermalization and transport of photoelectrons: A comparison of theoretical approaches, *J. Geophys. Res., 78,* 6709, 1973.

Fontheim, E. G., A. E. Beutler, and A. F. Nagy, Theoretical calculations of the conjugate predawn effects, *Ann. Geophys., 24,* 489, 1968.

Guiter, S. M., T. I. Gombosi, and C. E. Rasmussen, Two-stream modeling of plasmaspheric refilling, *J. Geophys. Res., 100,* 9519, 1995.

Hanson, W. B., Electron temperatures in the upper atmosphere, *Space Res., 3,* 282, 1963.

Hoegy, W. R., J.-P. Fournier, and E. G. Fontheim, Photoelectron energy distribution in the F region, *J. Geophys. Res., 70,* 5464, 1965.

Khazanov, G. V., and M. W. Liemohn, Non-steady-state ionosphere-plasmasphere coupling of superthermal electrons, *J. Geophys. Res., 100,* 9669, 1995.

Khazanov G. V., M. W. Liemohn, and T. E. Moore, Photoelectron effects on the self-consistent potential in the collisionless polar wind, *J. Geophys. Res., 102,* 7509, 1997.

Lemaire, J., Effect of escaping photoelectrons in a polar exospheric model, *Space Res., 12,* 1413, 1972.

Liemohn, M. W. and G. V. Khazanov, Non-steady-state coupling processes in superthermal electron transport, in *Cross-Scale Coupling in Space Plasmas, Geophysical Monograph 93,* AGU, 181, 1995.

Liemohn. M. W., G. V. Khazanov, T. E. Moore, and S. M. Guiter, Self-consistent superthermal electron effects on plasmaspheric refilling, *J. Geophys. Res., 102,* 7523, 1997.

Min, Q.-L., D. Lummerzheim, M. H. Rees, and K. Stamnes, Effects of a parallel electric field and the geomagnetic field in the topside ionosphere on auroral and photoelectron energy distributions, *J. Geophys. Res., 98,* 19223, 1993.

Nagy, A. F., and P. M. Banks, Photoelectron fluxes in the ionosphere, *J. Geophys. Res., 75,* 6260, 1970.

Tam, S. W. Y., F. J. Yasseen, T. Chang, and S. B. Ganguli, Self-consistent kinetic photoelectron effects on the polar wind, *Geophys. Res. Lett., 22,* 2107, 1995a.

Winningham, J. D., D. T. Decker, J. U. Kozyra, J. R. Jasperse, and A. F. Nagy, Energetic (>60 eV) atmospheric photoelectrons, *J. Geophys. Res., 94,* 15,335, 1989.

---

G. V. Khazanov and M. W. Liemohn: Space Sciences Laboratory, NASA/MSFC ES-83, Huntsville, AL 35812.

G. V. Khazanov is also at: Center for Space Physics, Aeronomy, and Astrophysics Research, Department of Physics, The University of Alabama in Huntsville, Huntsville, AL 35899.

# A New Magnetic Storm Model

Robert B. Sheldon [1] and Harlan E. Spence

*Boston University Center for Space Physics, Boston, Massachusetts*

Recent observations from the new perspective of the POLAR orbit have elucidated the crucial role that parallel electric fields and the ionosphere play in the development of a magnetic storm. During the main phase, we observed field-aligned beams of ionospheric ions that appear to be caused by deeply convecting plasmasheet ions. These ionospheric beams are a substantial fraction of the ring current density, suggesting a novel mechanism for ring current and Dst growth during the main phase of a magnetic storm. From these observations, we construct a new magnetic storm model that has implications for all aspects of a storm sequence, including ground, ionospheric and magnetospheric observations.

## 1. INTRODUCTION

The elements of this new magnetic storm model have all been presented before, but they have lacked a coherent, causal chain, that could connect them with the "standard ring current storm model"[ e.g., *Smith and Hoffman*(1974); *Gonzalez et al.*(1994)]. This "standard model" supposes that the generation of an enhanced cross-tail convection electric field during a storm convects in the near Earth plasmasheet into the ring current region on open drift paths. The abrupt cessation of this electric field strands the plasma in the ring current (RC), on now closed drift paths.

As attractive as this model is, it has a number of drawbacks. First, the injection of RC plasma is initially on open drift paths, so that the current contents of the RC are emptied as the new plasma arrives. This makes multiple injections difficult to accomplish, since as much RC is being lost as injected. Thus the only

---

[1] now at The University of Alabama in Huntsville, Huntsville, Alabama

way a Dst effect could be accomplished, would be if the plasmasheet density were higher than RC density. In general, the RC is in diffusive equilibrium with the plasmasheet [ *Sheldon*(1994a)], so that a "superdense" plasmasheet increase is required to produce a large Dst storm. This prerequisite (which is only weakly supported by the data) adds complication to the simple model above. Yet even with this included, models of the storm-time RC generation, [e.g., *Fok et al.*(1995); *Jordanova*(1998)] find that there remains a factor 2 discrepancy between simulated and actual Dst. Or if the boundary condition is adjusted to produce enough Dst, the RC during storm recovery is too large.

*Chen et al.*(1993) recognized that this abrupt cessation of electric field on a timescale shorter than the drift time would violate the 3rd invariant, leading to radial diffusion. They then explored multiple electric field pulses, showing that rapid diffusion could also generate the RC from a plasmasheet source. This had the advantage that multiple injections were cumulative, but the same disadvantage that the final RC energy content remained dependent on plasmasheet densities.

A second objection to the standard model came from composition analysis of the storm-time RC. *Hamilton et al.*(1988) measured the enhanced ionospheric oxygen content of great storms, far above the quiescent

$H^+/O^+$ RC ratio. This ratio was shown to be proportional to Dst [ *Grande et al.*(1997)], which again, suggests that the plasmasheet must be "prepared" for a great storm. Yet despite the work of *Daglis and Axford*(1996), no such mechanism operating on storm timescales was found.

Our model takes as its input the cross-tail potential at the inner edge of the plasmasheet and the plasmasheet density at this location, to predict the abrupt Dst change as well as the subsequent recovery and compositional changes in a complete storm sequence. With a suitable inner tail model, we should be able to use only the solar wind density and electric field, $Ey = Bz \times Vx$, to make these predictions.

## 2. DATA ANALYSIS

The POLAR spacecraft is in a polar, 9 x 2 Re orbit that on April 15, 1996, was in the noon-midnight meridian with perigee over the south pole. Thus it made two diagonal passes through the midnight and noon radiation belts. A typical pass shows an energetic particle population whose average energy is proportional to the $|B|$. This is the normal distribution for ions diffusing in L-shell from a source region in the plasmasheet. On this day the Comprehensive Energetic Particle and Pitch Angle Distribution (CEPPAD) [ *Blake et al.*(1995)] experiment detected two nearly monoenergetic bands superposed on the night side ring current (plate 1): a population of trapped ions at ~90 keV, and a beam of field-aligned ions at ~40 keV.

A nearly monoenergetic trapped population is possible when a strong cross-tail electric field drives ions against the $\nabla B$ and corotation $E \times B$ drifts deep into the magnetosphere [ *Smith and Hoffman*(1974); *Sheldon*(1994b)]. Such a "nose" event must be nearly 90° trapped particles because of the large increase in $|B|$ while convecting from the plasmasheet, which is the characteristic of the upper energy band in our data. Since nose events are highly correlated with storms, and storms are defined by Dst, but Dst is generally not immediately available, we turn to the preliminary Dst provided by the Kyoto University web site. After subtracting the ionospheric Sq contribution using the "quietest day of the month" (April 7/8) method, we find a moderate storm of at least -63 nT on the first hour of April 15. Additional support for the nose event identification came from the extensive GGS database.

Examination of WIND data [ *Ogilvie et al.*(1995)] for April 14 showed that there were several $Bz < 0$ periods lasting for 1 hour or less. Corresponding CU and CL de-

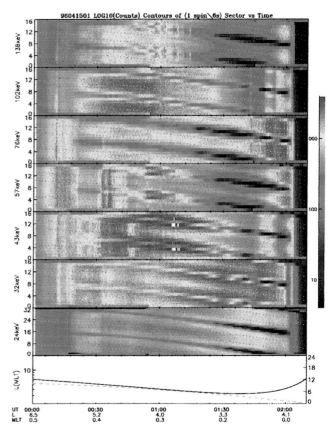

**Plate 1.** POLAR/CEPPAD/IPS data on April 15, 1996, displaying roll modulation of the counts in the 90° head in the energy bands from 24-138 keV. Dotted white and blue lines are 90° and 60° pitchangles.

rived from the CANOPUS array [ *Rostoker et al.*(1995)] show that these periods led to substorms with riometer absorption signatures at auroral latitudes. However the storm trigger appears to be the strong southward turning of $Bz < -10$ occurring at 2000 UT, accompanied by a jump in the solar wind speed from 450 km/s to 600 km/s, which produced an even larger $Bz$ in the compressed magnetic field of the magnetosheath. This period of strong southward $Bz$ lasted more than 3 hours, effectively saturating the ability of the tail to shield out the polar cap potential. The IZMIRAN model [ *Papitashvili et al.*(1994)] predicted in excess of 150 kV across the polar cap for these solar wind conditions.

The CANOPUS array detected a magnetic bay, a nearly equal response of AU and AL, suggesting that the current systems had moved equatorward, overhead of the magnetometers. Indeed, the Halley Bay magnetometer at L~4 [ *Dudeney et al.*(1995)] showed a large H deflection with almost no Z deflection, indicative of strong overhead currents. The ionosphere responded strongly at this time (private communication, J. Aarons, 1996) as ionospheric scintillations were enhanced.

While CANOPUS riometers at L>6 recorded very little activity, the riometer at L=4.4 as well as the Halley Bay riometer at L~4, (private communication A. Rodger, 1996) showed an extremely intense and narrow absorption feature at this time, indicating that precipitation had penetrated to low latitudes, deep in the magnetosphere and down to E-layer ionospheric depths. From these observations, as well as complementary ground observations, we surmise that after several intense substorms had pumped up the plasmasheet, a strong convection field injected the plasma to at least L=4.4, which POLAR/ CEPPAD observed as a 90-100 keV band.

The band of 40 keV field-aligned ions, however, are harder to explain. They have the wrong pitchangles to have convected from the plasmasheet, because the adiabatic decompression involved in backtracing them to their origin would place them in the plasmasheet loss cone. Nor would ions of this energy have had access to the plasmasheet simultaneously with higher energy ions, since the magnetosphere is a "notch filter" for only one energy. This implies that these ions are trapped on closed drift orbits. But if they undergo the same processes as the adiabatically energized ring current, which can be seen simultaneously with the banded distribution, they would not be monoenergetic, nor would they track the energy of the nose ions so precisely. That is, if they had resided for any length of time in the magnetosphere, the same convection that brought in the plasmasheet ions would disperse these ions as well. Nor do these ions show the energy dispersion associated with substorm injections. Thus we conclude that these field-aligned ions are *in situ* accelerated during the time of the measurement.

"Zipper" events, [ *Fennell et al.* (1981); *Kaye et al.* (1981)], have the same bimodal pitch-angle distributions, though at somewhat lower energy. They found that the "zippers" were rich in $O^+$, and concluded that they were observing beams coming from the ionosphere. Since beams are seen at auroral latitudes, they concluded that they were on flux tubes connected to auroral arcs. The composition experiment on POLAR, CAMMICE, was turned off during this period, so we cannot determine the $E > 30 keV$ $O^+$ content for April 15, however when it was switched back on, around April 19, it found an anomalously large amount of $O^+$ in the ring current (RC) in two energy bands centered at 40

and 100 keV. Because of the short lifetime of $O^+$ against charge exchange, such a measurement is consistent with a storm injection occurring only a few days previously, since between storms, CAMMICE detects no $O^+$ enhancements. A similar storm on March 21, 1996 showed that the IPS zippers are simultaneous with the double peaked $O^+$ spectrum observed by CAMMICE. And POLAR/ TIMAS [ *Shelley et al.*(1995)] (private communication W.K. Peterson, 1996) did detect $E < 20 keV$ $O^+$ beams for this day. So we conclude that on April 15 we are observing $O^+$ ions accelerated to ~30-40 keV by strong parallel electric fields in the ionosphere. Since these ions track ~50 keV below the nose ions, it appears there must be a causal connection.

*Whipple*(1977) argues that one can produce a field-aligned electric field if the electron and ion pitch angle distributions (PADs) are not identical. In our case, a hot monoenergetic nose ion superposed on a cold plasmasphere electron population will produce a parallel electric field of several $kT_e$ pointing away from the equator. That is, since a mirroring ion spends most of its time away from the equator, the electrons at the equator will experience a force pulling them toward higher latitudes. If the energetic ion density (or current density in a dynamic system) is greater than the plasmaspheric cold electron density, a second ion-dominated solution to the Whipple equations is possible that produces an ion space-charge potential at approximately the ion beam parallel energy. (Since the Whipple equations are derived for a neutral plasma, we must rederive these results using Poisson's equation for a non-neutral plasma. Intriguing evidence that such space charge is both possible and probable come from laboratory experiments on magnetized, trapped electron clouds *Hansen and Fajans*(1995)) This space charge potential would expel magnetospheric electrons to the equator and could only be neutralized somewhere earthward on the flux tube where the cold electron density would again exceed the hot ions. At this point, the potential would drop from kV to a few $kT_e$, forming a double layer. Naturally this space charge should be shielded by neighboring flux tubes, generating a locally perpendicular electric field that may be manifest in the ionosphere by a "polarization jet" or a subauroral ion drift (SAID).

By measuring the energy of these beams as the spacecraft crosses through this region, we can map out some features of this potential structure. We have fit the spectra with a sum of three peaks: a Chapman layer function appropriate for the ionospheric beam, a Gaussian for the nose ions, and a log-space Gaussian for the ring current ions, achieving remarkably good 9 parameter fits to 16 energy points. We note that an asymmetric Chapman layer function, $y = \exp(1 + x - \exp(x))$, is what one expects if the extraction potential is extended over some distance.

Figure 1 demonstrates how the three populations, beam, nose and RC, evolve during this orbit. The RC is adiabatically energized so that as the model equatorial magnetic field increases the energy increases, as noted by the squares and dashed line. A latitude effect in RC is apparent due to the more rapid loss of low energy ions at higher latitudes. This energization is not seen by the nose ions because they are not an equilibrium trapped population, rather we are seeing the nose ions that have access to this location, so that stronger B-field ions also began with smaller magnetic moment. A more detailed analysis of the pitch angles is needed to clarify the coupled energy-L dependence of these ions.

If the first solution of *Whipple* holds, then both the nose and the beam energies should move in concert as we move along the field line, and we should observe a constant difference between these two energies (circles). We find instead that the ratio (X) is more constant, that the beam ion energy tracks the nose ion energy with a nearly constant ratio of 1/2. This suggests that the beam energy is determined not by the electron thermal temperatures or anisotropies, but by the energy of the nose ions themselves, consistent with the second, ion-dominated solution to the *Whipple* equilibrium.

Such a parallel electric field should also accelerate magnetospheric electrons to 30 keV into the ionosphere, producing a riometer absorption event. Since the electric field is colocated with the nose plasma, the absorption must be a narrow strip in latitude but distributed in longitude, as seen in the CANOPUS data. This extended signature is reminiscent of SAIDs which have been previously identified with substorm injections. Our mechanism would identify a large SAID with a storm injection, and which may also account for the E-layer keV electron signature seen in subauroral riometer data.

We carefully distinguish between substorm and storm injections because the characteristic signatures of each are different. A substorm involves a reconfiguration of the magnetic field which produces a dB/dt electric field in the region between 6-10 Re near midnight. This produces two types of signatures in the POLAR data. The inductive electric field may *in situ* accelerate the entire plasma population to ~30 keV, which is observed in our instrument as an isotropic, dispersionless <50 keV enhancement over a restricted MLT and L-shell range. Or

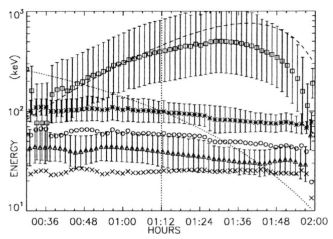

**Figure 1.** Peak fits to 96s averaged spectra of the 90° head. Ring current (□), nose ion (∗), beam ion (△) energies, difference (○) and 10× the ratio (×) of nose:beam are plotted. "Error" bars are FWHM from fits, since fitting errors are negligible. Overplotted with different scales are: one half the model equatorial B-field intensity (dashed); and on linear scale the magnetic latitude from -40 to 40 degrees (dotted), with the equator marked with a vertical dotted line.

it may bring in plasmasheet material with a distinctive, highly energy dispersed signature. A storm injection, on the other hand, is observed as a nose event, a monoenergetic band of ions penetrating to subauroral L-shells and existing over a broad range of MLT determined by the duration of the cross-tail field. Energization of low energy plasma is not seen, but adiabatic energization of the nose ions occurs as the ions convect toward stronger B. From an ionospheric viewpoint, substorms are in the auroral zone, whereas storms penetrate down to subauroral latitudes albeit in a restricted latitude band. Thus we argue that these beams cannot be substorm generated.

## 3. DISCUSSION

If our mechanism generates field aligned electric fields and oxygen rich beams for every storm injection, then we have elucidated a new model for magnetic storms. Since it is generally thought that magnetic storms are characterized by intense convection fields that bring plasmasheet plasma deep into the RC, then all storms should create parallel fields and extract oxygen from the ionosphere as well. Thus large storms should extract more oxygen than small storms, with a larger Dst effect. Similarly, Dst should change on minute bounce timescales, rather than the 10's of minutes convection timescales.

Several recent observations lend support for this theory. Analysis of the CRRES data set shows that there is a positive correlation between the magnitude of Dst and the $O^+$ content of the RC [ *Grande et al.*(1997)]. A Dst prediction filter [ *Gleisner et al.*(1996)] found that a neural network with one hidden layer, representing an unknown quadratic dependence on solar wind Vx,Bz, and density, could explain up to 84% of the variance in Dst, an improvement over the linear filter's 70%. If significant $O^+$ is extracted during the main phase as we predict, one would expect such a non-linear dependence of Dst with Ey. In addition, Kyoto 1-minute Dst show excursions much too rapid for convective timescales.

A study of storm-time PC1 pulsations by *Mursula* (COSPAR 1996 paper to be published in *Adv. Sp. Res.*) showed that the PC1 waves associated with ioncyclotron waves during the most intense part of the main phase of a storm occurred primarily at dusk and the PC1 frequency dropped down to below the oxygen gyrofrequency. These observations are suggestive of an oxygen rich plasma occuring at dusk, the location of the deepest penetration of the nose ions and also in agreement with our theory.

This extraction of ions and precipitation of electrons near midnight will generate an outward flowing current which then drifts westward with the bulk of the RC. We expect that the disappearance of the parallel field, occurring near the dusk terminator, will result in a downward current thus completing the loop of the partial ring current as measured by *Suzuki et al.*(1985) using Magsat data. They concluded that 1/3–1/4 of RC amperage was observed in the partial RC. If we assume that half of the partial RC is carried by upward flowing ions, then we estimate that 1/6 - 1/8 of the total RC is composed of ions of ionospheric origin.

Finally, if electrons are being precipitated into the ionosphere, one would expect the generation of X-rays. POLAR/PIXIE has observed an intense emission in a single spot near the dusk terminator that occurs midway through main phase of almost every major Dst storm [ *Imhof et al.*(1998)]. This global observation is again consistent in location and timing with our acceleration mechanism.

Thus we find that the mechanism described in the abstract not only provides a causal chain for the entire storm sequence, but has great predictive power in explaining many other observations not previously linked to storms.

## 4. CONCLUSIONS

We have attempted to construct a new model of magnetic storms that incorporates field-aligned electric fields as an intrinsic part of the storm development. A storm proceeds then with the following steps: 0) A critical density plasmasheet may be required for a significant magnetic storm, in which case precursor substorm activity or a solar wind density enhancement would be a prerequisite. 1) Large solar wind $Ey$ produces a large polar cap electric field, which if persistant for an extended period (>3hr) exceeds the ability of the tail to shield out the electric field. 2) This produces a strong cross-tail convection electric field that transports plasmasheet ions deep into the magnetosphere. 3) The opposing $\nabla B$ and $E \times B$ drifts act as a selective filter permitting a single energy the deepest penetration, a monoenergetic "nose" event. 4) This spatially narrow band of hot ions, compressed as it convects towards stronger B reaches its highest density near the dusk terminator where it can outnumber the local cold electron density at the equator and thus generate a field-aligned space-charge potential that attempts to confine the hot ions to the equator, a parallel electric field. 5) At some point down the field line earthward of the equator, the growing cold electron density exceeds the hot ions and the potential is neutralized to approximately the electron thermal temperature, forming a double layer that very quickly extracts ionospheric ions, including $H^+$ and $O^+$. 6) Ion cyclotron waves, generated by this unstable particle population, then perpendicularly heat these ion beams and scatter them out of the loss cone so that they become a permanent part of the ring current. 7) This entire structure is a "convection instability", a dynamic balance between the current of corotating cold electrons and convecting hot ions, which would rapidly decay if it were not for the continuous power input of the convection electric field. When the field switches off, the convection power source is removed, the parallel electric fields vanish, and the hot ions are trapped in the ring current and subsequently decay through charge exchange.

*Acknowledgments.* This study was supported by NASA contract NAS5-30368 and NSF grant ATM-9458424. The authors gratefully acknowledge K. Ogilvie, E. Shelley, J. Dudeney, G. Rostoker, B. Blake and T. Fritz for the use of their data.

## REFERENCES

Blake, J. B. et al. CEPPAD: Comprehensive energetic particle and pitch angle distribution experiment on POLAR. In C. T. Russell, editor, *The Global Geospace Mission*, pages 531–562. Kluwer Academic Publishers, 1995.

Chen, M. W., M. Schulz, L. R. Lyons, and D. J. Gorney. Stormtime transport of ring current and radiation-belt ions. *J. Geophys. Res.*, 98, 3835–3849, 1993.

Daglis, I. A. and W. I. Axford. Fast ionospheric response to enhanced activity in geospace: Ion feeding of the inner magnetotail. *J. Geophys. Res.*, *101*, 5047–5065, 1996.

Dudeney, J. R. et al. Satellite experiments simultaneous with antarctic measurements (SESAME). In C. T. Russell, editor, *The Global Geospace Mission*, pages 705–742. Kluwer Academic Publishers, 1995.

Fennell, J. F., D. R. Croley, Jr., and S. M. Kaye. Low-energy ion pitch angle distributions in the outer magnetosphere: Ion zipper distributions. *J. Geophys. Res.*, *86*, 3375, 1981.

Fok, M.-C., T. E. Moore, J. U. Kozyra, G. C. Ho, and D. C. Hamilton. Three-dimensional ring current decay model. *J. Geophys. Res.*, *100*, 9619–9632, 1995.

Gleisner, H., H. Lundstedt, and P. Wintoft. Predicting geomagnetic storms from solar-wind data using time-delay neural networks. *Ann. Geophys.*, *14*, 679–686, 1996.

Gonzalez, W. D., J. A. Joselyn, Y. Kamide, H. W. Kroehl, G. Rostoker, B. T. Tsurutani, and V. M. Vasyliunas. What is a geomagnetic storm? *J. Geophys. Res.*, *99*, 5771–5792, 1994.

Grande, M., C. H. Perry, A. Hall, J. Fennell, and B. Wilken. Survey of ring current composition during magnetic storms. *Adv. Space Res.*, *20*, 321–326, 1997.

Hamilton, D. C. et al. Ring current development during the great geomagnetic storm of february 86. *J. Geophys. Res.*, *93*, 14,343–14,355, 1988.

Hansen, C. and J. Fajans. Debye shielding and the dynamic response of a magnetized, collisionless plasma. In *Non-Neutral Plasma Physics II*, volume AIP Conference Proceedings 331, pages 87–91, New York, 1995.

Imhof, W. L., D. L. Chenette, D. W. Datlowe, J. Mobilia, M. Walt, and R. R. Anderson. The correlation of AKR waves with precipitating electrons as determined by plasma wave and x-ray image data from the POLAR spacecraft. *Geophys. Res. Lett.*, *25*, 289–292, 1998.

Jordanova, V. K. October 1995 magnetic cloud and accompanying storm activity: Ring current evolution. *J. Geophys. Res.*, *103*, 79, 1998.

Kaye, S. M. et al. Ion composition of zipper events. *J. Geophys. Res.*, *86*, 3383–3388, 1981.

Ogilvie, K. W. et al. Swe, a comprehensive plasma instrument for the wind spacecraft. In C. T. Russell, editor, *The Global Geospace Mission*, pages 55–77. Kluwer Academic Publishers, 1995.

Papitashvili, V. O. et al. Electric potential patterns in the northern and southern polar regions parameterized by interplanetary magnetic field. *J. Geophys. Res.*, *99*, 13,251, 1994.

Rostoker, G. et al. Canopus - a ground-based instrument array for remote sensing the high latitude ionosphere during the istp/ggs program. In C. T. Russell, editor, *The Global Geospace Mission*, pages 743–760. Kluwer Academic Publishers, 1995.

Sheldon, R. B. Ion transport and loss in the earth's quiet ring current 2. diffusion and magnetosphere-ionosphere coupling. *J. Geophys. Res.*, *99*, 5705–5720, 1994a.

Sheldon, R. B. Plasmasheet convection into the inner magnetosphere during quiet conditions. In D. N. Baker, editor, *Solar Terrestrial Energy Program: COSPAR Colloquia Series*, volume 5, pages 313–318, New York, 1994b. Pergamom Press.

Shelley, E. G. et al. The toroidal imaging mass-angle spectrograph (TIMAS) for the POLAR mission. In C. T. Russell, editor, *The Global Geospace Mission*, pages 497–530. Kluwer Academic Publishers, 1995.

Smith, P. and R. Hoffman. Direct observations in the dusk hours of the characteristics of the storm time ring current particles during the beginning of magnetic storms. *J. Geophys. Res.*, *79*, 966, 1974.

Suzuki, A., M. Yanagisawa, and N. Fukushima. Antisunward space current below the magsat level during magnetic storms, and its possible connection with partial ring current in the magnetosphere. *J. Geophys. Res.*, *90*(B3), 2465–2471, 1985.

Whipple, Jr, E. C. The signature of parallel electric fields in a collisionless plasma. *J. Geophys. Res.*, *82*, 1525, 1977.

---

Robert B. Sheldon, Boston University Center for Space Physics, rsheldon@bu.edu

# The Magnetospheric Trough

M. F. Thomsen, D. J. McComas, J. E. Borovsky, and R. C. Elphic

*Los Alamos National Laboratory, Los Alamos, New Mexico*

We review the history of the concepts of the magnetospheric cold-ion trough and hot-electron trough and conclude that the two regions are actually essentially the same. The magnetospheric trough may be viewed as a temporal state in the evolution of convecting flux tubes. These flux tubes are in contact with the earth's upper atmosphere, which acts both as a sink for precipitating hot plasma-sheet electrons and as a source for the cold ionospheric plasma, leading to progressive depletion of the plasma sheet and refilling with cold plasma. Geosynchronous plasma observations show that the rate of loss of plasma-sheet electron energy density is commensurate with the precipitating electron flux at the low-latitude edge of the diffuse aurorae. The rate at which geosynchronous flux tubes fill with cold ionospheric plasma is found to be consistent with previous estimates of early-time refilling. Geosynchronous observations further indicate that both Coulomb collisions and wave-particle effects probably play a role in trapping ionospheric material in the magnetosphere.

## 1. INTRODUCTION

### 1.1 Ion Trough

When the plasmasphere and plasmapause were discovered in the early sixties through lunar rocket measurements and ground-based whistler wave studies [e.g., *Gringauz et al.*, 1960; *Carpenter*, 1962a,b, 1966], the low-density region immediately outside the dense plasmasphere was termed the "plasma trough" [*Carpenter*, 1966]. Whistler measurements revealed not only the existence of the plasma trough, but the fact that the density in the nightside trough was typically lower than that in the dayside trough [*Angerami and Carpenter*, 1966]. With observations from the Lockheed light ion spectrometer on Ogo 5, *Chappell et al.* [1971] showed that in addition to the day/night asymmetry, there is a local time gradient in the dayside trough density, from low values near dawn to increasingly larger values through noon toward evening. They interpreted this gradient as a progressive filling of the flux tubes with outflowing ionospheric material as the flux tubes convect across the dayside magnetosphere. Figure 1 illustrates a) the morning-to-evening density gradient at L=5 [after *Chappell et al.*, 1971] and b) the convection/filling interpretation [from *Chappell*, 1972].

The convection pattern underlying the *Chappell et al.* explanation was based on the models proposed earlier by *Axford and Hines* [1961], *Carpenter* [1962a], *Nishida* [1966], and *Brice* [1967]. This basic pattern, illustrated in Figure 2, arose from the superposition of a dawn-to-dusk electric field imposed across the magnetosphere by the solar wind and a corotational electric field imposed by the ionosphere. The resultant cold plasma flow paths (i.e., the equipotentials of the total electric field) form two classes: paths that close around the earth, and open paths that enter from the night side and exit through the dayside magnetopause. The location of the separatrix between the two is determined by the strength of the imposed convection, being closer to the earth for stronger convection. This simple picture accounted nicely for the existence of the plasmasphere, including its dusk bulge [*Nishida*, 1966].

Geospace Mass and Energy Flow: Results From the International Solar-Terrestrial Physics Program
Geophysical Monograph 104
Copyright 1998 by the American Geophysical Union

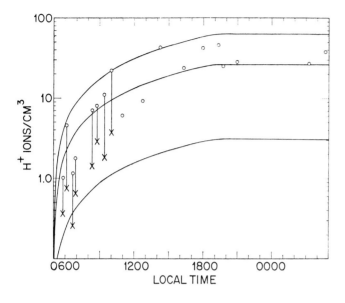

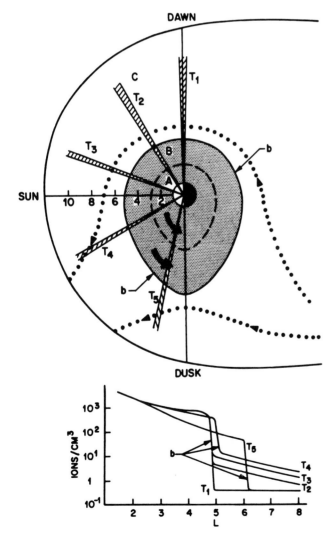

The second component of the *Chappell et al.* [1971] explanation of the dayside trough was that of flux tube filling from the ionosphere during the convection process. Flux tubes following closed drift paths around the earth remain exposed to the ionospheric source for long periods of time, resulting in the accumulation of large cold plasma densities (the plasmasphere). Flux tubes on open trajectories are exposed only for the time it takes them to move through the dayside magnetosphere. The ionospheric source was described theoretically in terms of a hydrodynamic model by *Banks et al.* [1971], based on a similar approach to describe the polar wind [*Banks and Holzer*, 1969]. The filling process, including mechanisms by which outflowing ionospheric material can be trapped in the magnetosphere, is a topic of considerable interest [c.f., the February, 1992, special section of *J. Geophys. Res.*, especially the introductory review by Horwitz and Singh].

*1.2 Electron Trough*

In the same time frame in which the morphology and properties of the cold ion trough were being investigated, interest was also focused on the nightside plasma sheet, including its inner (earthward) edge. *Frank* [1971] summarized the spatial relationship of several magnetospheric boundaries, including the inner edge of the nightside plasma sheet. Using Ogo 3 measurements, he found that in the midnight region the inner edge of the plasma sheet often abuts the plasmapause, but prior to local midnight there appears to be a significant gap between the two (see Figure 3, from *Frank* [1971]). The gap between the plasmapause and the earthward edge of the plasma sheet was termed the electron "trough." This gap was consistent with the earlier results of *Vasyliunas* [1968], who reported that in the local time range of 17-22 hours, "weak or no electron fluxes are found between the inner boundary of the plasma sheet and the outer boundary of the plasmasphere."

This region was revisited by *Fairfield and Viñas* [1984] using ISEE-1 low-energy electron measurements. They confirmed the earlier findings that the plasma sheet abuts the plasmapause post-midnight and that a gap between the two commonly exists in the premidnight region. The explanation they proposed for this gap is illustrated in

**Figure 1.** (a) Local time dependence of cold ion density at L=5 measured by the Lockheed Light Ion Spectrometer on OGO 5 The symbols indicate the two different procedures that were used to project the measurements to the magnetic equator, and the curves indicate different theoretical expectations [after Chappell et al., 1971]. (b) Schematic illustrating the progressive filling of flux tubes as they convect from the night side across the day-side magnetosphere [from Chappell, 1972].

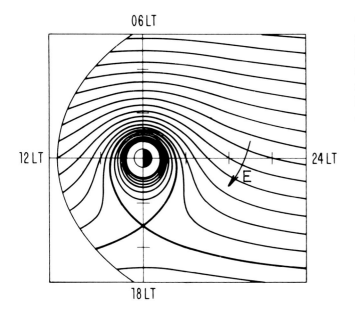

**Figure 2.** Equipotential contours in the magnetospheric equatorial plane for a superposition of the corotational electric field and a uniform, dawn-to-dusk electric field [after Lyons and Williams, 1984].

Figure 4. Comparison with Figure 2 shows that the explanation depends on the same flux-tube convection pattern responsible for ion trough filling. As flux tubes convect from the tail into and around the near-earth region, they begin to lose their load of hot plasma-sheet electrons through precipitation into the atmosphere. This precipitation is responsible for the diffuse aurora [*Fairfield and Viñas*, 1984] and results in significant depletion by the time the flux tubes reach the afternoon/evening sector. Fairfield and Viñas proposed that a spacecraft exiting from the plasmasphere in the evening sector could pass through a region of depleted flux tubes before entering the "fresh" plasma sheet on flux tubes that are newly arrived from the tail. In this picture, the evening "trough" between the plasmasphere and plasma sheet is simply a natural extension of the region of depleted electron fluxes in the afternoon sector. Hence, it would be appropriate to expand the use of the term "trough" to cover the full depleted sector, and this is the approach we adopt here.

In a later study based on DE-1 and DE-2 data, *Horwitz et al.* [1986] found no systematic evidence for a separation between the plasmapause and the plasma sheet in the evening sector and suggested that the gap found in the various earlier studies may have been the result of using 1-keV electrons rather than 100-eV electrons to mark the inner edge of the plasma sheet. *Horwitz et al.* [1986] thus called into question the existence of the original electron "trough," i.e., the extension of the afternoon depleted region into the evening sector beyond the plasmasphere bulge, as illustrated in Figure 4. The electron trough defined more broadly to apply to the entire afternoon region of lower electron fluxes is, however, well explained by the interpretation offered by Fairfield and Viñas (Figure 4).

*1.3 Synthesis*

The historical development sketched above brings us to the conclusion that the ion and electron trough regions, on the surface apparently different entities, are in reality physically the same. Moreover, the underlying physical picture suggests that the magnetospheric trough should be viewed not so much as a spatial region, but as a temporal state in the evolution of convecting flux tubes that are in contact with the earth's upper atmosphere. The atmosphere acts both as a sink for precipitating hot plasma sheet electrons and as a source for the cold ionospheric plasma, leading to progressive depletion of hot plasma sheet electrons and refilling with cold plasma. Figure 5 schematically summarizes this picture. The two panels (ion trough and electron trough) are essentially mirror images. In reality of course, this highly simplified picture is complicated by a number of factors, including a non-uniform convection electric field and temporal changes in the convection pattern and in the precipitation and filling rates. It is further modified by divergence of hot electron drift paths from the ExB drift paths of the cold plasma, due to gradient and curvature contributions to the drift. Nonetheless, Figure 5 provides a useful conceptual framework from which to examine the physics of the trough region.

As indicated in the historical discussion above, the trough has been observed with ground-based whistler measurements and numerous satellites. Elliptically orbiting satellites such as the Ogo satellites [e.g., *Taylor et al.*, 1965; *Vasyliunas*, 1968; *Frank*, 1971; *Chappell et al.*, 1971], the ISEE satellites [e.g., *Nagai et al.*, 1985; *Carpenter and Anderson*, 1992], DE [e.g., *Sojka et al.*, 1983], and Exos D [e.g., *Watanabe et al.*, 1992] provide a radial cut through the plasmasphere, plasmapause, and trough, with local time information accumulated through the seasonal evolution of the orbit. A complementary view is provided by geosynchronous satellites [e.g., *Lennartsson and Reasoner*, 1978; *Higel and Wu*, 1984; *McComas et al.*, 1993; *Moldwin et al.*, 1994], which sample all local times within 24 hours, but only at the radial distance of 6.6 $R_e$. In the following sections we present some illustrative examples of trough observations at geosynchronous orbit

358 THE MAGNETOSPHERIC TROUGH

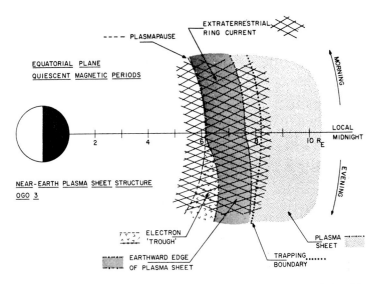

**Figure 3.** Schematic summary of the structure of the near-Earth plasma sheet and its relationship to the plasmapause in the midnight region, as observed by OGO 3 [from Frank, 1971].

and of the physical issues that can be addressed with those observations.

## 2. GEOSYNCHRONOUS OBSERVATIONS

As discussed above, the configuration illustrated in Figure 5 changes size and shape according to the strength of the convection electric field. A geosynchronous satellite will therefore find the system in different states over the course of time. Figure 6 shows our schematic configuration with three different circular orbits to illustrate the different sequences of regions that a geosynchronous satellite might observe, depending on the strength of the convection. A satellite travelling the orbit illustrated in the left panel would pass from the dayside trough region into the evening bulge of the plasmasphere and then out of the plasmasphere directly into the nightside plasma sheet. The center panel illustrates the case when the convection is strong enough that the plasmapause lies inside of the satellite orbit. In this case, the satellite would cross the afternoon separatrix from the dayside trough to the nightside plasma sheet without ever encountering the plasmasphere. Finally, the right panel illustrates the situation when an interval of strong convection is followed by an interval of quieting, in which a new separatrix forms beyond the old plasmapause, and flux tubes that had previously been part of the dayside trough can now drift on closed trajectories into the night side. Although not illustrated in Figure 6, another case that can occur is when the convection is weak for several days, allowing the plasmasphere to fill up to beyond geosynchronous orbit. In that case the satellite would remain within dense, cold plasmaspheric plasma throughout an entire day.

Plate 1 shows representative geosynchronous plasma observations for a case such as that illustrated in the left-hand panel of Figure 6. The data shown here and in subsequent figures were obtained by one of the Los

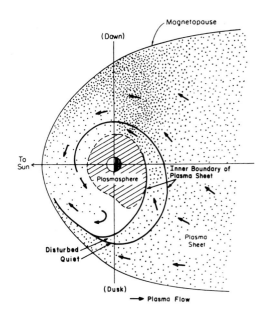

**Figure 4.** Schematic view of the inner edge of the plasma sheet in the magnetospheric equatorial plane as suggested by Fairfield and Viñas [1984].

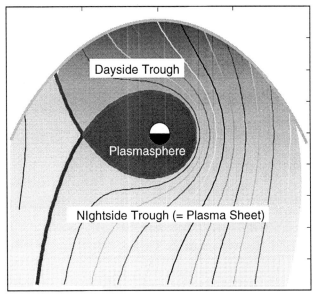

**Figure 5.** Synthesis view of the ion and electron trough regions. To lowest order the ion trough (right) and the electron trough (left) are colocated, representing a temporal state in the evolution of convecting flux tubes that are in contact with the earth's ionosphere. Convecting flux tubes progressively lose their electron plasma sheet content to auroral precipitation, while they progressively fill with cold plasma of ionospheric origin.

Alamos magnetospheric plasma analyzers (MPA) now operating in geosynchronous orbit [c.f., *McComas et al.,* 1993]. The MPAs are spherical sector electrostatic analyzers that measure the three-dimensional velocity-space distributions of ions and electrons over the nominal energy-per-charge range of ~1 V to 40 kV. Further details on the instrument are provided by *Bame et al.* [1993]. Plate 1 shows 24 hours of observations of ions (upper panel) and electrons (lower panel) from the MPA on spacecraft 1994-084 on June 17, 1996. Both panels present the color-coded logarithm of the instrument count rate (proportional to the measured energy flux) as a function of the logarithm of the particle energy on the vertical axis and time along the horizontal axis. The measurements shown in Plate 1 were obtained in the southward-viewing quadrant of the satellite spin.

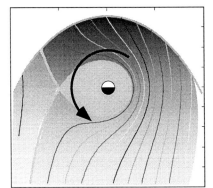 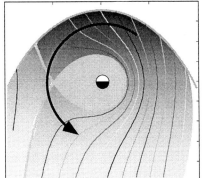 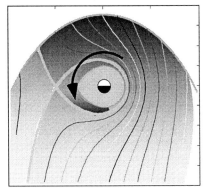

**Figure 6.** Schematic trough configuration with circular orbits illustrating possible sequences of regions observed by a geosynchronous satellite.

360 THE MAGNETOSPHERIC TROUGH

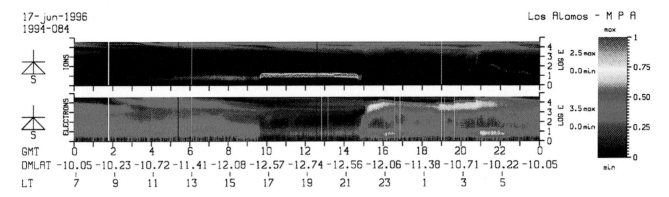

**Plate 1.** Ion (top) and electron (bottom) measurements made by the MPA onboard 1994-084 on June 17, 1996. Both panels show the color-coded logarithm of the instrument count rate as a function of the logarithm of the particle energy (on the vertical axis) and Universal Time (GMT) on the horizontal axis. The corresponding local times of the measurements are shown below the time axis.

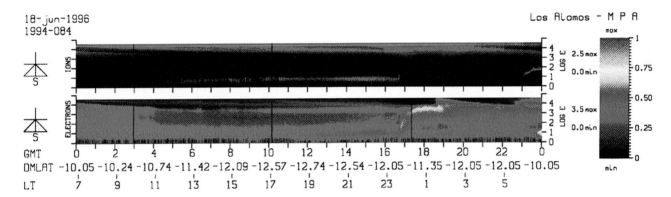

**Plate 2.** Same as Plate 1 for June 18, 1996.

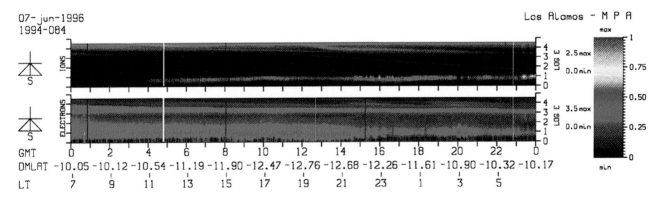

**Plate 3.** Same as Plate 1 for June 7, 1996.

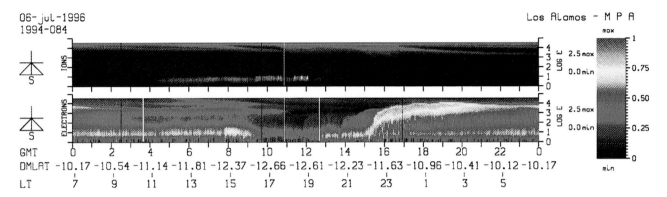

**Plate 4.** Same as Plate 1 for July 6, 1996, except the electron panel shows observations from the eastward-looking spin quadrant, and the color bar has been adjusted to bring out weaker fluxes.

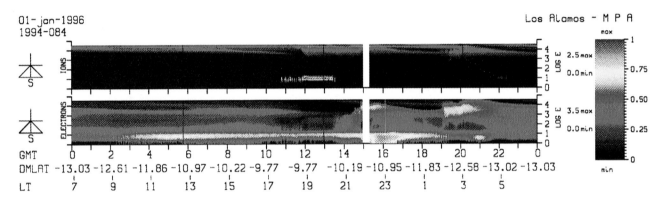

**Plate 5.** Same as Plate 1 for January 1, 1996.

Plate 1 illustrates the appearance of several different magnetospheric regions at geosynchronous orbit. In the top (ion) panel, the population at the highest energies (above ~10 keV) is the low-energy edge of the ion plasma sheet/ring current. At low ion energies between ~0930 and 1500 UT, there is a very dense, cold population that we identify as plasmaspheric plasma [c.f., *McComas et al.*, 1993; *Moldwin et al.*, 1994]. Just prior to this plasmaspheric interval, there are several hours of low-density, low-energy ions that appear to build up as the satellite traverses the day side. This is the classic appearance of the dayside ion trough at geosynchronous orbit. When the satellite leaves the dense plasmasphere just before 1500 UT, it enters immediately into a region of fairly dense hot electrons (~several keV), which is the nightside plasma sheet. This population seems to be refreshed several times as the satellite passes through the night hours. After local dawn (~2300 UT), the hot electron fluxes decline steadily on into the next day, as shown in Plate 2, which presents the MPA observations for June 18, 1996.

The measurements shown in Plate 2 are representative geosynchronous plasma observations for a case such as that illustrated in the center panel of Figure 6, in which the satellite enters the nightside plasma sheet directly from the dayside trough, without ever passing through the plasmasphere extension. The decline of hot electron fluxes and the gradual increase in cold ions in the dayside trough are similar to those seen on the previous day (Plate 1), but in Plate 2 the satellite enters into the fresh plasma sheet from the trough at ~1640 UT, without encountering any cold, dense, plasmaspheric ions. It is interesting to note that this plasma-sheet entry occurred near midnight local time, underlining the point made earlier that the shape and location of the separatrix may be more complicated than suggested by the simple tear-drop schematic of Figures 5 and 6.

The upper panel of Figure 7 shows the densities derived from the MPA cold-ion and hot-electron measurements shown in Plates 1 and 2. These densities represent numerical sums over the measured distribution functions from $E_i$~1 eV to ~100 eV and from $E_e$~30 eV to ~40 keV, respectively [cf., *Thomsen et al.*, 1996]. To draw attention to the dayside trough, horizontal bars have been added to show the range of UT during which the satellite travelled from 06 LT to 18 LT on these two days. The decrease of the hot electron density and the simultaneous increase of the cold ion density across the day side of the orbit can be seen on both June 17 and 18.

The lower panel of Figure 7 shows the energy density of the hot electron component for the same two days. The loss of plasma sheet energy density in the morning hours is

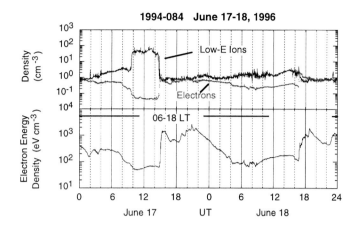

**Figure 7.** (Upper panel) Computed densities of low-energy ions (1 eV < $E_i$ < 100 eV) and hot electrons (30 eV < $E_e$ < 41 keV) for the 24-hour interval shown in Figure 7. (Lower panel) Corresponding hot-electron energy density (density × temperature). The horizontal bars indicate times when the satellite was between 06 and 18 LT.

a clear signature of the development of the electron trough. In the discussion below, we will compare the loss of plasma sheet energy with the energy deposited in the atmosphere by diffuse aurorae.

The cold ion densities plotted in Figure 7 are somewhat uncertain due to uncertainties in the potential of the satellite, including possible surface charging asymmetries (for a discussion of the factors affecting the precision of the measurements, see *Moldwin et al.* [1995]). A number of lines of evidence suggest that, especially at low densities (<10 cm$^{-3}$), our moments calculation may be underestimating the cold ion population. For example, an examination of the density measured as the satellite passed through eclipse on various occasions suggests that the density computed from measurements made in sunlit conditions may be underestimated by factors of ~1.2-4 for computed densities less than 10 cm$^{-3}$. A comparison between the median MPA cold ion densities measured as a function of local time for several months in 1990-91 and the median cold plasma densities measured by the GEOS 1 mutual impedance probe during 1977-78 [*Décréau et al.*, 1982] leads to a similar conclusion.

An example of data from the type of orbit illustrated in the right-hand panel of Figure 6 is presented in Plate 3. In this example, from June 7, 1996, the satellite never enters the fresh plasma sheet on the night side, but remains within the trough region of increasing cold ions and decreasing hot electron fluxes. The electron plasma sheet population trapped on these temporarily closed drift paths has been called the "remnant layer" by *Feldstein and Galperin*

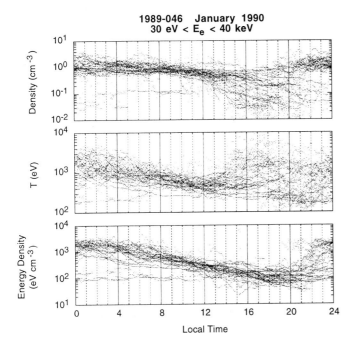

**Figure 8.** Local time dependence of the measured density, temperature and energy density of plasma sheet electrons (30 eV < $E_e$ < 40 keV) during the month of January 1990.

[1985], who related it to the region of diffuse aurorae equatorward of the discrete auroral oval.

The types of orbits described schematically in Figure 6 and illustrated in Plates 1-3 are quite commonly observed at geosynchronous orbit, confirming the general utility of the concepts summarized in Figure 5. There are occasions, however, when the geosynchronous obser-vations do not fit neatly into this basic framework. Plates 4 and 5 illustrate two such examples. Plate 4, from July 6, 1996, is quite similar to Plate 2 in that the satellite leaves the dayside cold-ion trough (at ~1200 UT) without encountering any dense, cold, plasmaspheric plasma. It would thus appear to be a case of crossing the afternoon separatrix, as in the center panel of Figure 6. However, unlike Plate 2 and unlike what one might expect on the basis of Figure 6, the satellite does not obviously observe fresh plasma sheet electrons upon crossing the separatrix. Rather, there is a 2-3 hour interval of trough-like electron fluxes, but with no corresponding cold ions, before the satellite clearly encounters the evening plasma sheet at ~14-15 UT. Close inspection of this trough-like interval, however, shows that it does contain a very tenuous, cool, plasma-sheet-like electron population (note that we have adjusted the color bar for the electrons to bring out these very low fluxes), and it is possible to reconcile this event with the expectations of Figure 6 if one assumes that the plasma sheet in this region was extremely tenuous at this time. One possible explanation for the low density might be that these plasma sheet flux tubes, while nominally on open drift trajectories (c.f., Figure 6), may have spent a long time in the dusk stagnation region and have consequently suffered appreciable losses to atmospheric precipitation. It may be that a similarly tenuous (depleted?) plasma sheet was responsible for some of the gaps between the plasmapause and the inner edge of the plasma sheet reported by *Vasyliunas* [1968] and by *Frank* [1971].

Plate 5 from January 1, 1996, shows another example that appears to be at odds with the predictions of Figure 6, namely, a case in which essentially no cold ion trough was seen across the entire day side of the orbit. Although there are possible concerns about a positive spacecraft potential "hiding" the cold ions [c.f., *Olsen*, 1982], a number of arguments lead us to conclude that this is probably not the case and that the cool ion density was indeed quite low throughout the interval shown in Plate 5. Such events can potentially shed light on what controls the upflow of ionospheric material into the equatorial magnetosphere.

## 3. TROUGH PHYSICS

### 3.1 *Plasma Sheet Electron Losses*

As suggested by *Fairfield and Viñas* [1984], the decline in the plasma sheet electron content as flux tubes drift from the near-earth tail on the night side around onto the day side is due to the precipitation of the plasma sheet electrons into the ionosphere, producing the morning-side diffuse aurora. We can estimate the rate of energy deposition into the ionosphere by measuring the typical rate at which the plasma sheet electron energy density declines as a function of local time. Figure 8 shows the local time dependence of the density, temperature, and energy density (defined as the density times the temperature) of electrons between ~30 eV and ~40 keV as measured by the MPA on 1989-046 during the month of January 1990. The decline of the energy density past midnight is quite clear. Figure 9 shows the local time dependence of the various percentile levels of the energy density in one-hour local-time bins for the same month of data. A linear fit to the 25th, 50th, and 75th percentile levels between 01 and 09 LT yields values of -95, -170, and -212 eV cm$^{-3}$ per hour of local time, respectively, for the slopes of the curves. To convert this spatial loss rate to a temporal loss rate, we use the azimuthal drift speed of 2 keV electrons, a typical plasma sheet temperature (Figure 8), which for corotation plus gradient and curvature drift in a dipole field is 1.43 hours of local time per hour. Thus the temporal loss rates for

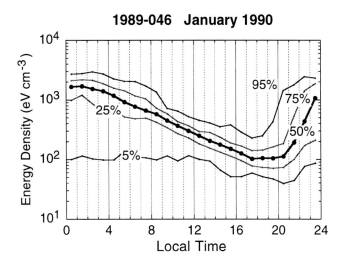

**Figure 9.** Local time dependence of the 5th, 25th, 50th, 75th, and 95th percentile values of the measured plasma sheet energy density in one-hour local-time bins for the month of January 1990.

plasma sheet electron energy density between 01 and 09 LT in Figure 9 are 135, 243, and 303 eV cm$^{-3}$ hr$^{-1}$ for the 25th, 50th, and 75th percentile levels, respectively.

The temporal loss rate of plasma sheet energy density, $\mathcal{E}$, is related to the energy flux into the ionosphere, $\mathcal{F}$:

$$(\partial/\partial t) \int \mathcal{E} \, d^3V = 2 \int \mathcal{F} \, dA$$

where the integral on the left-hand side is over the volume of a given flux tube, and the integral on the right-hand side is over the cross-sectional area of the same flux tube at the ionosphere. If we assume that the electron energy density is roughly a constant along the flux tube and approximate the field as a dipole, we have

$$\mathcal{F} = (1.05 \times 10^{12} \text{ cm}) (\partial \mathcal{E} / \partial t).$$

Thus, the 50th percentile level in Figure 9 implies a precipitating electron energy flux of 0.11 erg cm$^{-2}$ s$^{-1}$. For comparison, *Spiro et al.* [1982] found the quiet-time mean energy flux of precipitating electrons at low altitudes to be in the range 0.2-1.0 erg cm$^{-2}$ s$^{-1}$ between 01 and 09 LT at a geomagnetic latitude of 67°, the nominal invariant latitude for geosynchronous orbit in a dipole field. However, recent studies comparing geosynchronous plasma measurements with similar measurements made with DMSP at low altitudes [*Weiss et al.*, 1996; *Reeves et al.*, 1997] suggest that the typical ionospheric magnetic latitude to which geosynchronous orbit maps in the post-midnight region is generally lower than 67° by a few degrees. At a magnetic latitude of 64°, *Spiro et al.* [1982] found mean energy fluxes between 0.05 and 0.17 erg cm$^{-2}$ s$^{-1}$, in very good agreement with the estimate based on the geosynchronous observations.

### 3.2 Plasmasphere Refilling

One of the earliest objectives of studies of the cold ion trough was the question of how rapidly flux tubes could fill with upflowing ionospheric material [e.g., *Angerami and Carpenter*, 1966; *Chappell et al.*, 1971]. Because it is one of the most fundamental examples of ionosphere/magnetosphere coupling, this issue is still of high interest. The question of plasmaspheric refilling has two aspects: 1) What is the rate at which material flowing up from the ionosphere fills the flux tubes? and 2) What are the mechanisms responsible for trapping that material in the magnetosphere? Geosynchronous observations can help address both issues:

*3.2.1 Refilling rate.* Because the flux tubes of the dayside plasma trough are approximately corotating, the spatial gradients in cold plasma density that are typically observed at geosynchronous orbit across the day side can be converted to temporal filling rates. This approach has been used by several authors [e.g., *Park*, 1970, 1974; *Higel and Wu*, 1984; *Sojka and Wrenn*, 1985; *Song et al.*, 1988; *Farrugia et al.*, 1989]. During steady magnetospheric conditions, this can be done for an individual orbit (e.g., Figure 7); more generally, because of temporal variations in the convection pattern, this estimation of the filling rate can be done statistically. In Figure 10 we show the occurrence frequency distribution for the cold ion density measured at geosynchronous orbit by the MPA on spacecraft 1989-046 during the month of July, 1990. The data have been sorted by local time of observation, with the distribution for every other one-hour bin across the day-side filling region stacked vertically. For comparison, the distribution for the 05-06 LT bin is repeated as the light curve in each panel. The tick marks at the top of each panel mark the 25th percentile values of the measured density for each distribution. It is clear from Figure 10 that the cold plasma density increases systematically across the dayside trough, from values of less than 1 cm$^{-3}$ before dawn to roughly 1.5 cm$^{-3}$ in the afternoon.

In Figure 11 we plot the value of the cold-ion density at various percentile levels in each one-hour local time bin for the same month of data. For comparison, we also plot as the light dashed line the curve derived from the plasma trough model of *Carpenter and Anderson* [1992], evaluated at L=6.6. Since the Carpenter and Anderson values represent total electron density (cold plus hot), we also show as the heavier dashed line their curve reduced by 0.7 cm$^{-3}$, a representative value for the hot plasma sheet ion density.

An analysis like that illustrated in Figure 11 has been done for 11 different months of data, from different seasons and different parts of the solar cycle, and the resulting 25th and 50th percentile levels for the 12-13 LT bin are plotted in Figure 12. The 50th percentile levels tend to vary erratically from month to month because of the variable probability of encountering the dense plasmasphere near noon, but there is a remarkable steadiness to the 25th percentile level, which we therefore take to be a good measure of the true trough density,

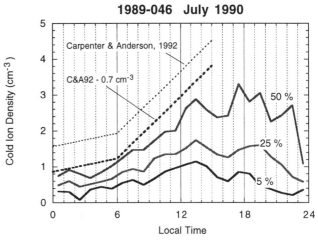

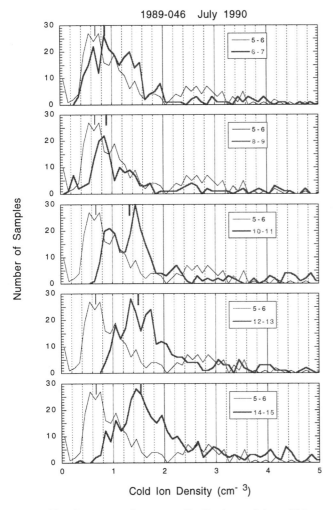

**Figure 11.** Local time dependence of the 5th, 25th, and 50th percentile values of the measured cold-ion density in one-hour local-time bins for the month of July 1990. The upper dashed line shows the the trough density at L=6.6 from the model of Carpenter and Anderson [1992]. The lower dashed curve is the Carpenter and Anderson trough density reduced by 0.7 cm$^{-3}$, to remove approximately the contribution from the hot ion population.

uncontaminated by dense intervals. These measured values tend to be lower than the 12 LT trough density given by the Carpenter and Anderson relation (see Figure 11) by factors of ~1-3. As discussed earlier, there is reason to believe that the computed MPA cold ion densities may be underestimated by a similar factor when the density is low (<10 cm$^{-3}$). With such a correction, the trough densities we observe are quite consistent with the Carpenter and Anderson relation.

To estimate the filling rate from the spatial gradient of the observed cold ion density, we use the lower percentile levels shown in Figure 11 since these presumably represent the "freshest" trough flux tubes, i.e., those which are executing their first pass across the dayside region. Under the assumption that the flux tubes are approximately corotating (i.e., traverse one hour of local time in one hour of actual time), the slopes of the 5th, 25th and 50th percentile levels in Figure 11 correspond to filling rates of 0.085, 0.124, and 0.218 cm$^{-3}$ hr$^{-1}$, respectively (where the slopes are taken between 04 and 14 LT). If the filling is assumed to occur only on the day side (i.e., when the ionospheric footpoints are illuminated), these correspond to approximate daily filling rates of 1.0, 1.5, and 2.6 cm$^{-3}$ day$^{-1}$, respectively. This can be compared to the values 0.287 cm$^{-3}$ hr$^{-1}$ or 3.4 cm$^{-3}$ day$^{-1}$ determined by *Carpenter and Anderson* [1992].

Table 1 summarizes hourly filling rates derived in this fashion from the 25th percentile levels for the eleven

**Figure 10.** Occurrence frequency distributions of the cold ion density (1 eV < $E_i$ < 100 eV) for one-hour local-time bins across the day side of the magnetosphere. The measurements were made by the MPA on geosynchronous satellite 1989-046 during the month of July 1990. The tick marks at the top of each panel indicate the 25th percentile values.

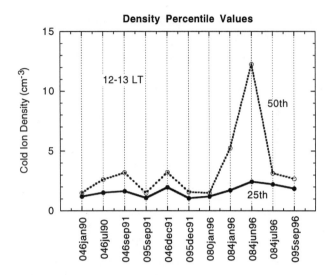

**Figure 12.** Twenty-fifth and fiftieth percentile levels of the cold ion density measured at geosynchronous orbit between 12 and 13 LT for various months covering different seasons and different parts of the solar cycle. The months are labelled according to the satellite that made the measurements (046=1989-046, 095=1990-095, 080=1991-080, 084=1994-084).

different months of observations. All the satellites had nearly complete coverage for the months included in this study, with the exception of 1990-095, which had 73% coverage in September 1996 and 71% coverage in September 1991 (during the latter month the sector from ~12 to 16 LT was essentially not covered). The 25th percentile level was chosen because it seems to correspond fairly well to the peak of the low-density end of the hourly distributions (see Figure 10) and appears to be uncontaminated by encounters with the dense plasmasphere (see Figure 12). The months included in Table 1 represent northern winter, summer, and fall conditions near solar maximum (1990-91) and near solar minimum (1995-96). These months provide a crude exploration of seasonal and solar cycle dependences [c.f., *Carpenter and Anderson*, 1992; *Rasmussen et al.*, 1993].

From Table 1 it appears that the refilling rate ranges from about 0.05 to 0.25 cm$^{-3}$ hr$^{-1}$, with no strong indication of either seasonal or solar cycle dependence. The average derived rate for the solar maximum months is 0.09 ± 0.03 cm$^{-3}$ hr$^{-1}$, while for solar minimum it is 0.16 ± 0.07 cm$^{-3}$ hr$^{-1}$. It is possible that the slightly higher rate at solar minimum is due to slightly lower convection speeds associated with lower magnetic activity (slower convection allows more refilling time). The final column in Table 1 shows the average value of the magnetospheric activity index Kp for each of the months examined. There is a weak negative correlation between the refilling rates and average Kp (R ~ -0.6). This is largely driven by the high rate found during June 1996, the quietest month in our study. With the increase of a factor of ~1.2-4 discussed above, the derived refilling rates in Table 1 are consistent with the Carpenter and Anderson value.

*3.2.2 Trapping mechanisms.* There are several mechanisms that might be operating to trap upflowing ionospheric material in the magnetosphere. One such mechanism is due to electromagnetic ion cyclotron waves driven by the hot ring-current ion temperature anisotropy; these waves have been observed to cause transverse heating of cool ions [*Mauk and McPherron*, 1980; *Young et al.*, 1981], thereby increasing their pitch angles and trapping them in the magnetospheric "bottle." In support of this mechanism, *Gary et al.* [1994, 1995] have shown that the geosynchronous hot-ion temperature anisotropy has an upper bound that obeys the scaling relationship expected for this instability on theoretical grounds. Moreover, the mechanism accounts for the warm (10 eV) ion population often seen in the afternoon trough region and characterized by strong $T_\perp > T_\parallel$ temperature anisotropies [*Olsen*, 1981; *Gary et al.*, 1997], quite in contrast to the field-aligned nature of cool ion outflows observed in the morning sector. The consequences of the presence of such warm, equatorially trapped ions for the

Table 1. Derived Refilling Rates

| Spacecraft | Month | Rate (cm$^3$hr)$^{-1}$ | Average Kp |
|---|---|---|---|
| 1989-046 | Jan 1990 | 0.049 | 2.62 |
| 1989-046 | July 1990 | 0.124 | 2.22 |
| 1989-046 | Sept 1991 | 0.100 | 3.01 |
| 1990-095 | Sept 1991 | 0.078 | 3.01 |
| 1989-046 | Dec 1991 | 0.136 | 2.66 |
| 1990-095 | Dec 1991 | 0.076 | 2.66 |
| 1991-080 | Jan 1996 | 0.070 | 1.88 |
| 1994-084 | Jan 1996 | 0.140 | 1.88 |
| 1994-084 | June 1996 | 0.250 | 1.27 |
| 1994-084 | July 1996 | 0.146 | 1.59 |
| 1990-095 | Sept 1996 | 0.184 | 2.52 |

further trapping of ionospheric upflows have been addressed by Singh and Chan [1992].

Another mechanism that can be important in trapping ionospheric upflows is Coulomb collisions [e.g., *Schulz and Koons*, 1972; *Lemaire*, 1989; *Wilson et al.*, 1992], which become increasingly important as a flux tube fills with plasma. The numerical simulations of *Wilson et al.* [1992] show that Coulomb collisions not only increase the filling rate, but they produce a characteristic two-stage filling process as seen at the magnetic equator: At early times, the plasma flowing up from the ionospheric footpoints of a given flux tube exhibits a large parallel temperature anisotropy ($T_\parallel > T_\perp$) all along the flux tube. As the density builds up in the flux tube, two regions of near-isotropy develop at the ionospheric ends, and these regions expand up along the flux tube until they meet at the equator. At this point, which at L=6 is after about 28 hours of refilling for the outflow parameters used by *Wilson et al.*, the plasma at the equator is nearly isotropic. Beyond this time the apparent refilling rate at the equator (dn/dt) increases markedly. In the L=6 simulation, *Wilson et al.* found an equatorial density of about 8 cm$^{-3}$ after the first 24 hours, with a subsequent increase to 30 cm$^{-3}$ in the next 24 hours.

There are at least two pieces of observational evidence from geosynchronous orbit that support this picture. First, there is the difference between the early-time (i.e., the first 12-15 hours) filling rates estimated above and by several previous authors based on dayside trough density observations, which are typically in the range of several particles per cm$^3$ per day, and the late-time (> 24 hours) filling rates estimated to be in the range of 30-50 cm$^{-3}$ per day [*Sojka and Wrenn*, 1985; *Lawrence et al.*, 1997]. Second, there is the fact that at low trough densities the low-energy ion distributions are typically field-aligned, becoming nearly isotropic at densities ~10-100 cm$^{-3}$ [e.g., *Nagai et al.*, 1985; *Sojka and Wrenn*, 1985; *Horwitz et al.*, 1990; *Carpenter et al.*, 1993]. In agreement with these earlier studies, Los Alamos geosynchronous observations indicate that this transition occurs at measured cold ion densities near 10 cm$^{-3}$.

## 4. SUMMARY

Recent geosynchronous plasma measurements complement and confirm earlier observations of the magnetospheric trough. The separately described cold-ion and hot-electron troughs are in reality the same region. The magnetospheric trough is not so much a spatial region as a temporal state in the evolution of convecting flux tubes that are in contact with the earth's upper atmosphere. The atmosphere acts both as a sink for precipitating hot plasma-sheet electrons and as a source for the cold ionospheric plasma, leading to progressive depletion of plasma sheet content and filling with cold plasma. The rate of depletion of hot electron energy density at geosynchronous orbit is commensurate with the precipitating electron flux at the low-latitude edge of the diffuse aurorae, and the measured rates of cold-ion density build-up are in general agreement with previous estimates of early-time (< 15 hours) flux-tube refilling. Geosynchronous observations further indicate that transverse heating of upflowing ionospheric ions by electromagnetic ion cyclotron waves produced by the hot ring-current temperature anisotropy may contribute to cold-ion trapping in the magnetosphere. In addition, at least two lines of evidence indicate that Coulomb collisions play a prominent role in trapping the new material.

*Acknowledgments.* We are grateful to D. L. Carpenter and to our colleagues at Los Alamos for numerous very helpful discussions and suggestions. We also thank J. L. Horwitz for his interest and encouragement, and we thank the reviewers for their helpful comments. This work was performed under the auspices of the U. S. Department of Energy with support from the NASA ISTP program.

## REFERENCES

Angerami, J. J., and D. L. Carpenter, Whistler studies of the plasmapause in the magnetosphere 2. Electron density and total tube electron content near the knee in magnetospheric ionization, *J. Geophys. Res.*, *71*, 711, 1966.

Axford, W. I., and C. O. Hines, A unifying theory of high latitude geophysical phenomena and geomagnetic storms, *Can. J. Phys.*, *39*, 1433, 1961.

Bame, S. J., et al., Magnetospheric plasma analyzer for spacecraft with constrained resources, *Rev. Sci. Instrum.*, *64*, 1026, 1993.

Banks, P. M., and T. E. Holzer, High-latitude plasma transport: The polar wind, *J. Geophys. Res.*, *74*, 6317, 1969.

Banks, P. M., A. F. Nagy, and W. I. Axford, Dynamical behavior of thermal protons in the mid-latitude ionosphere and magnetosphere, *Planet. Space Sci.*, *19*, 1053, 1971.

Brice, N. M., Bulk motion of the magnetosphere, *J. Geophys. Res.*, *72*, 5193, 1967.

Carpenter, D. L., The magnetosphere during magnetic storms; a whistler analysis, Ph.D. thesis, Tech. Rep. 12, Radiosci. Lab., Stanford Univ., Stanford, Calif., 1962a.

Carpenter, D. L., New experimental evidence of the effect of magnetic storms on the magnetosphere, *J. Geophys. Res.*, *67*, 135, 1962b.

Carpenter, D. L., Whistler studies of the plasmapause in the magnetosphere 1. Temporal variations in the position of the knee and some evidence on plasma motions near the knee, *J. Geophys. Res.*, *71*, 693, 1966.

Carpenter, D. L., and R. R. Anderson, An ISEE/whistler model of equatorial electron density in the magnetosphere, *J. Geophys. Res.*, *97*, 1097, 1992.

Carpenter, D. L., et al., Plasmasphere dynamics in the duskside bulge region: A new look at an old topic, *J. Geophys. Res., 98*, 19,243, 1993.

Chappell, C. R., Recent satellite measurements of the morphology and dynamics of the plasmasphere, *Rev. Geophys. and Space Phys., 10*, 951, 1972.

Chappell, C. R., K. K. Harris, and G. W. Sharp, The dayside of the plasmasphere, *J. Geophys. Res., 76*, 7632, 1971.

Décréau, P. M. E., C. Beghin, and M. Parrot, Global characteristics of the cold plasma in the equatorial plasmapause region as deduced from the GEOS 1 mutual impedance probe, *J. Geophys. Res., 87*, 695, 1982.

Fairfield, D. H., and A. F. Viñas, The inner edge of the plasma sheet and the diffuse aurora, *J. Geophys. Res., 89*, 841, 1984.

Farrugia, C. J., D. T. Young, J. Geiss, and H. Balsiger, The composition, temperature, and density structure of cold ions in the quiet terrestrial plasmasphere: GEOS 1 results, *J. Geophys. Res., 94*, 11, 1989.

Feldstein, Y. I., and Y. I. Galperin, The auroral luminosity structure in the high-latitude upper atmosphere: Its dynamics and relationship to the large-scale structure of the Earth's magnetosphere, *Rev. Geophys., 23*, 217, 1985.

Frank, L. A., Relationship of the plasma sheet, ring current, trapping boundary, and plasmapause near the magnetic equator and local midnight, *J. Geophys. Res., 76*, 2265, 1971.

Gary, S. P., M. B. Moldwin, M. F. Thomsen, and D. Winske, Hot proton anisotropies and cool proton temperatures in the outer magnetosphere, *J. Geophys. Res., 99*, 23, 1994.

Gary, S. P., M. F. Thomsen, L. Yin, and D. Winske, Electromagnetic proton cyclotron instability: Interactions with magnetospheric protons, *J. Geophys. Res., 100*, 21, 1995.

Gary, S. P., M. F. Thomsen, J. Lee, D. J. McComas, and K. Moore, Warm protons at geosynchronous orbit, *J. Geophys. Res., 102*, 2291, 1997.

Gringauz, K. I., V. G. Kurt, V. I. Moroz, and I. S. Shklovskii, Results of observations of charged particles observed out to R=100,000 km, with the aid of charged-particle traps on Soviet space rockets, *Astron. Zh., 37*, 4, 716, 1960.

Higel, B., and L. Wu, Electron density and plasmapause characteristics at 6.6 Re: A statistical study of the GEOS 2 relaxation sounder data, *J. Geophys. Res., 89*, 1583, 1984.

Horwitz, J. L., and N. Singh, Foreward, *J. Geophys. Res., 97*, 1047, 1992.

Horwitz, J. L., R. H. Comfort, and C. R. Chappell, A statistical characterization of plasmasphere density structure and boundary locations, *J. Geophys. Res., 95*, 7937, 1990.

Horwitz, J. L., et al., Plasma boundaries in the inner magnetosphere, *J. Geophys. Res., 91*, 8861, 1986.

Lawrence, D. J., J. E. Borovksy, D. J. McComas, and M. F. Thomsen, Late-time plasmaspheric refilling at geosynchronous orbit, *EOS Trans. Am. Geophys. Union, 78*, no. 17, p. S286, 1997.

Lemaire, J., Plasma distribution models in a rotating magnetic dipole and refilling of plasmaspheric flux tubes, *Phys. Fluids B, 1*, 1519, 1989.

Lennartsson, W., and D. L. Reasoner, Low-energy plasma observations at synchronous orbit, *J. Geophys. Res., 83*, 2145, 1978.

Lyons, L. R., and D. J. Williams, *Quantitative Aspects of Magnetospheric Physics*, D. Reidel, Dordrecht, Holland, 1984.

Mauk, B. H., and R. L. McPherron, An experimental test of the electromagnetic ion cyclotron instability with the earth's magnetosphere, *Phys. Fluids, 23*, 2111, 1980.

McComas, D. J., et al., Magnetospheric plasma analyzer: Initial three-spacecraft observations from geosynchronous orbit, *J. Geophys. Res., 98*, 13453, 1993.

Moldwin, M. B., M. F. Thomsen, S. J. Bame, D. J. McComas, and K. R. Moore, An examination of the structure and dynamics of the outer plasmasphere using multiple geosynchronous satellites, *J. Geophys. Res., 99*, 11475, 1994.

Moldwin, M. B., M. F. Thomsen, S. J. Bame, and D. J. McComas, The fine-scale structure of the outer plasmapause, *J. Geophys. Res., 100*, 8021, 1995.

Nagai, T., J. L. Horwitz, R. R. Anderson, and C. R. Chappell, Structure of the plasmapause from ISEE 1 low-energy ion and plasma wave observations, *J. Geophys. Res., 90*, 6622, 1985.

Nishida, A., Formation of plasmapause, or magnetospheric plasma knee, by the combined action of magnetosphere convection and plasma escape from the tail, *J. Geophys. Res., 71*, 5669, 1966.

Olsen, R. C., Equatorially trapped plasma populations, *J. Geophys. Res., 86*, 11, 1981.

Olsen, R. C., The hidden ion population of the magnetosphere, *J. Geophys. Res., 87*, 3481, 1982.

Park, C. G., Whistler observations of the interchange of ionization between the ionosphere and the protonosphere, *J. Geophys. Res., 75*, 4249, 1970.

Park, C. G., Some features of plasma distribution in the plasmasphere deduced from Antarctic whistlers, *J. Geophys. Res., 79*, 169, 1974.

Rasmussen, C. E., S. M. Guiter, and S. G. Thomas, A two-dimensional model of the plasmasphere: refilling time constants, *Planet. Space Sci., 41*, 35, 1993.

Reeves, G. D., L. A. Weiss, M. F. Thomsen, and D. J. McComas, A Quantitative Test of Different Magnetic Field Models Using Conjunctions Between DMSP and Geosynchronous Orbit, in *Radiation Belt Models and Standards*, vol. 97, edited by Lemaire, J. F., D. Heynderickx and D. N. Baker, p. 167, Am. Geophys. Union, 1997.

Schulz, M., and H. C. Koons, Thermalization of colliding ion streams beyond the plasmapause, *J. Geophys. Res., 77*, 248, 1972.

Singh, N., and C. B. Chan, Effects of equatorially trapped ions on refilling of the plasmasphere, *J. Geophys. Res., 97*, 1167, 1992.

Sojka, J. J., and G. L. Wrenn, Refilling of geosynchronous flux tubes as observed at the equator by GEOS 2, *J. Geophys. Res., 90*, 6379, 1985.

Sojka, J. J., R. W. Schunk, J. F. E. Johnson, J. H. Waite, and C. R. Chappell, Characteristics of thermal and suprathermal ions associated with the day-side plasma trough as measured by the Dynamics Explorer retarding ion mass spectrometer, *J. Geophys. Res., 88*, 7895, 1983.

Song, X. T., G. Caudal, and R. Gendrin, Refilling of the plasmasphere at the geostationary orbit: A Kp-dependent model deduced from the GEOS-1 measurements of the cold plasma density, *Adv. Space Res., 8*, 45, 1988.

Spiro, R. W., P. H. Reiff, and L. J. Maher Jr., Precipitating electron energy flux and auroral zone conductances - an empirical model, *J. Geophys. Res., 87*, 8215, 1982.

Taylor, H. A., Jr., H. C. Brinton, and C. R. Smith, Positive ion composition in the magnetoionosphere obtained from the Ogo-A satellite, *J. Geophys. Res., 70*, 5769, 1965.

Thomsen, M. F., D. J. McComas, G. D. Reeves, and L. A. Weiss, An observational test of the Tsyganenko (T89a) model of the magnetospheric field, *J. Geophys. Res., 101*, 24, 1996.

Vasyliunas, V. M., A survey of low-energy electrons in the evening sector of the magnetosphere with Ogo 1 and Ogo 3, *J. Geophys. Res., 73*, 2839, 1968.

Watanabe, S., B. A. Whalen, and A. W. Yau, Thermal ion observations of depletion and refilling in the plasmaspheric trough, *J. Geophys. Res., 97*, 1081, 1992.

Weiss, L. A., M. F. Thomsen, G. D. Reeves, and D. J. McComas, *Observational Testing of Magnetospheric Field Models at Geosynchronous Orbit*, p. 93, Space Environment Center, Boulder, Colorado, 1996.

Wilson, G. R., J. L. Horwitz, and J. Lin, A semikinetic model for early stage plasmaspheric refilling 1. Effects of coulomb collisions, *J. Geophys. Res., 97*, 1109, 1992.

Young, D. T., S. Perraut, A. Roux, C. de Villedary, R. Gendrin, A. Korth, G. Kremser, and D. Jones, Wave-particle interactions near $\Omega_{He+}$ observed on GEOS 1 and 2 1. Propagation of ion cyclotron waves in He$^+$-rich plasma, *J. Geophys. Res., 86*, 6755, 1981.

---

J. E. Borovsky, MS D466, Los Alamos National Laboratory, Los Alamos, New Mexico, 87545.

R. C. Elphic, MS D466, Los Alamos National Laboratory, Los Alamos, New Mexico, 87545.

D. J. McComas, MS D466, Los Alamos National Laboratory, Los Alamos, New Mexico, 87545.

M. F. Thomsen, MS D466, Los Alamos National Laboratory, Los Alamos, New Mexico, 87545.

# Data-Based Models of the Global Geospace Magnetic Field: Challenges and Prospects of the ISTP Era

Nikolai A. Tsyganenko

*Raytheon STX Corp., NASA GSFC, Greenbelt, Maryland*

The global magnetic field dominates many of the physical processes of geospace. It is critically important to predict and quantitatively model that field, given information on the state of the incoming solar wind. The ISTP project is an abundant source of data on the distribution and dynamics of the magnetic field and on that field's response to the solar wind variations. Together with earlier sets of spacecraft magnetometer data, the new ISTP data will provide an unprecedented pool of experimental information on the magnetic environment of our planet. Combining this wealth of data with the newest mathematical methods for representing the geomagnetic field will significantly advance our knowledge on the structure and dynamics of the Earth's magnetic environment.

## 1. INTRODUCTION

The geomagnetic field is one of the most important features of the near-Earth space environment. It links the interplanetary medium (and, hence, the solar atmosphere) with the Earth's ionosphere, guides energetic particles produced by solar activity, channels low-frequency electromagnetic waves and heat flux, confines the radiation belt and auroral plasma, and serves as a giant storehouse of solar wind energy, released during magnetic storms. All these aspects are closely related to the problem of forecasting conditions in the Earth's plasma environment, which is why magnetic field models play a central role in the recent US interagency "space weather" initiative.

A large database of spacecraft magnetometer measurements made by many missions of the last three decades has given us a wealth of information on the structure of Earth's distant magnetosphere and on its response to varying solar wind conditions. This information, however, is often hidden behind chaotic fluctuations, due to the complex nature of the solar-wind-magnetosphere interaction. To reveal the coherent structure underlying the data, one needs flexible mathematical representations for the principal magnetospheric field sources, parametrized by the characteristics of the solar wind and by available geophysical indices. These mathematical "modules", combined into a global model, are then fitted to the entire body of magnetospheric and solar wind data, making it possible to extract information on the global configuration of the geomagnetic field and its response to varying external conditions.

In recent years, significant progress was made in the data-based modeling of the magnetosphere. In contrast to earlier magnetic field representations [*Mead and Fairfield*, 1975; *Tsyganenko*, 1987, 1989], the new-generation models [*Tsyganenko*, 1995, 1996] have a realistic explicitly defined magnetopause, whose size is parametrized by the solar wind pressure, and an IMF-controlled interconnection field across the boundary. Another important new feature is the field from the Region 1 and 2 Birkeland currents [*Tsyganenko and Stern*, 1996]. The amplitude of the field-aligned currents was found from data and parametrized by solar wind characteristics [*Tsyganenko*, 1996].

The purpose of this paper is to give a brief review of the data-based approach, highlight the latest developments, and discuss the new prospects emerging from the availability of the freshly obtained data of the ISTP spacecraft.

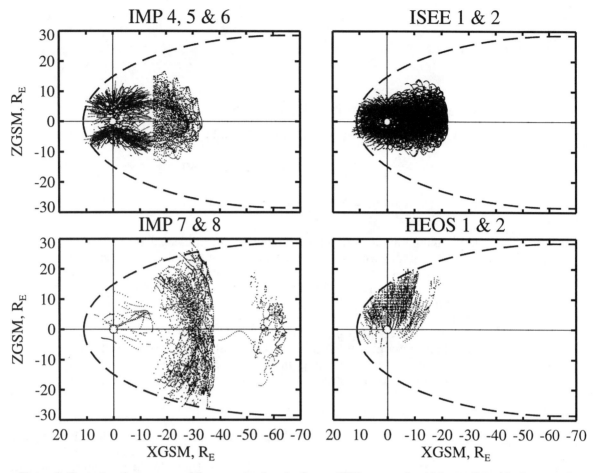

**Figure 1.** Illustrating the coverage of the magnetosphere by the pre-ISTP spacecraft, which contributed in the currently used data set by *Fairfield et al.* [1994]. The data points are displayed in projection on the noon-midnight meridian plane, and the average position of the magnetopause is shown by broken line.

## 2. DATA SETS FOR THE MODELING

Over the past three decades, an immense quantity of space magnetometer data has accumulated, collected by many spacecraft at different locations, seasons, and during different solar wind conditions and disturbancy levels. *Mead and Fairfield* [1975] compiled the first set of magnetic field data using four IMP spacecraft during 1966-72, and from that dataset they created an empirical model of the distant field, binned by the Kp-index. *Tsyganenko and Usmanov* [1982] extended the set of Mead and Fairfield by adding HEOS-1 and -2 data and developed a more realistic model with explicitly defined ring current and tail current sheet. Subsequently, the IMP-HEOS set of Tsyganenko and Usmanov was further extended by Tsyganenko and Malkov [see *Peredo et al.,* 1993] by adding ISEE 1/2 data from 1977-81 to the database, while Fairfield independently added HEOS observations along with additional IMP-6 data to the original Mead-Fairfield data base. Additional editing of these data sets and their merging into one large database was done by *Fairfield et al.* [1994], yielding the latest version of the modeling data set, used in the derivation of the new-generation global field model [*Tsyganenko,* 1996].

Each orbital pass contributing to the original high-resolution data was plotted, and a selection of the intervals inside the magnetosphere was made for it. To reduce the very large quantity of data to a manageable size, the values of the field components and of the spacecraft position were averaged, so that consecutive values corresponded to significantly different locations. Finally, the values of the field components were tagged by simultaneous values of the solar wind plasma and magnetic field parameters, compiled by J.H. King at NSSDC [*King,* 1977], as well as by the simultaneous values of the AE/Dst/Kp indices. The total number of the data points in the final set was 79,745, though for a significant part of the data ($\approx 40\%$) simultaneous solar wind plasma and IMF information was not available. Figures 1a-d show noon-midnight projections of the data point locations for the principal contributors to the database of *Fairfield et al.* [1994]: (a) the IMP-4,-5,-6 spacecraft, (b) ISEE-1 and -2, (c) IMP-7,-8 and Explorer-33,-35, and (d) HEOS-1 and -2 data.

Due to inaccuracies of spacecraft attitude determination, data taken closer than $\sim 4R_E$ were not included in the sets. In general, the coverage is quite non-uniform: the points are relatively dense at middle latitudes in the intermediate range of distances ($5 \leq R \leq 20R_E$), but become much sparser at higher latitudes. At larger distances a spread-out cloud of IMP-7 and -8 data covers the range $25 \leq R \leq 40R_E$, separated by a wide gap from a handful of Explorer-35 data points around the Moon's orbit.

In spite of the incomplete and non-uniform data coverage, the existing set proved to be a valuable source of information on the distant geomagnetic field. It not only allowed average static field configuration to be obtained, but also made it possible to reveal the response of individual magnetospheric current systems to changes in the solar wind, specifically, to its plasma flow pressure and the IMF conditions, as discussed in more detail in the next sections. At first glance, the size of the modeling data set seems quite large; however, considering the enormous spatial extent of the modeling region, gaps in the coverage, and the fact that each global parameter adds another dimension, it becomes clear that more data are urgently needed. To provide the necessary information about all current systems, the data should adequately sample all regions - not only in the (X,Y,Z) space, but also in the added dimensions of the geodipole tilt angle, the geomagnetic activity level (e.g., Dst-index), solar wind presure, and the IMF components.

The new ISTP observations can fill numerous gaps in the existing coverage. Figure 2 shows the distribution of the GEOTAIL positions, projected on the GSM X-Y plane and separated by 1-hour intervals.

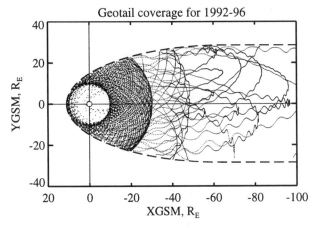

**Figure 2.** Coverage of the magnetosphere up to tailward distance of $100R_E$ by the Geotail observations. The data points are projected on the GSM equatorial plane.

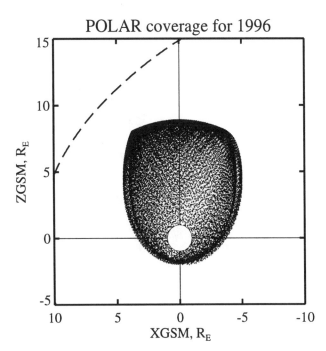

**Figure 3.** Coverage of the near-Earth magnetosphere by POLAR spacecraft in 1996.

Combined with simultaneous IMP-8 data on the solar wind parameters, the GEOTAIL magnetometer experiment will add to the existing database, as a rough estimate, about 6000 hours worth of data taken inside the magnetosphere.

Inclusion of WIND data (taken both in the solar wind and in the magnetosphere) will further extend the database. An additional advantage of the inclusion of the GEOTAIL data into the modeling data set will be a significant improvement of the equatorial coverage, owing to the relatively low inclination of the GEOTAIL orbit, in comparison with the family of IMP spacecraft.

Another important addition to the existing dataset will be data from POLAR. Due to its highly inclined orbit, that spacecraft will greatly improve the sampling of the high-latitude region, permeated by Birkeland currents. The urgent need for exploring the magnetic field in that region, relevant to the establishing of the large-scale structure of field-aligned currents and to their response to the solar wind parameters, was recognized long ago. The only systematic information on Birkeland currents, so far, has come from low-altitude satellites such as TRIAD and MAGSAT; however, it still remains largely unclear, where field-aligned currents (especially, those of Region 1) come from. Figure 3 displays a distribution of the POLAR positions for 1996, separated by 5-min intervals of time.

Another potential benefit of using the POLAR data can be the development of a more realistic model of the polar cusp region. It has been found in the past [e.g., *Fairfield*, 1991] that the diamagnetism of the polar cusp plasma reveals

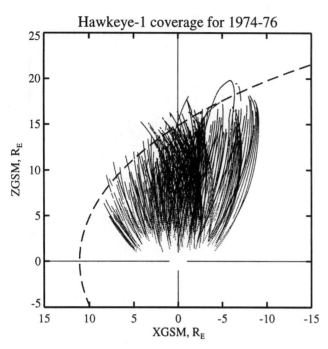

**Figure 4.** Coverage of the high-latitude magnetosphere by Hawkeye-1 spacecraft in 1974-76.

itself in the deviations of the model magnetic field from that measured inside the polar cusps.

The structure of the polar magnetosphere at higher altitudes was probed in the past by the high-apogee spacecraft HEOS-1, HEOS-2, and Hawkeye-1. The data from the first two were included in the existing database of *Fairfield et al.* [1994] (see Figure 1d above). Figure 4 shows the data cloud for the Hawkeye subset, which is being added to the database.

Another very important domain is the near-Earth equatorial magnetosphere, the region where substorm explosions are believed to be "ignited". New sets of the magnetic field data from that region have been compiled recently. The CRRES data (taken in 1990-91) span a relatively limited range of distances, extending to the spacecraft apogee at $R \approx 6.3 R_E$; nevertheless, the dataset is a valuable addition to the modeling data base, since it greatly improves the coverage of the near-equatorial region occupied by the radiation belt. The AMPTE/CCE data set is of unique importance due to the nearly equatorial orbit of the spacecraft, covering a wider range of distances than CRRES, up to the $R \approx 8.8 R_E$, and spanning a relatively long period of four years (1984-88). Figure 5 shows the distribution of CRRES and AMPTE/CCE data points. Finally, a huge amount of data on the dynamics of the inner magnetosphere is provided by the extensive set of geosynchronous measurements by GOES satellites, made during 1986-94. Although some of the data contain systematic errors due to magnetic fields generated aboard the spacecraft, most of them are of decent quality and are thus a rich additional resource for magnetospheric field modeling.

## 3. MODELING THE MAGNETOPAUSE AND DERIVATION OF THE SHIELDING FIELD

The essence of modeling lies in combining the data with flexible mathematical forms, providing a realistic representation of the main individual sources of the total observed field. In devising these forms, we make appropriate assumptions based on the physics (e.g., the magnetopause shielding fields are assumed as curl-free gradients of suitable scalar potentials) and employ all available independent observations (e.g., use a model of the magnetopause based on the direct crossings data).

Knowing the magnetopause position as a function of the upstream solar wind conditions plays a key role in modeling the field configuration inside the magnetosphere. The solar wind confines (shields) the fields produced by all internal sources, by adding to them Chapman-Ferraro field due the current flowing on the magnetopause. In a crude approximation, this results in a "closed" geometry of the total field with $B_n = 0$ on the boundary. Once the position of the magnetopause is known, it is in principle possible to calculate the shielding fields for all internal sources (dipole, ring currentm tail, etc.)

In actuality, $B_n$ is not always zero, which entails important consequences for the physics. The distribution of the normal component $B_n$ on the magnetopause defines the amount of the solar wind magnetic flux linking the geomagnetic field to the solar wind and, hence, the magnitude of the externally induced electric fields in the magnetosphere and ionosphere. That important feature can also be reproduced by the data-based models.

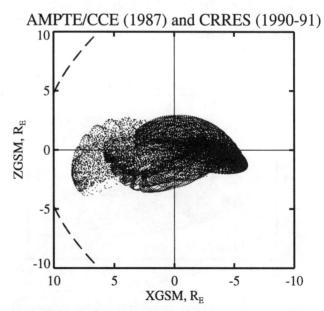

**Figure 5.** Coverage of the near-Earth equatorial magnetosphere by AMPTE/CCE and CRRES spacecraft.

In the deriving the earlier models [*Mead and Fairfield*, 1975; *Tsyganenko and Usmanov*, 1982; *Tsyganenko*, 1987, 1989] no direct information about the shape and position of the magnetopause was used. Neither did those models specify that shape and position explicitly. Instead, the magnetopause appeared as a "de-facto" surface that separated two families of field lines: those with no connection to Earth and those crossing Earth's surface at least once. In essence, this "de-facto" magnetopause represented an outward extrapolation of the model field beyond the region covered by measurements. On the dayside, its shape and size did not differ much from those observed. However, because of the poorer coverage of the high-latitude tail lobes, at larger distances the shape of the field lines became unstable and resulted in unrealistic features of the "de-facto" boundary.

This motivated us to look for a more rigorous mathematical treatment of the magnetopause and its shielding field. Significant progress in this area was made recently, based on two cornerstones, discussed below in more detail. The first one is the empirical modeling of the magnetopause shape and size using data of actual boundary crossings [*Sibeck et al.*, 1991; *Roelof and Sibeck*, 1993], and the second is an approximate method for the derivation of the shielding field for a wide class of general boundaries, using the least squares minimization of magnetic flux across the magnetopause [*Schulz and McNab*, 1987, 1996].

The first empirical models of the magnetopause shape, based on spacecraft crossing observations were developed by *Fairfield* [1971], *Howe and Binsack* [1972], and *Formisano et al.* [1979]. *Sibeck et al.* [1991] further developed that approach by extending the database of boundary crossings and *Roelof and Sibeck* [1993] introduced a bi-variate parameterization of the model boundary by both the solar wind ram pressure and IMF $B_z$.

In the latest version of the data-based magnetospheric field model [*Tsyganenko*, 1996] the magnetopause model of *Sibeck et al.* [1991] was adopted for the front of the boundary ($X_{GSM} > -50 R_E$) and was then smoothly extended tailward as a cylinder with the radius $R \approx 28 R_E$. No IMF dependence of the magnetopause shape and size was assumed in that model, while the effects of pressure variations were simulated by a self-similar compression/expansion. The scaling factor $\kappa$ was assumed as $\kappa = (p/p')^\alpha$, where the index $\alpha$ was a free model parameter. Its best-fit value $\alpha \approx 0.14$ was found to be quite close to $\alpha = 1/6$, given by simple theory [e.g., *Mead and Beard*, 1964].

The assumption of self-similar scaling of magnetopause dimensions by $\kappa(p)$ greatly simplifies the task of maintaining the shielding condition (or, more generally, keeping $B_n$ on the boundary under control) for different pressure values, since in that case re-calculation of the shielding field can be reduced to a simple scaling. However, as shown by *Roelof and Sibeck* [1993], self-similarity should only be considered as a rather crude approximation: in the limit of strong

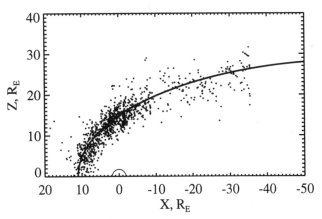

**Figure 6.** The shape of the model magnetopause for average solar wind conditions and the positions of the observed boundary crossings, rotated into the $X - Z$ plane.

northward and southward IMF $B_z$, the response of the magnetopause shape to variations of the pressure is substantially different from a self-similar change in size. The IMF-related change of the boundary shape does not seem to modify the internal field structure in any major way; however, it becomes important in the evaluation of the IMF interconnection with the geomagnetic field across the magnetopause, as will be discussed below in more detail.

Figure 6 shows the shape of the model magnetopause for average solar wind conditions, adopted in the data-based model by *Tsyganenko* [1996], and the cloud of boundary crossing points, compiled from the set of *Sibeck et al.* [1991] and further extended by adding crossings by Hawkeye-1 during 1974-75 [*Tsyganenko et al.*, 1996].

Once the analytically prescribed boundary is available, it is possible to derive a shielding field for any internal electric current system, by specifying a set of suitable curl-free fields with a sufficient degree of flexibility and combining those fields by least squares fitting, to make the resultant $B_n$ on the boundary as close as possible to its desired distribution. *Schulz and McNab* [1987] were first to implement that idea in developing their "source-surface" model of the magnetosphere. Subsequently, the method was extended for the derivation of shielding fields for all principal sources of the magnetospheric magnetic field in data-based models, including the ring current and the tail current sheet [*Tsyganenko*, 1995] and the Region 1/2 Birkeland currents [*Tsyganenko and Stern*, 1996]. With a slight modification, the same method can be used for the derivation of the IMF interconnection field inside the magnetosphere, based on a prescribed distribution of $B_n$ on the boundary [*Tsyganenko*, 1996].

As an example, Figure 7 displays a configuration of the magnetic field produced by a model ring current, confined inside the model magnetopause. In this case, the shielding field $\vec{B}_{sh} = -\nabla U$ was represented by a sum of 8 cylindrical harmonics and included a dependence on the geodipole tilt angle $\Psi$

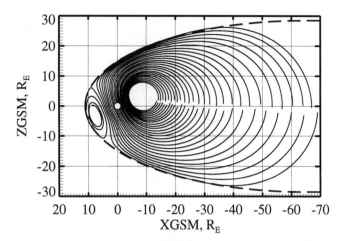

**Figure 7.** Lines of the magnetic field produced by the model ring current, fully confined within the model magnetopause by the shielding field (1). The north-south asymmetry is due to the tilt of the Earth's dipole, which affects the position of the ring current.

$$U = \cos\Psi \sum_{i=1}^{2} \sum_{m=1,3} a_{im} J_m(\rho/p_{im}) \exp(x/p_{im}) \sin m\phi$$
$$+ \sin\Psi \sum_{i=1}^{2} \sum_{m=0,2} b_{im} J_m(\rho/q_{im}) \exp(x/q_{im}) \cos m\phi \quad (1)$$

The coefficients $\{a_{im}, b_{im}\}$ and nonlinear parameters $\{p_{im}, q_{im}\}$ were found by an iterative algorithm, combining a standard linear least-squares fitting of $\{a_{im}, b_{im}\}$ with a simplex search for the optimal nonlinear parameters $\{p_{im}, q_{im}\}$ [*Press et al.*, 1992].

All other internal sources can be shielded in a similar way; the difference is in the number of terms necessary for satisfying the boundary conditions with sufficient accuracy. The tail current sheet and the Region 1 system of Birkeland currents require more terms in (1), because they extend to larger distances and create a strongly non-uniform field near the magnetopause.

The primary reason for explicitly including the shielding fields in the models is to control the degree of interconnection between the field of terrestrial origin and the IMF. Most of the recent progress in this area was made possible by the "modular approach", in which separate internal sources of the magnetospheric field were modeled by separate terms with their own amplitude coefficients, parameterized by appropriate combinations of solar wind characteristics [*Tsyganenko*, 1996]:

$$\mathbf{B} = \mathbf{B}_{Earth} + \mathbf{B}_{Tail} + \mathbf{B}_{RC} + \mathbf{B}_{Birk} + \mathbf{B}_{int} \quad (2)$$

The consecutive terms on the righthand side denote, respectively, the contributions from the Earth's main field, the tail current sheet, the ring current, Birkeland currents, and the interconnection term, representing the effect of non-zero $B_n$ and IMF penetration. It is implicitly assumed in (2) that each term on the right-hand side, except for the last one, includes its own "partial" magnetopause field, so that the net field from internal magnetospheric sources remains fully shielded inside the boundary, for any combination of amplitudes of the individual terms.

The last term in (2) represents the effect of the IMF penetration. Due to the full shielding of other terms, the corresponding normal component of the total field on the boundary is reduced to $B_n = (\mathbf{B}_{int} \cdot \mathbf{n})$, where the vector $\mathbf{B}_{int}$ was assumed to be proportional to the transverse component of the IMF $\vec{\mathbf{B}}_t = B_y \vec{e}_y + B_z \vec{e}_z$. The proportionality factor was determined from data.

Fitting several versions of the above model field to the data, tagged by the solar wind parameters, yielded persistently large values of the proportionality factor ($\sim 0.8$) [*Tsyganenko*, 1996]. It was also found that the penetration efficiency was almost independent of the direction of the transverse component of the IMF, that is, the penetration pattern just rotated around the Sun-Earth line, as if rigidly tied to that component.

More recent studies recognized, however, that a subtle effect must be taken into account here, or else a systematic overestimate of the penetration field may arise. So far, all derivations of $\mathbf{B}_{int}$ were based solely on magnetic field vectors observed inside the magnetosphere, many of them in the magnetotail. But the boundary of the tail, especially near Earth, spreads outwards – its diameter increases with distance from Earth. Lobe field lines, especially those far from the plasma sheet, tend to follow the boundary and therefore also diverge outwards, an effect which gives them a southward $B_z$, increasing when the flaring angle of the boundary grows larger.

On the other hand, one should expect that the magnetopause flaring angle also increases during times of southward IMF, due to an increased reconnection rate. If a larger southward $B_z$ is then observed inside the tail, it is hard to tell offhand what part of it is caused by the added penetration of the IMF and what part by greater flaring of the boundary.

This ambiguity cannot be resolved by models, in which the solar wind effects upon the magnetopause are reduced to the pressure-controlled expansion and compression, with no dependence of the boundary shape upon the IMF. On the other hand, introducing a magnetopause model with a bivariate dependence of shape and size, similar to that of *Roelof and Sibeck* [1993], meets with two problems. The first one is the above mentioned need to recalculate the shielding field parameters for any new combination of the solar wind parameters, since the corresponding changes in the boundary shape are no longer self-similar. The second problem is a sparsity of magnetopause crossing data tailward of $x \sim -10 R_E$ and the almost total absence of such data in the middle and far tail, which limits the reliability of magnetopause models in that region and, hence, makes it difficult to separate the effects of the tail flaring from those of the IMF penetration.

Observations of magnetopause crossings by the ISTP spacecraft (GEOTAIL, IMP-8, and WIND), made during recent years, will provide a valuable addition to the existing dataset and may help in obtaining more reliable estimates of the reconnection of the interplanetary magnetic flux. Combining that data with models of the magnetosheath plasma flow [e.g., *Spreiter and Stahara*, 1980] can allow us to derive the expected distribution of the solar-wind-induced electric field in the polar caps in a way similar to what was done by *Toffoletto and Hill* [1989, 1993]. The resulting electric potential pattern can then be compared to the one given by independent data-based models of the ionospheric convection [e.g., *Heppner and Maynard*, 1987].

## 4. MODELING THE RING CURRENT AND THE TAIL CURRENT SYSTEM

There exist several known methods to analytically represent the field produced by the ring current. *Tsyganenko and Usmanov* [1982] suggested a model, based on a simple mathematical modification of the vector potential of a dipole. It is probably the simplest possible model with only two parameters: the ring current amplitude and its characteristic radius. The principal deficiency of the model stems from its extreme simplicity: the electric current is excessively spread out in space, and there is no easy way to control its distribution. *Hilmer and Voigt* [1995] used a combination of two models of that kind with different scale sizes for modeling a more realistic ring current, including a zone of eastward $\vec{j}$ at the inner boundary of the radiation belt.

*Tsyganenko and Peredo* [1994] suggested a more flexible model, based on a superposition of analytical vector potentials for current disks with a controllable finite thickness. The radial distribution of the current density in that kind of model can be adjusted by an appropriate choice of the expansion coefficients and by scaling the field as a whole. That type of the ring current was used in the new global models, parameterized by the solar wind pressure, IMF, and Dst-index [*Tsyganenko*, 1995; 1996].

Here we describe another model, combining flexibility and realism with a very fast computational performance. The basic idea is to start from the vector potential for a circular current loop [e.g., *Smythe*, 1950]:

$$\vec{A} = \frac{(1 - k^2/2)K(k) - E(k)}{k\rho^{1/2}} \vec{e}_\phi, \quad (3)$$

where $E(k)$ and $K(k)$ are complete elliptical integrals of the 1st and 2nd kind, and modify the original form of the function $k(\rho, z)$ by adding a term $D^2$ in the denominator, so that

$$k^2 = \frac{4R\rho}{(R + \rho)^2 + z^2 + D^2} \quad (4)$$

This results in a spreading out of the originally infinitely thin current loop over all of space. However, most of the current remains localized in the vicinity of the loop, so that the current density peaks at $\rho = R$, $z = 0$, and rapidly falls off to zero with increasing distance from the loop. The variable $D$ specifies a characteristic thickness of the current, and can be made a function of position without violating the condition $\nabla \cdot \vec{B} = 0$. In particular, taking $D = D(x)$ makes it possible to model the day-night asymmetry of the ring current (or the dawn-dusk asymmetry, with $D = D(y)$). Combining two distributions with different characteristic radii $R$ allows a realistic ring current to be modeled, reversing to an eastward directed current in the inner magnetosphere, an effect of the pressure gradient. Figure 8 illustrates an example of such a model ring current, displaying radial profiles of the magnetic field disturbance (left panel) and the volume current density (right panel).

The tail current system has been recently modeled by using the above described approach with analytical spread-out current disks [*Tsyganenko and Peredo*, 1994; *Tsyganenko*, 1995]. To achieve a greater flexibility in modeling the extended current sheet on the nightside, the tail field was represented as a combination of several modes having different tailward variation scale distances and different weight amplitudes, with those parameters determined from fitting the model to data.

The analytical disk-like modes have an inconvenient feature: for large radial distances $r$, individual terms in the corresponding expansions for the tail field decrease as $r^{-2}$, while the actual far-tail lobe $B$ varies more slowly [e.g., *Slavin et al.*, 1985]. By combining several terms, it is in principle possible to obtain a slower decrease over a limited range of $X_{GSM}$ with a sufficiently steep inner edge of the current sheet. However, the number of terms necessary for maintaining a realistic profile becomes too large when one tries to extend the tail model to larger distances. For that reason, in the last release of the model [*Tsyganenko*, 1996] a disk-like mode was combined with an "asymptotic" mode, based on another model of the current sheet, suggested in our earlier work [*Tsyganenko*, 1987], in which the current flows along straight lines in the Y-direction, and its density in the limit $x \to -\infty$ tends to a finite value, rather than to zero. That term provided a smooth transition from the rapidly varying field in the near tail to the almost uniform $\mathbf{B}$ in the far tail, where the actual lobe field is nearly constant. Even with just a few hundred data points in the far tail, contributed by the Explorer-35 magnetometer, this approximation yielded quite realistic values of the best-fit quiet-time asymptotic magnetotail field $B_\infty$ and of the coefficient defining its response to changes in the solar wind pressure around its average value $P_0$:

$$B_\infty = 6.7 + 5.5\left(\sqrt{P/P_0} - 1\right) \quad (5)$$

This result suggests that adding new data of GEOTAIL to the

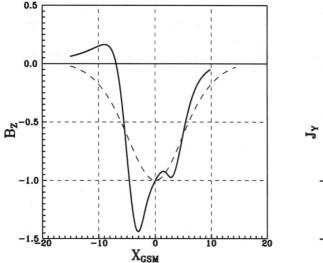 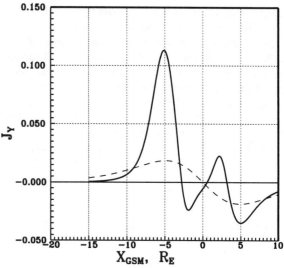

**Figure 8.** Variation along the Sun-Earth line of the $B_z$ component of the magnetic field (left) and the corresponding profile of the azimuthal electric current density (right) in the model ring current. Both quantities are plotted in arbitrary units.

already available data will extend the model to more distant regions.

## 5. INCLUSION OF BIRKELAND CURRENTS IN THE MODEL

One of the most recent breakthroughs in the data-based modeling of the magnetospheric field was the inclusion of the global field of the Region 1 and 2 field-aligned currents [*Tsyganenko and Stern,* 1996]. Long before that, the great role played by the Region 1 currents in the magnetic structure of the outer dayside magnetosphere was realized [e.g., *Maltsev and Lyatsky,* 1975; *Tsyganenko and Usmanov,* 1984]. The latest results, discussed in the next section, give more evidence for the strong effect of the global system of Region 1 Birkeland currents upon the position of the polar cusps.

A principal difficulty with Birkeland currents, not found in other current systems, is the lack of information on their actual global configuration. Little is as yet known on how far from Earth do the Birkeland currents remain field-aligned and on where they close. Existing observations [*Iijima and Potemra,* 1976; *Tsyganenko et al.,* 1994] and recent results of MHD simulations [*Tanaka,* 1995; *Janhunen,* 1996] suggest two important features: first, that Region 1 currents strongly depend on the IMF orientation both in their magnitude and geometry and, second, that most of them flow in the dayside and dawn-dusk magnetosphere, at least for southward IMF conditions.

The new model of the field-aligned current systems [*Tsyganenko and Stern,* 1996] is based on an Euler potential representation for the current density and on existing data on the geometry and distribution of Birkeland currents at iono-

spheric altitudes and at larger distances. The variation of the current strength along the ionospheric Region 1 and 2 ovals was specified to fit the statistical results of *Iijima and Potemra* [1976]. The parameters to be found from data were the coefficients relating the total current in both systems with the solar wind pressure and IMF. The principal results of the data-based modeling about field-aligned currents will be briefly discussed in the next section.

However, modeling the Birkeland currents is still in its infancy, and much remains to be done. The most important limitation of the existing model is the assumption of a fixed geometry of the current, regardless to IMF conditions. In future upgrades of the magnetospheric models, more flexibility will be allowed for the field-aligned current by splitting the term $\mathbf{B}_{Birk}$ into a sum of independent modes with variable IMF-dependent coefficients. As already mentioned, a substantial input is expected from the POLAR data, since they cover a relatively unexplored region and can provide a valuable information on field-aligned currents at intermediate distances.

## 6. SOLAR WIND EFFECTS IN GLOBAL MODEL FIELD CONFIGURATIONS

The optimal way of extracting from the data information on the solar wind control of individual current systems would be to specify an analytical dependence of the model field on all input variables (e.g., solar wind pressure, IMF components, and Dst) and then fit the model to the entire set of data. A simpler method, used in the earlier models, was to divide the data into several subsets, each covering a different interval of the model input parameter (e.g., Kp-index) and then to

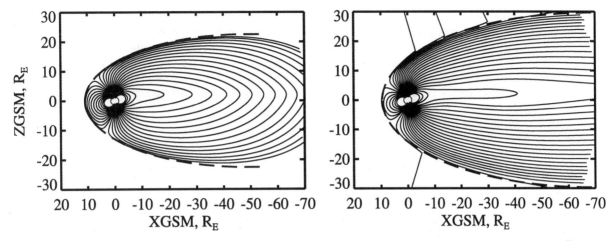

**Figure 9.** Configurations of the model magnetic field lines for opposite polarities of the IMF $B_z$-component: $B_z > +5nT$ (left) and $B_z < -5nT$ (right). In both cases, the same values were used for the solar wind pressure and the Dst-index. Note a striking difference in the amount of the magnetic flux both on the day and night sides.

separately fit the model to the subsets, generating separate sets of coefficients. However, with a larger number of input parameters, this approach becomes unfeasible due to rapidly growing number of bins in the multi-dimensional parameter space, leaving too few data points in each of the individual subsets.

In the last version of the global model [*Tsyganenko*, 1996], the input solar wind parameters and the Dst-index entered the field components as continuous variables and were therefore treated in the same way as the spatial coordinates $\{x, y, z\}$ and the geodipole tilt angle $\Psi$. That approach avoided the binning of the database into smaller sets and, hence, allowed more effective extraction of the information contained in the data. However, the price of this was sacrificing the ability of the model magnetopause to change its shape in response to changes of the IMF polarity. It was assumed that the only parameter controlling the size of the boundary was the solar wind pressure, which allowed fast recalculation of the shielding field, as already discussed in Section 3.

In this section, we discuss the results of the most recent global modeling study on the response of the magnetospheric field structure to changes in the solar wind state, for two opposite extremes of the polarity of the north-south component of the IMF [*Tsyganenko et al.*, 1996]. It was aimed at a more accurate estimate of the IMF interconnection effect, taking into account the IMF-induced changes of the magnetopause shape. For that reason, instead of a continuous parameterization by the IMF $B_y$ and $B_z$, we had to retreat to the old binning approach with regard to the IMF, having retained a continuous dependence on the wind pressure.

Two subsets were compiled from the entire modeling database, including the data taken during periods with strongly northward and strongly southward IMF. More specifically, the first subset contained the data points with IMF $B_z >$ $+5nT$, and the second one with $B_z < -5nT$. In both cases, an additional constraint $B_y < 5nT$ was imposed on the azimuthal IMF component, to make sure that effects of the north-south component were predominant. The numbers of data points in the two subsets were 1787 and 1722, respectively.

The next step was to construct pressure-driven magnetopause models for the two opposite IMF polarities. For simplicity, effects of the IMF and of the pressure $p$ upon the boundary position were reduced to two fundamental separate modes: (i) a self-similar compression/expansion, driven by $p$, and (ii) an IMF-driven change of the shape of the boundary, in which an earthward shift of the dayside magnetopause was accompanied by an increase of the tail radius. The individual response amplitudes of the two modes were represented as simple functions of $p$ and IMF $B_z$, evaluated by least squares fitting of the model boundary to the data set of the magnetopause crossings, in the same way as was done by *Roelof and Sibeck* [1993].

Using the two magnetospheric data subsets and magnetopause models, two sets of model coefficients were found by least squares, for IMF $B_z > +5nT$ and for IMF $B_z < -5nT$.

Figure 9 shows two configurations of the field lines in the noon-midnight meridian plane, for the same values of $p = 4nPa$ and Dst=0, but for opposite IMF polarities: $B_z > +5nT$ (left panel) and $B_z < -5nT$ (right panel).

As seen in the plots, there is a striking difference between the two configurations. First, changing the IMF from strongly northward to southward decreases the polar cusp footpoint latitude from $\sim 80°$ to $\sim 71°$. This shift is primarily due to the large increase of the total strength of the Region 1 Birkeland current, from nearly zero for northward IMF conditions to $\approx 2.4MA$ for southward IMF. Second, a huge difference can be seen in the amount of the magnetic flux closure across the tail plasma sheet. This is mostly due to a dramatic redistribution of the tail current: the reversal to strongly northward IMF $B_z$ is accompanied by a nearly complete disappearance of the far-tail current, while, at the same time, no significant change in the amplitude of the near-tail current was found. An additional contribution to this effect comes from the direct penetration of the IMF; however, the overall magnitude of the penetrating field was found to be much smaller than in the model with the IMF-independent magnetopause shape [*Tsyganenko, 1996*]. More specifically, for the strongly northward IMF, the penetrating field was virtually absent, while for strongly southward field, a penetration coefficient of only $\approx 0.17$ was obtained.

This finding differs significantly from the strong penetration ($\approx 0.8$) reported previously [*Tsyganenko, 1996*]. The difference is due to the inclusion of the effects of the IMF-dependent magnetopause, discussed in more detail in Section 3, and the strong correlation of the far-tail current with the IMF (no IMF-dependence of that current was assumed in the earlier model; see Equation 5 above).

The above results demonstrate that even with sparse data subsets and relatively crude time resolution, the data-based models can reproduce the expected global re-configuration of the magnetic field surprisingly well. At the same time, separating the subtle interconnection effects from the background of stronger variations is still a hard task with much uncertainty involved. Adding new data from the ISTP spacecraft and a re-creation of the older datasets with a higher time resolution will increase the reliability of the modeling.

## 7. MODELING OF THE NEAR-EARTH MAGNETIC FIELD

Many studies require an accurate model of the geophysical environment only within a limited region of the near-Earth space. In particular, the low-latitude region inside $R \approx 10 R_E$ is of special importance. From the viewpoint of physics, that is where the most dramatic events of particle injection and of magnetic field re-configuration take place during storms and substorms. In addition, studies of "space weather" effects on communication satellites need reliable, compact, and fast models of the magnetic field in the vicinity of geostationary orbit [e.g., *Rufenach et al., 1992*].

In some aspects, the local modeling of the near-Earth magnetic field is easier than its global representation. First, the data coverage of this region is relatively dense, accumulated over many years under different solar activity conditions, including a large database of GOES measurements. Second, due to the smaller spatial extent, local models can be made mathematically simpler and hence much faster for computations: for example, the magnetopause field can be represented by simple linear or quadratic expansions. Also, a simpler model for the field of Birkeland currents can be envisioned, because the behavior of the field components at large distances is no longer important. On the other hand, as one descends to smaller spatial scales and shorter time intervals, one can include here effects not covered by global models, for instance, large-amplitude effects of the substorm currents. Attempts were made in the past to model the substorm growth and recovery phases by a modification of the global tail current system [e.g., *Pulkkinen et al., 1994* and references therein]. However, a truly realistic and full description of the substorm disturbance field should include a feasible 3-D model of the current wedge on the nightside. First successful steps in this direction have been made recently [*Tsyganenko, 1997*].

On the dayside, even the existing global models, based on data with relatively crude time resolution, track the field observed at the geosynchronous orbit surprisingly accurately [*Lu et al., 1997*]. As an example, Figure 10 compares the components of the magnetic field measured there by GOES-5 with those computed using the global model [*Tsyganenko, 1996*]. The best agreement is observed for the $B_z$-component: in this example and in many other cases we found an almost perfect match between the model and the data in the pre-noon sector. Note that even short-scale excursions of $B_z$ are reproduced with remarkable accuracy. A closer inspection showed that those excursions were due mainly to variations in the IMF $B_z$, which controls the intensity of the model Birkeland currents and also contributes through the term for the directly penetrating IMF.

Another important feature, still awaiting its proper representation in models, is the dawn-dusk asymmetry of the storm-time equatorial field, conspicuous at geosynchronous distance [*Coleman and McPherron, 1976*]. ISTP data, first of all those of POLAR and GOES, will significantly improve the coverage of high- and low-latitude regions of the inner magnetospphere and will thus promote rapid progress in modeling the near-Earth magnetic environment.

## 8. SUMMARY

Global data-based modeling of the geomagnetosphere have seen significant progress during the last years. A large arsenal of mathematical methods has been developed recently for modeling all principal sources of the magnetic field. Realistic representations have been devised for the magnetopause, based on direct crossings data and dependent on solar wind parameters. Powerful methods were developed for the derivation of the magnetopause magnetic field, with an IMF-controlled interconnection of terrestrial and solar-wind magnetic fields across the boundary. Large sets of space magnetometer data for modeling have been compiled from the archived higher-resolution data.

A synthesis of the above elements has made it possible to create new-generation models, parameterized by solar wind characteristics and the Dst-index. In spite of a still incom-

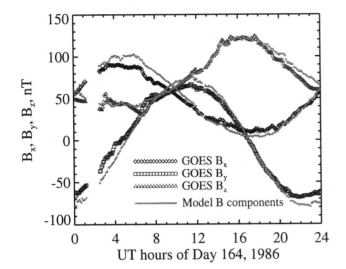

**Figure 10.** An example of comparison of the diurnal variation of the magnetic field components measured by geosynchronous spacecraft and those obtained from the data-based model.

plete coverage by data, these models reproduce the observed response of the magnetosphere to solar wind input surprisingly well. The newly obtained ISTP data will provide an abundant influx of fresh experimental information. Combining that data with the existing datasets and application of the newest methods will boost our progress in understanding the physical links between the state of the interplanetary medium and the near-Earth magnetic environment.

*Acknowledgments.* The author is grateful to David Stern for his numerous helpful comments on the manuscript. This work is supported by NASA grants NAS5-32350 and NSF Magnetospheric Physics Program grant ATM-9501463.

## REFERENCES

Coleman, P.J. and R.L. McPherron, Substorm observations of magnetic perturbations and ULF waves at synchronous orbit by ATS-1 and ATS-6, in: *The Scientific Satellite Programme During the International Magnetospheric Study*, ed. K. Knott and B. Battrick, p.345, D.Reidel, Dordrecht, 1976.

Fairfield, D.H., An evaluation of the Tsyganenko magnetic field model, *J. Geophys. Res., 96*, 1481, 1991.

Fairfield, D.H., N.A. Tsyganenko, A.V. Usmanov, and M.V. Malkov, A large magnetosphere magnetic field database, *J. Geophys. Res., 99*, 11319, 1994.

Heppner, J.P. and N.C. Maynard, Empirical high-latitude electric field models, *J. Geophys. Res., 92*, 4467, 1987.

Hilmer, R.V. and G.-H. Voigt, A magnetospheric magnetic field model with flexible current systems driven by independent physical parameters, *J. Geophys. Res., 100*, 5613, 1995.

Iijima, T. and T.A. Potemra, The amplitude distribution of field-aligned currents at northern high latitudes observed by Triad, *J. Geophys. Res., 81*, 2165, 1976.

Janhunen, P., H.E.J. Koskinen, and T.I. Pulkkinen, A new global ionosphere-magnetosphere coupling simulation utilizing locally varying time step, in *Proceedings of the ICS-3 Conference on Substorms* (Versailles, France, May 12-17, 1996), ESA SP-389, pp.205-210, 1996.

King, J.H., *Interplanetary medium data book*, NSSDC/WDC-A-R&S 77-04, Natl.Space Data Cent., Greenbelt, MD., 1977.

Lu, G., G.L. Siscoe, A.D. Richmond, T.I. Pulkkinen, N.A. Tsyganenko, H.J. Singer, and B.A. Emery, Mapping of the ionospheric field-aligned currents to the equatorial magnetosphere, *J. Geophys. Res., 102*, 14467, 1997.

Maltsev, Yu.P. and W.B. Lyatsky, Field aligned currents and erosion of the dayside magnetosphere, *Planet.Space Sci., 23*, 1257, 1975.

Mead, G.D. and D.B. Beard, Shape of the geomagnetic field solar wind boundary, *J. Geophys. Res., 69*, 1169, 1964.

Mead, G.D. and D.H. Fairfield, A quantitative magnetospheric model derived from spacecraft magnetometer data, *J. Geophys. Res., 80*, 523, 1975.

Peredo, M., D.P. Stern, and N.A. Tsyganenko, Are existing magnetospheric models excessively stretched ?, *J. Geophys. Res., 98*, 15343, 1993.

Pulkkinen, T.I., D.N. Baker, P.K. Toivanen, R.J. Pellinen, R.H.W. Friedel, and A. Korth, Magnetospheric field and current distributions during the substorm recovery phase, *J. Geophys. Res., 99*, 10955, 1994.

Roelof, E.C., and D.G. Sibeck, Magnetopause shape as a bivariate function of the interplanetary magnetic field $B_z$ and solar wind dynamic pressure, *J. Geophys. Res., 98*, 21421, 1993.

Rufenach, C.L., R.L. McPherron, and J. Schaper, The quiet geomagnetic field at geosynchronous orbit and its dependence on solar wind dynamic pressure, *J. Geophys. Res., 97*, 25, 1992.

Schulz, M., and M. McNab, Source-surface model of the magnetosphere, *Geophys. Res. Lett., 14*, 182, 1987.

Schulz, M., and M. McNab, Source-surface modeling of planetary magnetospheres, *J. Geophys. Res., 101*, 5095, 1996.

Sibeck, D.G., R.E. Lopez, and E.C. Roelof, Solar wind control of the magnetopause shape, location, and motion, *J. Geophys. Res., 96*, 5489, 1991.

Slavin, J.A., E.J. Smith, D.G. Sibeck, D.N. Baker, R.D. Zwickl, and S.-I. Akasofu, An ISEE 3 study of average and substorm conditions in the distant magnetotail, *J. Geophys. Res., 90*, 10875, 1991.

Smythe, W.R., *Static and dynamic electricity*, New York: McGraw-Hill, 1950.

Spreiter, J.R. and S.S. Stahara, A new predictive model for determining solar wind–terrestrial planet interactions, *J. Geophys. Res., 85*, 6769, 1980.

Tanaka, T., Generation mechanisms for magnetosphere-ionosphere current systems deduced from a three-dimensional MHD simulation of the solar wind-magnetosphere-ionosphere coupling processes, *J. Geophys. Res., 100*, 12057, 1995.

Toffoletto, F.R. and T.W. Hill, Mapping of the solar wind electric field to the Earth's polar caps, *J. Geophys. Res., 94*, 329, 1989.

Toffoletto, F.R. and T.W. Hill, A nonsingular model of the open magnetosphere, *J. Geophys. Res., 98*, 1339, 1993.

Tsyganenko, N.A. and A.V. Usmanov, Determination of the magnetospheric current system parameters and development of experimental geomagnetic field models based on data from IMP and HEOS satellites, *Planet. Space Sci., 30*, 985, 1982.

Tsyganenko, N.A. and A.V. Usmanov, Effects of field-aligned currents in structure and location of polar cusps, *Planet. Space Sci., 32*, 97, 1984.

Tsyganenko, N.A., A magnetospheric magnetic field model with the warped tail current sheet, *Planet. Space Sci., 37,* 5, 1989.

Tsyganenko, N.A., D.P. Stern, and Z. Kaymaz, Birkeland currents in the plasma sheet, *J. Geophys. Res., 98,* 19455, 1993.

Tsyganenko, N.A., and M. Peredo, Analytical models of the magnetic field of disk-shaped current sheets, *J. Geophys. Res., 99,* 199, 1994.

Tsyganenko, N.A. Modeling the Earth's magnetospheric magnetic field confined within a realistic magnetopause, *J. Geophys. Res., 100,* 5599, 1995.

Tsyganenko, N.A. and D.P. Stern, Modeling the global magnetic field of the large-scale Birkeland current systems, *J. Geophys. Res., 101,* 27187, 1996.

Tsyganenko, N.A., Effects of the solar wind conditions on the global magnetospheric configuration as deduced from data-based field models, in *Proceedings of the ICS-3 Conference on substorms* (Versailles, France, May 12-17, 1996), ESA SP-389, pp.181-185, 1996.

Tsyganenko, N.A., M. Peredo, S. Boardsen, and T.E. Eastman, Interconnection of IMF with the geomagnetic field: What can data based models tell us ?, *Eos Trans. AGU, 77,*(46), Fall Meet.Suppl., F638, 1996.

Tsyganenko, N.A., An empirical model of the substorm current wedge, *J. Geophys. Res., 102,* 19935, 1997.

---

N. A. Tsyganenko, Raytheon STX Corporation, Laboratory for Extraterrestrial Physics, Code 690.2, NASA Goddard Space Flight Center, Greenbelt, MD 20771. (e-mail: kolya@nssdca.gsfc.nasa.gov)

# Improvements to the Source Surface Model of the Magnetosphere

Vahé Peroomian

*Institute of Geophysics and Planetary Physics, UCLA, Los Angeles, CA 90095-1567*

Larry R. Lyons

*Department of Atmospheric Sciences, UCLA, Los Angeles, CA 90095*

Michael Schulz

*Space Sciences Department, Lockheed Martin Palo Alto Research Laboratories, Palo Alto, California*

In this paper we outline a technique developed for including large-scale current systems in magnetospheric models and apply it to the inclusion of Birkeland currents in the Source Surface Model (SSM) of the terrestrial magnetosphere. This model is ideal for studying the effects of Birkeland currents since it has a well-defined boundary between open and closed field lines, magnetopause shielding currents separate the geomagnetic and interplanetary magnetic fields in the absence of a specified connection between the two, and Birkeland currents are not included in the model either implicitly or explicitly. The Birkeland currents are modeled using a series of field-aligned, infinitely thin wire segments. The magnetic field produced by these currents is calculated from the Biot-Savart law and its normal component on the surface of the magnetopause is shielded (minimized) by image currents carried on wires placed outside the magnetosphere. This process is iterated until a convergent solution is obtained. We find that inclusion of Birkeland currents in the SSM results in the closure of previously open flux near midnight and in a sunward shift of the separatrix. Moreover, the sunward shift is found to increase with increasing Birkeland current strength by an amount that agrees with observations of the auroral oval.

In the second part of this paper, we describe modifications of the SSM undertaken to achieve a more realistic testbed for our Birkeland current studies. These modifications include an enlargement of the magnetotail radius, the modification of the tail current sheet in order to obtain a thinner current sheet in the model, and the reduction of the asymptotic lobe **B** field strength.

## 1. INTRODUCTION

Birkeland currents play an important role in magnetosphere-ionosphere coupling and in auroral and substorm dynamics and are thought to be responsible for momentum and energy transfer from the outer magnetosphere to the ionosphere and atmosphere. Birkeland currents are especially prominent during geomagnetic disturbances. Magnetic disturbances from field-aligned (Birkeland) currents and their variation as a function of geomagnetic activity have been the subject of numerous studies [*Birkeland*, 1908; *Sugiura and Chapman*, 1960; *Boström*, 1964; *Atkinson*, 1967; *Iijima and Potemra*, 1976a, 1976b; *Kamide*, 1982 and references therein]. The Birkeland current system consists of a Region I current (which flows into the high-latitude ionosphere on the AM side and out of the ionosphere on the PM side) and a slightly weaker Region II current, which flows at slightly lower latitudes ( into the ionosphere on the PM side and out of the ionosphere on the AM side) [*Iijima and Potemra*, 1976a, 1976b]. By now it is generally accepted that the Region II system closes via a partial ring current in the outer part of the near-equatorial trapping region [*Schield et al.*, 1969; *Vasyliunas*, 1970, 1972; *Schindler and Birn*, 1978].

Although much is now known about the distribution of Birkeland currents in the ionosphere, the global flow pattern and closure of Region I currents beyond the ionosphere remains largely unknown. The Region I currents have been postulated to flow along the separatrix between open and closed field lines and somehow connect in the tail region, either by diverting the cross-tail current or by connecting to the interplanetary magnetic field lines through the magnetopause [see discussion by *Stern*, 1994].

While magnetotail currents are an integral part of all contemporary magnetic field models, the only existing model that includes the contribution of Region I and Region II Birkeland currents is that of *Tsyganenko* [1995] and *Tsyganenko and Stern* [1996]. However, because *Tsyganenko*'s model is empirical in nature and the magnetic field is obtained via a least squares technique, it cannot be used to study the effects of Birkeland currents on magnetospheric configuration and magnetic field topology. *Donovan* [1993] used a method similar to that of *Kaufmann and Larson* [1989] to incorporate field-aligned currents into the *Tsyganenko* [1987] magnetic field model. Although *Donovan* [1993] iterated the location of current-carrying wires to a converging solution, he made no attempt to shield the excess field from the Birkeland currents within the magnetosphere and obtained a $\delta B_z$ profile that was independent of $x$ in the tail region.

This paper discusses three main elements of our research. First, we address the development of our quantitative model of the Region I and Region II Birkeland current systems, which is consistent with the above premises and shielded by additional magnetopause currents so that their effects are confined to the magnetosphere itself. Secondly, we consider inferences made regarding the consequences of the modeled Birkeland current systems on the configuration of the magnetosphere, with special emphasis on the auroral oval. Thirdly, we discuss the modifications of the source surface model used in this study to obtain a better and more realistic testbed for our Birkeland current studies. In Section 2, we present the source surface model used as a baseline field model in our studies. In Section 3, we outline the method used to include Region I currents, Region II currents, and their shielding currents in the model, discuss the procedure for iterating this process toward a convergent solution and show details of results obtained with the new model. In Section 4, we discuss several modifications to the original SSM. Finally, in Section 5 we summarize our findings and outline our plans for the future.

## 2. BASELINE FIELD MODEL

We have chosen the source surface model (SSM) of *Schulz and McNab* [1987, 1996] as the template for our Birkeland current studies. The SSM is a very versatile and simple model that is ideal for developing and testing systematic techniques for including large-scale magnetospheric currents because it is easy to generate from analytical expressions, has a well-defined boundary between open and closed field lines, and has a clean separation between magnetospheric and external fields. Moreover, because it is analytical in nature, the SSM does not include the **B** field produced by Birkeland currents in any way (even implicitly). Therefore, we can use it to study (without bias) the effects of adding these current systems individually.

The SSM is a prescribed magnetopause model. The magnetopause in the original SSM is a figure of revolution specified by

$$(\rho/\rho_\infty^*)^{2.280} = \tanh[(1.0854\rho_\infty^*)(x + b)] \quad (1)$$

where $\rho$ is the distance from the Earth-Sun line, $\rho_\infty^*$ is the asymptotic radial distance to the magnetopause, $x$ is the distance downstream from Earth, and $b$ is the value of $-x$ at the subsolar point on the magnetopause, chosen so that $b = 10\ R_E$ in the present model (Figure 1). For the original model, $\rho_\infty^*$ was chosen to be $1.7696b$. The model is implemented by minimizing the mean square normal component of the magnetic field **B** on the surface of the magnetopause and the mean square tangen-

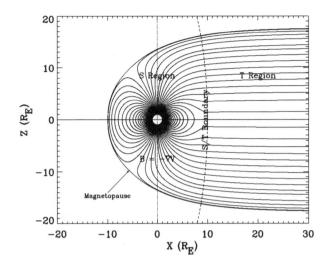

**Figure 1.** Field lines in the noon-midnight meridional plane calculated for the Source Surface Model. Field lines are calculated at $\lambda = 2°$ increments. The dashed curve ($x \sim 10\ R_E$) represents the location of the cross-tail normal surface, separating the near-Earth ("S") region from the tail ("T") region.

tial component of **B** along a cross-magnetospheric surface normal to the field lines in the magnetotail, given by

$$(1.0854\rho/\rho_\infty^*)^2 = 2.28\{\sinh^2[(1.0854/\rho_\infty^*)(x_0 + b)] - \sinh^2[(1.0854/\rho_\infty^*)(x + b)]\} \quad (2)$$

where $x_0$ is the value of $x$ at $\rho = 0$ ($x_0 = 10\ R_E$ in this case, dashed curve in Figure 1). The original SSM was current free in the near-Earth or S region, with **B** there derived from a scalar potential $V(r,\theta,\varphi)$ represented by a finite spherical-harmonic expansion of the form

$$V(r,\theta,\varphi) = (a^3/r^2)g_1^0 \cos\theta + (a^3/b^2)\sum_{n=1}^{N}\sum_{m=0}^{n}(r/b)^n \overline{g}_n^m P_n^m(\theta)\cos m\varphi \quad (3)$$

where $a$ is the Earth's radius and $P_n^m(\theta)$ are the associated Legendre functions with *Schmidt* [1935] normalization. The expansion coefficients $\overline{g}_n^m$ were determined by requiring a weighted linear combination of the mean-square normal component of **B** on the magnetopause and the mean-square tangential component of **B** on the cross-magnetospheric tail-entrance surface to be minimized. Field lines in the tail region were constructed geometrically so they would not intersect each other or the magnetopause. The resultant magnetic field in the tail region produced a neutral sheet in the $z = 0$ plane.

The original SSM did not contain the contributions to the magnetic field from any internal current systems, either implicitly or explicitly, other than from the cross-tail current. The original model was also deficient in several areas, most notably in its unrealistic magnetotail configuration.

## 3. INCORPORATION OF BIRKELAND CURRENTS IN THE SSM

### 3.1. Region I and Region II Currents

Region I field-aligned currents are believed to flow in the vicinity of the separatrix between open and closed field lines and somehow connect to the cross-tail current. Early models of the Region II currents [e.g., *Kirkpatrick*, 1952] placed these currents along dipolar field lines mapping out to several Earth radii and closed the circuit at the equator. Upon discovery of the ring current, this closure path for the Region II Birkeland currents eventually became identified as the "partial ring current" [*Schield et al.*, 1969; *Vasyliunas*, 1970, 1972; *Schindler and Birn*, 1978]. The problem of modeling these current systems is not a simple one. The surface on which Birkeland currents flow is rather complicated, and previous studies have used average configurations to approximate the global extent of these current systems [see e.g. *Stern*, 1993, 1995; *Tsyganenko*, 1991]. Our technique for including the Birkeland currents in our model is as follows: first, we follow the methodology outlined by *Olson and Pfitzer* [1974] and represent each wire by using a large number of finite-length elements. We place the footpoints of Region I current-carrying wires at each hour of local time along the separatrix between open and closed field lines (Figure 2). Region II wires are placed 3.5° equatorward of the Region I wires. The magnitude of $I_\parallel$ on each Region I wire is obtained by integrating $J_\parallel = 0.15\sin\varphi$ Amps/m over one hour of arc length along the separatrix in the ionosphere. Since Region II currents are about 80% as strong as Region I currents [*Iijima and Potemra*, 1976a, 1976b], we modeled the Region II $J_\parallel$ as $J_\parallel = -0.12\sin\varphi$ A/m. This gives a total Region I current of 0.42 MA per hemisphere and a total Region II current of 0.34 MA (oppositely directed) per hemisphere. Region I wires are then traced out along SSM field lines into the magnetosphere. Because the SSM does not contain the contribution of internal current systems (except for the cross-tail current), for zero dipole tilt the boundary between open and closed

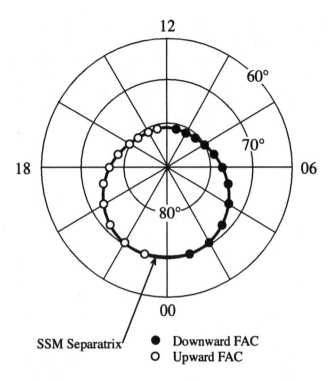

**Figure 2.** Projection of the SSM separatrix into the ionosphere above $\lambda = 60°$. Circles arrayed along separatrix represent footpoints of Region I current-carrying wires. Filled circles represent wires with downward current (current into ionosphere), and hollow circles represent wires with upward current (current out of ionosphere).

field lines maps to the $z = 0$ plane in the tail region (see Figure 1). Therefore, Region I current-carrying wires placed on the separatrix flow in the $z = 0$ plane in the tail region. *Kamide and Fukushima* [1972] have suggested that the closure of Region I currents occurs via a diversion of the cross-tail current. We model the closure of Region I currents with a dusk-to-dawn current-carrying wire connecting symmetric pairs of wires across midnight. We map the Region II currents to the equator (locus of minima in |**B**| along field lines) and close the circuit with wires placed along contours of constant |**B**| on this minimum-B surface. The minimum-B surface coincides with the equatorial plane on the nightside but not near the dayside magnetopause, where it bifurcates and deforms [*Shabansky*, 1971] to higher |z| values (see Figure 3).

### 3.2. Magnetopause Shielding

The original SSM confined the geomagnetic field inside the magnetosphere. The addition of fields produced by our Region I and Region II current systems violates this confinement by leaving a significant $\hat{\mathbf{n}} \cdot \mathbf{B}$ ($\approx \hat{\mathbf{n}} \cdot \delta\mathbf{B}$) on the surface of the magnetopause. In this study, we place "image" current-carrying wires outside the magnetopause at locations given by $\rho_b = \rho_\infty^2 / \rho_a$ (where $\rho_b$ is the radial distance to the image wire, $\rho_a$ is the location of the Region I wire in the tail, and $\rho_\infty$ is the magnetotail radius). Images placed at this location provide excellent shielding in the tail region. However, the shielding in the near-Earth region isn't perfect because of the deviation of the magnetopause from a cylindrical shape, the curvature of the Region I current-carrying wires, and the presence of Region II wires. We therefore have manually adjusted the image currents in order to compensate for these effects and to maximize shielding.

### 3.3. Iteration Toward Self-Consistent Topology

Having chosen to place our Region I currents on the separatrix between closed and open field lines (and our Region II currents 3.5° of invariant latitude equatorward from there), we should insist that these specifications (of currents parallel to **B**) apply in the resulting **B**-field configuration. Otherwise the resulting model would not be self-consistent.

To assure this consistency, we must iterate between the specification of Birkeland currents (plus closure and image currents) and the computation of $\Delta\mathbf{B}$. Figure 4

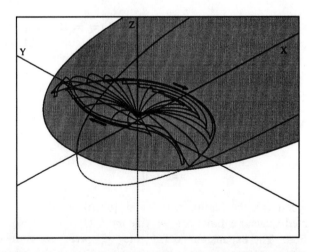

**Figure 3.** Placement of Region II current-carrying wires in the model. The Region II currents are closed via wires placed along contours of minimum B along field lines. These contours occur in the equatorial plane on the nightside and at higher |z| on the dayside. Although not shown, wires are placed in both hemispheres.

shows the separatrix between open and closed field lines at the Earth's surface, and illustrates the result of our iterative procedure for arriving at a final configuration. The dotted curve in this figure corresponds to the separatrix in the original SSM. The addition of Region I and II Birkeland currents to the SSM moves this boundary poleward on the night side by about 3° and equatorward on the day side by about 0.5° (to the dashed curve in Figure 4). The resulting polar cap is slightly smaller than in the original SSM because the Birkeland currents have added a positive $\delta B_z$ in the central portion of the tail. The positive $\delta B_z$ also means that Birkeland current-carrying wires no longer coincide with the $z = 0$ plane in the tail for the next iteration. Further iterations mainly change the locations of the wires between 03 and 06 MLT in the ionosphere. Our third and final iteration (thick solid curve) shows an intermediate position there.

Figure 5 illustrates the final magnetic field configuration in the new model. The field lines shown emanate

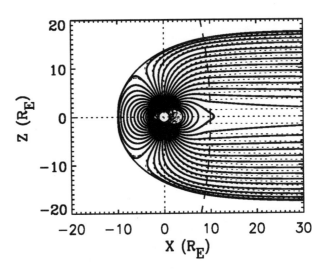

**Figure 5.** Field lines calculated at $\lambda = 2°$ intervals in the noon-midnight meridional plane for the third and final iteration of the model. Dotted curves represent field lines for the original SSM. The cross-tail normal surface is also shown (dashed line).

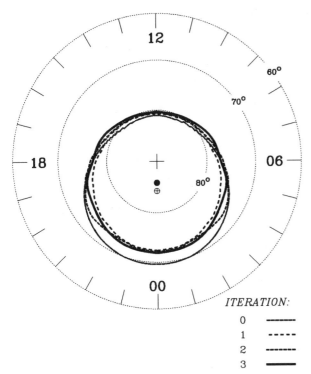

**Figure 4.** Location of the separatrix for the original SSM (dotted curve) and subsequent iterations projected onto the auroral zone. (dashed curve = first iteration, chain-dashed curve = second iteration, solid curve = final iteration). Circles of constant latitude above 60° are shown. The thick curve in this plot represents the final configuration of the model. The ⊕ symbol represents the center of the auroral oval for the SSM, and the • represents the center of the auroral oval for the final configuration.

from the Earth at 2° intervals of magnetic latitude in the noon-midnight meridian. The solid curves in this figure correspond to field lines of the third and final iteration. The dotted curves represent field lines in the original SSM. As noted above, the principal difference between the two sets of field lines occurs near the equatorial plane and near the dayside cusps. At the same time, field lines near the magnetopause converge toward the SSM field lines. This is because we have successfully shielded over 99% of the excess $\delta B$ from the Birkeland currents at the magnetopause.

One advantage of our model is its ability to isolate and study the effects of Birkeland currents explicitly. In the previous pages, we have seen the large-scale changes in magnetic field topology that occur when these currents are incorporated into the SSM. The strength of the Birkeland currents used in this study (0.42 MA for Region I) is on the lower end of observed stormtime values. This is because the strength of $\delta B_z$ in the equatorial plane is directly proportional to the strength of the Birkeland currents, and we found that the use of a stronger Birkeland current led to the closure of an unreasonably large portion of the magnetotail in our first iteration. In order to understand the effect of increasing the Birkeland current in our model (e.g., for application to substorms), we plot in Figure 6 the ionospheric mapping of the separatrix between closed and open field lines for six values of the Birkeland current. The solid curve in

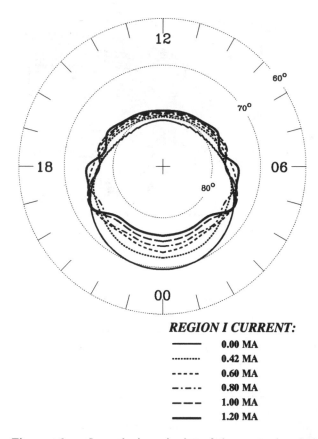

**Figure 6.** Ionospheric projection of the separatrix, similar in format to Fig. 4. The solid curve in this plot represents the separatrix of the final configuration of our model (0.42 MA Region I current). Successive curves (equatorward to poleward at midnight) represent higher magnitudes of Birkeland currents from 0.6 MA to 1.2 MA Region I current.

this plot represents the separatrix for the original SSM. The dashed curve just poleward represents the separatrix for the magnitude of the total current used throughout the paper (0.42 MA Region I current and -0.34 MA Region II current). Successive curves represent stronger currents up to the 1.2 MA Region I current (thick solid curve). These additional curves were obtained by simply increasing the magnitude of Birkeland current from the last iteration of our 0.42 MA Region I current model. No further iterations were carried out, and thus the results shown in Figure 6 are in this sense not self-consistent.

One notable feature of the separatrices shown in Figure 6 is the presence of corrugations for higher values of Region I current. These corrugations, the wavelength of which decreases with increased Birkeland current, suggest the occurrence of the Kruskal-Shafranov instability [*Kruskal and Schwarzschild*, 1954; *Shafranov*, 1956; *For-*

*slund*, 1970; *Hasegawa*, 1970] which leads to the 'buckling' of Region I current sheets. We saw in Figure 5 that the effect of adding the Birkeland currents was to shrink the region of open field lines from the night side and to shift the centroid of the auroral oval toward the day side. Whereas the offset between the magnetic pole and the center of the oval in the original SSM was about 6°, the addition of a 0.42 MA Region I and -0.34 MA Region II Birkeland current decreased this offset to about 4°. Here we see that an increased Birkeland current strength results in an increased shift of the polar cap toward the dayside. This result is in agreement with predictions by *Lyons et al.* [1994] and *Schulz* [1991] concerning the effect of Birkeland currents on the separatrix. In order to quantify this effect in the present context, we plotted in Figure 7 the offset between the magnetic pole and the center of the auroral oval calculated from the six curves shown in Figure 6 as a function of total Birkeland current. The "center" of the auroral oval was found by taking half the difference between the separatrix latitudes at noon and midnight. We see from Figure 7 that the offset is a linear function of the Birkeland current strength (correlation coefficient $R^2 \sim 1$ for the best fit curve), and ranges from 5.7° for very quiet conditions (0.0 MA Birkeland current) to about 1° for the most active conditions contemplated (1.2 MA Region I current). *Meng et al.* [1977] studied 50 DMSP photographs of the quiet auroral oval to obtain nominal auroral oval locations. They found that the offset between the magnetic pole and the center of the quiet-time auroral oval fell within a circular area of 3° radius centered at colatitude $\theta \sim 4.2°$ on the night side. *Schulz and McNab* [1996] found very good agreement between the original SSM

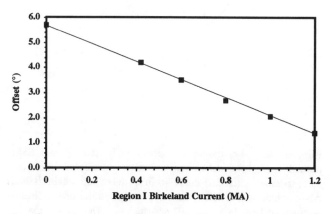

**Figure 7.** Plot of offset between the six auroral ovals shown in Fig. 15 and the magnetic pole as a function of Region I current strength (in MA). The solid curve represents a linear fit to the data and has a correlation coefficient of $R^2 = 0.999$.

auroral oval and one obtained during very quiet times by *Meng et al.* [1977]. However, the original SSM could not reproduce the observed auroral ovals at more active times. Observations during more active times by *Feldstein* [1963, 1969] have placed the offset at ~3°. Furthermore, *Feldstein and Starkov* [1967] found that the offset decreased with increased geomagnetic activity and could be less than 1° under very disturbed conditions. The results shown in Figure 7 show good agreement with *Meng et al.* [1977], *Feldstein* [1963, 1969], and *Feldstein and Starkov* [1967]. These studies and ours suggest a direct link between the auroral oval offset and the Birkeland current strength.

## 4. MODIFICATION OF THE SSM

*Schulz and McNab* [1996] identified several areas in which the original source surface model was deficient. These include the lack of a normal component of **B** across the tail current sheet at $x > 10\ R_E$, the absence of a ring current (or any other current system except the cross-tail current), a smaller than observed magnetotail radius, an asymptotic lobe **B** field that is too large, and a current sheet that is too thick in the $z$ direction in the tail. In the first part of this paper, we incorporated Birkeland currents in the SSM and obtained many important results. However, in order to be able to compare our results with observations, we require a baseline field model that is as realistic as possible, without sacrificing the elegant simplicity of the original SSM.

In this section, we outline our efforts to modify the original SSM to obtain a more realistic testbed for our Birkeland current studies. The modifications discussed below deal with the radial size of the magnetotail, the modification of the tail current sheet, and the reduction of the asymptotic lobe **B** field strength. Efforts to incorporate a ring current in the model and to add a normal component of **B** across the magnetotail current sheet are underway and will be published at a later date.

### 4.1. Magnetotail Radius

The original SSM places the asymptotic magnetotail radius at $\rho_\infty^* = 17.696\ R_E$ (or $1.7696b$). Observations place the nominal magnetotail radius at ~ $25\ R_E$ [see, for example, *Roelof and Sibeck*, 1993]. In order to increase the magnetotail radius in the SSM without changing the overall shape of the magnetopause, we follow the procedure outlined by *Schulz and McNab* [1996] and use Eq. (1) above with different values of $\rho_\infty^*$.

Figure 8 shows plots of the SSM magnetopause for four different magnetopause radii from $\rho_\infty^* = 1.7696b$

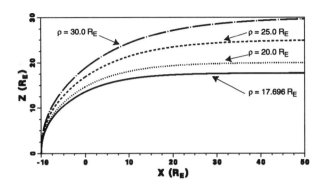

**Figure 8.** SSM Magnetopause for $\rho_\infty^* = 1.7696b$ (original SSM, solid curve), $\rho_\infty^* = 2.0b$ (dotted curve), $\rho_\infty^* = 2.5b$ (dashed curve), and $\rho_\infty^* = 3.0b$ (chain-dashed curve).

($17.696\ R_E$) (original SSM, shown with dotted curve) to $\rho_\infty^* = 3.0b$ ($30.0\ R_E$) (shown with chain-dashed curve). In each case, the enlarged magnetopause is used as the new boundary for the least-squares minimization technique and a new set of $\bar{g}_n^m$s is produced (see, for example, Figure 15 of *Schulz and McNab* [1996]). We have selected $\rho_\infty^* = 2.5b$ ($25.0\ R_E$) (shown by the solid curve) for the magnetotail radius in the modified SSM. Along with bringing the SSM magnetotail radius closer to agreement with observations, this change also increases the magnetotail volume by a factor of two. As we will see below, this change also significantly decreases the asymptotic lobe **B** field value.

### 4.2. Magnetotail Current Sheet and Lobe Field Strength

One of the shortcomings of the original SSM is the thickness of the magnetotail current sheet. Figure 9a shows contours of constant **B**, or current streamlines, for the original SSM taken at $x = 20\ R_E$. As can be seen in the figure, the current sheet in the model is very thick, and many current streamlines close upon themselves in the lobe in an unrealistic manner. Also, both the average lobe field (25.5 nT) and the maximum lobe field (54.2 nT) in the SSM are much higher than observations have shown. The dotted curve in Figure 10 is a plot of $B_x$ as a function of $z$ at $x = 20\ R_E$ and $y = 0$. The half-thickness of the current sheet (measured from $z = 0$ to the peak in $B_x$) in the SSM is $6.0\ R_E$, much thicker than current sheet thicknesses observed in the magnetotail.

The method utilized here for obtaining a thinner and more realistic current sheet in the SSM is to add an additional constraint to the variational principle used in obtaining the SSM expansion coefficients (see Section 2). To do this, we specify a constant (except for a change of sign across the current sheet) lobe field strength at $x =$

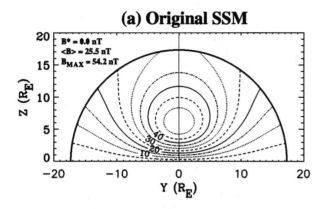

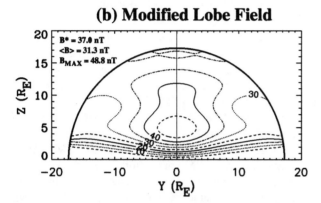

**Figure 9.** Contours of constant B in the $x = 20\ R_E$ plane for (a) the original SSM and (b) the model with a modified lobe field. Contours are shown at 5 nT increments.

+∞, and map this value along field lines to the normal surface of the SSM (the cross-tail boundary of the S region). The new expansion coefficients are then calculated using the new distribution of **B** on the normal surface. The selection of an asymptotic lobe field is not trivial, however. If the selected field value is too small, field lines in the S region are no longer confined to within the magnetopause. Similarly, if too large a value is chosen, the least-squares technique yields expansion coefficients which confine the **B** in the S region to a smaller region within the magnetopause. We are currently developing a model where the optimum lobe **B** field strength is found as an additional variable of the least squares technique. In this paper, however, we have used a "trial and error" method for determining the optimum **B** for each magnetotail configuration. For the SSM with a 17.696 $R_E$ magnetotail radius, the optimum asymptotic lobe **B** field (denoted by **B\*** henceforth) was found to be 37 nT. The resultant distribution of current streamlines is shown in Figure 9b. We see that the cur-

rent sheet is much thinner, and fewer current streamlines close upon themselves in the lobe. The solid curve in Figure 10 shows the $B_x$ versus z profile for this configuration. We see that although the current sheet thickness has been reduced to 4.8 $R_E$ and the maximum lobe field strength has decreased to 48.8 nT, the average lobe field has increased to 31.3 nT.

Increasing the SSM magnetotail radius to 25 $R_E$ does not change the unrealistic nature of the current streamlines in the model (Figure 11a). However, because of the significant change in the volume of the lobes, the average lobe field is reduced to 14.3 nT and the maximum lobe field is reduced to 43.0 nT (dotted curve in Figure 12). The current sheet thickness (~ 6.4 $R_E$) also remains large in this case. The result of applying the lobe field modification technique with **B\*** = 20 nT is shown in Figure 11b. Although there isn't a significant reduction in the thickness of the current sheet (5.6 $R_E$ in this case), the overall magnetic topology is much more realistic, and the maximum lobe field is notably reduced. The solid curve in Figure 12 shows that the modified model has a maximum lobe field strength of 34 nT and an average lobe field of 15.7 nT, both of which are in better agreement with observations.

One of the most important features of the SSM is the well defined separatrix between open and closed field lines in the model. Since the size and location of the separatrix are of paramount importance in our Birkeland current studies, we must ensure that the modifications carried out so far do not alter the shape and location of the separatrix in an unrealistic manner. Figure 13 is in a format similar to that of Figure 6 and shows the iono-

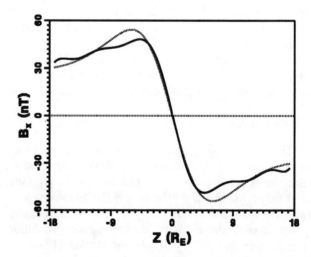

**Figure 10.** Plot of $B_x$ versus z at $x = 20\ R_E$ and $y = 0$. The dotted curve corresponds to the original SSM, and the solid curve to the SSM with the modified lobe field.

spheric projection of the separatrix in the original SSM (dotted curve) and the separatrix in the modified model (with $\rho_\infty^* = 25.0\ R_E$ and $B^* = 20$ nT, shown with solid curve). One of the deficiencies of the SSM was that it placed the dayside separatrix at too high a latitude compared to observations. We see that the modifications presented above have actually resulted in a separatrix that is ~4° equatorward of that of the original SSM on the dayside. We expect the difference between the dayside location of the modified SSM separatrix and the observed auroral oval to decrease even more when a ring current is incorporated into our model.

## 5. SUMMARY AND DISCUSSION

In this paper, we have outlined a newly-developed technique for including Birkeland currents in global magnetic field models and applied it to incorporate field-aligned currents into the Source Surface Model of

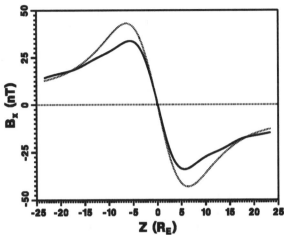

**Figure 12.** Plot of $B_x$ versus $z$ at $x = 20\ R_E$ and $y = 0$ for $\rho_\infty^* = 2.5b$. The dotted curve corresponds to the original SSM, and the solid curve to the SSM with the modified lobe field.

the magnetosphere [*Schulz and McNab*, 1987, 1996]. We found that:

1. The separatrix between closed and open field lines is shifted poleward on the night side by about 3° and equatorward by about 0.5° on the dayside after the addition of the Birkeland and image currents that are appropriate for quiet times (0.42 MA Region I current).

2. The entire auroral oval shifts toward the dayside when the total Birkeland current is increased. The oval in the original SSM was off-center by ~6° from the magnetic pole. Here, we have decreased this offset to ~4° by including a relatively weak (0.42 MA) Birkeland current in the model. Furthermore, we have found that the offset decreases linearly as the total Birkeland current is increased, so that the offset is only about 1° for a Region I current of 1.2 MA (Figure 7). Since the SSM provides an excellent description of the auroral oval for extremely quiet times, the decreasing offset between the center of the auroral oval and the magnetic pole as the total Birkeland current is increased clearly indicates that the presence of Birkeland currents can account for the observed decrease in the auroral oval offset under more disturbed conditions [*Feldstein*, 1963, 1969; *Feldstein and Starkov*, 1967]. Moreover, the corrugations seen in the separatrices for higher values of Birkeland current suggest the occurrence of the Kruskal-Shafranov instability in Region I current sheets.

3. Changing the asymptotic magnetotail radius of the SSM to $\rho_\infty^* = 25.0\ R_E$, along with bringing the model into better agreement with observations, increases the

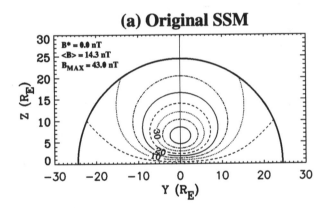

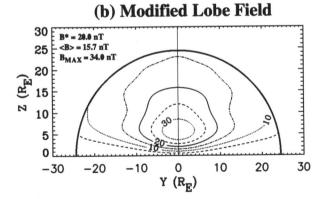

**Figure 11.** Contours of constant B in the $x = 20\ R_E$ plane for $\rho_\infty^* = 2.5b$. (a) Contours for the original SSM and (b) the model with a modified lobe field. Contours are shown at 5 nT increments.

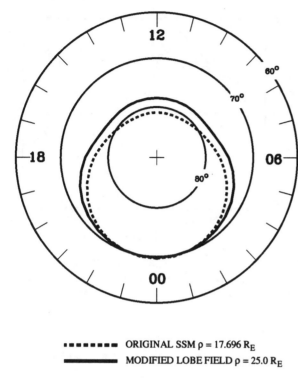

••••••• ORIGINAL SSM ρ = 17.696 $R_E$
——— MODIFIED LOBE FIELD ρ = 25.0 $R_E$

**Figure 13.** Ionospheric projection of the SSM separatrix, similar in format to Fig. 4, for the original model (dotted curve) and the model with $\rho^*_\infty = 2.5b$ and a modified lobe field (solid curve).

magnetotail volume by a factor of two. This results in a reduction of the average lobe field in the magnetotail.

4. By requiring the least-squares minimization to yield as constant a lobe field as possible, we have obtained a much more realistic lobe field profile. The corresponding decrease in the peak lobe B field value puts the modified SSM in closer agreement with observations.

While the technique outlined in this paper for incorporating Birkeland currents into the SSM has yielded several interesting results and the modifications outlined above have brought us much closer to a more realistic model, our work is far from finished. Our ultimate goal is to model the Birkeland currents using sheet and volume current representations of these current systems.

Before we do that, however, we will address and correct the remaining shortcomings of the original SSM. Our plans include the addition of a ring current to the model and the application of a stretch transformation to the near-Earth field lines to obtain a finite non-zero $B_z$ profile of the magnetotail.

*Acknowledgments.* The authors thank M. W. Chen for many helpful discussions. This work was supported by NASA grants NAG5-1100 and NAGW-4553 at UCLA as well as by NASA contract NAS5-30372 and NSF grant ATM-9119516 (both at Lockheed Martin). UCLA/IGPP publication number 4938.

## REFERENCES

Atkinson, G., The current system of geomagnetic bays, *J. Geophys. Res.*, 72, 6063, 1967.

Birkeland, K., *The Norwegian Aurora Polaris Expedition 1902-3*, vol. 1, *On The Cause of Magnetic Storms and the Origin of Terrestrial Magnetism*, H. Aschehoug, Christiania, Norway, 1908.

Boström, R., A model of the auroral electrojets, *J. Geophys. Res.*, 69, 4983, 1964.

Donovan, E., Modeling the magnetic effects of distributed magnetospheric currents, Ph.D.. thesis, University of Alberta, Edmonton, Alberta, 1993.

Feldstein, Y. I., Some problems concerning the morphology of auroras and magnetic disturbances at high latitudes. *Geomagn. i Aeronom.*, 3, 227, [English transl.: *Geomagn. Aeron.*, 3, 183] , 1963.

Feldstein, Y. I., Polar auroras, polar substorms, and their relationship with the dynamics of the magnetosphere, *Rev. Geophys.*, 7, 179, 1969.

Feldstein, Y. I., and G. V. Starkov, Dynamics of auroral belt and polar geomagnetic disturbances, *Planet. Space Sci.*, 15, 209, 1967.

Forslund, D. W., The Kruskal-Shafranov kink instability for field-aligned currents in the magnetosphere, *EOS, Trans. Amer. Geophys. Union.*, 51, 1970, p. 405.

Hasegawa, A., Theory of aurora band, *Phys. Rev. Lett.*, 24, 1162. 1970.

Iijima, T, and T. A. Potemra, The amplitude distribution of field-aligned currents at northern high latitudes observed by TRIAD, *J. Geophys. Res.*, 81, 2165, 1976a.

Iijima, T, and T. A. Potemra, Field-aligned currents in the dayside cusp observed by TRIAD, *J. Geophys. Res.*, 81, 5971, 1976b.

Kamide, Y., The relationship between field-aligned currents and the auroral electrojets: a review, *Space Sci. Rev.*, 31, 127, 1982.

Kamide, Y. and Fukushima, N., Positive geomagnetic bays in the evening high latitudes and their possible connection with partial ring current, *Rep. Ionos. Space Res. Japan*, 26, 79, 1972.

Kaufmann, R. L., and D. J. Larson, Electric field mapping and auroral Birkeland currents, *J. Geophys. Res.*, 94, 15,307, 1989.

Kirkpatrick, C. B., On current systems proposed for $S_D$ in the theory of magnetic storms, *J. Geophys. Res.*, 57, 511, 1952.

Kruskal, M., and M. Schwarzschild, Some instabilities of a completely ionized plasma, *Proc. Roy. Soc. (London)*, *A223*, 348, 1954.

Lyons, L. R., M. Schulz, D. C. Pridmore-Brown, and J. L. Roeder, Low-latitude boundary layer near noon: An open field line model, *J. Geophys. Res.*, *99*, 17,367, 1994.

Meng, C.-I., R. H. Holzworth, and S.-I. Akasofu, Auroral Circle--Delineating the poleward boundary of the quiet auroral belt, *J. Geophys. Res.*, *82*, 164, 1977.

Olson, W. P., and K. A. Pfitzer, a quantitative model for the magnetospheric magnetic field, *J. Geophys. Res.*, *79*, 3739, 1974.

Roelof, E. C., and D. G. Sibeck, Magnetopause shape as a bivariate function of interplanetary magnetic field $B_z$ and solar wind dynamic pressure, *J. Geophys. Res.*, *98*, 21,421, 1993.

Schield, M. A., J. W. Freeman, and A. J. Dessler, A source for field-aligned currents at auroral latitudes, *J. Geophys. Res.*, *74*, 247, 1969.

Schindler, K., and J. Birn, Magnetospheric physics, *Physics Reports*, *47*, 110, 1978.

Schmidt, A., *Tafeln der normierten Kugelfunktionen, sowie Formeln zur Entwicklung*, Engelhard-Reyer, Gotha, 1935.

Schulz, M., The Magnetosphere, in *Geomagnetism*, edited by J. A. Jacobs, vol. 4, p. 87, Academic, San Diego, 1991.

Schulz, M. and M. C. McNab, Source-surface model of the magnetosphere, *Geophys. Res. Lett.*, *14*, 182, 1987.

Schulz, M. and M. C. McNab, Source-surface modeling of planetary magnetospheres, *J. Geophys. Res.*, *101*, 5095, 1996.

Shabansky, V. P., Some processes in the magnetosphere, *Space Sci. Rev.*, *12*, 299, 1971.

Shafranov, V. D., The stability of a cylindrical gaseous conductor in a magnetic field, *Atomn. Energ.*, *1*, 38, 1956. English translation in *Soviet J. Atom. Energy*, *1*, 709, 1956.

Stern, D. P., A simple model of Birkeland currents, *J. Geophys. Res.*, *98*, 5691, 1993.

Stern, D. P., The art of mapping the magnetosphere, *J. Geophys. Res.*, *99*, 17,169, 1994.

Stern, D. P., A model of the Region I Birkeland field, presented at the 1995 AGU Spring Meeting, Baltimore, Maryland (*EOS*, *76*, S244, 1995).

Sugiura, M. and S. Chapman, The average morphology of geomagnetic storms with sudden commencements, *Abh. Deut. Akad. Wiss. Göttingen, Math. Phys. Kl.*, 4, 1960.

Tsyganenko, N. A., Global quantitative models of the geomagnetic field in the cislunar magnetosphere for different disturbance levels, *Planet. Space Sci.*, *35*, 1347, 1987.

Tsyganenko, N. A., Methods for quantitative modeling of the magnetic field from Birkeland currents, *Planet. Space Sci.*, *39*, 641, 1991.

Tsyganenko, N. A., Modeling the Earth's magnetospheric magnetic field confined within a realistic magnetopause, *J. Geophys. Res.*, *100*, 5599, 1995.

Tsyganenko, N. A., and D. P. Stern, Modeling the global magnetic field of the large-scale Birkeland current systems, *J. Geophys. Res.*, *101*, 27,187, 1996.

Vasyliunas, V. M., Mathematical models of magnetospheric convection and its coupling to the ionosphere, in *Particles and Fields in the Magnetosphere*, edited by B. M. McCormac, pp. 60-71, D. Reidel, Norwell, Mass., 1970.

Vasyliunas, V. M., The interrelationship of magnetospheric processes, in *Earth's Magnetospheric Processes*, edited by B. M. McCormac, pp. 29-38, D. Reidel, Norwell, Mass., 1972.

---

Vahé Peroomian, Institute of Geophysics and Planetary Physics, University of California, Los Angeles, CA 90095-1567. (e-mail vahe@igpp.ucla.edu).

Larry Lyons, Department of Atmospheric Sciences, University of California, Los Angeles. CA 90095.

Michael Schulz, Space Sciences Department, Lockheed Martin Palo Alto Research Laboratories, Palo Alto, California.